Pastoralism in Africa

Pastoralism in Africa

Past, Present and Future

Edited by

Michael Bollig, Michael Schnegg and Hans-Peter Wotzka

berghahn
NEW YORK • OXFORD
www.berghahnbooks.com

Published in 2013 by
Berghahn Books
www.berghahnbooks.com

Library of Congress Cataloging-in-Publication Data

Library of Congress Cataloging-in-Publication Data
Pastoralism in Africa: past, present and future / edited by Michael Bollig, Michael Schnegg, Hans-Peter Wotzka.
p. cm.
Includes bibliographical references.
ISBN 978-0-85745-908-4 (hardback: alk. paper) -- ISBN 978-0-85745-909-1 (ebook)
1. Pastoral systems--Africa. 2. Pastoral systems--Africa--History. 3. Herding--Africa. 4. Herding--Africa--History. 5. Land use--Africa. I. Bollig, Michael. II. Schnegg, Michael. III. Wotzka, Hans-Peter.
GN645.P185 2013
636.084'5096--dc23
2012032451

British Library Cataloguing in Publication Data

A catalogue record for this book is available from the British Library

Printed in the United States on acid-free paper

ISBN 978-0-85745-908-4 (hardback)
ISBN 978-0-85745-909-1 (institutional ebook)

Contents

Part II: Historical and Contemporary Dynamics of Pastoralism

Part III: Violence, Trade, Conservation and Pastoralism in Africa

List of Figures and Tables

Figures

Tables

List of Abbreviations

ACACIA	Arid Climate, Adaptation and Cultural Innovation in Africa
ADMADE	Administrative Management Design for Game Management Areas
ANR	Alliance Nationale pour la Résistance, Chad
BET	Borku-Ennedi-Tibesti, Chad's northern region
BOS	Besiedlungsgeschichte der Ost-Sahara (Colonisation History of the Eastern Sahara)
bp	before present
CAMPFIRE	Zimbabwean Communal Areas Management Programme for Indigenous Resources
CBC	Community-based Conservation
CBNRM	Community-based Natural Resource Management
CMA	Community Management Area
CNAR	Centre National d'Appui à la Recherche, Chad
CPA	Comprehensive Peace Agreement
CToT	Caloric terms of trade
DNA	Deoxyribonucleic Acid
DFG	Deutsche Forschungsgesellschaft (German Research Foundation)
DOP	Dakhleh Oasis Project
EIA	Early Iron Age
EUFOR	European Union Force
FUC	Front Uni pour le Changement, Chad
GA	Grazing Association
GEF	Global Environmental Facility
GOSS	Government of the Southern Sudan
GPS	Global Positioning System
GRAS	Geological Research Authority of Sudan, Khartoum
ICDP	Integrated Conservation and Development Projects
IDP	Integrated Development Planning, South Africa

IUCN	International Union for Conservation of Nature and Natural Resources
LAPC	Long-term Agricultural Policy Commission, Namibia
LBK	Linearbandkeramik
LSA	Late Stone Age
MAWF	Ministry of Agriculture, Water and Forestry, Namibia
MDTP	Maloti-Drakenberg Transfrontier Conservation and Development Project
MET	Ministry of Environment and Tourism, Namibia
MRA	Managed Resource Areas
MINURCAT	Mission des Nations Unies en République Centrafricaine et au Tchad
MOEST	Ministry of Education, Science and Technology, Kenya
MPS	Mouvement Patriotique pour le Salut, Chad
MRA	Managed Resource Association
MSA	Middle Stone Age
NAU	Namibian Agricultural Union
NCAM	National Corporation for Antiquities and Museums in Khartoum
NDVI	Normalised difference vegetation index
NPB	South African National Parks Board
NRMP	Natural Resource Management Project
NGO	Non-governmental Organisation
PIA	Pastoral Iron Age
PPF	Peace Park Foundation
RDL	Rassemblement pour la Démocratie et les Libertés, Chad
RMA	Range Management Areas
RMAP	Range Management Area Programme
RNP	Richtersveld National Park
SANParks	South African National Parks
SPLA	Sudan People's Liberation Army
SPN	Savannah Pastoral Neolithic
SST	Sea surface temperature
SWA	South West Africa
SWANLA	South West African Labour Organization
SWAPO	South West African People's Organisation
TBNRM	Transboundary Natural Resource Management
TFCA	Transfrontier Conservation Area
TGLP	Tribal Grazing Land Policy, Botswana
TLU	Tropical Livestock Unit

UN	United Nations
UNCED	United Nations Conference on Environment and Development
UNDP	United Nations Development Programme
UNESCO	United Nations Educational, Scientific and Cultural Organisation
USAID	United States Agency for International Development
WHS	World Heritage Site
WMBC	Walfish Bay Mining Company
WWF	World Wide Fund for Nature

Acknowledgements

It took several years to complete this volume and many years more to gather the data presented here. While the Collaborative Research Centre (CRC) 389 Arid Climate, Adaptation and Cultural Innovation in Africa (ACACIA) started its work in 1995, the project had several pertinent forerunners, some of them starting in the late 1980s with archaeological survey work in the Eastern Sahara. While the CRC was not directly focused on the development of pastoralism in Africa, its concentration on processes of adaptation to the manifold challenges of aridity in African landscapes entailed that pastoral lifestyles were central to a number of projects. This volume consists of sixteen contributions which directly originate from research done within the CRC. Contributions mainly come from anthropology and archaeology, two sciences rarely publishing in joint volumes these days. We hope that beyond a contribution to the field of African pastoralism this volume will inspire further cross-disciplinary thinking between anthropologists and archaeologists.

In order to bring the CRC's project to fruition, a comprehensive administrative structure was necessary both on the side of the financing agency GRC (the German Research Council) and at the University of Cologne. On behalf of many officials within the German Research Council contributing to the success of the CRC we would particularly like to thank Petri Winkler of the GRC desk for CRCs and the GRC people responsible for archaeology and anthropology: Ursula Far-Hollender, Wolfgang Rohe, Jochen Briegleb (†), Christina Petry, Susanne Anschütz (†), Ulrich Bienert and Corinne Flacke. Over the years a number of referees gave very valuable advice to the project, amongst them Ulrich Braukämper, Thomas Bierschenk, Gerd Spittler, Manfred K.H. Eggert, Horst Hagedorn, Pierre M. Vermeersch and Bernhard Streck, who all contributed to the project by discussing efforts and challenges with members of the collaborative research centre. At the University of Cologne, the project was carefully administered by Werner Schuck over many years, and it was his intimate knowledge of the research processes involved and the requirements on the ground which made the project go even to the most remote spots of the African continent (e.g. the interior of the Eastern Saharan Desert). Werner Schuck also accompanied this volume to its completion.

The research of the CRC was made possible by numerous African institutions that supported fieldwork as colleagues and hosts. In even larger numbers the people in these research areas have assisted us as patient informants and friendly hosts. We are deeply indebted to them and owe much more than communicating what resulted from our stays.

The work on the volume was accompanied by a number of technical experts and various student assistants. We would like to make particular mention of Lutz Hermsdorf-Knauth who did most of the graphics for the volume. As student assistants, Fabienne Braukmann, Ivanka Klein, Kerstin Lawan, Katja Metzmacher, Frederik Weck and Verena Werner also helped with the volume. They did an invaluable job and often worked overtime, especially during the last phase of the publication process when citation styles had to be adjusted, bibliographies polished up and CVs gathered.

INTRODUCTION
Specialisation and Diversification among African Pastoral Societies

Michael Bollig and Michael Schnegg

The Approach of this Volume

In 1995 the collaborative research centre Arid Climate, Adaptation and Cultural Innovation in Africa (ACACIA) started research on the complex inter-relationship between ecology, politics and culture in Africa's arid environments (Bubenzer, Bollig and Riemer 2006; Bollig et al. 2007; Bolten et al. 2009).[1] According to the original hypothesis of the ACACIA programme, the emergence of pastoralism in Africa as well as its spread were intricately connected to major climatic changes in the earlier part of the Holocene (Kuper and Kröpelin 2006). Once developed as a successful form of adaptation to arid and semi-arid environments, pastoralism made its way across the continent, and contemporary pastoral economies have developed characteristic strategies to cope with climatic perturbations and ensuing livestock losses. Pastoralism was generally portrayed as a livelihood strategy successfully meeting the challenges of low overall biomass productivity and environmental variability. Only during the last few decades have demographic, political and economic changes allegedly led to an increasing vulnerability of pastoralism in many African countries. This, in a nutshell, was the basic storyline the interdisciplinary project was to work on.

However, archaeological results generated within the project and produced by associated researchers clearly indicated that pastoral adaptations had been inherently unstable, vulnerable and diverse (see Galaty; Jesse et al.; Sadr, this volume). In contrast to earlier publications (e.g., Smith 1992), which rather emphasise the longevity and stability of pastoral adaptations to arid landscapes, change and instability are highlighted. Since the 1990s, diversity, vulnerability and dynamic adaptive processes have been discussed as key elements in contemporary pastoral adaptations as well. Some of these adaptive changes are small and have taken place within the pastoral sector without altering the fundamental structures of pastoral adaptation, such as changing herd composition, the introduction of new species and adjusted mobility strategies.[2] Furthermore, archaeologists, anthropologists but also historians have showed that pastoralists have crossed boundaries to other modes of production at all times: stockless or poor pastoralists have turned to foraging (Galaty 1993; Spear and Waller 1993; see also Bollig; Henrichsen; Lane, this volume), foragers have transformed into pastoralists (e.g., Cronk 2004) and small-scale irrigation farmers have taken to mobile livestock husbandry (Anderson 2002). In fact, in recent decades many East African pastoralists have turned to more diversified livelihoods: demographic growth and declining relations between livestock and people have motivated herders to take up dryland agriculture (McCabe 1997; Fratkin and Roth 2005; Galvin 2009; McCabe, Leslie and DeLuca 2010; see also Bollig and Österle, this volume) and the negative effects of drought and/or violence have led to more emphasis on trading and other livelihood strategies in quite a number of cases (Galaty 2005; Mkutu 2008; see also Little; Meerpohl, this volume). The commercialisation and commoditisation of pastoral activities as well as labour migration have given rise to more internal differentiation within pastoral societies (Hogg 1986; Ensminger 1992; Spencer 1998; Galvin 2009; see also Bollig and Österle; Galaty; Henrichsen; Little, this volume). Some of these transformations have led to the abandonment of pastoral strategies altogether, to their relegation to subsidiary livelihood strategies, but in some cases also to the strengthening of livestock-based strategies (e.g., McCabe, Leslie and DeLuca 2010) within a mixed economy. Paul Spencer (1998) has interpreted some of these changes in East Africa as processes of 'peasantisation': higher dependence on external markets, diversified economic strategies, absentee herd ownership through local elites, increasing internal stratification and incorporation into national societies are seen as hallmarks of such a development. While the concept of peasantisation captures some of these dynamics convincingly, it suggests a unilinear transformation towards market dominated economies. The peasantisation model, however, does not convincingly address many processes of change within pastoral systems which have not always led to more stratification, market orientation and incorporation into larger social and political structures.

It was here that the ACACIA programme focused its research efforts: How did change and diversity within pastoral adaptations and transitions between pastoralism and other livelihood strategies develop and what caused these changes? It was not so much the diversity of pastoral systems as such which captured our interest but rather processes leading to more or less diversity.[3]

The ACACIA programme was privileged to combine the research efforts of archaeologists, historians, anthropologists, linguists, geographers and botanists. This allowed on the one hand research into diachronic dynamics in pastoral systems, and on the other encouraged research on the much debated issue of the impact of various pastoral strategies on the environment. During the course of a twelve-year-long research programme, the complex political ecology of African pastoralism was unravelled with focal research areas in the south-west and in the north-east of the continent. Process and change in pastoral strategies was researched for Holocene communities as well as for contemporary communities in Chad and Namibia. In this volume original research from within the ACACIA programme is combined with parallel research by international scientists.

On Terminology and Definitions

There are numerous attempts to define pastoralism. Galaty defines pastoral societies as 'dependent on domestic animals for subsistence and predominantly occupied with herding on natural pasture' (Galaty 1996: 415); Ingold depicts pastoralism as 'a form of livelihood based upon the management of herds of domestic animals' (Ingold 2009: 710); Bonte describes it as 'l'élevage d'animaux herbivores vivant en troupeaux et se déplaçant à la recherche de leur nourriture. Il en résulte à degrés divers, de la transhumance saisonnière au nomadisme erratique, une mobilité des groupes humains associés aux troupeaux et des formes particulières d'organisation de l'espace' ('the breeding of herbivore gregarious animals that move in search of food. This results to different degrees, from the seasonal transhumance to the erratic nomadism, in a mobility of human groups associated with the herds and particular forms of their living space') (Bonte 1992: 561); and Salzman, in his concise overview of pastoralism, hierarchies and the state, defines pastoralism as 'the raising of livestock on "natural" pasture unimproved by human intervention' (Salzman 2004: 1). These definitions certainly help to differentiate pastoralism from other types of subsistence economy. However, they cannot (and do not want to) capture the diversity and inherently dynamic character of pastoral societies. It is symptomatic that edited volumes like Galaty and Johnson (1990) and Galaty and Bonte (1991), as well as review articles such as Dyson-Hudson and Dyson-Hudson (1980) and Fratkin (1997), desist from offering any definition. It is the heterogeneity and dynamic character of adaptations based on the use of livestock which has been emphasised consistently in edited volumes, review articles and encyclopaedia entries in recent years.

We propose here to characterise pastoralism through a set of indicators. This approach has three distinct advantages: it captures the diversity of ideological, social and economic aspects of pastoral livelihoods; allows us to determine to what degree an individual, a household or a community is pastoral; and sketches a framework within which dynamic transitions take place. Our approach rests on the assumption that these fields are interdependent but not isomorphic – that is, a society may consider itself pastoral and cherish pastoral values but may economically rather depend on a mixed set of resources. Before we can move on to define these dimensions and their indicators, we need to make clear what entities we aim to address: individuals, households or communities.

While it is useful and probably mandatory for comparisons to define entire communities as specialised or diversified, it is necessary to bear in mind that it is first of all individual actors who decide whether to invest or not to invest in diversification and/or specialisation. To label a pastoral community as specialised or diversified addresses a process that already needs explanation: Why does an entire community move in one direction? Under what circumstances does a community allow individual members to invest in pastoral strategies and sedentary agricultural strategies alike? When does a society prescribe some sort of specialisation and inhibit its members from diversifying (such as by putting up taboos against or even ridiculing agriculture)? It is tempting to address households as a major reference point. In fact, in most pastoral communities decisions about specialisation and diversification are taken at the household level, and many ethnographic studies of pastoralists take households as the basic unit of analysis.

The analysis of the case studies presented in this volume suggests three continua that need to be considered in order to determine the degree to which an individual or a larger social unit is pastoral: labour, capital and world-view. To what extent do individuals within a community invest labour in pastoral activities? If they invest major amounts of labour in their herds we could easily call a person 'pastoral' or a 'herder'. However, there is no community that we know of that invests all its labour in pastoral activity. In subsistence economies, supplementary food acquisition is of importance and in more market-orientated economies a number of people may derive income from activities not directly linked to herds. How actors invest different forms of capital is another continuum that describes process and change in pastoral economies. Whether all land is devoted to pastoral strategies or only parts of it, whether herds constitute the major form of wealth or are just one form of wealth is directly connected to our key question. Do these two dimensions – labour and capital – capture the continuum of pastoral specialisation and diversification fully?

We have been acquainted with 'pastoral communities' who present themselves very much as pastoral, but when looking at labour and capital investment they are not pastoral in a strict sense. The Nuer are a classic example: while they represented themselves as a truly pastoral community to Evans-Pritchard in the early 1930s, they mainly lived on millet, sorghum and fish, and invested large amounts of labour in their fields (Evans-Pritchard 1940: 57). In the contemporary world we also find this paradox: the communities of the Richtersveld in South Africa (Berzborn and Solich, this volume) and Fransfontein in Namibia (Schnegg, Pauli and Greiner, this volume) cherish a pastoral identity while at the same time the majority of the population is engaged in wage labour, and labour migration. In all four cases studied by Galaty (this volume), pastoralists have significantly diversified their assets and have put substantial amounts of labour into non-pastoral activities in recent years, thereby sustaining pastoral strategies in contemporary settings. All four communities conceptualise themselves as pastoralists despite these recent economic and social changes. Similar disjunctures between economy and ideology are suggested by data on prehistoric communities: rock art associated with middle Holocene herders of Wadi Howar in Sudan clearly hints at a pastoral ideology, while assemblages of artefacts rather suggest a lifestyle only partially dependent on herds. This points to world-view and identity as the third essential dimension to be captured.

There are only a few attempts in the literature to define specialisation in pastoral communities – perhaps a consequence of major cultural, social and economic differences among specialised herders in Africa (let alone in a global setting). Specialised pastoralism is often associated with entire ethnic groups (Maasai, Boran, Turkana), fractions or sections of such groups (Pokot, Sebei, Fulbe) or (more rarely) specialised professional and/or status groups within larger agro-pastoral societies (for example, Bedouin). Bates and Lees (1977) see a continuum from less specialised herders such as the Ugandan Karimojong, who incorporate some degree of rain-fed agriculture into an overall pastoral economy, to sectional specialisation as among the Pokot, among whom an agricultural faction employing irrigated agriculture and a specialised pastoral faction make up for a larger system.[4] It is especially this latter form of pastoralism – herding with employed herders, which is widespread in the Himalayas (Rao 1995), India (Rao and Casimir 2003), the Andes (Göbel 1997) and the Alps (Netting 1981) – which has been rare in Africa. However, recently, due to the sedentarisation of pastoralists and the emergence of absentee herd ownership, herding for herd owners living in centres or in towns has become more prominent.[5] Specialised pastoral systems engaging entire ethnic communities are more often found in environments where agricultural pursuits are highly risk-prone or virtually impossible. However, Bonte (1981) aptly warns that such a perspective easily leads to environmental determinism, and he focuses attention on factors like labour, accumulation and ideology to explain processes leading to specialised pastoralism.

Authors generally agree that in specialised herding economies the size of household herds is usually bigger than in agro-pastoral systems. Galaty and Bonte (1991) point out that specialised herders maintain ratios of people to livestock of between 1:10 and 1:20; meanwhile richer agro-pastoralists maintain ratios of around 1:3 or 1:4 and poorer ones tend to around 1:1 (Schneider 1979). As a consequence, more labour is devoted to the herding of livestock, and assets are allocated to the expansion of, or to the risk management of, pastoral herds. Bonte defines pastoral specialisation as 'production where the labour invested in animal domestication is dominant, i.e., labour involved in the transformation and reproduction of animals' (Bonte 1981: 33). While Bonte emphasises the role of labour allocation and capital accumulation, Bates and Lees (1977) highlight the importance of exchange as a major variable to explain pastoral specialisation. The following paragraphs attempt to depict the three continua on which we attempt to grasp processes of pastoral specialisation and diversification.

Labour

Pastoralists may allot different amounts of labour to livestock husbandry according to the degree of specialisation. That is, in cases where the entire community maintains a pastoral specialisation, most of the labour of women and men, elderly people and youths, is allotted to the herding of animals (entailing such tasks as migrating with herds, watering, searching for new pasture, protecting against predators), the processing of livestock products (such as milking, producing butterfat), the maintenance of a material culture linked to pastoral production (such as producing calabashes, milk pails) and the management of herds and livestock-based exchange systems (marketing, social exchange of livestock).[6] Even ritual is often intricately linked to livestock and conceptualised as pastoral labour. Labour is usually organised at a household level. In other cases only a section of the community maintains this kind of labour allocation, whereas other sections allocate labour to other pursuits, such as agriculture. This means that alternative sets of knowledge need to be developed. In yet other cases, it is a group of hired herders which allocates all its time to herding livestock, the owners of which pursue non-pastoral strategies.

Pastoralists are always occupied with the problem of controlling pastoral labour. Whereas agricultural labour frequently takes place near the homestead and labour performance can be easily monitored, much pastoral labour takes place far away from the main homestead, and whether a herder (related or hired) cheats on the herd owner is a matter of much discussion and dissent, and a recurrent topic in pastoral communities (e.g., Ensminger 1992).

There are major cross-cultural differences regarding labour allocation in pastoral communities. In the highly heterogeneous north-west Namibian Fransfontein community described by Schnegg, Pauli and Greiner (this volume), nowadays only some individuals (often hired herders) invest a lot of time in livestock, while a majority of the population is concerned with off-farm labour or does not own any livestock. Among the neighbouring Himba, a large proportion of the labour of all individuals is devoted to livestock. Nevertheless, during the rainy season, Himba women often invest considerably more time in their gardens than in their herds (Bollig, this volume). Communities also differ on time allocation historically. While the pastoral Pokot of the 1980s devoted labour mainly to livestock-related activities, only twenty years later the same community devoted much more time to agriculture and trade (see Bollig and Österle, this volume). In the cases explored in this volume it is often women, young men or poor herders who opt for alternative economic strategies. At first these are often just add-ons to pastoral strategies and fulfil risk-minimising functions, but gradually become more significant. John Galaty (this volume) compares four African pastoral communities (the Fulbe, Maasai, Nuer and Tswana) and shows how pastoral mobility (labour invested in the mobility of herds) has changed over time.

While archaeologists, of course, find it more difficult to develop ideas about labour allocation in prehistoric communities, there is ample evidence from excavations that allow inferences about the organisation of labour. The prehistoric herders of the Sudanese Wadi Howar apparently allocated considerable time to their livestock (Jesse et al., this volume). However, the great number of grinding stones found clearly indicates that the collection and processing of cereals must have also featured prominently in time budgets. The Late Stone Age hunter-herders Sadr (this volume) reports on for Southern Africa must have prioritised foraging strategies over herding activities. Their herds survived only for shorter periods and the archaeological record often shows major discontinuities of pastoral engagement whereas the foraging component is continuous.

In many pastoral societies, labour allocation is fairly uniformly distributed along the lines of gender and age. Typically, children are early on integrated into pastoral labour. Often they start their career as small-stock herders in the vicinity of the homestead. A lot of herding labour rests on male adolescents and young men managing highly mobile livestock camps. In many pastoral communities women are in charge of milking and milk processing (into ghee, butter or butterfat, for instance). Male seniors often act as herd managers within an environment in which decisions regarding mobility and intensity of grazing have to be taken in a context of communal access rights to pasture and water. It is of great interest to record how labour allocation and the control of labour changes when there is a distinct drive towards diversification, commoditisation and state integration (see Bollig and Österle; Galaty; Schnegg, Pauli and Greiner, this volume). The higher the degree of diversification, the more labour is devoted to non-pastoral efforts. Such non-pastoral activities are also gender or age specific: often it seems that women are the first to turn away from pastoral labour in order to invest more time in household gardens and small business. Is this because they can control wealth generated from their gardens better than wealth emanating from the herds? In many recent cases children are completely withdrawn from pastoral labour while they attend school. In diversified agro-pastoral systems it is often some male youths plus some hired herders who do most pastoral labour. It is of great interest to scrutinise moments of transition of labour allocation: What happens when more and more children attend school? Is sub-optimal herding accepted? Who replaces school going kids – the children of poorer relatives, neighbours or hired herders? What mechanisms of control are instituted when hired herders take over pastoral labour from sons, and how are gender roles redefined when women take to farmsteads while men are still active in distant livestock camps? How is control over labour, firmly vested in the herd owner in specialised pastoral systems, negotiated when economic strategies become more diversified and more difficult to monitor?

Capital

In broad terms, capital may be regarded as a factor of production, which is not needed for itself but for its ability to help in producing other goods (Mankiw 1998: 397). Economic capital – that is, any form of wealth potentially employable as a means of creating more wealth – is differentiated from social capital, which refers to the capacity of social linkages to be used to increase productivity, and from cultural capital, which comprises knowledge, skills, education and advantages that a person has to gain status within society. The way capital is structured, controlled and allocated is decisive. The more capital is allocated to livestock husbandry, the more a society is specialised – that is, if all economic capital is constituted by herds, all land put to the use of livestock husbandry and all material culture is geared towards livestock production, then a society is highly specialised. If social relations are sanctioned through the exchange of livestock and if cultural capital is mainly constituted by livestock-related knowledge, then all social exchanges constantly reproduce pastoral institutions and the accumulation of social capital is closely intertwined with pastoral specialisation. It is of key interest to investigate three issues. First, who controls the allocation of capital; that is, who decides which parts of the available land are used for what purposes, when livestock is sold and what kind of knowledge is accepted as relevant? Second, at what levels of political organisation are such decisions taken (household, section, state)? And third, how does capital allocation change when herders diversify or specialise? These questions are especially pertinent in the East African cases where a distinct trend towards more agriculture and more marketing is recently observable (Holtzman 1997; McCabe, Leslie and DeLuca 2010).

We will first turn to economic capital. On the one hand there are man made implements (such as calabashes) and infrastructure (such as wells, livestock enclosures) needed for herding. Then the herds themselves represent a crucial form of capital in specialised herder societies (Schneider 1979; Dyson-Hudson and Dyson-Hudson 1980). Diversification by necessity brings about a diversification of material culture as well. Hunting and gathering has its own technology, as has agriculture. In contemporary diversified pastoral systems, some may live in stone-built sedentary houses, with TV sets and radios, fridges and gas cookers, while some families may still cling to a more traditional material culture. When Bollig did research with the Pokot in the early 1990s, a survey of material culture yielded few inter-household differences – most items had some kind of function within the pastoral context; nowadays, such a survey would result in a number of items not related to pastoralism (Österle 2008), with bicycles, radios and 'urban' clothing being part and parcel of a more diversified pastoral economy.

Herds constitute perhaps the major form of capital in pastoral societies. In specialised pastoral communities, a household's property is mainly constituted by its livestock. In diversified herder communities, fields, ploughs or trucks may add to livestock capital. Herds often show highly complex ownership rights. Within one pastoral herd a great number of herd owners may be represented, each of them owning just a handful of animals. To the outside world such a herd may be claimed by the household head, but a closer look at the herd discloses a myriad of well-established rights in animals. There are even instances where single animals are owned by a number of people. Complex property rights in livestock with multiple ownership are common (Schlee and Khazanov 2012) and a hallmark of specialised pastoral systems.

The land managed by (and through) pastoralists must also be regarded as a form of capital asset. This capital is usually jointly managed by a community, more or less well defined. Anthropologists have forcefully pointed out that herders need labour (time, assets) in order to solve the problems of communal tenure. Landscapes utilised by pastoral communities bear the imprint of livestock husbandry. Generations of herders have moulded an environment favourable to large herds of herbivores: the selective burning of pastures, the cutting of unwanted bush, the establishment of networks of wells and the protection of specific pastures have formed the environment profoundly in many instances (see, e.g., Lane, this volume). Pastoral landscapes may be referred to as landesque capital (see Widgren 2007; Bollig 2009): investments in land with long-term returns within a pastoral context. Research into the recent transformation of pastoral landscapes in East Africa due to sedentarisation and agriculture shows how entire landscapes have been reshaped and such landesque capital is transformed: not just pastures with prime grasses but well ploughed and fertile fields, sturdy fences (to keep livestock away) and larger sedentary homesteads are becoming major markers of such transformed landesque capital.

Different forms of pastoral capital are characterised by different forms of access rights: while man-made implements are usually privately owned (by a person), man-made infrastructure (wells, enclosures, fences) are often owned by a household. Wells may also be owned by larger collectives, such as lineages (see, e.g., Helland 1982). Recent legal reforms in northern Namibia, for example, stipulate communal well ownership through formally constituted water-point associations (Bollig, this volume).

We may also differentiate to what extent pastoral capital is commoditised. Material culture used by pastoralists in many instances is not produced by herders themselves but by craftsmen (and women) belonging to other ethnic communities (Hodder 1982; Bollig 1987). The Kenyan Pokot obtained pottery from Marakwet or Il Chamus potters and spear blades from Samburu smiths (Kurita 1982; see also Bollig and Österle, this volume). The Himba of northwestern Namibia obtain pottery and ironware from itinerant Thwa craftsmen (Malan 1973; Bollig 2008), while Fulbe and Tuareg have traditionally obtained ironware and wooden utensils from the Laobe and the Inaden respectively, caste-like sub-groups (Bollig 1987). However, such material culture is rarely traded on – except, nowadays, to the odd tourist interested in exotic implements. Wells or enclosures are rarely regular trade goods. While Himba sometimes have their wells dug by employed labourers, such a well could not be sold on. Livestock, of course is (and probably always has been) easily exchangeable. Livestock trade has increased enormously in most pastoral systems in recent decades (Kerven 1992). The consumption needs of rapidly growing urban centres, improved transport facilities and road networks, and a general decline in transaction costs through such things as the introduction of weighing scales (Ensminger 1992), mobile phones (Mahmoud 2003) and so on have facilitated livestock trade. In two contributions to this volume – Little on Somali–Kenyan livestock trade; Meerpohl on livestock trade between Chad and Libya – the international dimension and enormous scale of contemporary pastoral livestock trade across African borders is elucidated. Despite sizeable trade barriers implemented by international boundaries, livestock trade across such boundaries is rapidly gaining in importance and offers new income opportunities for herders.

World-view

Specialised pastoral communities often have world-views in which livestock plays a key role. Ethnographers have developed different concepts to capture this aspect of pastoral specialisation. Herskovits's 'cattle complex', Bonte's 'bovine fetishism', Ferguson's 'bovine mystique', and Comaroff and Comaroff's allusion to 'gods on four legs' (Herskovits 1926; Bonte and Miller 1975; Bonte 1981; Ferguson 1985; Comaroff and Comaroff 1991) suggest that cattle are enshrined in symbolic significations and are attributed religious value. Herskovits had intended to define a cultural area as a complex of cultures which had similarities in many accounts. With Alfred Kroeber's list of culture elements in mind and Clark Wissler's method of delineating culture areas as a methodological base, Herskovits found that many cultures in north-eastern, eastern and Southern Africa display major similarities as cattle are central in their economic as well as in their symbolic systems. The concentration on cattle keeping, according to Herskovits, brought about an inclination to use a bovine idiom when expressing values and beliefs, passion and personal visions. According to Baxter, summarising Herskovits's lengthy article, cattle were regarded as 'the source of the images which express their social and imaginative lives', and 'most enduring social relationships were mediated through the loan, gift or exchange of cattle' (Baxter 2002: 113). Cattle, the providers of cherished food, of social relations and emotional comfort, were only to be slaughtered to mark the great transitional events of life, from birth to final funerary rites. In many of the societies covered by Herskovits's cattle complex, special honour was linked to individual head of cattle, with men taking their honorary names from their favourite oxen, songs being sung for renowned cows, bulls and oxen, and the character traits of cattle being fleshed out in oral lore.

Bonte (1981), drawing on Marx, employed the concept of fetishism to capture the structural effects of a particular mode of organising labour – that of capitalism (Wolf 1999: 34). Bonte applied the concept to East African pastoralists and argued that 'cattle fetishism is thought of and justified as reproducing the supernatural order' (Bonte 1981: 38–39). Peter Rigby, in an application of Bonte's concept to his data on the Parakuyo Maasai, argues that according to Bonte, 'the community level of production relations is inverted and mystified by the fetishism of cattle' (Rigby 1992: 49). In true Marxist fashion, Bonte argues that the cattle idiom is used to mask real differences in pastoral production within a community: 'L'utilisation religieuse du bétail, loin de s'opposer à ses autres formes d'utilisation, ressort d'un même fétichisme du bétail qui masque les conditions réelles de la vie sociale' ('The religious use of cattle, far away from being different from other forms of use, results from the same fetishism of cattle that also covers the real conditions of social life') (Bonte 1992: 561). In contrast to Bonte, Comaroff and Comaroff argue in rather Durkheimian fashion that 'livestock are first and foremost metaphors of social community' (Comaroff and Comaroff 1991: 34), and drawing on their material on the Barolong Tswana they argue that cattle establish 'the threads of which the social fabric is woven' and that cattle are the key to present social identities (ibid.: 45).

Rock art at the southern fringes of the Sahara tells of communities which cherished cattle far beyond their use value. Their images were laboriously marked through tattoos and their horns were meticulously manipulated (Smith 2005). Prehistoric herders in the Sudanese Wadi Howar buried cattle. Poetry and songs are devoted to cattle and they play key roles in all important rituals of a community. In many pastoral communities, livestock features significantly in rituals. There is no ritual among the pastoral Pokot or the pastoral Himba in which livestock are not slaughtered. Slaughtering takes place in a ritualised way. While among the Pokot a spear is thrust through the rips right into the heart, among the Himba the cow is strangled to death. In both communities the dying animal is positioned in a spatial context referring to ancestors. Among the Pokot, men rub themselves with the chyme of slaughtered animals in the course of a ritual. During marriages, Himba couples are smeared with butterfat. Probably one could extend the list of moments of ritual significance of livestock indefinitely.

Pastoral ideologies reflect both the world-view and ways of gaining prestige within a community. In many pastoral communities, the origins of the world, or of a specific ethnic community, is closely tied to the origins of livestock keeping. Famous are the Maasai myths of origin: people descend from the heavens together with their livestock. In other myths of origin, a god presents the founding father of the pastoral community with livestock, while he also presents a hoe to the forefather of neighbouring agriculturalists and a bow and arrow to the ancestors of hunters and gatherers. Pastoral specialisation gains in legitimacy as a divine act.

In the literature, other indicators have been used to specify the continuum of specialisation and diversification: nutrition has been frequently named as one indicator.[7] Communities depend to varying degrees on pastoral produce – that is, food which is either directly derived from herds (milk, meat, blood) or results from the exchange and/or marketing of livestock for other kinds of foodstuff. Comparative evidence shows that there are no contemporary pastoral communities which live of food only derived from their herds. It is also improbable that prehistoric herders were able to live entirely on pastoral food (Casimir and Bollig 1994). However, pastoral Turkana live during major parts of the rainy season mainly on milk from their cattle, camels and goats (Galvin and Little 1999). Throughout most of the year, however, maize bought with money derived from the sale of livestock features significantly. In many pastoral populations discussed in this volume, pastoral produce only makes up a minor part of the daily diet. Many pastoralists (perhaps most) thrive on food they obtain from agricultural producers and/or markets through the sale or barter of livestock. Many contemporary East African herders are switching more and more to a maize-based diet, bargaining on the fact that, in terms of weight, meat exchanged for maize results in six-fold the number of kilo-calories (Bollig and Österle, this volume). Nutrition hence does not seem to be a good indicator of pastoral specialisation.

Mobility is another aspect frequently mentioned as a defining indicator of pastoral specialisation. Livestock herding in arid and semi-arid environments necessitates mobility as resources are unevenly spread seasonally and between years (Asad 1964; Dyson-Hudson and Dyson-Hudson 1969; Salzman 1971). In order to make use of resources which are distributed unpredictably, herders have to be mobile. Some authors have also emphasised the relevance of political factors (such as conflict, violence, avoidance of state led encapsulation) to explain mobility patterns. McCabe et al. point out that 'despite a large number of studies of pastoral peoples, our understanding of pastoralism is inadequate' as 'most of the extensive literature studies present essentialist descriptions and typologies of patterns of mobility, and detailed information on patterns of mobility of specific individuals is lacking' (McCabe, Dyson-Hudson and Wienpahl 1999: 110). The mobility of herds and herders is rarely predictable and uniform but usually adapted to highly diverse household structures, distinct herd structures and annually varying resource distribution. In order to migrate with herds, households depend on able-bodied herders. Far-ranging migrations with small stock are especially troublesome. Goat kids often have to be carried many kilometres: Himba herders often tie some six or eight goats kids to their herding stick and carry them over long distances. Camel herds also need close supervision (Klute 1992; Spittler 1998) as animals tend to stray. When camel herds are taken to distant pastures, the daily work can be done by some children, while it needs at least one experienced herder who is able to check for spoors of straying animals (Spittler 1998). The herds of different households within a community may be fairly different: while one household may own many small stock (limiting mobility and tying the household to the close proximity of water holes), another household may have numerous non-milking stock (enabling mobility). Other households may own sizeable herds of camels, which enables them to utilise pastures far from water holes. There is also a number of highly idiosyncratic features of herders that determine mobility. McCabe (2004), in his study of four Turkana households, makes lucidly clear that some of the decisions taken are linked to personal character traits rather than to clearly identifiable environmental or social variables. Finally, resources in semi-arid areas are fairly unevenly spread from year to year. The lack of sufficient rainfall during a specific part of the year may seriously limit the growth of grasses, while bushes still sprout to give satisfactory grazing for camels and goats. In such years a herder may feel forced to take his cattle to distant places with sufficient grass while small stock and camels may stay. In yet other years rains may be abundant, and so a household can stay in one place throughout the year.

While it is certainly true that specialised herding communities often display higher degrees of mobility, pastoral specialisation and mobility do not directly overlap. Even herders with little propensity to move often may be highly specialised and may simply profit from favourable ecological circumstances. Hence, due to this high degree of intra-community variability, spatial mobility also does not seem to be a convincing indicator for describing specialisation in herder communities.

The Case Studies in this Volume

The case studies presented in this volume show prehistoric, historical and contemporary herders moving between specialisation and diversification. They describe under what conditions specialisation and diversification occur, thereby giving an idea of the diversity and the specific socio-ecological and cultural dynamics of pastoralism in Africa. On the basis of fresh empirical data, contributions by archaeologists, historians and ethnographers address changing patterns of pastoral land use. Phases of pastoral specialisation differed in length covering several decades to several centuries. This fact questions older visions of African pastoralism as a single mode of production or a clearly separable livelihood strategy. We will show that shifts into and out of specialised pastoralism have been rather frequent. Contributions to this volume explain what exactly drove and still drives these shifts. There are a number of likely candidates: aridification (Vansina 2004; Smith 2005), patterns of accumulation (Bonte 1981), patterns of exchange and changing rates of exchange (Bates and Lees 1977), the impact of colonial policies (Bollig, this volume) and the impact of larger markets (Henrichsen, this volume). A closer look at the duration of phases of pastoral specialisation raises questions about the resilience and duration of specialised pastoral systems.

The volume spans pastoral adaptations as diverse as prehistoric foragers husbanding minor numbers of small stock for limited periods of time in north-eastern Africa and Southern Africa to prehistoric pastoralists in the Libyan Desert who maintained a pastoral culture for several millennia (Jesse et al., this volume), to highly specialised pastoralists such as the Herero of Namibia in the late nineteenth century, among whom big men (often referred to as 'chiefs' in historical accounts) owned herds of several thousand animals (Henrichsen, this volume), to the pastoral Pokot of the mid twentieth century who scorned agricultural activities (Bollig and Österle, this volume) and to pastoralists rapidly adapting to globalised flows of finance, information and identities (Galaty; Meerpohl; Little, this volume). On the other hand, there are several case studies where pastoral strategies were integrated into a wider livelihood system: prehistoric South African foragers at times kept some livestock, as did prehistoric hunter-gatherers in East Africa; impoverished herders in north-western Namibia and Kenya at the end of the nineteenth century resorted to a great number to non-pastoral risk-buffering strategies which allowed for survival and the protection of emaciated herds. Contemporary absentee herd owners in north-western Namibia (Schnegg, Pauli and Greiner, this volume) and South Africa's Namaqualand and Lesotho (Berzborn and Solich, this volume) earn their major income as migrants but constantly invest in livestock in order to emphasise a pastoral identity. In a longitudinal comparison of four African pastoral societies, Galaty (this volume) shows how former mobility patterns are adapted to contemporary conditions in a globalised world.

Of course, heterogeneity in pastoral adaptations is not entirely new to anthropologists and archaeologists. Melville Herskovits emphasised a long time ago that the pastoral communities he sampled under the label 'cattle complex' showed a lot of diversity (Herskovits 1926). More recent volumes on pastoralists (Khazanov 1994; Galaty and Johnson 1990; Galaty and Bonte 1991; Fratkin and Roth 2005) also hint at the heterogeneity of communities conventionally labelled pastoralists. Our volume directly addresses pastoral diversity and sheds critical light on concepts such as 'pastoral culture' and 'pastoral mode of production'. Contributions to this volume show that such cultures or modes of production have a lifetime of their own, one which is often astonishingly short. It is possible to specify which processes drive a move to a more specialised pastoral culture or mode of production. We may also deduce mechanisms stabilising a specific livelihood and can name drivers leading to a more diversified mode of production and more diverse symbolic orientations. Moves into a more specialised mode of production or out of it are theoretically interesting. Clearly it is the accumulated decisions of individual actors which underpin such moves. But how does individual behaviour eventually translate into a cultural pattern? Such shifts often seem to be abrupt and not the result of long-term gradual changes. While a cultural pattern is reiterated and elaborated for some time, a kind of threshold (how can such a threshold be meaningfully conceptualised?) is crossed and a new set of rules, values and institutions emerges.

The Emergence and Spread of Pastoralism

Five archaeological contributions present data on the emergence of pastoralism in various regions of the African continent. Kuper and Riemer deal with prehistoric foragers dwelling in the Eastern Sahara between 8000 and 5000 bp. They included some livestock in their livelihood portfolio, but seemingly were reluctant to change to a more specialised pastoral mode of production. Livestock were probably utilised in an effort to diversify a hunting-and-gathering mode of existence. Jesse et al. describe prehistoric pastoral populations in the Sudanese part of the Libyan Desert. While the herding populations they describe are definitely 'more' pastoral than the pastro-foragers of the Egyptian Desert discussed by Kuper and Riemer, these populations still relied on gathered plant food to a significant degree. Despite their reliance on grass seeds and other gathered food though, these people may have felt themselves to be herders. Cattle featured large in their symbolic world as rock art from the wider region impressively shows. Burials of entire bodies of dead cattle also hint at the religious significance of cattle. This pastoral system was extremely resilient and – as archaeological sources indicate – remained fairly unchanged for about 2,000 years.

Lane shows for East Africa how pastro-foragers and pastoralists with different specialisations co-existed for protracted periods of time. Only later on, in the northern lowlands of Kenya by around 4200 bp, by about 3800 bp in eastern Kenya and by about 3400 to 3000 bp in highland central, southern and western Kenya, more specialised modes of pastoralism relying on cattle and small stock developed, apparently in conjunction with the specialisation of agricultural livelihoods. Niche specialisation in this context seems to be interdependent: under similar conditions, specialising either in pastoralism or agriculture seems to have been opportune (and communities profit from interdependent specialisation through exchange and markets, as Bates and Lees 1977 would argue), whereas under other circumstances a widely diversified portfolio of different livelihood strategies is more appropriate. Lane also describes how pastoralists have changed the landscapes they dwelled in. Specialised pastoralists may have significantly contributed to the establishment of grasslands through applying fire to wooded stretches of land.

The situation again seems to have been fairly different in West Africa. Linseele analyses archaeozoological data from northern Burkina Faso and from the Bama Deltaic Complex and the *firgi* area of north-eastern Nigeria. The data indicate that there was considerable regional variation, despite some general trends. By the second millennium BC, the first domestic animals – cattle, sheep and goats – had reached certain parts of the West African savannah. They apparently originated from north-eastern Africa. Also around the second millennium BC, the first crops were being grown in the area. However, full farming only appeared at the beginning of our era. This seems to have allowed the development of economically specialised groups, including pastoral nomads. The latter also profited from the appearance of new types of domestic animals.

In contrast to the prehistoric development of pastoralism in western and eastern Africa, pastoral strategies remained peripheral until about 1,000 years ago south of the Zambezi, as Sadr discusses. In Southern Africa domesticated livestock was first introduced about 2,000 years ago, but it took a long time – possibly until the early colonial period, Sadr argues – before pastro-foragers became pastoralists. While there is a distinct increase in livestock holdings in south-eastern Africa in the sixth and seventh century (Hall 1987: 46–60), this intensification of pastoral strategies occurred within an agricultural context. Very clearly, an increasing accumulation of cattle was linked to increased stratification within these societies, and the rulers of Great Zimbabwe seem to have held extensive herds of cattle (Garlake 1978). Some authors have even explained the demise of Mapungubwe (Voigt 1984) and Great Zimbabwe (Beach 1980) as being to do with severe overgrazing problems. Sadr clearly elaborates that such pastoral intensification has not taken place in south-western Africa: the few indications of the intensification of pastoralism are transient moments until the sixteenth century.

Henrichsen as well as Bollig and Österle show how even over shorter time spans than those conventionally addressed by archaeologists, the significance of livestock for subsistence differs within a population. Processes of pastoralisation and repastoralisation have occurred repeatedly within a period of two hundred years in quite a number of African pastoral societies. While the ancestors of the pastoral Pokot of Kenya were a highly heterogeneous set of foragers, agriculturalists and pastoralists until the early nineteenth century (and did not form an ethnic unit according to their own recollections), they rapidly developed a highly specialised pastoral economy in the second part of the nineteenth century, along with clearly demarcated ethnic boundaries. This was elaborated during the twentieth century, but at the turn to the twenty-first century, the pastoral Pokot not only integrated sedentary agriculture into their livelihood portfolio at an amazing pace but also widened their value system to include non-pastoral orientations. Henrichsen reports how Herero herders were forced to abandon pastoral strategies for some decades due to constant raiding from other, better-armed communities from the 1830s to the 1850s. When pastoral options became feasible again, another form of pastoralism developed, characterised by the ownership of large herds by a few big men and the articulation of pastoral livestock husbandry and larger markets. The great rinderpest epidemic of 1897/98 and the genocidal war of the German Empire against the Herero brought an end to this phase of pastoralisation.

States and the Dynamics of Pastoral Specialisation and Diversification

The advent of the colonial state and its successor, the independent nation-state, has probably been the most influential determining factor of the dynamics of pastoral economies and societies. Not only were pastoral communities encapsulated in reserves, pastoral districts, homelands and the like, but spatial mobility was also severely curtailed. Independent states seamlessly continued the project of the containment of mobile communities, of the enforcement of boundaries and of governmental control over pertinent assets. In some instances the colonial state instituted a new kind of specialised pastoralism. Commercial ranching – a topic rarely dealt with in anthropological accounts of pastoralism – developed fairly different patterns of pastorally orientated lifestyles in the twentieth century in Africa's (former) settler colonies. Although there are a number of sizeable differences with other forms of pastoral adaptation (such as pasture held in private property, herding through employed shepherds), there are also parallels (susceptibility to drought and market fluctuations, the problem of controlling herding labour). It was mainly state subsidies which transformed fairly diversified small-scale settler farm enterprises into cattle ranches in Namibia in the mid twentieth century. Botha argues that a drive towards pastoral specialisation was carefully planned by the administration of the South African Apartheid regime. Major investments in cattle ranching – in fences, boreholes, improved breeds, veterinary drugs and so on – were made possible through government subsidies. Dieckmann concentrates on settler pastoralism in north-western Namibia's Outjo district. She shows how settler farms started off as small-scale peasant farms applying a number of highly diversified economic strategies. Dieckmann also describes to what extent such farms were dependent on cheap African labour. In the Outjo district in the 1950s there was a distinct switch towards more livestock husbandry and pastoral specialisation – then with a distinct focus on meat production – furthered by state-led subsidies and extension programmes.

Bollig shows how herders in north-western Namibia were forced out of the pastoral niche through violent encounters with raiders in the late nineteenth century before the advent of state control. Their turn towards foraging was enforced by outside forces beyond their control. Oral traditions recall that ancestors occasionally slaughtered their remaining animals in order to escape being the target of raiders. As soon as the threat of violence was contained by the emergent colonial state, they again invested labour and capital in pastoral strategies. However, it was mainly the drilling of hundreds of boreholes in the second half of the twentieth century which contributed to pastoral intensification in the region and led to the abandonment of foraging strategies.

Schnegg, Pauli and Greiner, focusing on Nama- and Damara-speaking communities in Fransfontein, western Namibia, show how land loss to settlers (the very settlers Dieckmann deals with) and reserve policies by the Apartheid government lead to impoverishment and the continued need to diversify. Due to continued land loss to commercial farms and the increasing control of the colonial administration, hunting soon had to be discarded and space for a comprehensive expansion of agricultural strategies became extremely limited. Labour on white-owned farms and on mines became not only an option but a necessity. Successful labour migrants regularly invested in livestock back home. Their herds were herded by relatives and progressively by hired herders from northern Namibia and/or southern Angola (Greiner 2008). The Fransfontein case hints at the challenges of our specialisation/diversification trajectory: only a minor segment of the population nowadays invests a large amount of labour in pastoral strategies – these are either relatives or labourers of wealthy absentee herd owners. The majority of the population invests labour in and allocates capital to a variety of off-farm activities, despite the fact that most (men) dream of a pastoral future. Food for the entire population is mainly store-bought and pastoral produce only makes up a minor percentage of food. Capital accumulation and allocation has a pastoral bias only among the few wealthy households of the community. They invest in wells, fences, cars, veterinary medicine, troughs and enclosures and will invest gains in more livestock while many simply cannot afford livestock beyond a few head of small stock.

In rare cases the state is dominated by a pastoral community – which then may lead to an expansion of pastoralism. Jánszky and Jungstand report that violent strategies in Chad have led to an expansion of pastoral activities. The Chadian Zaghawa have expanded violently into Tama territory, taking advantage of the politics of state tutelage, clientelism and tribalism: since 1990, Chad has had a Zaghawa president who has succeeded in putting close relatives in key political offices. The Zaghawa migrate seasonally with their flocks into the land of the Tama. Increasingly they also settle more permanently there. Backed by state authorities, the Zaghawa do not fear claims for compensation. The fact that they are better equipped with weapons has protected them from vigilante justice on the part of the Tama. State-condoned violence in the Chadian case has led to an expansion of pastoralism into lands previously held by agriculturalists. The recent pattern of conflicts in Darfur, Sudan, suggests a similar scenario: Arab-speaking pastoralists have violently gained ground in the region due to state protection and state investment in pastoral militias (De Waal 1989).

Beyond the State: Pastoralism in the Age of Globalisation

Since the early 1990s the state has been in retreat in many pastoral regions. The total demise of state power in pastoral Somalia is the most extreme case. In many other instances, however, planned decentralisation, violence and conservation led by international NGOs have led to the decline of governmentality in pastoral regions. In other cases states willingly declined to control the international trade activities of pastoralists or were simply too weak to effectively scrutinise and tax them. Trading livestock across international boundaries became an important livelihood strategy in many pastoral populations during the past decade. While in some instances traditional caravan routes were rejuvenated, in other cases international trade routes were newly developed. Of course, trade in livestock is age-old and barter with livestock has been documented in many pastoral settings. Cattle were exchanged for blades and spearheads, pottery, beads and so on by Maasai herders. East African pastoralists as much as West African herders regularly paid itinerant craftsmen with livestock for their services (Bollig 1987: 185–95). In much of East and West Africa, inter-regional trade in livestock was facilitated by colonial regimes. Kerven (1992) has shown that the colonial impact at least in northern Kenya was ambiguous: at times livestock trade by Somali traders was encouraged and at times it was discouraged. In general though, we see an increase in cash-based livestock trade throughout the twentieth century. The direction of this trade is unidirectional: whereas pastoralists were selling and buying livestock in the early twentieth century (for example, Pokot herders exchanged cattle with Somali merchants for sheep), since the 1930s pastoralists were generally suppliers of livestock to meat markets. Ensminger (1992) and Bollig (2006: 95) report that the development of pricing livestock during the twentieth century has been generally favourable for pastoral producers in Kenya: the livestock: maize exchange ratio improved consistently throughout the century.

During the past two decades inter-regional and inter-state trade in livestock increased significantly throughout Africa: both Little and Meerpohl deal with aspects of large-scale trade in livestock across international boundaries in two distinct settings. While Little describes the growth and organisational capacity of the cattle trade between Somalia and Kenya, Meerpohl deals with the camel trade across the Chadian–Libyan boundary. In Little's case, Somali livestock traders bargain on differential prices for cattle on both sides of the boundary: per head of cattle they gain considerably more on Nairobi's livestock market than on any livestock market in Somalia. The material presented by Little shows how cross-border livestock trade drives the economies of the Horn of Africa's border regions and areas well into the interiors of Ethiopia, Somalia and Kenya. This trade is highly relevant not only for inhabitants of the border region but also for consumers and producers located several hundred kilometres from the borders. In Meerpohl's case, camels are driven in large caravans from Abéché in eastern Chad across one of the most inhospitable deserts of the world to southern Libya, where they are sold to provide meat for Libya's and Egypt's urban populations.

The gains from the sale of camels are invested in sugar which is then transported back to Chad on lorries and enters markets there. In both cases the development of complex international trade networks contributes significantly to pastoral livelihoods. They provide the necessary cash to buy food and consumer items at local markets. In this way emergent larger-scale markets sustain pastoral specialisation in both cases – that is, without such markets local herders would have to invest more time in off-farm income-generating strategies. While increasing inter-regional and international trade is opening up new options for pastoralists, conflicts work in the opposite direction.

Pastoralism in southern and eastern Africa has been heavily influenced by state-led national conservation policies, discussed in the chapter by Berzborn and Solich. In general, pastoralism has been regarded as environmentally problematic and state policies have sought to control pastoral land use through restricting mobility and/or fixing carrying capacities. African pastoralists were also considered to be detrimental to wildlife, even if game had been severely reduced by European hunters (e.g., Adams and McShane 1996). As a consequence, herders were in many instances evicted from nature conservation areas. In other settings they were intentionally contained together with wildlife in emergent conservation areas, as many romantic conservationists firmly believed that game and traditional herders could exist in harmony. In this way the Maasai were moved from their aboriginal rangelands into what are today the Amboseli and Maasai Mara parks. Due to international pressure and increasing local resistance since the 1980s, models of 'fortress' conservation made way for community-based programmes, which sought the participation of local herders in conservation initiatives. Recently endorsed Transboundary Natural Resource Management (TBNRM), which led to the emergence of major park areas, again tended to ignore the needs of local farmers.

In a final comparative chapter, John Galaty discusses the 'modernity' of strategies pursued by pastoralists as they respond to the constraints imposed on them by modern nation-states and the possibilities brought about by globalisation. Focusing on the institutions of pastoral territoriality, he compares four pastoralist societies from across the African continent. While he first looks at the interrelation between seasonality, local ecology and socio-political institutions, he moves on to compare the impacts of the state and global dynamics on pastoral mobility and territoriality. In this second part of the chapter, Galaty shows how sedentarisation, population growth and encroachment on pastoral rangelands by non-pastoral populations, the growth of small towns, increasing social stratification and the appropriation of key resources by elites, the spread of arms and low-intensity war and a diverse set of development projects change the perspectives of African pastoralists on pastoral diversification and specialisation. The intensification and commercialisation of pastoral strategies have often been the aim of state-led tenure changes favouring privatisation over communal forms of tenure, encouraging marketing structures and the commoditisation of livestock husbandry. Galaty's contribution impressively shows that pastoral strategies, both specialised and diversified, have a place in Africa's modernity.

Specialisation and Diversification

The contributions to this volume show that pastoral livelihoods are more diverse and undergo more rapid and profound changes than generally assumed. Highly specialised pastoral economies often only last a number of decades before herders diversify their livelihood strategies once again and homogeneous land use patterns are replaced by more fragmented land use. Specialised forms of pastoral production develop within a few decades if incentives are stable. Our cases show frequent transitions from diversified to specialised and vice versa, and we will attempt to sketch some few commonalities here.

Why Specialisation?

In Bollig's study of the development of pastoralism among the Himba of northern Namibia (this volume), he shows that livestock husbandry underwent significant changes during the last century. While Himba combined foraging and pastoral strategies at the beginning of the period studied, a sole focus on livestock emerged when, first, colonial forces ended a period of severe raiding at the fringes of the Cape-based trade system; second, when the colonial administration severely curtailed hunting; and third, when external exchange of livestock was inhibited and rich herders were forced to invest in internal exchange and redistribution. Through encapsulation and the curtailment of other livelihood strategies (foraging, trading), the colonial state clearly set a framework which worked in favour of pastoral specialisation. Despite inhibiting exchange with agricultural producers and/or larger markets, specialisation took place and wealth was progressively documented via the accumulation of herds. Throughout most of the twentieth century the Himba gained most of their economic resources through pastoralism, maintained social ties via the exchange of cattle and represented symbolic capital in a bovine idiom. While many women maintained gardens during the short rainy season, the major part of labour went into livestock husbandry. Capital investment was also mainly allocated in ways that improved the household herd or facilitated herd management. Pastoral specialisation and intensification was accelerated when the Apartheid government started a large-scale campaign to drill more than 400 boreholes in the area from 1954 onwards. The availability of year-round water allowed a much more intense pastoral use of the land and allowed much higher stocking rates than before. Pastoral specialisation in this case study developed within a framework in which the state established conditions that encouraged pastoral specialisation: the state-led interventions provided new land and water at very low cost, established rudimentary veterinary services and (in early colonial times) brought an end to raids by militarily superior groups targeting Himba herds.

Dag Henrichsen's contribution to this volume on the history of (re)-pastoralisation of the Namibian Herero in the nineteenth century also shows that diversification and specialisation are not always self-selected strategies but sometimes adaptations to violence and external drivers. The Herero were forced into diversification in the mid nineteenth century when raiders stole their cattle. For two to three decades many Herero lived as foragers and some turned to agriculture around mission stations. They often combined these subsistence strategies with some temporary livestock husbandry. Only when the Herero gained the upper hand militarily in these conflicts in the 1860s did pastoralisation pay off once again. As before, the accumulation of wealth through amassing livestock became an option, and exchange with larger markets a highly lucrative option. Within just two decades Herero had switched from a diversified livelihood to a more specialised one. In this situation increasing pastoral specialisation was a way to reclaim resources (land and wells), to access markets and to entertain social ties and political dominance. As in the East African cases raiding is linked to pastoral specialisation; however, as long as Herero and Himba were less well armed than their adversaries they suffered sizeable losses and raiding discouraged any form of pastoral specialisation. As soon as the Herero gained in military strength, however, raiding contributed to rapid pastoral expansion. Apart from the political situation, pastoral specialisation was further supported by their integration into regional trading networks that converted cattle into horses and guns, which were used to build and support a power base. This in turn allowed for the accumulation of ever more cattle through raiding. Pastoral specialisation in this case was conditioned by external forces (better access to weapons, emergence of livestock markets, high demand for cattle) and internal factors (easy access to labour by cattle-owning big men, a pastoral ideal of accumulating livestock and expressing relations of dominance through livestock exchanges).

Pastoral specialisation among the Herero lasted some three or four decades and then abruptly came to an end when rinderpest killed up to 95 per cent of Herero cattle in 1897 (Gewald 2000). The Herero settling in the reserves in the 1920s still maintained some herds but a major part of the population at that time was already living from migrant labour (Werner 1998). The phase of extreme pastoral specialisation was short and only lasted from the 1860s to the 1890s. As among the Himba, pastoral land became available at relatively low cost: land could be obtained through force relatively easily. However, only when all three factors (low cost of land, access to markets via the exchange of cattle, and accepted accumulation pathways of wealth through herds) historically coincided, did a massive move towards specialisation take place. This made specialisation for more and more households attractive and led herds to grow considerably.

Among the commercial ranchers in Namibia described by Dieckmann (adopting a local perspective) and Botha (adopting a national perspective focusing on the development of commercial pastoralism in Namibia), the circumstances were rather different. The colonial state had paved the way towards pastoral specialisation for settlers. While the economy of settler households was rather diversified during the first half of the century, including gardening and milk production, and in Dieckmann's case also hunting, the state fostered pastoral specialisation through grants and subsidies.[8] Massive support in infrastructure, including farm roads, fences and boreholes, and exploitative labour legislation, facilitated the use of privately owned land for the rearing of large herds of cattle. In addition the state provided cheap money and risk-reducing mechanisms that allowed for intensified farming. Even though the state heavily supported and fostered the shift towards specialisation in beef farming, the latter never became the sole strategy, especially as long as farmers still had to pay off the farms or other loans, or in times of drought. Again, land was provided at very low cost which paved the road to specialisation. Additionally, specialised farmers were guaranteed preferential market access and could exchange livestock at favourable rates.

Bollig and Österle show for the Pokot of Kenya how pastoralism emerged as a dominant strategy from a much more diverse bundle of economic strategies including foraging and small-scale dryland agriculture in the first half of the twentieth century. The Pokot which Bollig worked with during the 1980s were almost entirely dependent on livestock. Colonial encapsulation between the early 1900s and early 1960s contributed to maintaining this type of specialisation rather than to its change! In search of the causes of pastoral specialisation in the early nineteenth century, the authors point to recent results from climate research that has identified an inter-decadal drought between 1760 and 1840 for eastern Africa (Verschuren 2004). The inter-decadal drought made cultivation in the catchment of Lake Baringo almost impossible, and the severe desiccation of the region may have encouraged the use of rangelands with mobile herds. Galaty (1993) for the Maasai and Lamphear (1988) for the Turkana observe similar specialisation dynamics for roughly the same period. Under conditions of desiccation, mobile pastoral strategies offered advantages over sedentary agro-pastoral strategies. Competition between herders led to a strong emphasis on ethnic boundaries. In the East African cases, pastoral specialisation according to oral accounts is inextricably linked to raiding. Raids at the same time offer a chance to spur the growth of herds, while on the other hand they present a risk: the bigger the herd, the more they are likely to be the target of raiders. Successful raiders spur the process of pastoral specialisation. Pastoral specialisation apparently was concomitant with the advent of irrigated agriculture – that is, more specialised forms of agriculture – in the region. The chapter by Jánszky and Jungstand on the expansion of Zaghawa pastoralism in Chad also exemplifies this argument. Violent strategies and military superiority (here guaranteed by a clientelistic state) lead to a rather rapid increase in pastoral specialisation. Agro-pastoral communities lose access to resources to militarily superior pastoral communities.

The comparative analysis of the case studies reveals that in all cases where people started to specialise in pastoral production land had become accessible either through institutional, technological or climatic change or through warfare. In the case of the Himba it was boreholes that opened new lands at still relatively constant population rates; among the Herero military superiority paved the way to accessing new pastures in the 1860s; among European settlers the massive technological support of the state made land accessible; among the Pokot an inter-decadal drought made farming almost impossible and prompted people into pastoral specialisation. In several cases, pastoral specialisation is linked to improved exchange options: livestock may be exchanged at favourable rates with other groups. In all the cases, new circumstances (altered relative prices) made specialised herding attractive to households, and only in one case are environmental processes apparently linked to a move towards more specialisation.

What are the costs of specialisation? In general, highly specialised pastoral communities are vulnerable to a number of factors. Droughts may destroy herds within a few months. Livestock diseases may finish entire livestock populations within short periods of time (as rinderpest did in the 1890s), and droughts may have profoundly negative effects on herds. Raiders may steal great numbers of livestock at a stroke and push households and/or communities out of pastoral livelihoods. Unlike farmers, who can harvest from their fields again if the rains are good the next year, herders who have lost significant parts of their herds are hampered for a number of years. Is pastoral specialisation therefore economically irrational as the risks are too high? Probably not: under benevolent conditions – military superiority or state administration guaranteeing peaceful conditions or patronage, favourable rates of exchange and pathways of accumulation linked to herds and the absence of long-term drought conditions – cattle herds grow exponentially. If land is available at low cost and exchange systems are in place, pastoral specialisation is a rapid means of amassing personal property.

Why Diversification?

Diversification has been pointed to as an effective strategy for buffering risk or reducing the vulnerability of the individual or the household. Galvin (2009), in a recent review article on pastoral development, also showed the downside of diversification: increasing compartmentalisation of land use and increasing inequality. The case studies collected in this volume show a wide range of examples for diversification and also show that diversification is not only about risk management.

In their chapter, Schnegg, Pauli and Greiner show convincingly how colonisation created very different livelihood opportunities for European settlers and indigenous populations. Indigenous people were forced to live in reserves that were well known to be too small to feed their populations from livestock production alone. This forced people into the labour market and made cheap manpower available to the industries supported by the Apartheid state (such as ranching). However, the symbolic and cultural relationship with their home communities and the value of cattle led (and still leads) many urban migrants to invest in livestock in the Fransfontein area. This put increasing stress on the ecosystem and led to a situation where those left behind became more and more vulnerable. The main cause behind the economic diversification process described in the chapter is often state politics, resulting in sizeable land shortages and a delimitation of pastoral mobility.

Bollig and Österle show for the Pokot of Kenya that after a period as specialised pastoralists, livelihoods have diversified over the last few decades. Today the Pokot economy includes small-scale gardening, wage labour and migration. The main drivers behind diversification were demographic growth (and restricted space or the lack of opportunity to expand spatially). This changed the relative price of land making it more valuable. At the same time, the market created new opportunities to engage in trade.

Meerpohl shows in her chapter that the livelihoods of the Zaghawa of Chad were dominated by cattle pastoralism till the 1960s, when a series of severe droughts between 1968 and 1985 killed most of their livestock. This had two major effects. On the one hand, many people shifted from cattle to less vulnerable goats and camels; on the other hand, many families started to migrate to urban areas where they worked as wage labourers. This created new economic opportunities and allowed restocking after the earlier drought-induced crisis. While in the past cattle were used in subsistence farming, camels now became a good traded between Chad and Libya. Meerpohl shows that the diversification of pastoralism was triggered by critical climatic events. At the same time, new trade opportunities have created markets in Libya in which camels from Chad and other goods are traded. Pastoralism in this case has thus become a link in the global network of goods that brings sugar from the Rhine via Libya's Mediterranean ports in exchange for camels from Chad.

In combination, the cases studies show that diversification often occurs if land falls short. The trigger may be demographic growth, as in the case of the Pokot, or a shortage of land due to political processes and land allocations, as in Fransfontein. Increasing inequalities are also linked to diversification and land use fragmentation in all the case studies. In addition to these factors, severe crises have the potential to force larger parts of the population out of the pastoral mode of production, leading to new economic strategies at the household level. This is what we see among the Zaghawa.

In sum, we find no case in which diversification occurred without some factors pushing people into new modes of production. Diversification does not seem to be a strategy pastoralists choose if they do not have to. At the same time, many specialise as soon as circumstances present them with the opportunity. This may again hint at the fact that pastoralism is a risky but also a profitable economic strategy, with the exponential growth of herds being a potential pathway to accumulation in times of plenty.

Notes

1. For more on ACACIA, see: http://www.uni-koeln.de/sfb389/.
2. See Dyson-Hudson and Dyson-Hudson (1980), Galaty and Johnson (1990: 1), Fratkin (1997) and Galaty (this volume) for contemporary societies.
3. The diversity of pastoral systems is well documented in Galaty and Bonte (1991).
4. Bates and Lees (1977) give some non-African examples.
5. See, e.g., Greiner (2008) for north-western Namibia, and Ensminger (1992) for the Kenyan Orma.
6. On pastoral labour, see Beck and Klute (1991); for ethnographic accounts of the complexity of pastoral labour, see Klute (1992) and Spittler (1998).

7. For a review of this issue, see Casimir (1991).
8. For a case from southern Namibia, see Silvester (1998).

References

Adams, J.S.G., and T.O. McShane. 1996. *The Myth of Wild Africa: Conservation without Illusion*. Berkeley: University of California Press.

Anderson, D.M. 2002. *Eroding the Commons: The Politics of Ecology in Baringo, Kenya, 1890s–1963*. Oxford: James Currey.

Asad, T. 1964. 'Seasonal Movement of the Kababish Arabs of Northern Kordofan', *Sudan Notes and Records* 45: 45–58.

Bates, D., and S. Lees. 1977. 'The Role of Exchange in Productive Specialization', *American Anthropologist* 79: 824–41.

Baxter, P.T.W. 2002. 'Cattle Complex', in A. Barnard and J. Spencer (eds), *Encyclopedia of Social and Cultural Anthropology*. London: Routledge, p.113.

Beach, D.N. 1980. *The Shona and Zimbabwe, 900–1850: An Outline of Shona History*. London: Heinemann.

Beck, K., and G. Klute. 1991. 'Hirtenarbeit in der Ethnologie', *Zeitschrift für Ethnologie* 116: 91–124.

Bollig, M. 1987. 'Ethnic Relations and Spatial Mobility in Africa: A Review of the Peripatetic Niche', in A. Rao (ed.), *The Other Nomads: Peripatetic Minorities in Cross-cultural Perspective*. Cologne: Böhlau, pp.179–228.

——— 2006. *Risk Management in a Hazardous Environment: A Comparative Study of Two Pastoral Societies*. New York: Springer.

——— 2008. 'Peripatetic Production', in J. Middleton and J.C. Miller (eds), *New Encyclopedia of Africa*. Detroit: Tomson and Gale, pp.252–57.

——— 2009. 'Kinship, Ritual, and Landscape Amongst the Himba of North-west Namibia', in M. Bollig and O. Bubenzer (eds), *African Landscapes: Interdisciplinary Approaches*. New York: Springer, pp.327–52.

Bollig, M., et al. 2007. *Aridity, Change and Conflict in Africa*. Cologne: Heinrich-Barth-Institut.

Bolten, A. et al. 2009. 'Towards a Reconstruction of Land Use Potential: Case Studies from the Western Desert of Egypt', in M. Bollig and O. Bubenzer (eds), *African Landscapes: Interdisciplinary Approaches*. New York: Springer, pp.57–77.

Bonte, P. 1981. 'Marxist Theory and Anthropological Analysis: The Study of Nomadic Pastoralist Societies', in J.S. Kahn and J.R. Llobera (eds), *The Anthropology of Precapitalist Societies*. London: Macmillan, pp.22–56.

——— 1992. 'Pastorals (Sociétés)', in P. Bonte and. M. Izard (eds), *Dictionnarie de l'Ethnologie et de l'Anthropologie*. Paris: Presses Universitaires de France, pp.561–62.

Bonte, P., and J.M. Miller. 1975. 'Cattle for God: An Attempt at a Marxist Analysis of the Religion of East African Herdsmen', *Social Compass* 22(3/4): 381–96.

Bubenzer, O., M. Bollig and H. Riemer. 2006. 'Kultur- und Landschaftswandel im Ariden Afrika', in H. Gebhardt et al. (eds), *Geographie und Humangeographie*. Munich: Elsevier, pp.550–57.

Casimir, M.J. 1991. *Flocks and Food: Biocultural Approach to the Study of Pastoral Foodways*. Cologne: Böhlau.

Casimir, M.J., and M. Bollig. 1994. 'Milchverarbeitung und Nahrungsmittelknappheit bei Afrikanischen Pastoralisten', in M. Bollig and F. Klees (eds), *Überlebensstrategien in Afrika*. Cologne: Heinrich Barth, pp.227–53.
Comaroff, J., and J.L. Camaroff. 1991. '"How Beasts Lost their Legs": Cattle in Tswana Economy and Society', J. Galaty and P. Bonte (eds), *Herders, Warriors, Traders: Pastoralism in Africa*. Boulder, CO: Westview Press, pp.33–61.
Cronk, L. 2004. *From Mukogodo to Maasai: Ethnicity and Cultural Change in Kenya*. Boulder, CO: Westview Press.
De Waal, A. 1989. *Famine that Kills: Darfur, Sudan*. Oxford: Clarendon Press.
Dyson-Hudson, R., and N. Dyson-Hudson. 1969. 'Subsistence Herding in Uganda', *Scientific American* 220(2): 76–80.
——— 1980. 'Nomadic Pastoralism', *Annual Review of Anthropology* 9: 15–61.
Ensminger, D. 1992. *Making a Market: The Institutional Transformation of an African Society*. New York: Cambridge University Press.
Evans-Pritchard, E.E. 1940. *The Nuer: A Description of the Modes of Livelihood and Political Institutions of a Nilotic People*. Oxford: Clarendon Press.
Ferguson, J. 1985. 'The Bovine Mystique: Power, Property, and Livestock in Rural Lesotho', *Man* 20: 647–74.
Fratkin, E. 1997. 'Pastoralism: Governance and Development Issues', *Annual Review of Anthropology* 26: 235–61.
Fratkin, E., K.A. Galvin and E.A. Roth (eds). 1994. *African Pastoralist Systems: An Integrated Approach*. London: Lynne Rienner.
Fratkin, E., and E. Roth. 2005. *As Pastoralists Settle: Social, Health, and Economic Consequences of the Pastoral Sedentarization in Marsabit District, Kenya*. New York: Springer.
Galaty, J. 1993. 'Maasai Expansion and the New East African Pastoralism', in T. Spear and R. Waller (eds), *Being Maasai: Ethnicity and Identity in East Africa*. London: James Currey, pp.61–86.
——— 1996. 'Pastoralists', in A. Barnard and J. Spencer (eds), *Encyclopedia of Social and Cultural Anthropology*. London: Routledge, pp.415–16.
——— 2005. 'Double-voiced Violence in Kenya', in V. Broch-Due (ed.), *Violence and Belonging: The Quest for Identity in Post-colonial Africa*. London: Routledge, pp.173–94.
Galaty, J., and P. Bonte (eds). 1991. *Herders, Warriors and Traders: Pastoralism in Africa*. Boulder, CO: Westview Press.
Galaty, J., and D. Johnson (eds). 1990. *The World of Pastoralism: Herding Systems in Comparative Perspective*. London: Guilford Press.
Galvin, K. 2009. 'Transitions: Pastoralists Living with Change', *Annual Review of Anthropology* 38: 185–98.
Galvin, K., and M. Little. 1999. 'Dietary Intake and Nutritional Status', in M. Little and P.W. Leslie (eds), *Turkana Herders of the Dry Savanna: Ecology and Biobehavioural Response of Nomads to an Uncertain Environment*. Oxford: Oxford University Press, pp.125–46.
Garlake, P. 1978. 'Pastoralism and Zimbabwe', *Journal of African History* 19: 479–93.
Gewald, J. 2000. 'Colonisation, Genocide and Resurgence: The Herero of Namibia 1890–1933', in M. Bollig and J. Gewald (eds), *People, Cattle and Land: Transformation of a Southwest African Pastoral People*. Cologne: Köppe, pp.187–226.
Göbel, B. 1997. '"You Have to Exploit Luck": Pastoral Household Economy and the Cultural Handling of Risk and Uncertainty in the Andean Highlands', *Nomadic Peoples* 1(1): 37–53.
Greiner, C. 2008. *Zwischen Ziegenkraal und Township: Migrationsprozesse in Nordwestnamibia*. Berlin: Dietrich Reimer Verlag.

Hall, M. 1987. *The Changing Past: Farmers, Kings and Traders in Southern Africa.* Cape Town: David Phillip.

Helland, J. 1982. 'Social Organization of Water Control among the Borana', *Development and Change* 13: 239–58.

Herskovits, M. 1926. 'The Cattle Complex in East Africa', *American Anthropologist* 28(1): 230–72.

Hodder, I. 1982. *Symbols in Action: Ethnoarchaeological Studies of Material Culture.* Cambridge: Cambridge University Press.

Hogg, R. 1986. 'The New Pastoralism: Poverty and Dependency in Northern Kenya', *Africa* 56: 319–33.

Holtzman, J.D. 1997. 'Gender and the Market in the Organization of Agriculture among Samburu Pastoralists', *Research in Economic Anthropology* 18: 93–113.

Ingold, T. 2009[1985]. 'Pastoralism', in A. Kuper and J. Kuper (eds), *The Social Science Encyclopedia.* London: Routledge, pp.710–11.

Kerven, C. 1992. *Customary Commerce: A Historical Reassessment of Pastoral Livestock Marketing in Africa.* London: Overseas Development Institute.

Khazanov, A.M. 1994. *Nomads and the Outside World.* Madison: University of Wisconsin Press.

Klute, G. 1992. *Die Schwerste Arbeit der Welt: Alltag von Tuareg-Nomaden.* Munich: Trickster.

Kuper, R., and S. Kröpelin. 2006. 'Climate-controlled Holocene Occupation in the Sahara: Motor of Africa's Evolution', *Science* 313: 803–7.

Kurita, K. 1982. 'A Market on the Boundary: The Economic Activities of the Pokot and the Marakwet in Kenya', *Africa Studies Monographs* (supplementary issue) 1: 71–103.

Lamphear, J. 1988. 'The People of the Grey Bull: The Origin and Expansion of the Turkana', *Journal of African History* 29: 225–43.

McCabe, J.T. 1997. 'Risk and Uncertainty among the Maasai of the Ngorongoro Conservation Area in Tanzania: A Case Study in Economic Change', *Nomadic Peoples* 1(1): 54–65.

——— 2004. *Cattle Bring Us to Our Enemies: Turkana Ecology, Politics, and Raiding in a Disequilibrium System.* Ann Arbor: University of Michigan Press.

McCabe, J.T., R. Dyson-Hudson and J. Wienpahl. 1999. 'Nomadic Movements', in M. Little and P.W. Leslie (eds), *Turkana Herders of the Dry Savanna: Ecology and Biobehavioural Response of Nomads to an Uncertain Environment.* Oxford: Oxford University Press, pp.109–21.

McCabe, J.T., P.W. Leslie and L. DeLuca. 2010. 'Adopting Cultivation to Remain Pastoralists: The Diversification of Maasai Livelihoods in Northern Tanzania', *Human Ecology* 38(3): 321–34.

Mahmoud, H. 2003. 'The Dynamics of Cattle Trading in Northern Kenya and Southern Ethiopia: The Role of Trust and Social Relations in Market Networks', PhD dissertation. Lexington: University of Kentucky.

Malan, S. 1973. 'Double Descent Among the Himba of South West Africa', *Cimbebasia* 2: 81–112.

Mankiw, N.G. 1998. *Principles of Economics.* Fort Worth, TX: Dryden.

Mkutu, K.A. 2008. *Guns and Governance in the Rift Valley: Pastoral Conflicts and Small Arms.* Oxford: James Currey.

Netting, R. 1981. *Balancing on an Alp: Ecological Change and Continuity in a Swiss Mountain Community.* Cambridge: Cambridge University Press.

Österle, M. 2008. 'Innovation und Transformation bei den Pastoralnomadischen Pokot (East Pokot, Kenia)', PhD dissertation. Cologne: Universität zu Köln.

Rao, A. 1995. 'From Bondsmen to Middlemen: Hired Shepherds and Pastoral Politics', *Anthropos* 90: 149–67.
Rao, A., and M. Casimir. 2003. *Nomadism in South Asia.* Oxford: Oxford University Press.
Rigby, P. 1992 *Cattle, Capitalism and Class: Ilparakuyo Maasai Transformations.* Philadelphia, PA: Temple University Press.
Salzman, P.C. 1971. 'Movement and Resource Extraction Among Pastoral Nomads: The Case of the Yarahmadzai Baluch', *Anthropological Quarterly* 44(3): 185–97.
——— 2004. *Pastoralists: Equality, Hierarchy, and the State.* Boulder, CO: Westview Press.
Schlee, G., and A.A. Khazanov (eds). 2012. *Who Owns the Stock? Collective and Multiple Property Rights in Animals.* Oxford: Berghahn.
Schneider, H.K. 1979. *Livestock and Equality in East Africa: The Economic Basis for Social Structure.* Bloomington: Indiana University Press.
Silvester, J.G. 1998. 'Beasts, Boundaries and Buildings: The Survival and Creation of Pastoral Economies in Southern Namibia 1915–1935', in P. Hayes et al. (eds), *Namibia under South African Rule: Mobility and Containment 1915–1946.* Oxford: James Currey, pp.95–116.
Smith, A.B. 1992. *Pastoralism in Africa: Origins and Development Ecology.* London: Hurst.
——— 2005. *African Herders: Emergence of Pastoral Traditions.* Walnut Creek, CA: AltaMira Press.
Spear T., and R. Waller. 1993. *Being Maasai: Ethnicity and Identity in East Africa.* London: James Currey.
Spencer, P. 1998. *The Pastoral Continuum: The Marginalisation of Tradition in East Africa.* Oxford: Oxford University Press.
Spittler, G. 1998. *Hirtenarbeit. Die Welt der Kamelhirten und Ziegenhirtinnen von Timia.* Cologne: Rüdiger Köppe Verlag.
Vansina, J. 2004. *How Societies Are Born: Governance in West Central Africa Before 1600.* Charlottesville: University of Virginia Press.
Verschuren, D. 2004. 'Decadal and Century-scale Climate Variability in Tropical Africa During the Past 2000 years', in R.W. Battarbee, F. Gasse and C.E. Stickley (eds), *Past Climate Variability through Europe and Africa.* Dordrecht: Springer, pp.139–58.
Voigt, E.A. 1984. 'The Faunal Remains from Magogo and Mhlopeni: Small Stock Herding in the Early Iron Age of Natal', *Annals of the Natal Museum* 26: 95–104.
Werner, W. 1998. *'No One Will Become Rich': Economy and Society in the Herero Reserves in Namibia, 1915-1946.* Basel: Schlettwein.
Widgren, M. 2007. 'Precolonial Landesque Capital: A Global Perspective', in A. Hornborg et al. (eds), *Rethinking Environmental History: World-system History and Global Environmental Change.* Lanham, MD: AltaMira Press.
Wolf, E. 1999. *Envisioning Power: Ideologies of Dominance and Crises.* Berkeley: University of California Press.

Part I

The Prehistory of Pastoralism in Africa

Chapter 1
Herders before Pastoralism
Prehistoric Prelude in the Eastern Sahara

Rudolph Kuper and Heiko Riemer

This chapter is dedicated to Fred Wendorf, who, through the groundbreaking work of his Combined Prehistoric Expedition, has been a source of permanent encouragement in our own research.

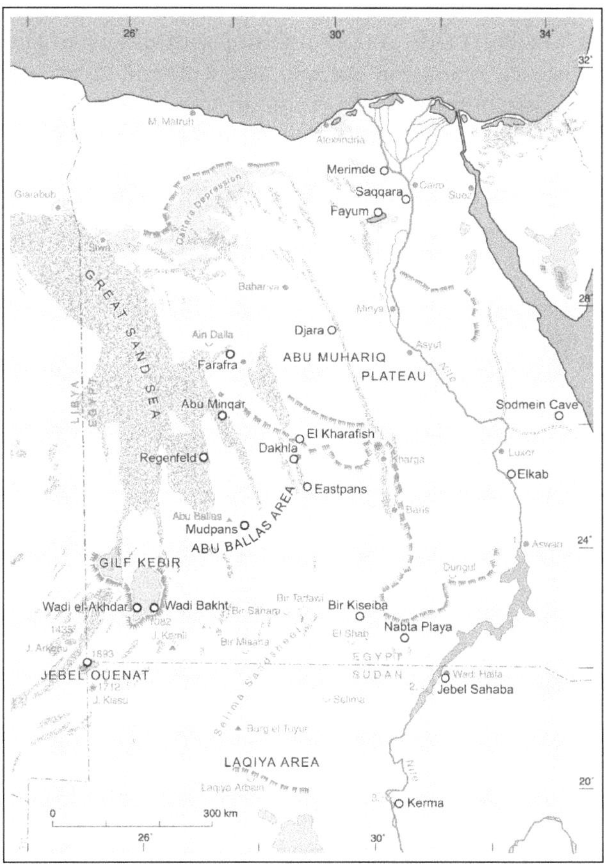

Figure 1.1: Map of North-east Africa, showing sites and regions mentioned in the text.

Introduction

Looking at the stunning assemblage of hundreds of cattle skulls around the second millennium BC tombs at Kerma (see Figure 1.2) (Chaix 2001; Bonnet 2004) or the early Dynastic bucrania from Saqqara (Emery 1954), and feeling the pride of intimate relations between humans and cattle expressed by countless much older rock paintings from all over the Sahara, it becomes evident that cattle pastoralism with its related ideological background must have played an important role in arid North-east Africa for thousands of years. To turn this general observation into a more anthropologically founded statement it should be useful at the outset to briefly consider what is understood by 'pastoralism' in the present study. Basic definitions regard pastoralism as a mode of subsistence involving either year-round or seasonal livestock herding, often combined with a nomadic way of life (e.g., Wendrich and Barnard 2008). Anthropological and archaeological classifications of specific human societies or cultural complexes as 'pastoral' are frequently based on this notion of nomadic pastoralism in which domestic animals play an important or dominant role in subsistence (Cribb 1991).

Pastoralism based on cattle, often combined with sheep and goats, is still the prevailing subsistence strategy in the arid and semi-arid zones of Africa today. The origins and development of this cultural phenomenon are far beyond the chronological reach of cultural anthropology, and it is a strength of archaeology to be able to record cultural behaviour over long periods of time and to differentiate between changes of various time depths. On the other hand, prehistoric evidence is scarce due to the frequently poorly developed state of archaeological

Figure 1.2: Bucrania deposited around a tumulus of a Nubian sovereign of the Classic Kerma phase, Kerma, Nubia (from Bonnet 2004: 70).

research in Africa and the mostly bad preservation of faunal remains at open-air settlement sites in arid regions especially because of the effect of wind erosion. On this weak base, neither questions regarding when and where the symbiosis between humans and cattle began, nor those relating to subsequent developments into specific types of specialisation in economic or ideological terms (such as the 'cattle complex' proposed by Herskovits 1926), can be answered satisfactorily. For instance, there is ethnographic evidence indicating that in East Africa such specialised forms of pastoralism might only have emerged 200 to 300 years ago. However, below we will discuss a much deeper time range with the intention of showing that, far from a purely evolutionary development, different types of pastoralism have existed side by side and have changed through time and space.

Radiocarbon dates for the earliest domesticates in Africa clearly prove the north-eastern part of the continent to be the place where animal husbandry obviously took off (MacDonald and MacDonald 2000; Close 2002), reaching West Africa in 2300 BC (Breunig, Neumann and Van Neer 1996; Linseele 2007), East Africa in 3000 BC (Marshall 2000), and the southern end of the continent about 2000 years ago (Smith 2000, 2005; Sadr, this volume).[1] As a review of the radiocarbon-dated sites in north-eastern Africa shows (Riemer 2007a), unequivocal evidence for cattle keeping in this area starts at around 6000 BC. From then onwards it was present throughout the Neolithic/Predynastic period of Egypt, until it found during the third millennium BC marvellous expression in fascinating illustrations, such as those furnishing the Old Kingdom tombs of Ti, Mereruka and other nobles in Saqqara, vividly reflecting the daily life of the farmers and testifying to close human–animal relations (Figure 1.3). Outside the Nile Valley the same intimacy is mirrored in numerous Saharan rock paintings that, despite only approximate dating, unequivocally demonstrate to what extent cattle dominated the life of the artists (see Figure 1.4).

Figure 1.3: Birth of a calf, Mastaba of Ti, Saqqara, Egypt, Fifth Dynasty (from Wild 1953: pl. 124).

Figure 1.4: Painting from the 'cattle period', when pastoralism had become a basis of human life throughout the Sahara. Lhote Fund, Musée de l'Homme, copyright and courtesy of the Musée National d'Histoire Naturelle, Paris (see cover).

Looking for the origins of this way of life one needs to follow up the Holocene human occupation of the Eastern Sahara and its dependence on changing environmental conditions with regard to different geographical settings and latitude in the various sub-areas. In the following considerations we will first present a general outline of the Holocene development as reflected in more than 700 radiocarbon dates from the Eastern Sahara and the Nile Valley. This framework comprising four stages of Holocene human occupation, from 'early' over 'middle' and 'late' to 'final' (see Figure 1.5), will be then used as a background for discussing the different phases in the evolution of pastoral life in north-eastern Africa.

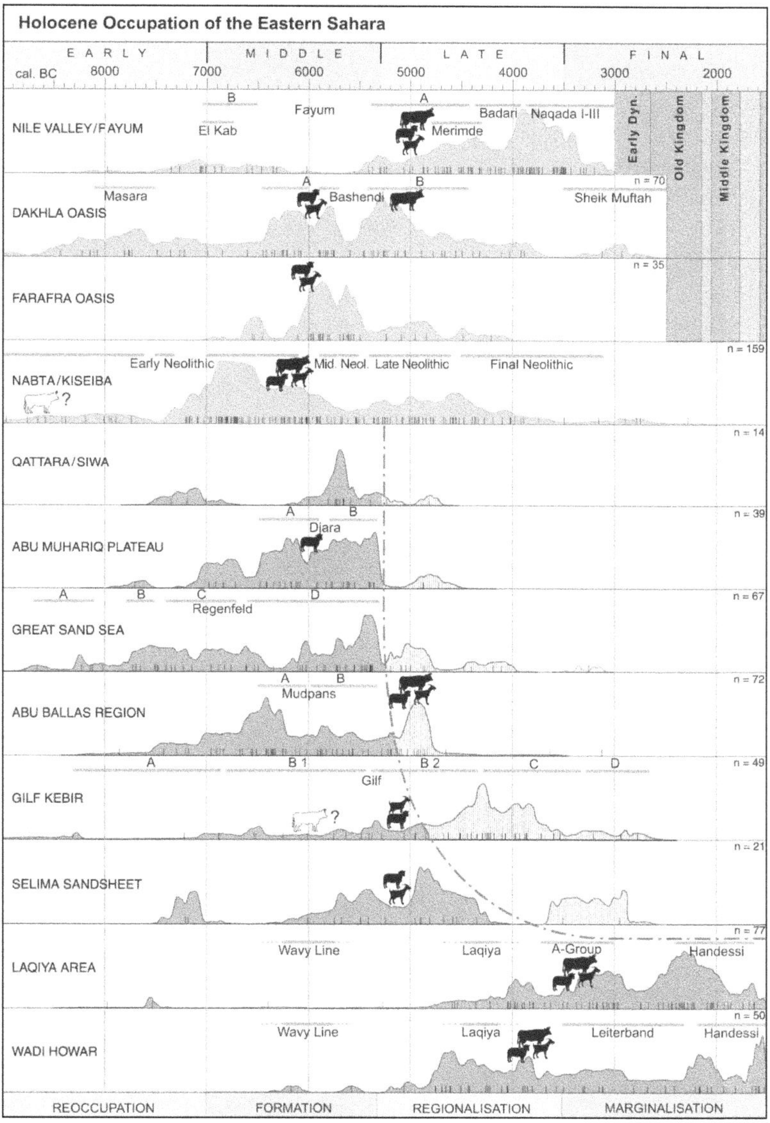

Figure 1.5: Earliest appearance of domesticated cattle, sheep and goats in the regional sequences of the Holocene occupation of the Eastern Sahara, arranged along a north–south gradient. Outlined cattle represent uncertain identification or dating. Sequences are represented by cumulative curves of calibrated radiocarbon dates from the BOS and ACACIA projects (bottom rows, Qattara/Siwa to Wadi Howar regions) and from other research conducted in the Nile Valley, the Egyptian oases and the region of Nabta Playa / Bir Kiseiba (top rows). The dashed line indicates the latitudinally distinct times of human abandonment of rain-dependent core desert areas; dates to the right of the line are from sites located close to permanent water or in extra-zonal ecologically favoured areas, such as the Gilf Kebir.[2]

New Life in the Savannah: From Reoccupation to Formation

As a result mainly of large-scale interdisciplinary research into the landscape archaeology of the Eastern Sahara, started in 1980 by the Besiedlungsgeschichte der Ost-Sahara (BOS) project into the settlement history of the Eastern Sahara at the University of Cologne, and continued between 1995 and 2007 by its ACACIA (Arid Climate, Adaptation and Cultural Innovation in Africa) collaborative research centre, a chronological framework for the Eastern Sahara is now available (see Figure 1.6), covering an area from Siwa in the north to the Wadi Howar more than 1,500 kilometres further south (Kuper and Kröpelin 2006). Far from being complete and first of all apt to reveal gaps of knowledge, especially regarding environmental and economic data, it can nevertheless serve as a background to a scenario of Holocene human occupation in the desert areas.

For a period of at least 20,000 years during the terminal phase of the Pleistocene, before the onset after 9000 BC of monsoonal rains that opened the Holocene in northern Africa, the Sahara desert extended about 400 kilometres further south than it does today (Kröpelin 1999: 491). This is clearly reflected by a lack of late Pleistocene sites in the desert regions. In contrast, in the Nile Valley a cluster of late Palaeolithic sites can be observed, especially in the area now covered by the Aswan Lake where significant archaeological research and rescue work took place in the early 1960s. However, at the end of the Pleistocene, the Nile Valley experienced major changes in the river's course triggered by increased rainfall in the African highlands. Obviously due to a 'Wild Nile' (Butzer 1980: 272), living conditions along the river became harsh and may have caused competition for space and food resources. This is, for instance, suggested by drastic evidence of violence within the cemetery of Jebel Sahaba on the eastern bank of Aswan Lake, dated to between approximately 16000 and 13000 BC (Wendorf 1968). At this Sudanese site, near Wadi Halfa and close to the Egyptian border, many of the sixty-one individuals (including nine children) were found with flint projectile points in their decayed bodies or embedded in their skeletons. Nearly half of the individuals died violently.

At the beginning of the Holocene, with the arrival of monsoon rains in the ninth millennium BC, a savannah like environment made the Eastern Sahara habitable. When the first people reoccupied the area is still a matter of speculation. Nile dwellers might have left the then inhospitable river valley, while groups from the south, already adapted to savannah ecology, would just have followed their traditional way of life. Their Epipalaeolithic tool kit as well as archaeozoological findings evince them to have been hunter-gatherers, who – following Wendorf's arguments (Gautier 2001, 2007: 81; Wendorf et al. 2001: 648–75) – possibly already practised some cattle husbandry in the Nabta/Kiseiba region (see Figure 1.5). While the evidence for such an early 'pastro-foraging' economy is inconclusive, the presence of well-made pottery bearing wavy-line decorations is a general achievement of the ninth millennium BC, also found in other parts of the southern Sahara between the Niger and the Nile rivers (Jesse 2003: 197, fig. 40). This reoccupation of the Eastern Sahara must have proceeded rather quickly up to the far north-west of Egypt, where Epipalaeolithic settlement in the Regenfeld area in the central Great Sand Sea proves suitable living conditions already existed before 8000 BC in what is today the Libyan Desert's most barren part (Kuper 2002: pl. 5; Riemer 2003).

Most striking in the overall distribution of sites from this period is an almost complete lack of evidence of occupation in the Egyptian Nile Valley, where only the site of El Kab testifies to human presence during that time (Vermeersch 1978). This corresponds to the Wadi Howar area in northern Sudan, which also shows a lack of dates before 6000 BC. Although one cannot exclude the possibility that this pattern may be related to insufficient research or to sites being irretrievably buried under metre-thick river sediments, most probably conditions in the Wadi Howar were too marshy and hazardous for human settlement due, amongst other things, to frequent inundations, mosquitoes and crocodiles. Therefore hunters and gatherers obviously preferred the less wooded grasslands in the northerly zones of the Eastern Sahara to the regularly flooded and densely wooded environments further south.

After 7000 BC, human settlement became well-established all over the Libyan Desert by virtue of economic and technological adaptation to the different regional ecological requirements. In the north, within the reach of the Egyptian limestone plateau, bifacial lithic technology, obviously rooted in the Levant, caused a complete change in the stone tool kit which later can be followed up into the Predynastic cultures of the Nile Valley (Kindermann 2004; 2010). Rocker-stamped pottery of Sudanese tradition is well represented as far north as the Egyptian oases region and the Great Sand Sea (Kuper 1988: 134, fig. 5; Hope 2002: 50; Riemer and Jesse 2006). However, the most important achievement of this formative phase was the adoption and rapid spread of livestock herding after the introduction of domesticated sheep and goats from the Near East (see Figure 1.6). On the other hand, cattle might have been domesticated locally and are well documented at places such as Nabta Playa (Wendorf et al. 2001: 648–75; Gautier 2007) where they mark the beginning of herding – the ultimate origin of African pastoralism. However, due to regional factors the economic development differed substantially from one region to another, such as indicated from the Abu Ballas region where rich faunal material from several sites at Mudpans, dated to between 6500 and 5500 BC, revealed no evidence of domestic livestock (Van Neer and Uerpmann 1989: 328; Kuper 1993).

While neither the dates themselves nor the radiocarbon curves indicate any rupture in the climatic and cultural developments during this phase, at some localities, for instance Djara and Mudpans, a faint change in pottery and stone artefacts is detectable before 6000 BC, suggesting a break between two phases (A and B) that obviously coincides with the appearance of small livestock, fully domesticated cattle (Gautier 2001) and the beginning of the so-called Middle Neolithic in the Nabta-Kiseiba area. Some cultural changes may consequently

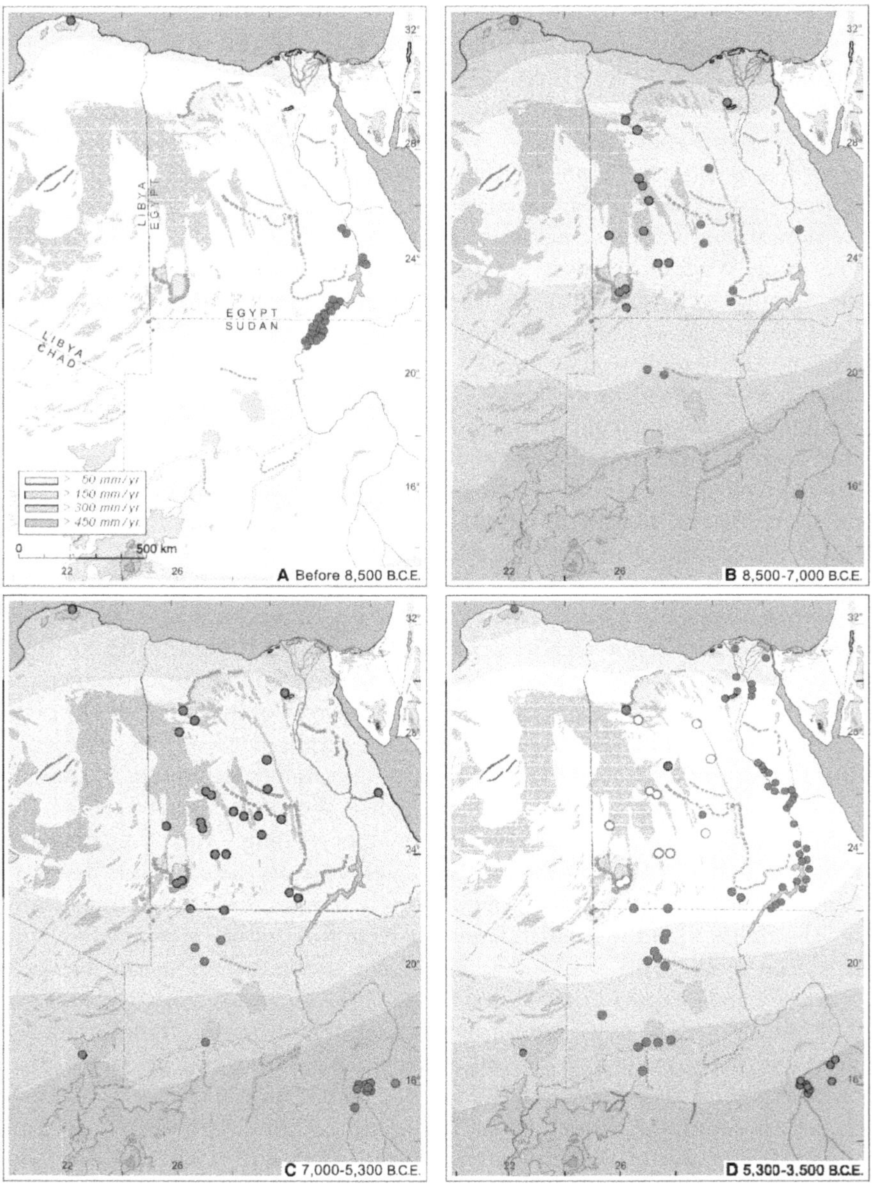
LIBYA
EGYPT
EGYPT
SUDAN
LIBYA
CHAD
> 50 mm / yr
> 150 mm / yr
> 300 mm / yr
> 450 mm / yr
0
500 km
A Before 8,500 B.C.E.
B 8,500-7,000 B.C.E.
C 7,000-5,300 B.C.E.
D 5,300-3,500 B.C.E.

Figure 1.6 (facing page): Climate-controlled occupation in the Eastern Sahara during the main phases of the Holocene (Kuper and Kröpelin 2006). Dots indicate major occupation areas; circles indicate isolated settlements in ecological refuges and episodic transhumance. Rainfall zones are delimited by best estimate isohyets on the basis of geological, archaeozoological, and archaeobotanical data: A – During the last Glacial Maximum and terminal Pleistocene (20000 to 8500 BC) the Saharan desert was void of any settlement outside of the Nile Valley. B – With the abrupt arrival of monsoon rains at 8500 BC, the hyper-arid *(continued)* desert was replaced by savannah-like environments and was swiftly inhabited by hunter-gatherer groups. C – After 7000 BC human settlement became well-established all over the Egyptian Sahara, fostering the earliest adoption of domesticated animals (by circa 6000 BC) by hunter-gatherer groups. D – Retreating monsoon rains caused the onset of desiccation of the Egyptian Sahara at 5300 BC. Pastro-foragers were forced to migrate to the Nile Valley, to ecological refuges, or to the Sudanese Sahara where they developed fully fledged pastoralism.

have occurred beyond local climate control. Finally, at the end of this period around 5300 BC, when the regular monsoon rains ceased to reach the Egyptian Sahara, a significant decline of data in the core desert of the Great Sand Sea and on the Abu Muhariq Plateau indicates a termination of settlement. Comparing, for example, the radiocarbon dates from Djara with those from the Fayum and Dakhla oasis, emigrations from the desert into the latter areas become obvious.

The few dates after 5300 BC (see Figure 1.5) belong to sites relatively close to permanent water and – in the case of Djara – might reflect episodic visits by small stock herders from the Nile Valley. Alternatively, such evidence of short term settlement may be linked to locally available groundwater, which attracted occasional grazing trips from the oases region, for example at Eastpans near the eastern end of the Abu Ballas region where cattle are well documented in the archaeozoological record around 5000 BC (Gehlen et al. 2002: 96–8). Generally, it was not before the end of the Formation phase at 5300 BC that herding appears to have become a vital human subsistence strategy in the Egyptian Sahara, while at the same time the first farming communities developed in the Fayum area.

First Steps towards Pastoralism: Pastro-foragers of the Sixth Millennium BC

Dating the Introduction of Domesticated Animals

It is commonly accepted that goats and sheep were introduced from the Near East, as wild progenitors do not exist on the African continent (Gautier 1980; Close 2002). Although initial domestication of ovicaprines took place in the Near East around 8500 BC at the latest (Bar-Yosef 1984; Legge 1996; Moore, Hillman and Legge 2000; Watkins 2007), it was probably around 6000 BC when Neolithic activities shifted into previously marginal areas to the south of the Levant. It is reasonable to suggest that the diffusion of sheep and goats through the Sinai Peninsula into Africa can be linked to this expansion of the Levantine Neolithic (Rosen 2008).

Preservation of bone material is notoriously poor on Saharan camp sites dating to the time of the Holocene humid phase which lasted from around 8500 to 5300 BC in the Western Desert of Egypt (see Figure 1.5). This always limits the possibilities of tracing the emergence and exact dating of domesticated animals in northern Africa. However, the last decades have seen an increase in archaeological data which now allows a more precise reconstruction of the timing and expansion of early livestock. The temporal and spatial distribution of radiocarbon dates associated with early domesticated animals does not reveal a long-lasting movement of domesticated animals into and throughout North-east Africa, but instead document initial appearance and rapid spread by about 6000 BC, the above-mentioned period of other cultural changes. The very earliest dates for sheep and/or goats at Sodmein Cave in the Eastern Desert around 6000 BC (Vermeersch et al. 1996) are nearly contemporaneous with those at Nabta Playa/Bir Kiseiba in the south of Egypt (Wendorf et al. 2001), at Farafra (Barich and Hassan 2000; Barich and Lucarini 2002), Djara (Kindermann et al. 2006) and possibly Dakhla, deep in the Western Desert (McDonald 1998; cf. McDonald 2009: 25). Dates from the Nile Valley are still lacking for this period of time, a fact that might correlate with the then unpleasant humidity in the valley which was only coped with by some groups from the desert during the dry season when the rains had stopped.

Herding Meets Hunting and Gathering

In principle, the spread of domesticated animals into North-east Africa, where there was no incipient domestication of ovicaprines, can either be seen as a process of the diffusion and adoption of herding elements by hunter-gatherers from their neighbours or as a displacement of hunter-gatherers by the expansion of pastoralist or farming communities. Looking at the Western Desert it becomes clear that sheep and goats first appear within hunter-gatherer communities. The material culture at the archaeological sites shows a general continuity before and after the appearance of domesticates, which can be recognised in the lithic traditions and, where present, the ceramic traditions as well. This suggests that domestic stock intruded into hunting and gathering subsistence, which then continued to exist during the sixth millennium BC until the beginning of the climatic drying. There is nothing to suggest that Levantine herders, or even their possible neighbours on the east bank of the Nile, might have displaced Western Desert groups. The donor society and the place where goats and sheep were exchanged are, however, difficult to identify – a fact not only due to the small number of known sites, but also to the problem of identifying possibly Near Eastern traits in Egypt.

The primary subsistence strategies in the Western Desert, namely hunting wild animals and gathering plant material, likewise continued. Wherever sufficient bone material is available to allow the calculation of species percentages, wild game clearly dominates the inventories, if domesticated animals are present at all. For example, in the bone record from Djara, two fragments of sheep make up less than 1 per cent of the mammal bones (Kindermann et al. 2006; Pöllath 2009). Even at Nabta Playa and Bir Kiseiba, where bones of domesticated animals are most frequent compared to other Western Desert sites, wild animals outnumber domesticates (Gautier 2001; cf. Smith 2005: 87). The opposite is true of the fully fledged Neolithic societies of the early Predynastic period of the fifth millennium BC, when domesticated animals clearly dominate the bone spectra (Driesch and Boessneck 1985; Brewer 1989; Boessneck and Driesch 1990, 1994; Vermeersch, Van Neer and Hendrickx 2004; Linseele and Van Neer 2009).

A good indicator of the importance of hunting, even where bone material did not survive, is the percentage of arrowheads among the retouched stone tools (Riemer 2007a). Although the Western Desert was clearly separated into a northern and a southern sphere during the mid Holocene – as expressed in a dichotomy of arrowhead traditions characterised by microlithic insets in the south as opposed to stemmed or leaf-shaped points, often with bifacial modification, in the north (Riemer 2007b) – arrowheads always make up a substantial proportion of the tools, though the percentages per site can vary between extremes of 10 per cent and 50 per cent. Even at sites where bones of domesticated animals have been found, arrowheads exist in great numbers, and it seems there was no significant change along with the introduction of sheep, goats or cattle. Again, the converse is indicated for Predynastic contexts where hunting devices are extremely rare (Riemer 2007a).

Subsistence of Late Hunter-gatherers in the Western Desert

What appears in the archaeological records of Holocene Western Desert sites around 6000 BC is the picture of a typical egalitarian foraging society, well adapted to the unpredictable arid environment. The Western Desert during the climatic optimum was far from being a paradise. The amount of annual rainfall has been estimated to a maximum of 100 mm during the Holocene (Neumann 1989), slightly higher in mountainous regions such as the Gilf Kebir (Linstädter and Kröpelin 2004), indicating a dry savannah with episodic rains and the patchy and unpredictable availability of surface water and related resources such as vegetation and wild animals. These factors created living conditions of high risk and stress for the foragers. Highly variable logistical and residential mobility patterns in space and through time are likely to be adaptive expressions of risk minimisation in the Western Desert. During the dry season when the rains had stopped, water pools dried up and prehistoric groups were forced to return to the wells and spring mounds of the oases (Kindermann and Bubenzer 2007). The rains were, however, highly variable in their distribution over time and space, which is a general characteristic of arid regions. The prehistoric adaptation to the risk of finding no water was a strategy of high mobility and the flexibility of seasonal movements that could cover several hundreds of kilometres during the course of a year; archaeological evidence of such behaviour is available through the distribution of pottery, exotic raw materials, mollusc shells and the like (Riemer 2007b). Archaeological evidence for an inter-regional network of contacts is also given by stylistically homogeneous lithic and ceramic traditions, such as bifacial lithic technology in northern Egypt (Kindermann et al. 2006; 2010) and Khartoum-style pottery, associated with microlithic technologies, in southern Egypt and northern Sudan (Riemer and Jesse 2006).

There was no potential to exploit aquatic resources which would have increased productivity and allowed a more sedentary way of life. The only exceptions were some areas in relatively close proximity to the Nile Valley, such as Djara where shells of the Nile bivalve *Chambardia rubens* (formerly *Aspatharia*) have been found (Kindermann et al. 2006; Kindermann and Bubenzer 2007). But even in this case, the resource was strictly seasonal and could only complement the terrestrial diet during a short period of the year. The main sources of the diet were small gazelles (*Gazella dorcas* and *G. leptoceros*), antelopes (oryx and addax), hare and fennek and the gathering of plant foods (Van Neer and Uerpmann 1989; Kindermann et al. 2006; Pöllath 2009). Grinding stones begin to appear sporadically during the early Holocene, but increase steadily in frequency through the seventh and sixth millennia BC. The latter implies the processing of hard seeds. However, remains of domesticated plants, such as those known for the early Predynastic groups, are still lacking in the archaeobotanical material, and the exploitation of wild grass seeds appears to be the only explanation for the presence of heavy grinding implements.

A foragers' way of life is also reflected in camp site patterns. Camps were only temporarily used, displaying nothing of sedentary subsistence (Kindermann and Bubenzer 2007). Storage pits, which would refer to some kind of surplus aggregation on the sites, do not exist.

Towards Complex Hunter-gatherers

There are some faint indications during the sixth millennium BC which might indicate the beginning of more complex socio-economic relations among Western Desert communities. Around the time when early domesticated animals were introduced to forager societies, the number of grinding implements constantly increased. Intensifying plant-food exploitation may have gone along with higher logistical mobility, as part of the group, probably women, could spend more time in the base camps to process plant food while male hunting parties were underway. In fact, camp site patterns continuously shifted towards logistical mobility patterns with the establishment of seasonal residential camps represented in large, multiply occupied sites at the few water pools in the desert. Short-term camps often devoted to special tasks, such as hunting or flint exploitation, can be observed in the vicinity of such base camps, illustrating short-term logistical operations related to the exploitation of important resources. Seasonal long-distance residential moves of over 100 kilometres between base-camp sites still occurred, but became less frequent throughout the year, although this was highly variable, depending on individual landscape settings and local resource availability.

Table 1.1: The forager–pastoralist transition in various regions of the Libyan Desert, based on interpretations of archaeological data by means of ethnographic parallels.

MATERIAL CULTURE	ECONOMY	IDEOLOGY & SOCIAL RELATIONS
	Libyan Desert, c. 6000–5300 BC	
No significant change in forager material culture (10–50% arrowheads among stone tools), but: • Increasing diversity in lithic tools • Increasing number of grinders • Rare pottery	**Pastro-foraging** • Earliest introduction of domesticated animals without significant change of traditional subsistence based on hunting and, increasingly, gathering of wild cereals • Fewer long-distance residential moves (>100 km), whereas base-camp size and logistical mobility increase	**Early complex foragers** • Food sharing plus marginal stock keeping • Increasing diversity in new objects; at the end of this phase elaborate flint knives and cosmetic palettes may have been earliest status items
	Nabta, Late Neolithic, c. 5600–4600 BC	
• Small number of arrowheads (<10% in early phase; 1.5% in late phase) • New pottery style	**Multi-resource pastoralism** • Reliance on domesticated animals (cattle, goats, sheep; c. 35% of fauna); hunting (c. 65%) and gathering continue	**Cattle-centred** • Cattle burials and cattle cult indicate important social and religious roles of cattle
	Dakhla Oasis, Bashendi B, c. 5600–4000 BC	
(Forager) material culture continues, but: • Decrease in arrowheads (<10% of stone tools) • Increase in pottery	**Early multi-resource pastoralism** • Reliance on domesticated animals (sheep, cattle); hunting and gathering of wild cereals persist • Higher residential mobility, but decrease in distances of residential moves (<100 km) as a result of reduced territory for climatic reasons	**Complex foragers?** • Increasing use of (status) objects of exotic character
	Ouenat/Gilf Kebir, c. 4400–3000 BC	
• 'Pastoral Style' in rock art: Vivid paintings of cattle husbandry and nomadic life • Change in cultural tradition (new pottery style; informal lithic technology; few arrowheads)	**Multi-resource pastoralism** • Reliance on domesticated animals (cattle, goats); hunting probably a special (prestige?) task	**Cattle-centred** • Emphasis on cattle herding, reproduction and individuality in rock art
	Dakhla Oasis, Sheikh Muftah, c. 4000–2200 BC	
• Change in cultural tradition (new pottery style; informal lithic technology; <1% arrowheads) • Increasing quantity of pottery (esp. cooking pots), while grinders diminish (porridge cooking?)	Multi-resource pastoralism • Heavy reliance on domesticated animals (cattle, goats: >50% of fauna); opportunistic hunting emerges as special (seasonal) task • Oasis-centred residential mobility • Donkey-based desert travel (exchange?)	• Lack of grave goods and burial monuments may point to 'small egalitarian groups' (McDonald 2001)

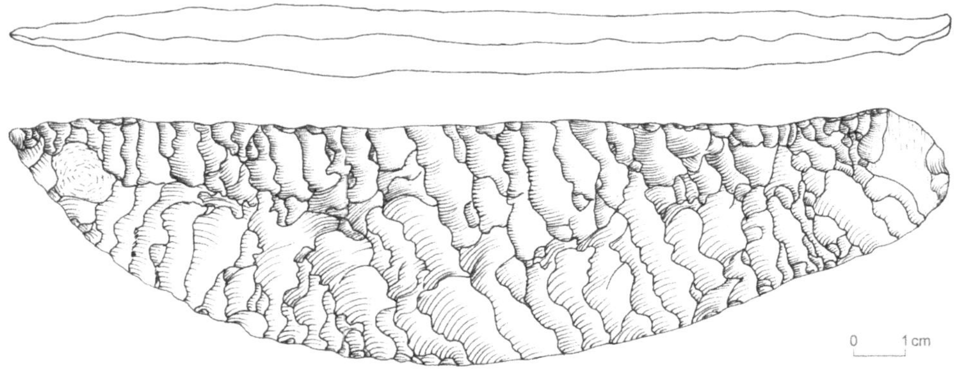

Figure 1.7: Fine pressure-flaked flint knife from Djara, Abu Muhariq Plateau, Egypt (circa 6600 to 5200 BC): early expression of incipient social inequality among the pastro-foragers? One end of the blade was truncated, probably to accommodate a wood or leather haft (Kindermann 2010: 404).

Taking a closer look at the timing of the processes mentioned above, the beginning of the shift towards more complex subsistence strategies and the development of a more sophisticated 'Mesolithic technology' (Hayden 1992) can be traced back to the seventh millennium BC or even to early Holocene times, predating the introduction of domesticated animals around 6000 BC; yet it possibly represents the substrate on which stock keeping – as the first expression of food storage, or, to use the term proposed by Woodburn (1988), 'delayed return foraging' – could grow. Around 5500 BC, thus postdating the introduction of livestock, the bifacial modification of stone tools in northern Egypt developed into a kind of decorative element in lithic production. Although most of the stone tools are clearly devoted to certain primary functions – such as arrowheads, scrapers and borers – a small amount of tools tend to represent rather non-functional personal accessories. This is manifest in finely pressure-flaked knives (see Figure 1.7). The elaborate character of the flaking and the fact that these knives cannot be re-sharpened do not suggest a highly functional tool but rather a kind of symbolic item representing the status of its owner. These knives, as well as cosmetic palettes, ear plugs, and other jewellery of exotic stones or molluscs, might reflect 'prestige technologies' and the beginning of a shift towards 'socio-economic inequality' (McDonald 2008: 101), but they do not principally change the picture of a hunter-gatherer society.

From all this it becomes clear that there is nothing to support the suggestion that the earliest herders adopted a fully fledged pastoral subsistence. The groups of the Western Desert clearly continued hunting game and gathering seeds and tubers of wild plants. It is therefore adequate to characterise the early herders of the sixth millennium BC as 'pastro-foragers' (Table 1.1).

Hunter-gatherer Conservatism

Finally, we need to discuss why the hunter-gatherers of the Western Desert did not develop a fully pastoral economy. Two arguments can be levelled against the spontaneous adoption of a pastoral way of life. First, it is unlikely that climatic and environmental conditions allowed for a fundamental change in subsistence strategies. In the light of the unpredictable availability of resources and a constant need for drinking water, herds could not be moved far out into the desert regions. The few fragments of sheep bones found at Djara have been explained by prevailing winter rains when low temperatures reduced water consumption to a level that offered the opportunity to cover larger, waterless distances with small livestock. Likewise, the risk of drought and consequent loss of herds was buffered by success in the traditional subsistence strategies of hunting and gathering. Evidence of domesticated animals elsewhere comes from sites in the oases or their surroundings, such as Dakhla or Farafra, where the high water table could be used throughout the year. It is, therefore, reasonable to suggest that risk minimisation was a major reason why livestock was at first only adopted in small proportions. Desert areas dominated by summer rains and far away from groundwater resources, such as the inner Great Sand Sea, have probably never seen the introduction of livestock because the water requirements of domesticated animals could not be met.

The second argument is that hunters cannot easily adopt herding. Using ethnographic data, Smith (1990, 1992) has argued convincingly that the socio-economic structure of hunter-gatherers is diametrically opposed to that associated with a pastoral way of life. In particular, it is the immediate consumption and sharing of food (Lee and DeVore 1968) in egalitarian societies that creates an ideology opposed to that of food and livestock accumulation in pastoral societies (Ingold 2000). Even the keeping of livestock is a concept of future orientation alien to hunter-gatherers. In turn, domesticated animals are not just food in the context of herders, but become a source of wealth, power and prestige (Sadr 1998).

In adjusting the two arguments mentioned above, it can be suggested that both unequally contributed to the behaviour of the Western Desert hunter-gatherers. There is no doubt that environmental conditions constituted limiting factors, in particular where water was lacking, but eventually it was the traditional hunter-gatherers ideology that prevented any ground-breaking economic change.

Despite the conservatism in ideology, the Western Desert example illustrates that the introduction of domesticates did occur. Furthermore, it is reasonable to suggest that this earliest food aggregation started to compete against and, in the long run, corrupted the reciprocity (food sharing) ideology, in particular in cases of food shortage. In fact, there is the beginning of a development towards more complex foragers that goes along with the earliest domesticated animals during the sixth millennium BC. But as long as the same environmental conditions persisted there was no ground-breaking shift to surmount the foragers' ideology completely – and perhaps no opportunity to adopt herding to any significant extent.

The End of Pastro-foraging

The insight that the hunter-gatherers of the Western Desert did not shift (quickly and completely) from their cost-efficient strategies to herding suggests that other factors, such as climatic and demographic stress, had to occur jointly before these societies started intensifying their productivity. This actually happened with the beginning of the climatic drying up of the Sahara by around 5300 BC that affected most of the foragers' territories and resources (see Figures 1.5 and 1.6). Then the groups of the Western Desert were forced to migrate into the remaining refuges, such as the oases and the Nile Valley. The loss of most desert territories for human habitation and the resulting population agglomeration in the Nile Valley can be seen as the primary cause for changes in traditional socio-economic concepts. What followed along the Nile during the fifth millennium BC was the development of the early Predynastic cultures which can be characterised as the first fully Neolithic societies featuring a sedentary life, a food-producing economy based on agriculture, stock keeping and fishing, and a rapidly developing social complexity that finally led to the emergence of the early Egyptian kingships (Riemer and Kindermann 2008).

The Return of the Desert: Regionalisation as Adaptation

The '5300 BC Event'

The retreat, after 5300 BC, from regions becoming more and more arid into areas with permanent water, into extra-zonal ecological niches such as the Gilf Kebir Plateau or further south to the Sudanese plains where rainfall and surface water were still sufficient (see Figure 1.6), fostered more regionally diverse socio-cultural adaptations. A small number of radiocarbon dates from the western fringes of the Great Sand Sea and the Abu Muhariq Plateau reflect sporadic occupation, whereas the eastern Abu Minqar and Abu Ballas areas lie within the range of transhumance from the Farafra and Dakhla oases (see Figure 1.5). In the valleys dissecting the Gilf Kebir, ceramic traditions developed that can be later detected in the Laqiya area of northern Sudan. In general, the previously widespread 'wavy-line complex' (or 'Khartoum-style complex') was replaced by new regional developments in pottery and other cultural components. This includes, in particular, the rise of cattle pastoralism as it is reflected in the rock art of Jebel Ouenat and Gilf Kebir and in the later settlement remains of the Laqiya area and the Wadi Howar (Keding 1997; Jesse, Keding, Lenssen-Erz and Pöllath, this volume).

Comparing this Saharan pathway into food-producing economies with the traditional Near Eastern model of 'Neolithisation', it turns out to be a specific North African variant of this basic change in cultural evolution: instead of the western Asian transition from mobile hunter-gatherers to sedentary, yet pre-ceramic farmers and stock keepers, we see pottery-producing hunter-gatherers, pastro-foragers as well as hunter-fishers replaced by nomadic cattle herders in the desert or dry savannah environments of the Sahara. With the exception of the

Nile Valley and the oases, cereal farming does not seem to have been a constituent of this Saharan mode, given that even the best of environmental settings were insufficient for agriculture and the mid-Holocene savannah still provided ample wild-growing grains, fruits and tubers.

Paradoxically, in certain landscapes the decreasing trend in annual precipitation may have been associated with an increase in the vegetation cover because of a change in seasonality. Geoarchaeological evidence from the Gilf Kebir suggests that the intense daytime summer monsoon rains of the early Holocene pluvial resulted in less grass growth than the quantitatively lesser winter rains of the terminal humid phase, which presumably fell at night (Linstädter and Kröpelin 2004). These favourable circumstances may have maintained the rich culture of cattle keepers who left their art in Jebel Ouenat and Gilf Kebir during the second half of the fifth and the fourth millennium BC.

Saharan Exodus

A comparison between the distribution of human settlement around 6000 BC with that around 4000 BC (see Figure 1.6) suggests that the exodus from the Sahara coincides with the rise of the first settled communities in the Nile Valley (Kuper 2002: pl. 7). With regard to their economic base, namely the Old World cereals wheat and barley, these first mixed-farming groups, which started with fully developed agriculture in the Fayum and at Merimde around 5000 BC, are clearly rooted in the Near Eastern Neolithic. However, on the other hand, not only the above-mentioned elements of their stone technology, but also essential aspects of their social and cognitive world might be traced back to western origins. This especially concerns the role of Saharan cattle herders and their spiritual heritage in the Neolithic of the Nile Valley. Recently it has been stressed that the Badari Culture on the middle Nile is mainly represented by cemeteries (Wengrow 2003). Occupation sites mostly consist only of ash layers, cultural debris and animal droppings, and they recall African stock pans. Together with poor evidence for permanent dwellings and domestic architecture, this may suggest a semi-sedentary existence during the early Predynastic of the second half of the fifth millennium BC. In the recent past the excavation of the seasonal hunting-and-fishing site of Mahgar Dendera 2 confirmed the 'temporary character of Badarian settlements' (Hendrickx and Vermeersch 2000: 43; Hendrickx, Midant-Reyes and Van Neer 2001), but also illustrate that hunting and fishing still continued, the first in elite and religious behaviour, the second to complement the diet (Linseele and Van Neer 2009).

This may go along with the practice of cattle burials, known also from the Badari Culture (Brunton and Caton-Thompson 1928: pl. 10.6), obviously a religious custom also recently reported from the Western Desert. In the Nabta area, 'Late Neolithic' burials of complete animal bodies were excavated, including one interred in a carefully built burial chamber with a wooden superstructure and dated to 5400 BC (Wendorf and Schild 2004: 20). Thus Saharan cattle traditions might well have become an essential part of Neolithic life on the fringes of the Nile Valley and may have contributed practically and spiritually to the long-lasting and complex process of the rise of Pharaonic civilisation.

Multi-resource Pastoralism during the Fifth to the Third Millennium BC

The close of the Holocene humid phase at the end of the sixth millennium BC and the resulting depopulation of the core zones of the Western Desert marks the beginning of a regionalisation of Eastern Saharan cultures (Kuper 2006). While the network of contacts began to concentrate on specific refuges and corridors, cultural and socio-economic elements started to establish independent expressions and concepts adapted to the individual configurations of their environments. The disjunction between the Western Desert cultures is clearest in the forms and styles of ceramic production and in the development of rock art. To some extent the same can be said of the development of pastoralism. Individual regions exhibit divergent and decoupled patterns of herding which, although here subsumed under the economic pattern of multi-resource pastoralism, widely differed in economy, ideology and social relations.

Early Pastoralists at Dakhla Oasis

At the time when the Predynastic cultures of the Nile Valley developed a fully sedentary way of life on the basis of agriculture, stock keeping and fishing, the first expressions of nomadic pastoral existence can be observed in the Western Desert. At Dakhla Oasis, one of the largest refuges with an extent of nearly 100 kilometres, archaeological work by the Dakhleh Oasis Project (DOP) has created a cultural sequence for late prehistory which offers a good insight into the economic development of the oasis (McDonald 1999; Hope 2002). While domesticated animals had already been introduced into the oasis in small amounts during the humid phase (McDonald 1998), it is the Bashendi B cultural unit during the second half of the sixth and the fifth millennium BC which appears for the first time to have heavily relied on herding (Churcher 1999; McDonald 2002). This may be concluded not only from a higher number of domesticates (cattle and sheep) in the bone material, but also from a decrease in arrowheads (McDonald 1998, 1999). Grinding implements occur, indicating a continuation in the gathering of wild cereals. Following McDonald (2009), exotic stone and shell might qualify as prestige items, suggesting emerging inequality in social relations. Bashendi B reveals smaller short-term camp sites and more frequent residential moves (McDonald 2002) compared to the preceding Bashendi A cultural unit; however, the distances of residential moves appears to be restricted to the oasis and its vicinity (no more than 100 kilometres) due to the climatically caused diminishing of territories. Although the drying trend was well underway during this period, arid conditions had not reached the present level of hyper-aridity. This offered groups the possibility of setting off to the vicinity of the oasis depression for herding and hunting activities. Archaeological sites, such as Eastpans 95/2 (Gehlen et al. 2002), yielded animal remains featuring mixtures of sheep, cattle, gazelles and antelopes, indicating a kind of early multi-resource pastoralism. Although the material culture of the Bashendi B unit does not differ significantly from that of the earlier period (McDonald 1999; Hope 2002), pottery, exotic stone and jewellery appear to be more frequent on the sites. It is therefore suggested that this earliest kind of pastoralism originated from a complex forager culture, or at least represents complex foragers in transition to a pastoral society and economy.

In Dakhla it was not before the emergence of the Sheikh Muftah cultural unit during the fourth and third millennia BC that pottery became a prevailing element at nearly all the sites (Hope 2002; McDonald 2002). The ongoing process of increasing aridity obviously restricted the territory of the people to the oasis depression, but there is no evidence of permanent settlements. Faunal remains are dominated by the bones of domesticated cattle and goats (Churcher 1999; McDonald 2001, 2002), but there is again evidence of hunting (gazelle, hare, hartebeest), in particular in the near-desert vicinity north of the oases. At El Kharafish on the Egyptian Limestone Plateau, approximately 30 kilometres north of Dakhla oasis, a multiply occupied seasonal camp site of the Sheikh Muftah people illustrates the character of a late-winter or early-spring camp when gazelles in the last weeks of their pregnancy were hunted (Riemer et al. 2008; Pöllath 2011). Although one cattle bone and a mass of dung pellets from ovicaprines attest to livestock, the bulk of bones speak of a camp especially devoted to hunting gazelles and wildfowl. The techniques applied to the hunting of mammals seem to differ fundamentally from earlier periods, since only a very few arrowheads have been found at the site. Alternative strategies, such as snaring and chasing, can be inferred.

Along with the development of the pastoral component, the material culture appears to have undergone a number of changes. Large cooking pots are frequent at the sites, indicating increased transport capacities. Cattle and donkeys, the latter evidenced in a Sheikh Muftah context in Dakhla (Churcher and Kleindienst 2006), may have served as pack animals. Judging from the maturity of most specimens, cattle were not primarily kept for meat, but perhaps for milk and transportation (Churcher 1999; McDonald 1999). Spouted bowls, well-known in Egyptian contexts for the distribution of milk, occur at Sheikh Muftah sites (Riemer 2011: 61–62). Grinding equipment is rare on Sheikh Muftah sites. With regard to the great number of cooking pots, it can be suggested that porridge cooking was the preferred way of processing plant material. It is as yet too difficult to tell whether plant remains represent gathered wild cereals or whether agricultural produce was additionally exchanged by means of small donkey 'caravans' through the desert. So far, no cultivated species have been found among the plant remains recovered from excavations (McDonald 2002). Surprisingly, this applies also to the latest Sheikh Muftah phase when Pharaonic Egypt had occupied Dakhla, around 2700 to 2600 BC, and irrigation and agriculture were introduced. Moreover, the lithic industry changed to informal tool production, a phenomenon that can likewise be observed in pastoral societies elsewhere (Smith et al. 1991).

For McDonald, 'the lack of noticeable burial monuments or of elaborate grave goods reinforces the picture of small, egalitarian groups' (McDonald 2002: 118). In fact, Sheikh Muftah societies did obviously not develop a cattle centrism in ideology and cult, nor do we see any traces of social complexity, such as illustrated in the following examples of pastoral societies.

A Cattle Cult at Nabta Playa

In many societies featuring cattle as the primary herding animals, cattle occupy a central position not only in an economic sense but also in the collective and individual tradition and ideology expressed in social relations, mythology and ritual behaviour. In the following it is shown that cattle ideology was central to the tradition of some pastoral groups in the Western Desert, which can be followed into the Predynastic and Dynastic periods of Nile-based Egyptian culture. Cattle ideology is expressed in cattle cults which are well attested in the Kerma Culture, where prominent burials yielded hundreds of skulls from sacrificed cattle (see Figure 1.2) (Chaix 2001). In the Egyptian Predynastic period, which was probably heavily influenced by pastoral traditions on both sides of the Nile, cattle became a symbol of power and kingship, not only in offerings but in a wide range of iconographic expressions (Wengrow 2001; Hendrickx 2002) which, among other things, culminated in the early Dynastic bucrania from Saqqara (Emery 1954). However, it remains an open question whether the desert pastoralists of the time practised an economic specialisation exclusively centred on cattle.

In the Western Desert the earliest expression of a cattle cult is represented at Nabta Playa by a number of cattle burials, one of which contained the skeleton of a productive cow. The intentional burial of whole skeletons or disarticulated cattle bones is a development that can be observed elsewhere in the Sahara since the end of the sixth millennium BC (Brass 2007), but what makes Nabta Playa so extraordinary is the dating evidence. The earliest radiocarbon date from cattle burials at Nabta E–94–1n is 5400 BC, falling into the (second half) of the so-called Late Neolithic phase as defined in the chronological sequence of Nabta. Dates from other sites at Nabta suggest that the tradition lasted approximately until the middle of the fifth millennium BC (Wendorf et al. 2001: 468–88). Nabta has been interpreted by the excavators as a ceremonial and political centre (ibid.: 671–4), and the burials are regarded by Brass (2007: 16) as ceremonial locales displaying the establishment of ritual and political authority within dominant lineages and leadership positions, although solid evidence for a hierarchical society and lineages is still lacking (cf. Wendorf et al. 2001: 672). Nevertheless, the cattle cult of Nabta Playa illustrates the central position cattle occupied in the pastoral belief system during this period, which probably developed into more elaborate symbolic forms along the Nile during the following centuries.

The economic basis of the Late Neolithic, however, is much more diversified. Plant remains testify to the gathering of wild cereals. In the faunal material goats and sheep make up a large part of the livestock, and the proportion of domesticates, around 35 per cent, is outweighed by that of hunted wild animals (Gautier 2001; Linseele and Van Neer 2009).

Cattle Pastoralism in the Rock Art of Jebel Ouenat and Gilf Kebir

Apart from Nabta, archaeological evidence for the establishment of cattle-centred pastoralism is meagre in the Western Desert. A prominent exception is the region of Jebel Ouenat and the southern Gilf Kebir, which lies at the intersection of the borders of Libya, Sudan and Egypt. As no certain indication is yet given by archaeozoological material for this region, the evidence is largely based upon the rich painted rock art (Rhotert 1952; Van Noten 1978; Le Quellec, de Flers and de Flers 2005; Riemer 2009; Zboray 2009). Pastoral themes dominate the images, as is widespread in Saharan pastoral rock art (Muzzolini 2000).

The pastoral 'style' of Ouenat and Gilf Kebir, in which cattle occupy a central position, is often scenic and narrative (see Figure 1.8), although small livestock also occur. Cattle not only tend to dominate the images, but the vivid depiction of cattle skin pattern and colours (see Figure 1.9) also illustrate the intimate relations between herders and their animals (cf. Smith 1992: 154–60; 2005: 194–5).

Individuality was also expressed in the diversity of horn forms. Moreover, depictions of calving cows (Figure 1.10) and other elements of reproduction indicate an emphasis on genealogy and cattle breeding (cf. Jesse, Keding, Lenssen-Erz and Pöllath, this volume).

Figure 1.8: A family scene with a cow suckling two children, illustrating intimate relations between herders and their animals. Karkur Talh, Jebel Ouenat, Sudan (photograph by R. Kuper; cf. Van Noten 1978: fig. 171).

Figure 1.9: The depiction of horn forms, skin pattern and colours of individual herd animals. Note the tent scene and the herders with bows. Ain Dua B, Jebel Ouenat, Libya (photograph by R. Kuper).

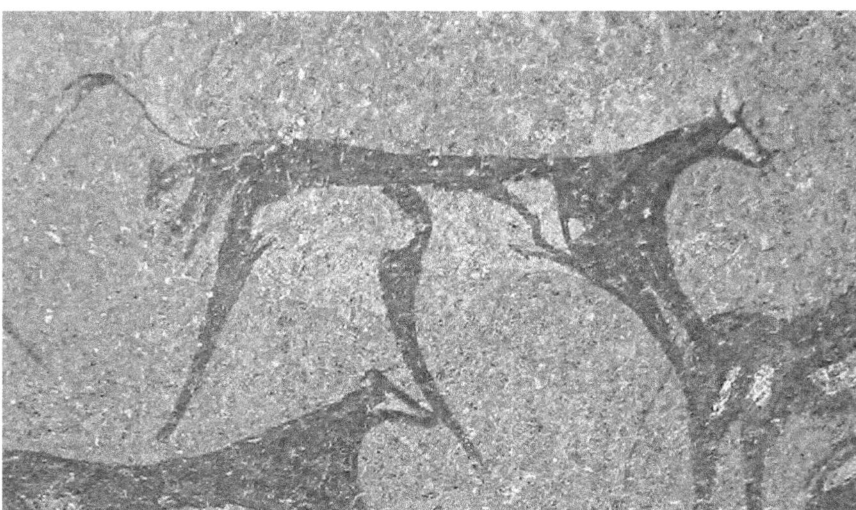

Figure 1.10: Birth and reproduction in pastoral rock paintings: a calving cow. Karkur Idris, Jebel Ouenat, Libya (photograph by R. Kuper).

An expression of pastoral territoriality can be seen in marked, small-size territorial units of (competing) pastoral groups (cf. Jesse, Keding, Lenssen-Erz and Pöllath, this volume) discernible by different rock-art styles (see Figure 1.11). Small-scale clusters of differing art styles representing territories of individual

Figure 1.11: Combat for cattle? Karkur Talh 23, Jebel Ouenat, Sudan (Almásy 1936: pl. vii).

pastoral groups have recently been suggested for Ouenat and the southern Gilf Kebir by Le Quellec, de Flers and de Flers (2005) and Zboray (2005).

Dating rock art is notoriously difficult, and this is also true of the pastoral rock art of Ouenat and Gilf Kebir. Although direct dating evidence as well as absolute dates from rock-art contexts are still lacking in the region, there is some evidence from local climatic and occupational history which provides clues as to when the pastoral style may have been practised. The first time-marker is represented by the occurrence of pastoral elements in the region. Due to a lack of sufficient archaeozoological data it cannot be securely determined exactly when the region was first occupied by pastoralists, but a dominance of microlithic arrowheads in Wadi Bakht and Wadi el-Akhdar in the Gilf Kebir during the period around 6600 to 4400 BC suggests that hunting ranked first in the contemporary mode of subsistence. By about 4400 BC the material culture indicates a fundamental change in ceramic and lithic traditions (Linstädter 2005). While a new type of thin-walled pottery often decorated by impressed or incised designs emerges, the lithic material changes towards a rather informal tool production. The end of this phase is defined by the breaking of the playa-lake dam in Wadi Bakht and the subsequent depopulation of the area around 3500 BC (Linstädter and Kröpelin 2004); subsequently the ceramic tradition persisted in more southerly areas until about 3000 BC. The collapse of the dam can be linked to a progressing trans-Saharan drying trend evident in Saharan and marine climatic archives elsewhere around the same time. However, this was not connected with a complete abandonment of the Gilf Kebir, as indicated by a cattle bone from Wadi el-Akhdar, dated to between 3200 and 2800 BC (Cziesla 1996: 229). If this indicates at least sporadic pastoral exploitation of the Gilf, then the same and more can be presumed for Jebel Ouenat, situated more than 100 kilometres to the south and, due to its altitude, favoured by more rainfall. Therefore pastoral rock art might have continued into the third millennium BC, illustrating the central role of cattle, whose presence in Jebel Ouenat was even reported as late as the 1950s.

Specialised Pastoralism or Cattle-centred Ideology?

In contrast to multi-resource pastoralism, cultural anthropologists define specialised pastoralism as a type of subsistence strategy involving (near) exclusive economic reliance on animal husbandry (Galaty 1996; Smith 2005: 4–6). With these two concepts in mind, it should be illuminating to try and classify the subsistence and wider cultural modes associated with the four archaeological case studies of pastoral people from the Western Desert of Egypt mentioned above. First, the example of the Bashendi B people in Dakhla Oasis (roughly between 5400 and 3900 BC) may be subsumed under the rubric of early multi-resource pastoralism; in this particular instance hunting was dominant, and there was the gathering of wild cereals and a clear forager tradition in material culture. However, the Sheikh Muftah economy (around 3900 to 2200 BC) was dominated by domesticates, wild animals contributing only secondarily to subsistence; it may therefore be conceptualised as developed multi-resource pastoralism. The example of the Nabta Late Neolithic (about 5400 BC) displays a kind of cattle specialisation in religious and social concerns which led Wendorf et al. (2001: 671) to stress the role of Nabta in the 'early development of the African Cattle Complex'. A closer look, however, indicates that wild animals outnumber domesticated species regarding the number of bones recovered from habitation sites. Consequently, in economic terms the Nabta Late Neolithic followed the multi-resource type and would hardly fit into the category of specialised pastoralism. Yet the important role cattle evidently played in ritual behaviour marks a clear step towards a pastoral ideology, and it is suggested that this type of pastoralism should be labelled 'cattle-centred' (see Table 1.1).

The vivid depictions of cattle of Jebel Ouenat (from around 4400 to 3000 BC) illustrate a pastoral ideology, but any assessment of the economic base of their creators founders on a lack of sufficient bone material. Even under the premise that it might have been focused nearly exclusively on cattle (which would be tantamount to ignoring that some depictions also show ovicaprines), questions concerning the plant food components of the human diet would remain. In communities of specialist pastoralists, the demand for plant food is usually met by systematic relationships with other specialised groups, in particular with farmers from whom the herders acquire primary agricultural products (Sadr 1991). Although protein rich food is inevitable in a specialist pastoral society, a large part of the diet consists of plant foods normally acquired by trade and exchange with sedentary agriculturalists. Another means of obtaining access to plant foodstuffs is raiding, threatening or offering 'protection' in exchange for tribute in the form of goods (Lees and Bates 1974). In the case of most of the early Saharan cattle herders, such as those around Ouenat, such relations could hardly have been maintained considering the long distances that existed between the potential partners. And certainly there would not have been any necessity to do so if one takes into account the rich yields of wild grains that the Sahelian environment still provides today. There are records from the Tuareg in Mali, where 'in good years' one man can harvest 50 kg of grass seeds per day (Klute 1991: 205–7); and from northern Darfur in Sudan it is reported that among the Mahria nomads the women of a camp of thirty-three members harvested about 375 kg per season (Holter 1994: 259). Clearly, pastoralists satisfying their demand for plant food by gathering wild cereals do not conform to the above-cited ethnographic definition implying (near) exclusive economic reliance on cattle, even if cattle bones clearly dominate their faunal records. In this light, the Ouenat evidence probably represents a multi-resource economy associated with a cattle-centred pastoralist ideology.

The discussion above supports the conclusion that cattle pastoralism may have a wide range of possible expressions that can vary from economic specialisation to a social and ideological obsession with cattle. No doubt pastoral ideologies will never be fully separate from pastoral economies, the latter supporting cattle-centred world-views and vice versa. Ouenat and Nabta illustrate that cattle orientation has its basis in cattle keeping but is not necessarily linked to specialised pastoralism.

Likewise, the bull's head developed into a transcendent symbol of power of the ruling elites during the process of Egyptian state formation, but probably had its roots in the economic strengths of the former pastoral communities.

Epilogue: Herders at the Margins of Ancient Egypt

It was the intention of the present chapter to assemble the relevant archaeological data on the early development of pastoralism in North-east Africa, in particular in Egypt's Western Desert. Although it remains an unresolved question whether cattle were originally domesticated in Africa, possibly already during the early Holocene, the spread of domestic cattle over northern Africa did not take place before around 6000 BC, along with the introduction of sheep and goats from the Near East. The earliest domesticated animals – goats, sheep and cattle – appear in hunter-gatherer contexts where hunting for desert game and the gathering of wild seeds clearly dominated subsistence activities. These 'pastro-foragers', though indicating a faint shift towards 'complexity' in material culture, mobility and possibly social relations, do not display a fundamental break in the hunter-gatherer tradition. Assuming a general conservatism in foragers' subsistence activities and ideology, and considering the harsh environmental conditions of the dry savannah, it is reasonable to suggest that the early 'experiments' in stock keeping at Nabta Playa, Djara, Farafra and elsewhere represent no more than a marginal contact phenomenon at the periphery of Near Eastern pastoralism. Nevertheless, pastro-foraging did emerge and possibly formed the substrate out of which more pronouncedly pastoral economies evolved during the following millennia.

It was not before the onset of the Saharan drying trend, at the end of the sixth millennium BC, that communities who had largely retreated to the refuges and margins of the desertified Sahara started to develop truly pastoral ways of life, both economically and ideologically. Around the same time, stock keeping appeared as one of the fundamental economic strategies of the early Predynastic cultures along the river Nile, accompanied by early agriculture and sedentarism. In the Western Desert, cattle, goats and sheep were herded in nomadic contexts, but hunting and gathering still persisted. The cases from the Western Desert, though subsumed here under 'multi-resource pastoralism', indicate a wide range of different pastoral expressions: an early cattle-centred ideology at Nabta Playa at the end of the sixth millennium BC, where cattle burials have been found, combined with a mode of subsistence apparently dominated by game hunting. A similar focus on cattle is in evidence at Jebel Ouenat between around 4400 and 3000 BC, here expressed by lively rock paintings depicting cattle herds and domestic life. At Dakhla Oasis, the society associated with the Sheikh Muftah cultural unit during the third millennium BC would seem to have been largely engaged in stock keeping judging from their faunal remains and a completely changed material culture, in particular higher amounts of domestic pottery and informal lithic production. However, hunting continued, and there is nothing to suggest an elaborate cattle orientation in ideology such as has been claimed for Nabta Playa and Jebel Ouenat.

The story of pastoralism in the Eastern Sahara is a book with many missing pages. The finale, however, is obvious: a happy ending that found its stage in Ancient Egypt and in many other regions of arid Africa. The very beginnings of the close symbiosis between humans and cattle still lie in the dark, and it remains a matter of discussion whether the introduction of small livestock around 6000 BC and the positive experience of having 'emergency rations on the hoof' stimulated the invention of the 'walking larder', or whether – perhaps some millennia earlier – it was just joyful experimentation which, after being successful, was recognised as also providing some comfort. To what extent this economic step then gradually changed the life and conceptions of hunter-gatherers, and whether they might be classified as 'pastro-foragers' or as 'multi-resource pastoralists', depends on regionally and temporally specific ecological adaptations, and is perhaps even more a matter of academic classification. Anyhow, climate can be regarded as the driving force behind the whole story, especially when economic pressures triggered by the increase of the aridity of western Egypt after 5000 BC caused the breakthrough of herding as an important subsistence strategy. The different directions of withdrawal from the expanding Western Desert obviously led to various transformations of pastoral elements, for instance to agro-pastoralism in the Nile Valley where cattle and cereals formed the backbone of the Egyptian state; to multi-resource pastoralism at Dakhla; and to a cattle-centred ideology among the pastoralists of Nabta Playa and Jebel Ouenat.

In these latter areas, far distant from the Egyptian Nile Valley, cattle keepers up to the second millennium BC practised lifestyles very different from those in the Pharaonic period. This has only left sparse traces in the written records of Ancient Egypt, in spite of growing evidence from recent archaeological discov-

Figure 1.12: Girl riding a bull in *galop volant*. Rock painting from Tamrit, Tassili. (Photograph by A. Zboray, enhanced with dStretch).

eries in the deep desert. For instance, the latter comprise some thirty outposts between Dakhla and the Gilf Kebir, clearly indicating the first trans-Saharan trail into the interior of the continent during the Old Kingdom (Kuper 2003; Förster 2007), and this is also confirmed by a hieroglyphic rock inscription at Jebel Ouenat mentioning the Middle Kingdom Pharaoh Mentuhotep II receiving products from the country of Yam to the south (Clayton, Trafford and Borda 2008). This oldest Saharan caravan traffic could not yet rely on camels, which were only introduced during the last centuries BC. The donkey had to replace the ox, which for a long time had served as a pack and riding animal (see Figure 1.12). Therefore, extraordinary logistic skills were required and a profound geographical knowledge of the area, especially of its water resources. Apparently this was rooted in age-old intensive contacts between cattle-herding societies all over the Holocene Saharan savannah that had continuously been reduced to only a few linear, but never fully abandoned, lines of communication.

With all this in mind it seems worth indicating the fact that this speculative encounter between history and prehistory includes more than these simply practical ties. There seem to be many elements of artistic and spiritual imagery in Saharan rock art in general, and in the herders' art in particular, that might have arguably influenced the ideological and artistic traditions in the Nile Valley (Leclant and Huard 1980; Almásy 1997: 332–6). In spite of fundamental transformations that these elements have surely undergone over millennia on the Nile, the goddess Hathor (see Figure 1.13) perhaps would not feel too much like a stranger among her beautiful cousins at Jebel Ouenat.

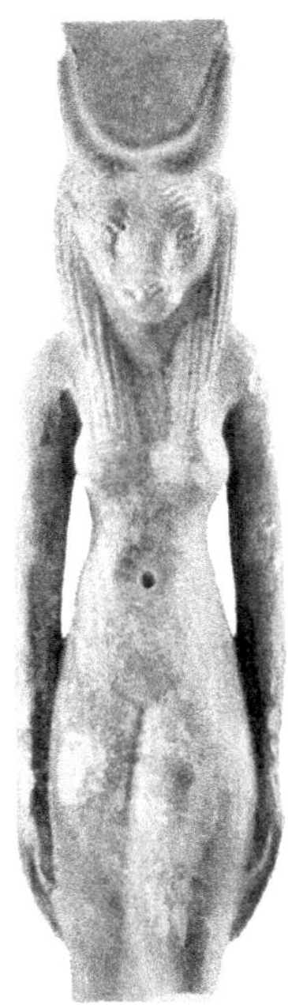

Figure 1.13: The Egyptian cow-goddess Hathor. Twenty-sixth Dynasty faience statuette (Hornung, Loeben and Wiese 2005: 111).

Notes

1. All dates in this contribution are calibrated BC.
2. Date calibration by means of the Cologne Calibration and Palaeoclimate Research Package CalPAL (Weninger and Jöris 2004).

References

Almásy, L.E. 1936. *Récentes explorations dans le désert libyque.* Cairo: Société royale de géographie d'Égypte.

——— 1997. *Schwimmer in der Wüste: Auf der Suche nach der Oase Zerzura.* Innsbruck: Haymon Verlag.

Barich, B.E., and F.A. Hassan. 2000. 'A Stratified Sequence from Wadi el-Obeiyd, Farafra: New Data on Subsistence and Chronology of the Egyptian Western Desert', in L. Krzyżaniak, K. Kroeper and M. Kobusiewicz (eds), *Recent Research into the Stone Age of North-eastern Africa.* Poznań: Poznań Archaeological Museum, pp.33–46.

Barich, B.E., and G. Lucarini. 2002. 'Archaeology of Farafra Oasis (Western Desert, Egypt)', *Archéo-Nil* 12: 101–8.

Bar-Yosef, O. 1984. 'Seasonality among Neolithic Hunter-gatherers in Southern Sinai', in J. Clutton-Brock and C. Grigson (eds), *Animals and Archaeology,* Vol. 3: *Early Herders and their Flocks.* Oxford: British Archaeological Reports, pp.145–60.

Boessneck, J., and A. von den Driesch. 1990. 'Tierreste aus der vorgeschichtlichen Siedlung von el-Omari bei Heluan/Unterägypten (Appendix III)', in F. Debono and B. Mortensen (eds), *El Omari: A Neolithic Settlement and Other Sites in the Vicinity of Wadi Hof, Helwan.* Mainz: Philipp von Zabern, pp.99–107.

——— 1994. 'Tierknochenfunde aus dem negadezeitlichen Fundplatz MA21/83 bei Armant', in B. Ginter and J.K. Kozlowski (eds), *Predynastic Settlement near Armant.* Heidelberg: Heidelberger Orientverlag, pp.183–89.

Bonnet, C. 2004. 'The Kerma Culture', in D.A. Welsby and J.R. Anderson (eds), *Sudan, Ancient Treasures: An Exhibition of Recent Discoveries from the Sudan National Museum.* London: British Museum Press, pp.70–7.

Brass, M. 2007. 'Reconsidering the Emergence of Social Complexity in Early Saharan Patoral Societies, 5000–2500 BC', *Sahara* 18: 7–22.

Breunig, P., K. Neumann and W. Van Neer. 1996. 'New Research on the Holocene Settlment and Environment of the Chad Basin in Nigeria', *African Archaeological Review* 13: 111–45.

Brewer, D.J. 1989. *Fishermen, Hunters and Herders: Zooarchaeology in the Fayum, Egypt (ca. 8200–5000 bp).* Oxford: British Archaeological Reports.

Brunton, G., and G. Caton-Thompson. 1928. *The Badarian Civilisation.* London: British School of Archaeology in Egypt.

Butzer, K.W. 1980. 'Pleistocene History of the Nile Valley in Egypt and Lower Nubia', in M.A.J. Williams and H. Faure (eds), *The Sahara and the Nile: Quaternary Environments and Prehistoric Occupation in Northern Africa.* Rotterdam: Balkema, pp.253–80.

Chaix, L. 2001. 'Animals as Symbols: The Bucrania of the Grave KN 24 (Kerma, Northern Sudan)', in H. Buitenhuis and W. Prummel (eds), *Animals and Man in the Past: Essays in Honour of A.T. Clason.* Groningen: Arc-Publicaties, pp.364–70.

Churcher, C.S. 1999. 'Holocene Faunas of the Dakhleh Oasis', in C.S. Churcher and A.J. Mills (eds), *Reports from the Survey of the Dakhleh Oasis, Western Desert of Egypt, 1977–1987.* Oxford: Oxbow, pp.109–15.

Churcher, C.S., and M.R. Kleindienst. 2006. 'A Predynastic Ass (*Equus asinus*) from the Sheikh Muftah Cultural Horizon of the Dakhleh Oasis, Western Desert, Egypt', in K. Kroeper, M. Chłodnicki and M. Kobusiewicz (eds), *Archaeology of Early North-eastern Africa.* Poznań: Poznań Archaeological Museum, pp.425–35.

Clayton, J., A. Trafford and M. Borda. 2008. 'A Hieroglyphic Inscription Found at Jebel Uweinat Mentioning Yam and Tekhebet', *Sahara* 19: 129–34.

Close, A.E. 2002. 'Sinai, Sahara, Sahel: The Introduction of Domestic Caprines to Africa', in T. Lenssen-Erz, U. Tegtmeier, S. Kröpelin et al. (eds), *Tides of the Desert/Gezeiten der Wüste: Contributions to the Archaeology and Environmental History of Africa in Honour of Rudolph Kuper.* Cologne: Heinrich-Barth-Institut, pp.471–83.

Cribb, R. 1991. *Nomads in Archaeology.* Cambridge: Cambridge University Press.

Cziesla, E. 1996. 'Der Fundplatz Wadi el Akhdar 80/14', in W. Schön, *Ausgrabungen im Wadi el Akhdar, Gilf Kebir (SW-Ägypten).* Cologne: Heinrich-Barth-Institut, pp.141–278.

Driesch, A. von den, and J. Boessneck. 1985. *Die Tierknochenfunde aus der neolithischen Siedlung von Merimde Benisalâme am westlichen Nildelta.* Munich: Uni-Druck.

Emery, W.B. 1954. *Great Tombs of the First Dynasty*, Vol. 2. London: Egypt Exploration Society.

Förster, F. 2007. 'With Donkeys, Jars and Water Bags into the Libyan Desert: The Abu Ballas Trail in the Late Old Kingdom/First Intermediate Period', *British Museum Studies in Ancient Egypt and Sudan* 7: 1–36.

Galaty, J.G. 1996. 'Pastoralists', in A. Barnard and J. Spencer (eds), *Encyclopedia of Social and Cultural Anthropology.* London: Routledge, pp.415–16.

Gautier, A. 1980. 'Contributions to the Archaeozoology of Egypt', in F. Wendorf and R. Schild (eds), *Prehistory of the Eastern Sahara.* New York: Academic Press, pp.317–44.

——— 2001. 'The Early to Late Neolithic Archaeofaunas from Nabta and Bir Kiseiba', in F. Wendorf, R. Schild and associates (eds), *The Archaeology of Nabta Playa: Holocene Settlement of the Egyptian Sahara, 1.* New York: Kluwer Academic/Plenum, pp.609–35.

——— 2007. 'Animal Domestication in North Africa', in M. Bollig, O. Bubenzer, R. Vogelsang and H.-P. Wotzka (eds), *Aridity, Change and Conflict in Africa.* Cologne: Heinrich-Barth-Institut, pp.75–89.

Gehlen, B., et al. 2002. 'The Holocene Occupation of the Eastern Sahara: Regional Chronologies and Supra-regional Developments in Four Areas of the Absolute Desert', in T. Lenssen-Erz, U. Tegtmeier, S. Kröpelin et al. (eds), *Tides of the Desert/Gezeiten der Wüste: Contributions to the Archaeology and Environmental History of Africa in Honour of Rudolph Kuper.* Cologne: Heinrich-Barth-Institut, pp.85–116.

Hayden, B.D. 1992. 'Models of Domestication', in A.B. Gebauer and T.D. Price (eds), *Transitions to Agriculture in Prehistory.* Madison, WN: Prehistory Press, pp.11–19.

Hendrickx, S. 2002. 'Bovines in Egyptian Predynastic and Early Dynastic Iconography', in F.A. Hassan (ed.), *Droughts, Food and Culture: Ecological Change and Food Security in Africa's Later Prehistory.* New York: Kluwer Academic/Plenum, pp.275–318.

Hendrickx, S., B. Midant-Reynes and W. Van Neer. 2001. *Mahgar Dendera 2 (Haute Egypte), un Site d'Occupation Badarien.* Leuven: Leuven University Press.

Hendrickx, S., and P. Vermeersch. 2000. 'Prehistory: From the Palaeolithic to the Badarian Culture (*c.* 700,000–4000 BC)', in I. Shaw (ed.), *The Oxford History of Ancient Egypt.* Oxford: Oxford University Press, pp.17–43.

Herskovits, M.J. 1926. 'The Cattle Complex in East Africa', *American Anthropologist* 28: 633–64.

Holter, U. 1994. 'Nomadenfrauen in der Dürre: Das Beispiel der Mahria Kamelnomaden (Nördliche Rizeigat) in Norddarfur, Sudan', in M. Bollig and F. Klees (eds), *Überlebensstrategien in Afrika.* Cologne: Heinrich-Barth-Institut, pp.255–73.

Hope, C.A. 2002. 'Early and Mid Holocene Ceramics from the Dakhleh Oasis: Traditions and Influences', in R. Friedman (ed.), *Egypt and Nubia: Gifts of the Desert.* London: British Museum Press, pp.39–61.

Hornung, E., C.E. Loeben and A. Wiese. 2005. *Immortal Pharaoh: The Tomb of Thutmose III.* Madrid: Factum Arte and United Exhibits Group.

Ingold, T. 2000. *The Perception of the Environment: Essays in Livelihood, Dwelling and Skill.* London: Routledge.

Jesse, F. 2003. *Rahib 80/87: Ein Wavy-Line-Fundplatz im Wadi Howar und die früheste Keramik in Nordafrika.* Cologne: Heinrich-Barth-Institut.

Keding, B. 1997. *Djabarona 84/13: Untersuchungen zur Besiedlungsgeschichte des Wadi Howar anhand der Keramik des 3. und 2. Jahrtausends v. Chr.* Cologne: Heinrich-Barth-Institut.

Kindermann, K. 2004. 'Djara: Excavations and Surveys of the 1998–2002 Seasons', *Archéo-Nil* 14: 31–50.

——— 2010. *Djara. Zur mittelholozänen Besiedlungsgeschichte zwischen Niltal und Oasen (Abu-Muharik-Plateau, Ägypten).* Cologne: Heinrich-Barth-Institut.

Kindermann, K., and O. Bubenzer. 2007. 'Djara: Humans and Their Environment on the Egyptian Limestone Plateau around 8,000 Years Ago', in O. Bubenzer, A. Bolten and F. Darius (eds), *Atlas of Cultural and Environmental Change in Arid Africa.* Cologne: Heinrich-Barth-Institut, pp.26–9.

Kindermann, K., et al. 2006. 'Palaeoenvironment and Holocene Land Use of Djara, Western Desert of Egypt', *Quaternary Science Reviews* 25: 1619–37.

Klute, G. 1991. *Die schwerste Arbeit der Welt: Alltag von Tuareg-Nomaden.* Munich: Trickster.

Kröpelin, S. 1999. 'Terrestrische Paläoklimatologie heute arider Gebiete: Resultate aus dem Unterem Wadi Howar (Südöstliche Sahara/Nordwest-Sudan)', in E. Klitzsch and U. Thorweihe (eds), *Nordost-Afrika, Strukturen und Ressourcen: Ergebnisse aus dem Sonderforschungsbereich 'Geowissenschaftliche Probleme in ariden und semiariden Gebieten'.* Weinheim: Wiley-VCH, pp.446–506.

Kuper, R. 1988. 'Neuere Forschungen zur Besiedlungsgeschichte der Ost-Sahara', *Archäologisches Korrespondenzblatt* 18: 127–42.

——— 1993. 'Sahel in Egypt: Environmental Change and Cultural Development in the Abu Ballas Area, Libyan Desert', in L. Krzyżaniak, M. Kobusiewicz and J. Alexander (eds), *Environmental Change and Human Culture in the Nile Basin and Northern Africa until the Second Millennium* BC. Poznań: Poznań Archaeological Museum, pp.213–23.

——— 2002. 'Routes and Roots in Egypt's Western Desert: The Early Holocene Resettlement of the Eastern Sahara', in R. Friedman (ed.), *Egypt and Nubia: Gifts of the Desert.* London: British Museum Press, pp.1–12.

——— 2003. 'Les marches Occidentales de l'Égypte: Dernières Nouvelles', *Bulletin de la Société Francaise d'Égyptologie* 158: 12–34.

——— 2006. 'An Attempt at Structuring the Holocene Occupation of the Eastern Sahara', in K. Kroeper, M. Chłodnicki and M. Kobusiewicz (eds), *Archaeology of Early North-eastern Africa.* Poznań: Poznań Archaeological Museum, pp.261–72.

Kuper, R. (ed.). 1978. *Sahara: 10.000 Jahre zwischen Weide und Wüste.* Cologne: Museen der Stadt Köln.

Kuper, R., and S. Kröpelin. 2006. 'Climate-controlled Holocene Occupation in the Sahara: Motor of Africa's Evolution', *Science* 313: 803–7.

Leclant, J., and P. Huard. 1980. *La Culture des Chasseurs du Nil et du Sahara*. Algiers: Centre de recherches anthropologiques prehistoriques et ethnographiques.

Lee, R.B., and I. DeVore. 1968. 'Problems in the Study of Hunters and Gatherers', in R.B. Lee and I. DeVore (eds), *Man the Hunter*. Chicago: Aldine, pp.3–20.

Lees, S.H., and D.G. Bates. 1974. 'The Origins of Specialized Nomadic Pastoralism: A Systemic Model', *American Antiquity* 39: 187–93.

Legge, A.J. 1996. 'The Beginning of Caprine Domestication in Southwest Asia', in D.R. Harris (ed.), *The Origins and Spread of Agriculture and Pastoralism in Eurasia*. London: University College London Press, pp.238–62.

Le Quellec, J.-L., P. de Flers and Ph. de Flers. 2005. *Du Sahara au Nil: Peintures et Gravures d'avant les Pharaons*. Paris: Soleb-Fayard.

Linseele, V. 2007. *Archaeofaunal Remains from the Past 4000 Years in Sahelian West Africa: Domestic Livestock, Subsistence Strategies and Environmental Changes*. Oxford: Archaeopress.

Linseele, V., and W. Van Neer. 2009. 'Exploitation of Desert and Other Wild Game in Ancient Egypt: The Archaeozoological Evidence from the Nile Valley', in H. Riemer, F. Förster, M. Herb and N. Pöllath (eds), *Desert Animals in the Eastern Sahara: Status, Economic Significance and Cultural Reflection in Antiquity*. Cologne: Heinrich-Barth-Institut, pp.45–78.

Linstädter, J. (ed.). 2005. *Wadi Bakht: Landschaftsarchäologie einer Siedlungskammer im Gilf Kebir*. Cologne: Heinrich-Barth-Institut.

Linstädter, J., and S. Kröpelin. 2004. 'Wadi Bakht Revisited: Holocene Climate Change and Prehistoric Occupation in the Gilf Kebir Region of the Eastern Sahara, SW Egypt', *Geoarchaeology* 19: 753–78.

MacDonald, K.C., and R.H. MacDonald. 2000. 'The Origins and Development of Domesticated Animals in Arid West Africa', in R.M. Blench and K.C. MacDonald (eds), *The Origins and Development of African Livestock: Archaeology, Genetics, Linguistics and Ethnography*. London: University College London Press, pp.127–62.

McDonald, M. 1998. 'Early African Pastoralism: View from Dakhleh Oasis (South Central Egypt)', *Journal of Anthropological Archaeology* 17(2): 124–42.

——— 1999. 'Neolithic Cultural Units and Adaptations in the Dakhleh Oasis', in C.S. Churcher and A.J. Mills, *Reports from the Survey of the Dakhleh Oasis 1977–87*. Oxford: Oxbow, pp.117–32.

——— 2001. 'The Late Prehistoric Radiocarbon Chronology for Dakhla Oasis within the Wider Environmental and Cultural Settings of the Egyptian Western Desert', in M. Marlow (ed.), *The Oasis Papers, 1: Proceedings of the First International Symposium of the Dakhleh Oasis Project*. Oxford: Oxbow, pp.26–42.

——— 2002. 'Dakhleh Oasis in Predynastic and Early Dynastic Times: Bashendi B and the Sheikh Muftah Cultural Units', *Archéo-Nil* 12: 109–20.

——— 2008. 'Emerging Social Complexity in the Mid-Holocene Egyptian Western Desert: Site 270 and its Neighbours in South-eastern Dakhleh Oasis', in M.F. Wiseman (ed.), *The Oasis Papers, 2. Proceedings of the Second International Conference of the Dakhleh Oasis Project*. Oxford: Oxbow, pp.83–106.

——— 2009. 'Increased Sedentism in the Central Oases of the Egyptian Western Desert in the Early to Mid Holocene: Evidence from the Peripheries', *African Archaeological Review* 26: 3–43.

Marshall, F. 2000. 'The Origins and Spread of Domestic Animals in East Africa', in F.A. Hassan (ed.), *Droughts, Food and Culture: Ecological Change and Food Security in Africa's Later Prehistory*. New York: Kluwer Academic/Plenum, pp.191–221.

Moore, A.M.T., G.C. Hillman and A.J. Legge (eds). 2000. *Village on the Euphrates: From Foraging to Farming at Abu Hureyra*. Oxford: Oxford University Press.

Muzzolini, A. 2000. 'Livestock in Saharan Rock Art', in R.M. Blench and K.C. MacDonald (eds), *The Origins and Development of African Livestock: Archaeology, Genetics, Linguistics and Ethnography*. London: University College London Press, pp.87–110.

Neumann, K. 1989. 'Zur Vegetationsgeschichte der Ostsahara im Holozän: Holzkohlen aus prähistorischen Fundstellen', in R. Kuper (ed.), *Forschungen zur Umweltgeschichte der Ostsahara*. Cologne: Heinrich-Barth-Institut, pp.13–181.

Pöllath, N. 2009. 'The Prehistoric Game Bag', in H. Riemer, F. Förster, M. Herb and N. Pöllath (eds), *Desert Animals in the Eastern Sahara: Status, Economic Significance and Cultural Reflection in Antiquity*. Cologne: Heinrich-Barth-Institut, pp.79–107.

——— 2011. 'Hunting and Fowling at El Kharafish 02/5', in H. Riemer, *El Kharafish: The Archaeology of Sheikh Muftah Pastoral Nomads in the Desert around Dakhla Oasis*. Cologne: Heinrich-Barth-Institut, pp.335–48.

Rhotert, H. 1952. *Libysche Felsbilder: Ergebnisse der XI. und XII. Deutschen Inner-Afrikanischen Forschungs-Expedition (DIAFE) 1933/1934/1935*. Darmstadt: L.C. Wittich.

Riemer, H. 2003. 'The "Re-conquest" of the Great Sand Sea', in Z. Hawass and L. Pinch Brock (eds), *Egyptology at the Dawn of the Twenty-first Century*. Cairo: American University in Cairo Press, pp.408–15.

——— 2007a. 'When Hunters Started Herding: Pastro-foragers and the Complexity of Holocene Economic Change in the Western Desert of Egypt', in M. Bollig, O. Bubenzer, R. Vogelsang and H.-P. Wotzka (eds), *Aridity, Change and Conflict in Africa*. Cologne: Heinrich-Barth-Institut, pp.105–44.

——— 2007b. 'Mapping the Movement of Pastro-foragers: The Spread of Desert Glass and Other Objects in the Eastern Sahara during the Holocene "Humid Phase"', in O. Bubenzer, A. Bolten and F. Darius (eds), *Atlas of Cultural and Environmental Change in Arid Africa*. Cologne: Heinrich-Barth-Institut, pp.30–3.

——— 2009. 'Prehistoric Rock Art Research in the Western Desert of Egypt', *Archéo-Nil* 19: 31–46.

——— 2011. *El Kharafish: The Archaeology of Sheikh Muftah Pastoral Nomads in the Desert around Dakhla Oasis*. Cologne: Heinrich-Barth-Institut.

Riemer, H., and F. Jesse. 2006. 'When Decoration Made Its Way: The Northern Extent of Khartoum-style Pottery in the Eastern Sahara', in I. Caneva and A. Roccati (eds), *Acta Nubica: Proceedings of the X International Conference of Nubian Studies, Rome 9–14 September 2002*. Roma: Università di Roma 'La Sapienza', pp.63–72.

Riemer, H., and K. Kindermann. 2008. 'Contacts between the Oasis and the Nile: A Résumé of the Abu Muhariq Plateau Survey 1995–2002', in B. Midant-Reynes, Y. Tristant, J. Rowland and S. Hendrickx (eds), *Egypt at Its Origins, 2: Proceedings of the International Conference Origin of the State*. Leuven: Peeters, pp.609–33.

Riemer, H., et al. 2008. 'El Kharafish: A Sheikh Muftah Desert Camp Site between the Oasis and the Nile', in B. Midant-Reynes, Y. Tristant, J. Rowland and S. Hendrickx (eds), *Egypt at Its Origins 2. Proceedings of the International Conference Origin of the State*. Leuven: Peeters, pp.585–608.

Rosen, S.A. 2008. 'Desert Pastoral Nomadism in the Longue Durée', in H. Barnard and W. Wendrich (eds), *The Archaeology of Mobility: Old World and New World Nomadism*. Los Angeles: Cotsen Institute of Archaeology, University of California Los Angeles, pp.115–40.

Sadr, K. 1991. *The Development of Nomadism in Ancient North-east Africa.* Philadelphia: University of Pennsylvania Press.

——— 1998. 'The First Herders at the Cape of Good Hope', *African Archaeological Review* 15: 101–32.

Smith, A.B. 1990. 'On Becoming Herders: Khoikhoi and San Ethnicity in Southern Africa', *African Studies* 49: 51–73.

——— 1992. *Pastoralism in Africa: Origins and Development Ecology.* Athens: Ohio University Press.

——— 2000. 'The Origins of the Domesticated Animals of Southern Africa', in F.A. Hassan (ed.), *Droughts, Food and Culture: Ecological Change and Food Security in Africa's Later Prehistory.* New York: Kluwer Academic/Plenum, pp.221–38.

——— 2005. *African Herders: Emergence of Pastoral Traditions.* Walnut Creek, CA: Altamira.

Smith, A.B., et al. 1991. 'Excavations in the South-western Cape, South Africa and the Archaeological Identity of Prehistoric Hunter-gatherers within the Last 2000 Years', *South African Archaeological Bulletin* 46: 71–91.

Van Neer, W., and H.-P. Uerpmann. 1989. 'Palaeoecological Significance of the Holocene Faunal Remains of the B.O.S.-Missions', in R. Kuper (ed.), *Forschungen zur Umweltgeschichte der Ostsahara. Africa Praehistorica 2.* Cologne: Heinrich-Barth-Institut, pp.307–41.

Van Noten, F. 1978. *Rock Art of the Jebel Uweinat (Libyan Sahara).* Graz: Akademische Druck- und Verlagsanstalt.

Vermeersch, P.M. 1978. *Elkab II, L'Elkabien: Epipaléolithique de la Vallée du Nil Égyptien.* Leuven: Universitaire Pers.

Vermeersch, P.M., W. Van Neer and S. Hendrickx. 2004. 'El Abadiya 2, a Naqada I Site near Danfiq, Upper Egypt', in S. Hendrickx, R.F. Friedman, K.M. Ciałowicz and M. Chłodnicki (eds), *Egypt at Its Origins.* Leuven: Peeters, pp.971–89.

Vermeersch, P.M., P. Van Peer, J. Moeyersons and W. Van Neer. 1996. 'Neolithic Occupation of the Sodmein Area, Red Sea Mountains, Egypt', in G. Pwiti and R. Soper (eds), *Aspects of African Archaeology.* Harare: University of Zimbabwe Publications, pp.411–19.

Watkins, T. 2007. 'Der Naturraum in Anatolien: Ein Zusammenspiel von Klima, Umwelt und Ressourcen', in Badisches Landesmuseum Karlsruhe (ed.), *Vor 12.000 Jahren in Anatolien: Die ältesten Monumente der Menschheit.* Stuttgart: Theiss, pp.37–47.

Wendorf, F. 1968. 'Site 117: A Nubian Final Palaeolithic Graveyard near Jebel Sahaba, Sudan', in F. Wendorf (ed.), *The Prehistory of Nubia.* Dallas, TX: Southern Methodist University Press, pp.954–95.

Wendorf, F., and R. Schild. 2004. 'The Western Desert during the 5th and 4th Millennia BC: The Late and Final Neolithic in the Nabta-Kiseiba Area', *Archéo-Nil* 14: 13–30.

Wendorf, F., et al. 2001. *Holocene Settlement of the Egyptian Sahara,* Vol. 1: *The Archaeology of Nabta Playa.* New York: Kluwer Academic/Plenum.

Wendrich, W., and H. Barnard. 2008. 'The Archaeology of Mobility: Definitions and Research Approaches', in H. Barnard and W. Wendrich (eds), *The Archaeology of Mobility: Old World and New World Nomadism.* Los Angeles: Cotsen Institute of Archaeology, University of California Los Angeles, pp.1–21.

Wengrow, D. 2001. 'Rethinking "Cattle Cults" in Early Egypt: Towards a Prehistoric Perspective on the Narmer Palette', *Cambridge Archaeological Journal* 11: 91–104.

——— 2003. 'Landscapes of Knowledge, Idioms of Power: The African Foundations of Ancient Egyptian Civilization Reconsidered', in D. O'Connor and A. Reid (eds), *Ancient Egypt in Africa.* London: University College London Press, pp.21–35.

Weninger, B., and O. Jöris. 2004. 'Glacial Radiocarbon Age Calibration: The CalPal Program', in T. Higham, C. Bronk Ramsey and C. Owen (eds), *Radiocarbon and Archaeology*. Oxford: University School of Archaeology, pp.9–15.

Wild, H. 1953. *Le Tombeau de Ti*, Fascicule 2, Part 1: *La Chapelle*. Cairo: Institut français d'archéologie orientale.

Woodburn, J. 1988. 'African Hunter-gatherer Social Organization: Is It Best Understood as a Product of Encapsulation?', in T. Ingold, D. Riches and J. Woodburn (eds), *Hunters and Gatherers*, Vol. 1: *History, Evolution and Social Change*. Oxford: Berg, pp.31–65.

Zboray, A. 2009. *Rock Art of the Libyan Desert (2nd edition)*, DVD. Newbury: Fliegel Jezerniczky Expeditions.

CHAPTER 2

'I Hope Your Cattle Are Well'

Archaeological Evidence for Early Cattle-centred Behaviour in the Eastern Sahara of Sudan and Chad

Friederike Jesse, Birgit Keding, Tilman Lenssen-Erz and Nadja Pöllath

Introduction

'I hope your cattle are well' (*abere chemegi tuga chekug*) is the greeting phrase of pastoral Kalenjin in Kenya, and reflects their special relationship with cattle.[1] In many African societies cattle are not only of great economic importance but also play an important role in the social and ritual domains. They are prestige items and 'companions for life', often occupying central positions in ritual and mythology. The use of animals as a medium for complex symbolism is a widespread phenomenon, probably above all since 'natural species are chosen not because they are "good to eat" but because they are "good to think"' (Lévi-Strauss 1962: 89).

The importance of cattle in the various aspects of the herding way of life is not a recent occurrence but was already highly visible in ancient societies, as can be seen in a wide area in northern Africa stretching from the Nile Valley to the Atlantic.[2] Rather striking expressions of cattle-centred behaviour with obvious regional differentiations have been found in the Eastern Sahara of Sudan and Chad during archaeological research by the University of Cologne carried out over the last three decades (e.g., Kuper 1981, 1995; Keding 1997b, 1998; Lenssen-Erz and von Czerniewicz 2005; Jesse and Keding 2007; Keding et al. 2007; Lenssen-Erz 2007). The present chapter is based on these discoveries.

For societies in which cattle are the most important animals, not only in the economic but also in the other above mentioned domains, the term 'cattle complex' introduced by Melville Herskovits (1926) immediately comes to mind. Although not explicitly defined, Herskovits's 'cattle complex' was mainly based

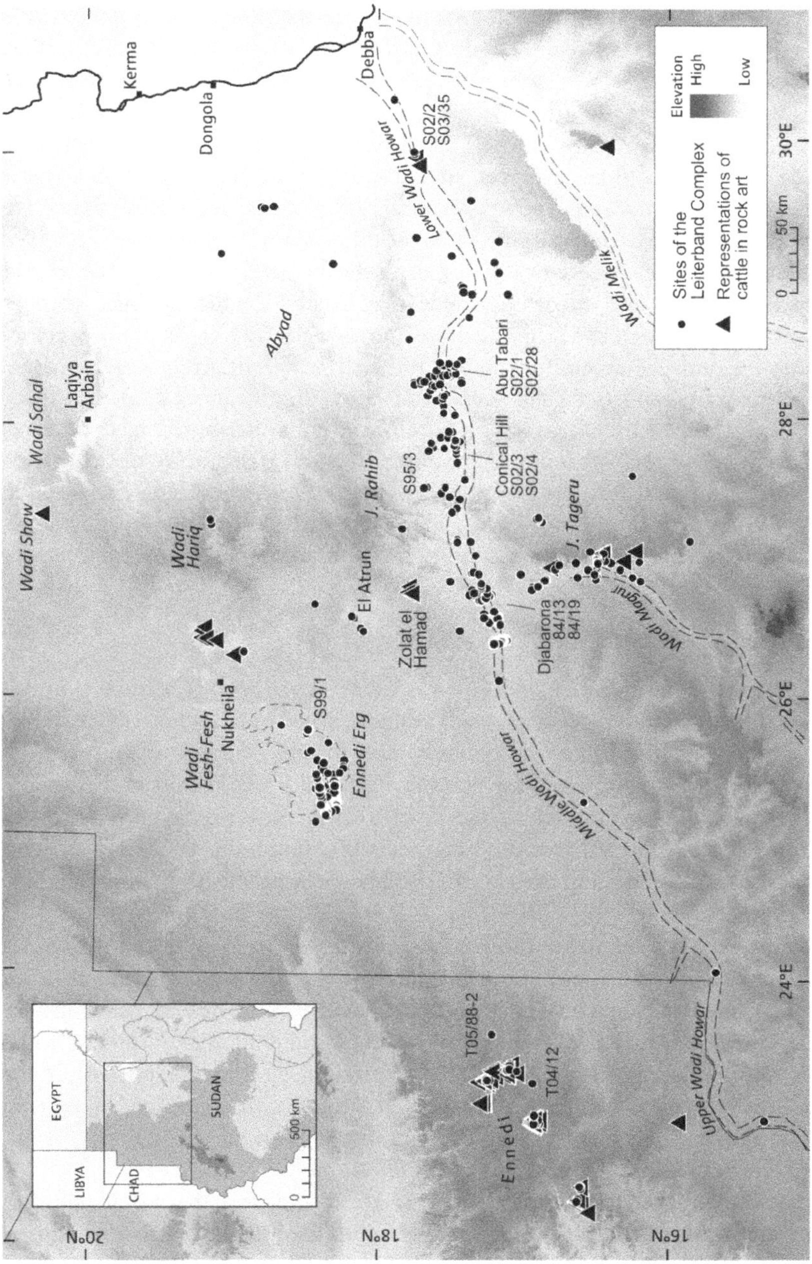

Figure 2.1: The study area in the Eastern Sahara of Sudan and Chad. Mapped are the sites of the Leiterband phase (fourth and third millennia BC) and representations of cattle in rock art. Sites discussed in the text are also indicated.

on observations of eastern and southern African pastoralists among whom cattle were used in many fields of ritual, social and economic life. Despite controversial debates among cultural anthropologists (e.g., Mtetwa 1982: 18–30; Mair 1985; Baxter 1996: 91–92; see also Comaroff and Comaroff 1990: 212), the concept is occasionally still used to describe cultural contexts where cattle are more than just a 'walking larder' (Clutton-Brock 1989) or provisions on the hoof.[3] Here the more general term 'cattle-centred behaviour' will be used to describe a special form of pastoralism[4] that developed in the southern Eastern Sahara during the middle and late Holocene. The main sources of evidence used here come from excavations, intensive surveys and rock-art studies. Numerous and well-preserved animal bones from archaeological sites permit detailed archaeozoological analysis. Prehistoric artefacts and, more importantly, archaeological features such as thousands of pits filled with cattle bones and/or pottery clearly suggest an importance of cattle beyond the economic sphere and help differentiate cultural entities. Complementary aspects of the social and ritual significance of cattle can be deduced from rock art. These sources are not evenly distributed in the area of research: Whereas rock art is abundant in the Ennedi Highlands of north-eastern Chad, archaeological features are much more numerous and better preserved in the Wadi Howar region of northern Sudan (Figure 2.1). This discrepancy determines the following discussion, structured by aspects of archaeology and rock art.

The Research Setting

The Eastern Sahara of Sudan and Chad is today hyper-arid. It features a multitude of different landscapes with wadi systems, highlands such as Jebel Rahib, Jebel Tageru and the Ennedi Highlands, and former lake and swamp regions such as Ennedi Erg and the West Nubian Palaeolake (cf. Pachur and Altmann 2006). The region's central axis is the Wadi Howar. Running from west to east over more than 1,050 kilometres, it connects the mountainous regions of Jebel Marra (Sudan), Borkou and the Ennedi Highlands (Chad) with the Nubian Nile Valley (e.g., Kröpelin 1993, 2007).

For the last 10,000 years, the Eastern Sahara has been subject to dramatic climate changes with humid and arid phases of varying intensity and regional impact. This is reflected in the Holocene cultural sequence of the southern Eastern Sahara (Figure 2.2), which covers the last 7,000 years. Three cultural phases can be distinguished. They are characterised by different pottery design styles, lithic tools and subsistence patterns. The economic sequence begins with a hunter-gatherer lifestyle, followed by a number of pastoral adaptations (for different scenarios, see Kuper and Riemer, this volume).

According to palaeoclimatic, archaeozoological, and (archaeo-)botanical studies, the area offered suitable ecological conditions for human occupation during the more humid periods of the early and middle Holocene.[5] During this time, long sections of the Wadi Howar were marked by chains of large, deep fresh-

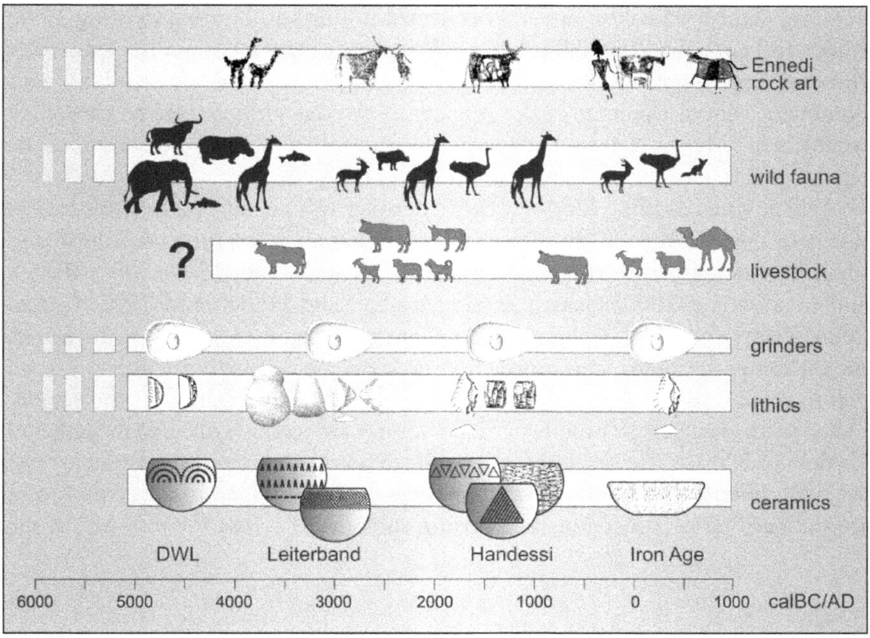

Figure 2.2: The Holocene cultural sequence of the Wadi Howar and Ennedi regions.

water lakes supporting a wide variety of fish species. Green bush and thorn savannah bordered the bodies of water, and offered good living conditions for various animal species. In the late Holocene (around 2500 BC) increasing aridity, progressing from north to south, finally reached the Wadi Howar. With the beginning of aridification, the ecological setting changed and resources became more and more patchy and unpredictable, forcing people to modify their ways of life. However, the Wadi Howar and the Ennedi Highlands, favoured by geographical position and hydrological conditions, remained refuges for humans and animals for hundreds, if not thousands, of years.

Cattle Pastoralism in the Wadi Howar Region: The Archaeological Evidence

Beginnings

The earliest evidence for domestic cattle in the Wadi Howar region (comprising Wadi Howar, the Jebel Tageru and the Ennedi Erg) are bones dated to the late fifth millennium BC (Table 2.1 and Figure 2.3). These finds, ranging in date between circa 4200 and 4050 BC, come from three localities near Jebel Rahib and the Middle Wadi Howar. The prehistoric pastoral groups, of which different variants are present from the beginning, practised a sort of multi-resource pastoralism based on herding, plant-food collecting, and occasional fishing and hunting (Keding 1998; Kuper and Kröpelin 2006: 804, Fig. 1). Already the earliest pastoralists apparently chose to largely confine themselves in their ritual, social and economic life to a dominant, central element: cattle. Apparently various groups fostered cattle-first strategies of different character as a means to concisely symbolise and communicate regulative mechanisms of social, economic and ritual life.

At that time, around 4200 BC, cattle herding was already well-established in northern Africa. At least from the seventh millennium BC onwards, cattle and small livestock are known from a number of sites (Gautier 1987, 2001; Blench and MacDonald 2000; Hassan 2002; Marshall and Hildebrand 2002; Garcea 2004; Kuper and Riemer, this volume). Compared to other regions of the world, the earliest pathways to food production in Africa are distinctive. Cattle occurred before crops and the first food producers were thus, depending on the region, either pastro-foragers or mobile herders, but not sedentary farmers (Marshall and Hildebrand 2002). The reasons for the shift to food-producing economies are not yet fully understood. Various ideas have been presented, including the notion of animal herding as a reaction to enduring subsistence crises, for instance in the

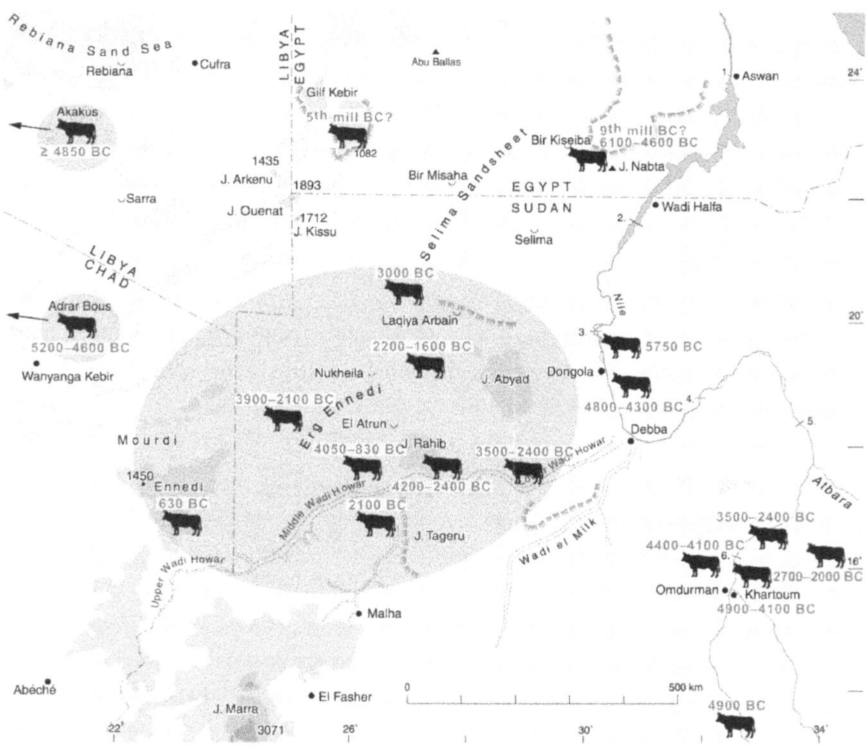

Figure 2.3: The appearance of domestic cattle in north-eastern Africa as indicated by the available radiocarbon evidence (see Table 2.1).

Table 2.1: Radiocarbon dates for domestic cattle remains found in the Wadi Howar region and in the Ennedi Highlands (see Figure 2.3).

Region[1]	Site	Lab. no.	bp	cal BC[2]	Material	$\delta^{13}C$	Remarks	Reference
MWH	84/19-3	KN-3417	5240±260	4060±280	bone	-11,9	cattle probable*	Keding 1997a: 34
MWH	84/19-3	KN-3522	4810±110	3570±130	bone	-11,9	cattle probable*	Keding 1997a: 34
MWH	84/25	KN-4669	4456±178	3160±230	ostrich eggshell	-9,95	cattle present	Jesse et al. 2007: Fig. 4
MWH	84/12-2	KN-3521	4420±130	3130±180	bone	-11,9	cattle present	Keding 1997a: 34
MWH	84/19-5	KN-3554	4410±120	3120±170	bone	-11,9	cattle probable*	Keding 1997a: 34
MWH	84/13-5	KN-3405	4210±200	2820±280	bone	-11,9	cattle	Keding 1997a: 34
MWH	84/13-3	KN-3408	4120±130	2680±170	bone	-11,9	cattle	Keding 1997a: 34
MWH	80/86-6	KN-2940	3740±120	2170±180	bone	-11,9	cattle present	Keding 1997a: 34
MWH	84/13-5	KN-3416	3580±110	1950±160	bone	-11,9	cattle	Keding 1997a: 34
MWH	S96/5-2	KIA-12419	3335±300	1660±390	bone	–	cattle**	Jesse et al. 2007: Fig. 4
MWH	96/119	UtC-5941	3320±60	1610±80	bone	-10,0	cattle probable	Jesse et al. 2007: Fig. 4
MWH	84/1	KN-3523	3250±60	1540±70	bone	-11,9	cattle***	Jesse et al. 2007: Fig. 4
MWH	84/9	KN-3889	2690±130	840±180	bone	-11,6	cattle present at site	Jesse et al. 2007: Fig. 4
LWH	84/24-1	KN-3945	5760± 65	4610±80	bone	-14,2	cattle present at site	Jesse 2003: 325
LWH	95/3-1	UtC-5583	5400±50	4240±80	charcoal	-25,2	fireplace; cattle bones	Jesse 2003: 330
LWH	95/3-1	KIA-14900	5275±35	4120±80	charcoal	-27,91	fireplace; cattle bones	Jesse et al. 2007: Fig. 4
LWH	84/24-1	KN-3946	5000±150	3810±150	bone	-8,4	cattle present at site	Jesse 2003: 325
LWH	02/52-1	KIA-26358-IV-2	4760± 30	3570±50	bone	-4,40	cattle****	–
LWH	03/31	KIA-26360	4740±35	3520±100	bone	-12,38	cattle*****	Jesse et al. 2007: Fig. 4
LWH	02/52-1	KIA-26358-IV-1	4685± 35	3470±70	bone	-3,61	cattle****	–
LWH	02/52-1	KIA-26359-III-2	4645± 30	3450±50	bone	-1,30	cattle probable****	–
LWH	84/24-1	KN-3944	4600±250	3300±310	bone	-13,0	cattle present at site	Jesse 2003: 325
LWH	84/24-1	KN-3943	4600±150	3310±210	bone	-10,0	cattle present at site	Jesse 2003: 325
LWH	02/52-1	KIA-26359-III-1	4530± 25	3240±100	bone	-2,82	cattle probable****	–
LWH	02/28-1	KIA-26362	4450±180	3150±230	bone	-23,23	cattle******	Jesse et al. 2007: Fig. 4
LWH	02/28, EF 462	KIA-24507-V	4110± 30	2720±100	bone	-2,24	cattle****	–
LWH	02/28, EF 462	KIA-24507-VI	4060± 45	2640±110	bone	-1,11	cattle****	–

Table 2.1: *continued*

Region[1]	Site	Lab. no.	bp	cal BC[2]	Material	δ13C	Remarks	Reference
LWH	K146	Hv-15588	3915±210	2410±300	bone	–	cattle	Kröpelin 1993: 212
LWH	02/4	KN-5561	3895±105	2370±150	bone	-10,14	cattle	Godhoff 2005: 17
LWH	03/13, EF 221	KIA-24508	2845± 25	1010±50	tooth	2,43	cattle****	–
Rahib	K154	Hv-15589	5350±275	4170±300	bone	–	cattle	Kröpelin 1993: 212
WH	?	Hv-14439	3540±180	1910±240	bone	–	cattle	Pachur 1999: 431
J. Tageru	western edge	Hv-15568	3665±250	2090±330	bone	–	cattle	Pachur 1999: 431
Laqiya	82/38-2	KN-3014	4350±320	2980±420	bone	–	cattle	Cziesla 1986: 144
Ennedi Erg	99/1-3	KIA-12420	5080±30	3880±60	charcoal	-22,63	pit with cattle bones among other contents	Holst 2003: 207
Ennedi Erg	99/1-7	KIA-12421	4875±30	3670±30	charcoal	-21,03	pit with cattle bones among other contents	Holst 2003: 207
West Nubian Palaeolake	–	–	3733±81	2150±120	bone	–	cattle*******	Pachur and Wünnemann 1996: 27
Wadi Hariq	97/7-1	KN-5447	3785±40	2220±70	charcoal	-25,0	cattle present at site	Jesse et al. 2004: 132, Tab. 2
Wadi Hariq	97/7-1	KN-5327	3675±40	2060±70	charcoal	-25,28	cattle present at site	Jesse et al. 2004: 132, Tab. 2
Wadi Hariq	01/4-1	KIA-17508	3430±25	1740±30	charcoal	-19,97	cattle present at site********	Jesse et al. 2004: 132, Tab. 2
Wadi Hariq	01/1-1	KIA-17543	3385±25	1690±40	faeces	-25,9	cattle present at site	Jesse et al. 2004: 132, Tab. 2
Wadi Hariq	01/1-1	KIA-17510	3355±25	1650±40	charcoal	-25,23	cattle present at site	Jesse et al. 2004: 132, Tab. 2
Ennedi	T 05/88-2	Poz-16595	2480 ± 30	630 ± 100	bone	–	cattle (carbonate fraction)	Keding et al. 2007: 38

1 LWH – Lower Wadi Howar; MWH – Middle Wadi Howar; WH – Wadi Howar.
2 Calibrations by CalPal (Version March 2007), using the IntCal04 calibration data set.
* Cattle bones largely dominate the faunal remains from Djabarona 84/19, including those from the pits excavated in trenches 3 and 5.
** The collagen fraction (0.06 mg C) was dated to 3340 +300 / –290 bp (d13C = –21.90 ± 0.05); the bone rest fraction (0.06 mg C) was dated to 3170 +300 / –290 bp (d13C = –23.17 ± 0.25) (letter by P.M. Grootes, 15-11-2001).
*** The bones for the radiocarbon sample were taken in square 12/21b-14, from a context with cattle bones (written communication W. Van Neer, 2006).
**** Apatite fraction; the sample was solubilised by hydrochloric acid HCl (letter by M. Hüls, 7-7-2009).
***** The collagen fraction (0.12 mg C) was dated to 9670 +360 / -340 bp (d13C = –26.23 ± 0.77); the bone rest fraction (1.0 mg C) was dated to 4740 ± 35 bp (d13C = –12.38 ± 0.18). The age of the collagen fraction is much too old and indicates sample contamination. The age of the bone rest fraction is in good accordance with archaeological expectation, but the date cannot be regarded as reliable (letter by P.M. Grootes, 26-07-2005).
***** The collagen fraction (0.10 mg) was dated to 10230 +450 / -430 bp (d13C = –27.67 ± 0.39); the bone rest fraction (0.14 mg C) was dated to 4450 ± 180 bp (d13C = –23.23 ± 0.72). The age of the collagen fraction is unreliable and indicates contamination; the age of the bone rest fraction appears more probable archaeologically, but the date cannot be regarded as reliable (letter by P.M. Grootes, 26-07-2005).
****** "Three dates of *Bos taurus* (identified by Boessneck, München) averaged 3733 ± 81 yr BP" (Pachur and Wünnemann 1996: 27).
******* The sample yielded two rather different results: 3430 ± 25 bp (fulvic acids, 2.2 mg C) and 1610 ± 35 bp (humic acids, 0.6 mg C). Interpretation is difficult but the age of the fulvic acids seems to be a better indication of the true site age (written communication P.M. Grootes, 10-11-2004).

Libyan Tadrart Acacus (Holl 1998: 84) or as a possibility to display wealth and social status (Garcea 2004: 125). Marshall and Hildebrand favour the improved predictability of food as a key argument: 'Domestication is linked to broad social or ecological processes, proximate contexts, in which predictability becomes crucial, and manipulation practices meant to restore or enhance predictable access and scheduled consumption' (Marshall and Hildebrand 2002: 105). Even if the exact origins of the pastoral economy in the Wadi Howar region have not as yet been elucidated, different scenarios are conceivable: the displacement of local foraging groups by foreign cattle herders; the assimilation of local hunter-gatherers by foreign herders; or the diffusion or adoption of cattle herding by local groups through contact with cattle keepers elsewhere.

Since the archaeological record of the Wadi Howar region shows no indications of displacement, the diffusion or adoption of cattle herding seem to be most probable (Keding 2006). In contrast to the northern areas of the Eastern Sahara, there is no evidence for a precedence of pastro-foraging (see Kuper and Riemer, this volume); instead, the earliest regional herding was tantamount to fully developed pastoralism. The available data suggest two potential areas of origin (see Figure 2.3). The first, and more probable (Keding 1997a), is the Nile Valley, where cattle remains are known from the beginning of the fifth millennium BC, or even earlier, at Rabak, Esh Shaheinab and Kadero, as well as in the Dongola Reach;[6] during the middle Holocene pastoralism with large and small livestock was practised in the Nile Valley, with cattle always dominant among domesticates (Gautier 1989). The other possible source area is the Central Sahara, where cattle herding was present from around 5000 BC, for example, at Adrar Bous (Niger) and in the Tadrart Acacus (e.g., Marshall and Hildebrand 2002: 110, Fig. 2; Gifford-Gonzalez 2008).[7]

In the Wadi Howar region different pastoral adaptations existed at the time of the beginning of pastoralism, that is, around 4000 BC: In the area west of Jebel Rahib, at sites of the Leiterband Complex such as Djabarona 84/19 and Ennedi Erg S99/1, cattle largely dominate the faunal spectrum of the excavated areas (Günther 1995: 46; Keding 1997a; Berke 2001). The sites are characterised by concentrations of animal bones and/or pottery, which turned out to be remains of former pits (see below). By contrast, prehistoric groups in the Lower Wadi Howar based their livelihood on cattle pastoralism, but also used the mosaic of different habitats around the sites for fishing and hunting; their sites are devoid of pits.

Examples of this form of multi-resource pastoralism have been documented at the Conical Hill S95/3 site where undecorated pottery and a few sherds of pottery of Leiterband type (Keding 1997b: 36) were found with cattle bones and a few bones of both goat and wild game.[8] Another example comes from the large open air site S02/1 in the Abu Tabari region, dated to around 3800 BC (Jesse 2006: 45–46). Domestic mammals (predominantly cattle) constitute 53 per cent of the faunal assemblage. Wild mammals (gazelle, antelope, warthog, elephant) make up 15 per cent of the faunal remains, and 32 per cent are constituted by a variety of aquatic species (Pöllath and Peters 2007: 65). Cattle were the most important source of food besides aquatic resources and, judging from the large amounts of grinding material, probably wild plants, whereas hunting played a minor role in subsistence. That cattle also played a role beyond the purely economic sphere might be reflected in their representation as small figurines made of clay. One complete piece of a probable representation of a cow with a length of 4.9 cm and several fragments of legs or horns has been found (Jesse 2006: 45–46).

The Lower Wadi Howar sites suggest contacts with Neolithic sites in the middle Nile Valley indicated by similar ceramics and by the size and robustness of the cattle bones. The proximity to the Nile Valley and the possibility of exchanging livestock and ideas certainly best explain this regional development of cattle-centred behaviour contrasting with the Leiterband Complex further west.

Pastoral Climax: The Leiterband Complex

In the Middle Wadi Howar, the Jebel Tageru and the Ennedi Erg, a distinctive and rather remarkable form of pastoralism, the Leiterband Complex, existed from the regional beginnings of pastoralism, around 4000 BC, and prospered during the fourth and third millennia BC. It may be regarded as the most expressive occurrence of cattle-centred behaviour in the southern Eastern Sahara.

More than 800 Leiterband sites are presently known. They are mainly distributed west of Jebel Rahib, in the region stretching from the Middle Wadi Howar to the Ennedi Highlands (see Figure 2.1) (Keding 1997a: 204–16; Jesse and Keding 2007). From the beginning of the Leiterband Complex, the Middle Wadi Howar emerged as the most important centre of occupation due to its favourable ecological conditions. In the Ennedi Erg to the north-west, wide plains and the shores of the Nubian Palaeolake offered good pasture and water (Hoelzmann et al. 2001). The highlands, Jebel Tageru to the south, and Ennedi to the west, in particular their forelands, were also integrated in transhumant cycles as were parts of the Lower Wadi Howar (Keding 2004; Jesse and Keding 2007; Keding et al. 2007).

Features

Most sites may be classified as camps of different sizes and functions; burials are rare (Jesse and Keding 2002). The settlement pattern of the Leiterband Complex is characterised by a high proportion of (semi-)permanent medium-sized camps (circa 70 per cent), numerous small sites (15 per cent), which are mainly interpreted as stock posts, and very large sites (circa 15 per cent). Sites of the first and second categories are located in different topographic positions, but are linked to a large extent to water sources for watering livestock. In contrast, most large camps are prominently situated on elevations such as dunes, with a wide view over pasture and land. They may be subdivided into at least two categories: first, settlements with a wide variability of artefacts such as pottery sherds, lithic artefacts and grinding material; and second, sites featuring countless (eroded) pits filled with pottery and cattle bones, which probably played a role in the social-ritual sphere.

These (eroded) pits are the most striking and widespread type of feature of the Leiterband Complex, appearing at about 50 per cent of the sites presently known. They are found at sites of all sizes and in comparable quantities, but probably had different functions. There are three different categories of pits, all apparently of the same age: First, pits containing only (nearly) complete vessels, which seem to have fallen apart *in situ*; the latter were probably deposited upside down for future use – for example, site Ennedi Erg S99/1. Second, rubbish pits filled with butchered remains of cattle – for example, Ennedi Erg S99/1–3 (Berke 2001: 245) – and/or discarded stone artefacts (Keding 1997a: 222–23), pottery and sometimes fragments of grinding tools. Third, pits featuring cattle bones apparently intentionally buried. Such features underline an important ritual rank of cattle. Due to erosion, the original contents of the pits, consisting mainly of pottery and cattle bones, now form dense, compacted concentrations of finds on the surface (Figure 2.4). Beneath some of these round accumulations, measuring about one metre in diameter, the remains of steep-walled pits are preserved, up to two metres deep, filled with a fine-grained grey sediment mixed with cattle bones, sometimes almost whole cattle skeletons or large parts of cattle, and an abundance of nearly complete pottery vessels. Though the main components of the fillings are uniform, their compositions appear to be non-standardised. Thus, numbers of pots and animals vary, occasionally bones are burnt, sometimes (fragments of) grinding tools are included, and, in rare cases, some bones of wild game are also found.

Figure 2.4: Djabarona 84/13 in the Middle Wadi Howar, a large site with numerous concentrations of bone and/or pottery.

Eight extremely large sites in the Middle Wadi Howar and the western foreland of Jebel Tageru are characterised by dozens, hundreds, or even more, pit remnants of the last mentioned type. The most impressive example is Djabarona 84/13 (Figure 2.4), with more than 1,000 pits, representing a chronological sequence of at least 800 years (e.g., Keding 1997a, 1998). Generally, these sites cover several hectares, being situated on dunes or slight elevations and covered by a dense artefact scatter of all types of finds (see above). Some, for instance Djabarona 84/13 (Keding 1997a: 29–30, Fig. 4) are spatially subdivided into areas with or without pits. In no case were above-ground pit covers or markers observed.

Material Culture

Typical Leiterband finds include pottery, large numbers of bones, numerous grinding tools, stone artefacts, axes, and ostrich eggshell beads (Figure 2.5) (Keding 1997a).

The pottery vessels (Figures 2.5.1 and 2.5.2) consist mainly of well-made, neck-less pots of different sizes and shapes. Many are decorated over the entire exterior with rocker stamp impressions, among which the 'Leiterband Decoration' is the most prominent pattern. According to the find contexts, pottery was

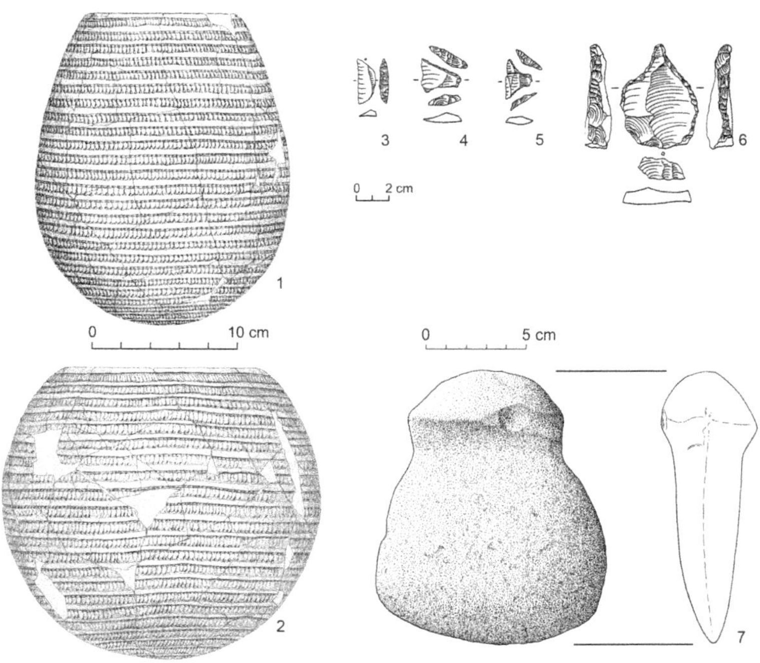

Figure 2.5: Material culture of the Leiterband phase. 1–2: Leiterband pottery; 3: segment; 4–5: transverse microliths; 6: borer; 7: axe of Darfur type.

used in the domestic as well as in the ritual spheres. While the early stages of Leiterband decorations reveal supra-regional connections with the east, especially with the Khartoum Neolithic (or Shaheinab Culture) of the Sudanese Nile Valley, dated to the fifth millennium BC, the later phases largely coincide with the pottery of Hohou type found in the Ennedi Highlands to the west of the Wadi Howar (Keding 1997a: 166–81; 1998: 5–9; Keding et al. 2007).

The stone artefacts (Figures 2.5.3 to 2.5.6) are characterised by a microlithic industry and are dominated, depending on site function, by a wide variety of transverse microliths, mega-segments, thumbnail scrapers and borers (Keding 1997a: 369, Pl. 77; Holst 2003). Besides the use of local quartz for artefacts, tools are mainly made from often non-local chalcedony (e.g., Keding 1997a: 200, Fig. 81). According to ethnographic sources, there is no typical herder's tool kit (Cribb 1991: 69), but some of the implements are certainly connected with the exploitation of cattle. Thumbnail scrapers might indicate the processing of (cattle) leather (e.g., Schön 2002: 316), and mega-segments, with their large cutting edge, the butchering of animals.[9] The transverse microliths were possibly used to pierce the veins of domestic animals to extract blood (Keding 1997a: 199–201). The 'bleeding' of cattle by means of comparable tools is still a widespread practice among cattle keepers in East Africa, where blood is a nutritional supplement (e.g., Dahl and Hjort 1976: 172).

Axes of the Darfur type (Figure 2.5.7) suggest the importance of wood working, though their function as prestige items cannot be excluded (e.g., Keding 1997a: 366, Pl. 74). Their distribution indicates north–south exchange cycles and, more dominant, east–west orientated networks.

Grinding tools, probably used for plant processing, are numerous. While lower grinding stones may take the form of flat slabs or basins, most of the upper grinders are characterised by an oval outline and section.

In total, the preserved Leiterband material culture is characterised by three significant traits: Firstly, by its relative homogeneity over a period of nearly two thousand years. This may be related to the apparent significance of material goods, especially pottery and its decoration, as potential media of identity.[10] Secondly, Leiterband material culture lacks definite prestige items or signs of (nascent) social complexity (e.g., Cribb 1991). Such items are neither found at habitation sites in the Wadi Howar or Ennedi regions, nor in ritual or cultic features such as graves (Jesse and Keding 2002), nor at the extraordinary sites featuring hundreds of pits. Cattle probably fulfilled the role of a status symbol, just as they are regarded as the sole expression of wealth in many recent societies (e.g., Herskovits 1926: 264). Thirdly, Leiterband material culture occurs in massive accumulations of artefacts, especially at medium sized to large sites, which include numerous breakable (such as pots) and/or heavy items (such as grinding tools). This indicates the restricted mobility of a large part of the pastoral society and repeated, planned use of the specific locations (see Cribb 1991: 75–80).

Subsistence

The Leiterband Complex is associated with people who practised a developed food-producing economy mainly based on cattle. Most sites yielded well-preserved cattle bones usually amounting to more than 90 per cent of the faunal inventories of excavated sites (Van Neer and Uerpmann 1989: 332, Table 5; Jesse et al. 2007: 46, Fig. 1). Nevertheless, the role of cattle in subsistence is difficult to determine unambiguously. According to the different uses of domestic animals, pastoralists can be divided into 'meat only' and 'meat plus milk' livestock keepers. Today, cattle herders' subsistence in comparably dry environments is essentially based on milk products at least during some time of the year (e.g., Dyson-Hudson and Dyson-Hudson 1970: 170; Dahl and Hjort 1976: 141–81). Usually beef is only consumed in the context of social, ceremonial or ritual events, and thus does not play an important role as an energy source (e.g., Dahl and Hjort 1976: 142; Smith 1980: 474; compilation of literature in Keding 1997a: 232–35). On such occasions, cattle are slaughtered and their meat is distributed within the community. Nowadays, such events are often periodic and mainly take place during the dry season (e.g., Dahl and Hjort 1976: 162).

Milking and the processing of milk are difficult to trace in the archaeological record. One important indicator, namely culling patterns of livestock, did not produce clear results for the Wadi Howar assemblages. Their mixture of juvenile, sub-adult and adult animals, with an explicit dominance of young animals, indicate a meat economy (Van Neer, pers. comm. 2003; see also Davis 1987: 157–58). However, since the bones come generally from sites and pits that are interpreted as ritual, rather than settlement sites with consumption refuse, the underlying criteria for slaughtering cattle might not reflect economic patterns. In sum, clear archaeological evidence for milking and milk processing is missing, but rock art and linguistic results suggest the use of milk in this region at least since the third millennium BC (Bechhaus-Gerst and Keding 2007).

Other faunal remains indicate some small stock keeping, fishing and, on a smaller scale, hunting. The bones of wild fauna are mainly gazelles (*Gazella dorcas*, *G. dama*) and a few hares, but also giraffe, elephant and hippopotamus (Van Neer and Uerpmann 1989; Berke 2001). Fishing was mainly restricted to species capable of surviving short periods of adverse living conditions, such as Tilapiini and Clariidae (Van Neer 1989: 62–66). Hunting and fishing probably only contributed on a small scale to daily food consumption and other requirements. However, the ideological value of both strategies might have been much more significant.

However, ethnographic studies of present-day societies shows that even specialised cattle pastoralists rely substantially on plant food (e.g., Johnson 2002). Botanical remains are rare at Leiterband sites, and derive mostly from charcoal. The known species list therefore comprises mainly trees, such as *Acacia nilotica*, *Ziziphus* sp., cf. *Grewia tenax* and Capparidaceae (Neumann 1989: 140). They all yield fruits and/or seeds, which were also consumed in the Old Egyptian Kingdom, and today are mainly prepared by livestock keepers into different kinds of porridge (e.g., Tubiana and Tubiana 1977; Germer 1985; von Maydell 1990). Impressions of Paniceae grass seeds were found in pottery sherds at sites Ennedi Erg S98/24 and Jebel Tageru S98/6.[11] Wild grass seeds were abundant in the region (Neumann 2005: 266) and are still used today by pastoralists either for their own nourishment or for that of their animals. Numerous grinding stones prove an intensive exploitation of wild plants by Leiterband groups.

Mobility Patterns

The Leiterband societies reveal different mobility levels and group structures during an annual subsistence cycle. Shifting centres of occupation and an obvious variability in site sizes reflect large and small scale mobility, seasonal uses of diverse sub-regions as well as environmental changes in the long term.

Within the territory of the Leiterband cattle keepers, the Middle Wadi Howar and, to a lesser extent, the Ennedi Erg were focal points characterised by medium-sized to large dry season camps, used by sedentary or at least semi-sedentary segments of a pastoral society. The limited distribution of the extremely large sites with their abundance of pits underlines the role of the Middle Wadi Howar, the Ennedi Erg and the Jebel Tageru as the core area of the Leiterband Complex.

The high share of (probably) semi-permanently or permanently occupied sites (around 75 per cent) attests to a low degree of mobility for a large segment of the pastoral society. Furthermore, the furnishing of sites with 'site fixtures' or 'site furnitures' (Binford 1980; Cribb 1991: 68) such as grinding tools, very large vessels, as well as features such as pits, which are all recorded at least for more than half of the sites, indicate planned repeated access of special locations and thus fixed subsistence cycles. In consideration of the environmental features of the areas and some slightly pronounced regional cultural expressions of the Leiterband Complex, east–west as well as north–south transhumance cycles with distances of 180 to 220 kilometres are conceivable.

The Ideological Role of Cattle

Cattle seem to have not only played an essential part in the diet but also a key role in the ideological or ritual sphere, as is shown by the extremely large sites with countless pits in the Middle Wadi Howar and Jebel Tageru. The accumulation of finds and features, as well as their special appearance and the time depth, suggest an interpretation in terms of regularly frequented sites which might have been associated with mythological or historical events remembered by a group (see Tilley 1994: 18).

The enormous amount of bones indicates that cattle played an important role in these special, probably ritual, social activities (Keding 1997a, 1998). Starting at around 5500 BC in different places in the Sahara, cattle cults, or the use of cattle in activities which seem to have had a ritual background, became a widespread phenomenon in North Africa (Figure 2.6).[12] Since the fourth millennium BC

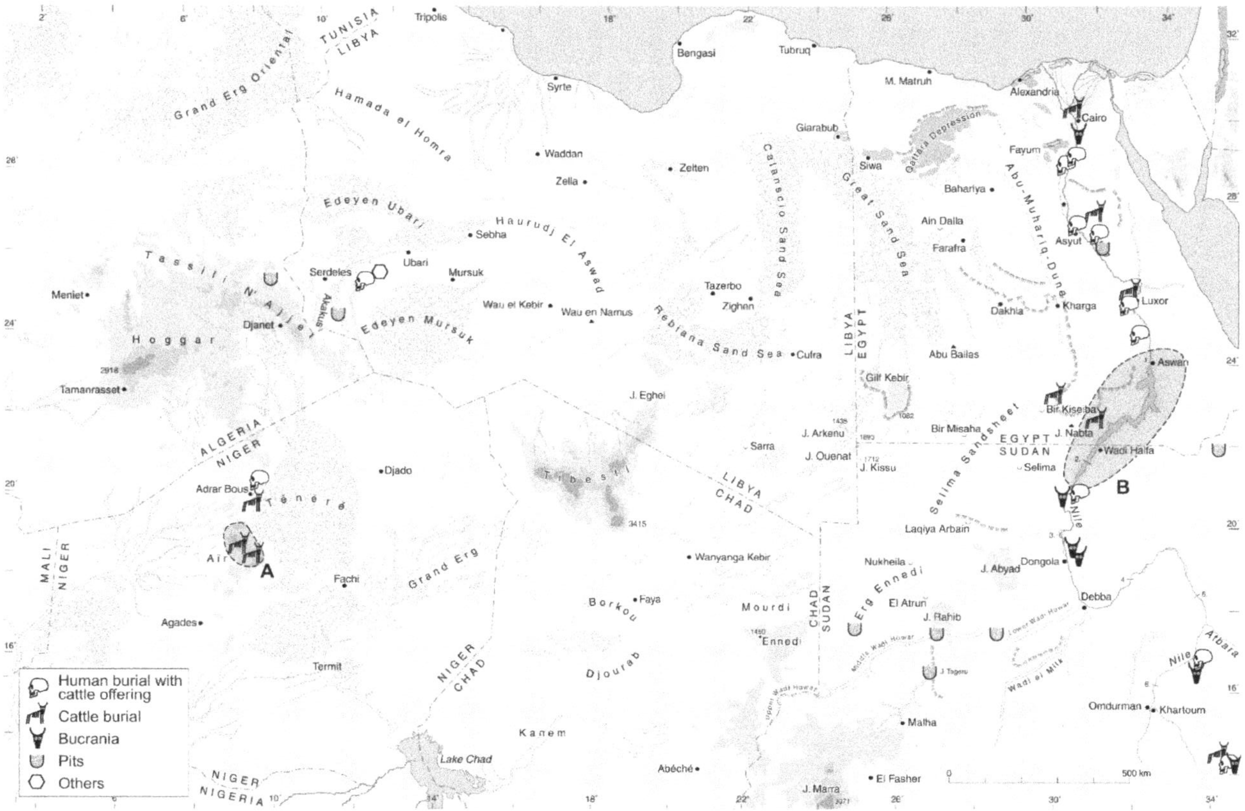

Figure 2.6: Regional aspects of cattle cults in northern Africa. Note cattle burials are frequent in regions A and B. A – Talak–Timersoï region; B – A-Group and C-Group region. Based on Keding (1997a), Paris (1997), Applegate, Achilles and Duncan (2001) and di Lernia (2006).

these features are known from a wide belt, stretching from Ancient Egypt and the Nubian area to the western Sahara (e.g., Ghoneim 1977: 196–216; Ferhat, Striedter and Tauveron 1996; Keding 1997a: 224–27; Paris 1997; Wendorf and Schild 2001: 668; di Lernia 2006). Apparently they took a variety of regional cultural expressions. Some represent without doubt cattle burials, occasionally even forming cemeteries, such as reported for the Bisharin Beja in eastern Sudan (Murray 1926). They are partially designated by above ground markers and/or are equipped with grave goods, such as Adrar Bous Site 1 (Paris 1997). Other cattle features are associated with human burials, for example, at Chin Tafidet in Niger (ibid.: 115). According to their find context, these cattle features are interpreted as sacrifices, burials of 'holy animals' or as offerings in human graves (e.g., Behrens 1963; Gratien 1985). Of special interest is the site of Mankhor in Algeria, dated to the second half of the fifth millennium BC. Here a necropolis of more than thirty cattle burials in pits was studied. All cattle had been buried at a young age, their bones having been placed in anatomical order or piled up in quarters (Ferhat, Striedter and Tauveron 1996; Ferhat 2003: 79). The presence of pits with cattle remains at Mankhor is a good analogue of the cattle features of the Leiterband Complex.

At present, no conclusive interpretation of the Leiterband pits is possible. However, ethnographic sources suggest a wide range of possible reasons for sacrificing cattle. Here only three feasible interpretations will be mentioned. The first conceives cattle sacrifices as offerings for deceased humans due to their economic or prestige value and/or emotional links between the deceased and special animals. This hypothesis is supported by the occurrence of pits filled with cattle bones together with human inhumations at Ennedi Erg S98/I31. However, in general, human graves are nearly unknown in the Leiterband Complex. Thus, it is conceivable that sacrifices of cattle were part of ritual mortuary practices accompanying archaeologically invisible human burials. The dead may have been buried elsewhere, or the mortuary practices may have left no surviving traces of the deceased (Jesse and Keding 2002).

A second line of interpretation sees cattle sacrifices as a reaction to poor conditions in the natural or social environment. Today, in traditional societies, imbalances or changes in climate and/or natural environment are often attributed to the spiritual, moral, social or political sphere (e.g., De Garine and Koppert 1988; De Waal 1989; Shipton 1990: 361). Rituals and sacrifices are often a means of coping. In their analyses of behaviour patterns among recent pastoralists, Dahl and Hjort state:

> apart from slaughter in cases of emergency, the most frequent type of slaughter is ... for ceremonial purposes. These may relate to life-cycle ceremonies, yearly thanksgiving or prayer for rain, etc. It should be remembered that such sacrificial feasts do not involve the destruction of the sacrificed animal, but rather public ceremonies with communal and communial sharing of food ... [and] many such feasts, such as appeasements for better times, occur during famines when milk production is low. (Dahl and Hjort 1976: 162)

In this vein, di Lernia (2006: 60) suggests prehistoric features displaying traces of 'ritual' slaughtering of cattle in the Eastern and Central Sahara to be partial relics of rainmaking ceremonies in areas characterised by drying environments. Since the period of the Leiterband Complex (4000 to 2200 BC) is characterised by a decrease of rainfall and biomass, the exceptional sites with hundreds of cattle pits would fit well in this scenario.

Last but not least, the prominent pit sites also reveal traits of 'competitive feasting' locations. During such social events, individuals, clans, or groups try to increase their prestige and strengthen their rank, power or authority by squandering luxury goods. Valuables are withdrawn from economic cycles without direct equivalents. Archaeological indicators might be 'architectural elements, food refuse, suites of serving vessels, and symbols or motifs associated with those vessels' (Arnold 2000: 29). Extensive slaughtering of cattle and the deposition of pottery vessels in great quantity might have been the pertinent media during the Leiterband phase.

In sum, it must remain a matter of speculation whether the Leiterband pits filled with cattle bones bear any relationship to ancient cults practised in other North African regions. However, regardless of their former actual use, these pits and sites document the enormous significance beyond the subsistence level that cattle must have had during the Leiterband phase.

Beyond the Leiterband Complex

The Leiterband Complex, with its main centre in the Middle Wadi Howar, had an influence on developments in neighbouring areas. This becomes obvious in the Lower Wadi Howar. Whereas at the beginning of the fourth millennium BC cattle only played a dominant role in areas to the west of Jebel Rahib, by the late fourth millennium BC cattle were also significant to the east of Jebel Rahib. Cattle bones on sites in the region of Conical Hill and Abu Tabari make up between 95 per cent and nearly 100 per cent of the identifiable faunal assemblages.[13] Pits have also been found on sites up to the region of Abu Tabari (see Figure 2.1), even if they are here not so much characterised by a predominance of concentrations of pottery and/or bones on the surface. The large site Abu Tabari S02/28, dating to around 3000 BC, is the best studied example. The ceramics found at this site display a variety of design styles, including incised herringbone patterns and Leiterband decorations (Jesse 2006). The pits here are explained by the disposal of the bones after butchering and consumption, but in some cases also with ritual practices with different forms of expression. In one pit, for example, the remains of at least five cattle were found and many bones were burnt. This bone assemblage might either be the result of a single event, say a big feast, or of refuse collected over some time and buried in one event. In another pit, pottery sherds, numerous ostrich eggshell beads, animal bones and human remains were found mixed together. Cattle were so important that they were even buried in graves like humans as is best indicated by the complete skeleton of a sub-adult cow carefully laid down on its right-hand side at Abu Tabari S02/28–24. The animal was orientated along a north–south axis, with the head to the north. No burial pit was recognisable (Jesse 2006: 48).

Increasing aridity during the fourth millennium BC and corresponding changes in the ecosystem in the Lower Wadi Howar (Pöllath and Peters 2007: 65) probably stimulated the growing importance of cattle in economic and social life, and therefore led to the adoption of features such as pits previously known only west of Jebel Rahib in the Leiterband Complex. The recurrence of cattle and the appearance of elements of cattle cults, such as cattle burials, can then be seen as a social response to environmental changes (di Lernia 2006). However, the sites of the late fourth and third millennium BC in the Lower Wadi Howar cannot simply be incorporated into the Leiterband Complex as they show originality reflected especially in the different pottery design styles present. Besides Leiterband patterns, incised herringbone patterns are present which clearly indicates strong contact with the A-Group and pre-Kerma culture in the Nubian Nile Valley around the third and second Cataracts (Keding 2000: 92; Jesse 2006: 49). Affiliation with the Nile Valley is also reflected in the archaeozoological record: the cattle bones found in the Lower Wadi Howar overlap in size with those of the Egyptian and Sudanese Nile Valley (cf. Pöllath and Peters 2005; Jesse et al. 2007), indicating the exchange of livestock and/or the adoption of efficient strategies of husbandry as they had been developed, for example, in Egypt since the Old Kingdom (Laudien 2000: 102–8). The obvious mixture of different cultural traits in the Lower Wadi Howar led to the development of local forms of cattle-centred behaviour at the periphery of the Leiterband Complex.

The Finale

At the end of the third millennium BC a cultural break becomes visible in the Wadi Howar region, triggered by the consequences of climatic change (see Figure 2.2). This settlement phase, the Handessi Horizon, mainly dates to the second millennium BC and is marked by pottery with geometric decorations (Francke 1986a, 1986b; Keding 1998: 10–11; Jesse 2007; Jesse and Keding 2007). In contrast to the preceding Leiterband Complex, a broader territory of the southern Eastern Sahara is exploited by the Handessi groups, including the Wadi Howar region and the Ennedi Highlands, but also areas more to the north such as Wadi Hariq and the Laqiya region (Jesse 2007; Jesse and Keding 2007: 42). In the Wadi Howar region, the limiting influence of aridification restricted settlement as well as herding, gathering and hunting activities to some favourable areas such as the Middle Wadi Howar and the Jebel Tageru. Probably as a reaction to the decreasing availability of usable land, a broader spectrum of resources was exploited. This included the addition of goats and sheep on a larger scale to livestock herds, as well as more intensive gathering and hunting. Cattle were still very important in subsistence (Jesse et al. 2007: 46, Fig. 1). However, compared to the preceding Leiterband phase, their role changed. This is borne out by a new site layout, now lacking pits filled with cattle bones and pottery. In the Wadi Howar region there are no traces to define the ideological value of cattle during this period, but already the presence of cattle in the light of the arid environment may indicate the high esteem in which cattle were still held. It definitely indicates the pastoral skills of the Handessi groups in managing the constraints of marginal environments, for example, by digging wells or the adjustment of their transhumant cycles to benefit from the effects of the rainy season in areas outside the Nile Valley (Jesse 2007). On the other hand, in the Nubian Nile Valley, sites of C-Group and Kerma Culture affiliation feature ample evidence of the high value of cattle, such as hundreds of bucrania found in burial contexts and countless small clay figurines of cattle (Chaix and Grant 1992; Bonnet 1996: 58–59; Kuper and Riemer, this volume, Fig. 1.2). The ideological importance of cattle is also manifest in the rock art of the Ennedi Highlands.

From about 1100 BC onwards, permanent use of the Wadi Howar region was no longer possible, but there are spots (such as watering troughs around wells) which still indicate its importance as a thoroughfare and even a habitat for pastoralists.

Pastoral Rock Art

Pastoral livelihood is a material condition defined by the characteristics elaborated above, but which cannot exist without adequate cognitive, ideological and social concomitants. These evanescent elements find expression, for instance, in the treatment of dead cattle, but are more tangibly reflected in rock art, thus providing clues for a deeper understanding of pastoralists' attitudes towards their livestock. If we postulate that the early herders already had cattle cults or a kind of 'infatuation' (Robertshaw 1990: 208) with their beasts, then a sophisticated communication medium such as rock art would mirror such an intimate coupling of economic organisation and epistemic structures.

For the region in question, the southern Eastern Sahara, two heterogeneous bodies of art will be considered here: the rock engravings of the Wadi Howar region and the paintings and engravings of the Ennedi Highlands (see Figure 2.1).[14] Both main research areas are subdivided into disconnected sub-regions, all of which have their own particular landscape setting.

The Wadi Howar Region

The richest rock-art site of the Wadi Howar region is Zolat el Hamad, located some 30 kilometres north of the Middle Wadi Howar (Figure 2.1). It comprises a roughly round array of numerous rocky hills with a maximum diameter of six kilometres. The region's rock art was first discovered by Newbold (1924, 1928), who made hasty records, as did the Frobenius expedition of 1933, so that no systematic documentation exists (Rhotert 1952: 79). After visits by research teams from Cologne University in 2003–2005, more rock art was reported (Kröpelin 2004), almost exclusively engravings. The bulk of the motifs show domestic cattle, with giraffes being the second most frequent motif (roughly a third of the number of cattle depictions), followed by ostriches (just above a tenth of the cattle motifs). Game animals representative of a more humid climate (elephants and rhinoceroses) are very rare, and it is a peculiarity of Zolat el Hamad that camels are missing entirely. The same situation seems to prevail in the rock art of the nearest oasis, Nukheila, about 120 kilometres to the north-north-west (Newbold 1928: 279).

Another hilly region loosely linked to the Middle Wadi Howar, though located 100 kilometres south, is Jebel Tageru. While no systematic records of this mountain's rock art are available, preliminary reports and photographs indicate that the engravings are very similar to those of Zolat el Hamad (ibid.: 277).[15] It appears that here the focus on cattle is even stronger since very few giraffe depictions are to be found and none of other game animals. It is surprising that here, too, depictions of camels seem to be absent.

In the Lower Wadi Howar, roughly 100 kilometres west of the Nile, rock engravings were found mainly on flat or slightly erect outcrops on the banks of the wadi (Kröpelin 2004; Jesse 2005). Three large sites (Figure 2.1) were detected in an 8.5 kilometre stretch along the wadi. Though relatively near to each other, the three sites are of rather different character as regards the motifs to be found there. While the record of engravings in the Lower Wadi Howar may as yet be incomplete, it seems to be rather certain that, in contrast to Zolat el Hamad and Jebel Tageru, there is no clear dominance of cattle in the art. At site S03/35 (Jesse 2005), where giraffes are the most frequent representational motif, cattle are engraved on a vertical wall in a way that they can be seen at a distance in the wadi.

In line with the familiar problem that rock art is difficult to date precisely, here the chronological aspects can only be dealt with in a sweeping sense, linking the art to general knowledge about climate development and archaeological sequences. However, at the central site in the Lower Wadi Howar, S02/2, a radiocarbon date was obtained on charcoal from the sediment covering an engraved giraffe. These were dated to no later than about 1200 to 1300 BC (ibid.: 33).

The Ennedi Highlands

The highlands of the Ennedi seem to have been a favourable landscape over long periods of the Holocene, not only in the early periods but even today.[16] Located in the Sahara–Sahel transition zone (Andigue, Bonnaventure and Adoum 2006: 13), the Ennedi Highlands are sparsely settled. The western foothills differ from the rest of the highlands not only in terms of hydrology, ecology and topography, but also as regards rock-art styles and motifs. Therefore different developments seem to have unfolded despite the close vicinity of the regions, which are echoed in patterns of use even today. Large herds of sheep and goats and even occasional horses can be seen in the vicinity of Archei (approximately 16°50' north of the Equator), a sight unseen in the inner parts of the highlands at the same or even more southerly latitudes. On the south-western outskirts of the Ennedi, near Bechikele (16°30'), one can encounter the northernmost cattle herds of Chad (cf. Clanet, Lemoalle and Dara 2006: 35). On the other hand, camels are omnipresent.

Though archaeological sites with stone artefacts and more or less well-preserved pottery are numerous in the Ennedi Highlands, their potential for reconstructing subsistence patterns or the ideological role of cattle is very restricted.[17] Classifiable faunal remains are limited to one cattle jaw, which was found in a disturbed archaeological context at site Mornou T 05/88–2 and has been dated to around 630 BC (Table 2.1).

The range of the mainly painted rock-art motifs, which is as narrow in the Ennedi as it is in the Wadi Howar (cf. Sauvet et al. 2006), is focused on cattle (and later also camels) in such a way that it does not seem to reproduce the true quantitative relations of domestic animals. Sheep and goats, the bones of which have been found in the Wadi Howar region (Van Neer and Uerpmann 1989; Berke 2001), and which are still present in the Ennedi today, are conspicuously absent from engravings in both regions, although they do appear in the paintings of earlier periods. Their absence from the engravings seems to be linked to a symbolic incongruence, for instance a taboo, of ovicaprids with engraving techniques, or it may indicate that painters were different from engravers in the value they accorded sheep and goats.

Judging from the distribution of cattle motifs in rock art, seemingly the entire Ennedi Highlands once used to be cattle-herder terrain, where people at any given time appear to have followed rather uniform subsistence patterns (Keding et al. 2007). Nevertheless, rock art is far from being homogeneous, certain sub-regional peculiarities indicating lively synchronic and diachronic cultural dynamics (ibid.: 28–32).

The Development of Ennedi Pastoral Art

There is a general trend in the pastoral art of the Ennedi by which, after their introduction into the region, cattle become more and more 'independent' of people, until they are no longer represented as an aid for work but as an indicator of wealth and status. Another way of expressing the significance of cattle was opened to the pastoralists through their immersion into the symbolism of livestock. This provided them with an expressive repertoire which allows the articulation of, among other things, identities in a cattle idiom.

The pastoralist's occupation with cattle, as expressed in art, underwent a development which mirrors an obviously narrowing focus – or maybe a growing infatuation – parallel to the ongoing climatic degradation. There are no hard data by which the beginning of so-called 'Bovidien' or pastoral art production in the Ennedi can be determined exactly, but general archaeological data indicate a more or less intense use of the highlands from the introduction of domestic stock in the fourth millennium BC (Figure 2.2).[18]

During this period, a remarkable rise in the number of archaeological camp sites is noticeable (Keding et al. 2007: 36, 39–40). These are mostly characterised by Hohou pottery (Bailloud 1969), which corresponds to the Leiterband ceramics used by cattle pastoralists in the Wadi Howar region (Keding 1997a; 1998: 9,

Fig. 4).[19] One radiocarbon date places Hohou ceramics at around 3000 BC (Bailloud 1969: 40). Thus the Ennedi Highlands were probably incorporated into a pastoral system that seems to have originated, or had its main base, about 300 kilometres to the east, in the area of the Leiterband Complex. In contrast to the Wadi Howar region, pits and bones were not registered in the Ennedi Highlands, the latter due to erosion.

For the art of the earliest Bovine Period, or Bovidien Ancien, Bailloud (1997: 26) observed that human figures were captured in lively action which generally goes together with a lateral view. They are mainly monochrome and do not seem to be highly standardised stylistically. Depictions of cattle are also largely monochrome and without patterning, so that there is no explicit individualisation of cattle. Activities imply attending on cattle, for instance in a milking scene from site T 04/12 in the Shekitiye region (Figure 2.7.1); images of cattle wearing what may be taken to be a harness (Figure 2.7.2) or pulling a rack (Figure 2.7.3) also fall in this period and seem to indicate that the symbolism of domestic stock implied, certainly among other things, pragmatic aspects of pastoral life, and that metaphorism to some extent anchored in everyday human–stock interaction.

Later the motifs become bichrome as well as polychrome. Human figures, for example, are frequently depicted in two or more colours, and feature elaborate details of clothing and ornament. While in the western foothills humans are still shown in distinct activities, in the central highlands the lateral aspect is replaced by a static frontal aspect unsuited to rendering distinct activities. Accordingly, depictions of humans interacting with their stock are no longer produced. The cattle begin to stand by themselves, symbolising their significance through their mere presence. At the same time they become individualised through the depiction of distinctive patterns of the coat. This feature is echoed in the work of contemporary pastoral groups, where the individuality of cattle is not only expressed by a multitude of terms applied to patterns of the coat, horns and other features (e.g., Haberland 1983; Galaty 1990; Eckl 2000; Smith 2005),[20] but also by individual names (Haberland 1983: 185).

The process of individualising cattle continued while aridity reached its peak and the Neolithic gave way to the Iron Age. The latter is well represented by numerous archaeological sites. The first use of iron is dated to around AD 400 but could have been earlier (Keding et al. 2007: 37–38, 40–42). Although evidence for iron use and technology is rare, sites with typical Iron Age ceramics, known as 'Djoki' pottery, are numerous (Bailloud 1969). This homogeneous pottery follows an entirely new concept. It includes high percentages of undecorated red-slipped sherds with thick, blackened rims (Keding, Lenssen-Erz and Pastoors 2007: 38, Fig. 8). The few decorations include shallow geometric incisions under the slip and mat impressions on the body and bottom (Keding et al. 2007: 39, Fig. 9). While connections to the east (Wadi Howar and the Nile Valley) are rare, some painted sherds may reflect western influence. This is also indicated by iron artefacts, including various types of awls and bead, as well as arrow-, lance- and spearheads (Keding et al. 2007: 40, Fig. 10) which were probably traded in from

Figure 2.7: Rock art from the southern Ennedi Highlands, presumably from early herding periods, showing scenes of pastoral cattle use. 1: a person milking a cow (length of cow, 9 cm); 2: a person attending to a cattle wearing a harness (height of human, 19 cm); 3: a cow pulling a rack. Tracings from photographs by T. Lenssen-Erz.

the west as prestige items. Although finds of metal points are extremely rare, numerous representations of them in rock art suggest the high social value of metal, itself probably a sign of prestige. In rock art, metal points are frequently associated with depictions of cattle accompanied by static sentinel-like figures of the 'Libyan warrior' type (e.g., Muzzolini 2001), equipped with lance and shield (see Figure 2.8) and often with rich head decorations. They are widespread in the inner highlands, also being an indication of western relationships since, for instance, these figures abound in the engravings of the Aïr mountains of Niger (Dupuy 1998). The shields are often decorated in ways recalling the unrealistic patterns of cattle markings (cf. Smith 2004: 46). Andrew Smith (2005: 14) describes South African pastoral Zulu warriors using cow-hide shields of a particular colour for each regiment. Another link of cattle hide pattern and group identity is reported for the Ethiopian Bodi, where each clan is identified with a specific hide pattern and deceased elders are buried in the hide of a cow with this pattern (ibid.). A similar association between clans and hide patterns can be seen among pastoral Herero of Namibia (Eckl 2000: 425–28). As for the Iron Age pastoral groups of the Ennedi Highlands, it may be concluded that when the climate had turned entirely arid they manifested their affiliation and identity through a symbolism closely linked to their cattle, and which they advertised in part through the designs on their shields.

Figure 2.8: Men presenting shields and lances with iron tips: three males are mounted on camels, juxtaposed with domestic cattle. Site T 03/26, southern central Ennedi (height of leftmost man, 40 cm). Photograph by T. Lenssen-Erz.

Figure 2.9: A cow with her calf, the pattern of the coat showing indisputably that the calf is her offspring. Site T 05/501, north-eastern Ennedi (length of cow, 22 cm). Photograph by A. Pastoors.

Another aspect of cattle centrism is addressed by the rather frequent depiction of cows with their calves. Often the painters made an effort to mark certain calves as being the obvious direct offspring of particular cows by giving both very similar coat patterns (Figure 2.9). This shows a concern with the descent and genealogy of cattle, which is also known from modern cattle pastoralists (e.g., Galaty 1990: 227; Eckl 2000: 424; Smith 2005: 14). Moreover, it seems to focus on reproduction, which in the context of cattle herding is tantamount to an emphasis on the growth of wealth and reputation. In accordance with the practices of many modern pastoral groups, cattle in the Ennedi Highlands were probably a means to advertise status and wealth, and through rock art this material as well as symbolic capital became meta-symbolic capital – that is to say that when living cattle were of symbolic value beyond their economic worth, then art provided an option to add a higher level of symbolism: if cattle were symbolic capital, among other things, then rock art allowed to save it in an even more abstract symbolic form. This may be an indication for the motivation for this particular rock art: It provided the means to petrify, as it were, once and forever the value cattle had for the owner. Moreover, in the periods under investigation, climatic development had gradually depleted the carrying capacity of the land and the usable space for pastoral economies, and therefore the tendency towards a manifestation of distinct identities and group affiliations would have increased.

Figure 2.10: An engraved, almost life-size cow (120 cm long) with deep grooves resulting from intense grinding on the navel, udder, neck, tail tuft and lower hind legs. Site T 05/560, north-eastern Ennedi. Photograph by T. Lenssen-Erz.

According to archaeological evidence, the Early Iron Age in the Ennedi Highlands probably began in the second century AD when the horse and camel may have added other options for pastoralist livelihoods, as is suggested by some depictions of camels juxtaposed with cattle in the inner highlands (Figure 2.8). But, unlike camels, horses seem to be entirely incompatible with cattle: there are almost no joint depictions. Horses display a uniform pattern of use, being largely ridden by armed people in a 'flying gallop' (74 per cent). Maybe they stand for a specific 'class' or 'caste' not involved in the ordinary working processes of the pastoralist environment. Brass (2007) suggests an emergence of elites in Saharan pastoral societies at a time even before the Ennedi was colonised by pastoralists. His argument is mainly founded on cattle and human burials under tumuli in the Acacus Mountains of Libya and the Egyptian Nabta Playa region, but he also draws on depictions of 'prominent individuals' in the rock art of the Libyan Messak Settafet and Tassili n'Ajjer in Algeria (ibid.: 17).

A final remark should be made about the difference between paintings and engravings. Typically, engravings are on flat rocks or on vertical walls. Unlike paintings, they are usually not found at inhabited sites, their location seeming to be determined by landscape properties rather than qualities linked to everyday livelihoods. Moreover, since they comprise no scenes and contain only a few depictions of humans, and exhibit many non-representational motifs, the whole context of engravings removes them from everyday life and from the experiences and causalities of the conscious, material world in comparison to paintings. This may be the reason why the 'infatuation' of veritable cattle centrism seems to be more comprehensively detectable in paintings than in engravings. With its narrative and metaphoric potential, painted art may be more open to the intricacy of 'cattle-first strategies' with their all-embracing influence on everyday life. Engravings, by contrast, with their largely symbolic function, lack the anchoring in the empirical, mundane sphere, and generally seem to be linked to a 'higher' sphere of a more abstract religiousness. This may be corroborated by large engravings of cattle, where the navel and in some instances also the udder or other body parts have been rubbed so intensively that deep hollows have been formed (Figure 2.10), which sometimes also show clear traces of red pigment. This bears witness to a particularly intense, probably repeated and ritual interaction with the finished pictures, being related to a special meaning of the animal which is not found among paintings.

Modelling the Mental Maps of Ennedi Cattle Pastoralists

Based on the analyses carried out in the Ennedi Highlands, where rock art was recorded comprehensively in relation to landscape properties, some general hypotheses as to the difference between engravings and paintings can be set forth, thus giving meaning to their parallel existence in the pastoral landscape and linking them to different mental maps. These concepts seem to have become increasingly distinct from the Neolithic to the Iron Age, without suggesting clear chronological thresholds or any cogent climatic causality. The dichotomy of

Table 2.2: Two ways of representing the significance of cattle in pastoral rock art.

Pastoral rock art – engravings	**Pastoral rock art – paintings**
Sites • sign posts (few traces of habitation) • motifs with signal character (long-range visibility) • landscape-related properties narrowly patterned	**Sites** • paintings usually at inhabited sites • paintings normally inside shelters • landscape-related properties variant
Art sample • 773 figures at 54 sites • 14.3 figures per site • thin scatter of sites over landscape • some areas without engravings	**Art sample** • 5795 figures at 104 sites • 55.7 figures per site • clear local concentrations • evenly distributed over highlands
Main motif • cattle are the first vehicle to transport notions of the extraordinary • engraved cattle are offerings to the landscape, not being referential to humans and their mundane actions • engravers had a relatively moderate 'infatuation' with cattle • they neglect goats and sheep in the art • this may indicate that the pastoral means of production were not suited per se to express meaning	**Main motif** • the symbolic capital constituted by cattle was 'saved' in the landscape • the landscape formed the stage on which to advertise one's 'infatuation' with cattle • paintings of cattle were self-referential, apparently having served to transcribe the value and meaning of cattle onto the landscape • this perpetuated the dominance of cattle in, and its intertwining with, everyday life
Hypothesis: 'Passive' approach to landscape • the entire landscape is charged symbolically • marked spots define the sacred landscape, beyond the everyday human life-world • places of symbolic significance were always there, even before art • configurations of the landscape were of importance in their own right, irrespective of the use made of them • the pre-existent significance of places was made visible by engravings	**Hypothesis: 'Active' approach to landscape** • the landscape is marked in the context of its mundane and ritual use • art sites are places of everyday routines (notwithstanding the existence of specialised ritual sites) • people appropriated the landscape through use in the context of their normal livelihood • when choosing a site they had no particular configuration and symbolism of the landscape in mind • first criterion of choice was a site's suitability for dwelling purposes • at a chosen site paintings completed the appropriation of the place by ritual means

representing the significance of cattle at the interface with the landscape may be summarised as in Table 2.2.

These hypotheses corroborate the findings from other disciplines and reach out beyond the Ennedi case study: although there are no paintings in the Wadi Howar, the characterisation of the relationship between landscape and engravings applies there, too. Zolat el Hamad is a peculiar landscape feature marked with engravings, and in the Lower Wadi Howar (S03/35) depictions of cattle were made in such a way that they attain a signalling effect over a considerable distance. Obviously, groups living in the wider area shared very similar economic strategies and a cattle-centred ideology but were clearly distinct in other cultural traits even though their territorial units may have partly overlapped.[21] With this trend towards regionalisation, expressed in different conceptualisations of landscape (that is, the mental maps of painters and engravers), with the rather narrow demarcation of spatial entities by the exclusiveness of certain motifs in special areas and also with different idioms to communicate within and through the landscape (paintings versus engravings), it seems plausible that these groups may also have spoken different languages. While the ecological conditions seem to have imposed a narrow choice of economic strategies, it was perhaps the growing scarcity of resources which favoured a parallel development of cultural multivocality and distancing.

Conclusion

Cattle-centred societies, based on multi-resource pastoralism, had already developed in the fifth millennium BC in the Eastern Sahara of Sudan and Chad and can be traced through to the beginning of the Iron Age in the first half of the first millennium AD. Cattle were seen as capital, not only in an economic but also in an ideological and social sense. Cattle, providing milk and meat, assured people's livelihood and served as a means of exchange, but their incorporation in rituals also allowed societies to cope both physically and mentally with vulnerable conditions such as climate change. Various cattle-first strategies developed as a means to concisely condense and communicate regulative mechanisms of social, economic and ritual life. Especially during the second half of the Holocene, this strong relationship seems to have counterbalanced for a long time the criticality of ecological conditions that deteriorated to a state where animals better adapted to arid ecotopes could possibly have yielded a more resilient source of subsistence. But what at first appears like an unsustainable strategy may have been an efficient feedback loop. The cognitive fixation on cattle might have tolerated no alternative. Through this focus the whole potential of all possibilities to keep cattle was creatively exploited. This allowed herding under conditions that from a strictly ecological perspective would seem completely unsuited. However, comparable attitudes can still be found, for example, among the BaSotho of Lesotho, where Ferguson (1985: 651–52) has pertinently documented a pastoralist's cattle strategy in the face of severe drought.

All sets of data investigated in this study – that is, artefacts, bones, and rock art – allow a consensual interpretation pointing towards a fixation with cattle, distributed over various landscapes and habitats of the southern Eastern Sahara. A variety of pastoral adaptations were present in different areas, such as the Ennedi Highlands or the Wadi Howar region, and also through time.

However, a set of characteristics of cattle-centred behaviour can be established, including some or all of the following features:

- cattle bones clearly dominate the faunal remains;
- pits filled with cattle bones and often with broken pottery; occasionally even large parts of still articulated or entire carcasses were deposited in these pits; these pits indicate various uses of cattle in secular and ritual contexts;
- large sites with hundreds of such pits, showing the interlinked special meaning people imposed on localities and animals; in some cases this particular significance, likely of a ritual and certainly of a social character, lasted over several centuries;
- less frequent features include: individual cattle burials; small cattle figurines of clay; bones from outstandingly large beasts;
- pastoral strategies of herd management such as the recurrence of different breeds or the installation of watering troughs;
- in the rock art, cattle depictions are predominant;
- the individualisation of cows in representations, with special efforts put into the production of cattle pictures and occasional tactile interaction with finished pictures;
- a passive role for humans in painted scenes involving cattle is characteristic of Iron Age rock art;
- in general, the fixation on cattle is maintained for a considerable period of time, even when climatic conditions deteriorate to a state where the herding of more resilient species (sheep, goats, camels) would seem to have been less risky and these species could be expected to gain in symbolic significance.

Cattle-centred behaviour functioned for rather a long time in the southern Eastern Sahara: ubiquitous during the Leiterband Phase (4000 to 2200 BC), it was more strongly focused on the Ennedi Highlands in the ensuing periods up to the beginning of the Iron Age in the first half of the first millennium AD.

The value of cattle is rooted in times long before their domestication, as indicated by evidence of final Pleistocene sites in Egypt and Algeria. For instance, at site 8905 in the Tushka area of Egypt, the skulls (or parts thereof) of wild cattle (*Bos primigenius*) were found associated with human burials, dated to between around 16000 and 12000 BC (Wendorf 1968: 872–73; Wendorf and Schild 2001: 657). At Columnata in Algeria, the horncores of five or six wild cattle were found in a stone setting associated with a human burial of Ibéromaurusian date (about 12000 BC) (Hachi and Roubet 2003: 17).

In the late middle Holocene, finds of cattle integrate sites of the southern Eastern Sahara into a large northern African pastoral sphere. Since their first appearance, cattle were of outstanding importance in the way of life of the prehistoric groups of this area. This was not restricted to subsistence but also included ritual aspects, as manifested in different forms in the archaeological record (see Figure 2.6). Despite cattle being a widespread phenomenon throughout northern Africa, regional differences become obvious.[22]

This general trend towards regionalisation becomes especially apparent when taking a look at phenomena with a high potential to communicate symbolic meanings, such as pottery styles and rock art. There are overarching divisions in the research area based on chronological and environmental features (most pronounced: the Ennedi Highlands versus Wadi Howar) that still allow for small-scale cultural expressions within these sub-divisions. In the rock art of the Ennedi and the Lower Wadi Howar there are regional specialisations which seem to circumscribe territorial entities no larger than a few dozen square kilometres. On the other hand, there are large supra-regional phenomena, such as the distribution of certain decorative patterns of pottery and types of lithic artefact. Obviously, lithic artefacts and pottery operated as expressions of identity on a different level than rock art. Pastoralism as the dominant form of subsistence could be the uniting but also the dividing factor for this. While on a broad scale the value and use of cattle was shared, from another perspective the ideological process of individualising cattle may have been coupled with the phenomenon in the social environment where identities, too, showed a tendency towards small-scale entities. Occasional social events and meetings between different pastoral groups could explain the far-flung spread of certain traits, whereas a more or less constant neighbourhood as well as competition for scarce resources may have advanced the need to advertise distinct identities. As a consequence, people sharing the same landscape and the same pastoral subsistence strategy may have developed different intensities of cattle-centred behaviour.

An archaeological study that provides evidence for cattle-centred behaviours over a vast landscape and over many centuries can neither elucidate the psychological aspect of the fascination with cattle, nor the extent of the cognitive effort to explain, order and understand the world. But the early appearance of such behaviours and their occurrence in the present may indicate that even emotional factors constitute a form of cultural capital that strengthens resilience under precarious ecological conditions and secures survival despite strong environmental and cultural changes. These changes may be of minor importance if only 'the cattle are well'.

Notes

Between 1980 and 1993, field research was carried out by a team from the University of Cologne as part of the Besiedlungsgeschichte der Ostsahara (BOS) project. Since 1995, research has continued for twelve years under the aegis of the ACACIA (Arid Climate, Adaptation and Cultural Innovation in Africa) collaborative research centre. Both projects owe their funding to the Deutsche Forschungsgemeinschaft. We would like to thank our archaeological and geological counterparts in Sudan and Chad, the National Corporation for Antiquities and Museums (NCAM) and the Geological Research Authority (GRAS), both in Khartoum, and the Centre National d'Appui à la Recherche (CNAR) in N'Djamena for their fruitful cooperation during this time. For useful critical comments on earlier drafts of this chapter we would like to thank Heiko Riemer, Michael Bollig and Hans-Peter Wotzka.

1. We are grateful to E. Ndiema (personal communication) for bringing this greeting phrase to our attention.
2. For the ritual aspect of cattle, see, e.g., Keding (1997a: 224–27), Paris (1997, 2000), Applegate, Achilles and Duncan (2001) and di Lernia (2006).
3. For an archaeological perspective on this matter, see, e.g., Wendorf and Schild (2001: 657, 671–73).
4. In the anthropological literature there is a wide range of definitions of pastoralism, the lowest common denominator being an economic dependence on the herding of domesticated animals (Johnson 2002: 152). Along the two axes of mobility and subsistence strategies it is possible to distinguish various forms of pastoralism (cf. Sadr 1991: 4; Johnson 2002: 166).
5. For palaeoclimatic studies, see Pachur and Kröpelin (1987), Kröpelin (1993), Hoelzmann et al. (2001), Kuper and Kröpelin (2006), Pachur and Altmannn (2006) and Kröpelin et al. (2008); for archaeozoological studies, see Van Neer (1989), Van Neer and Uerpmann (1989), Berke (2001), Peters and Pöllath (2004) and Pöllath and Peters (2007); for (archaeo-)botanical studies, see Neumann (1989) and Nussbaum, Kröpelin and Darius (2007). For the subdivision of the Holocene, see Kuper and Kröpelin (2006).
6. Recent research in the Kerma Basin has provided evidence for domestic cattle at least from the early sixth millennium BC. At El-Barga, a Neolithic grave had been covered by the skull of a domestic cow for which two radiocarbon dates clustering around 5750 BC are available. A sondage at a site about 5 kilometres from El-Barga yielded bones of domestic cattle in association with fragments of ostrich eggshell and shells which gave radiocarbon dates of around 7000 BC (Honegger 2005: 247). At Rabak, south of Khartoum, bones of domestic cattle were found in level six of the excavation. A shell sample found in the same level was dated to 4980 ± 140 BC (6050 ± 100 bp; T-5133). However, the fauna found at the site shows a heavy emphasis on fishing, with some hunting of various animal species (El-Mahi and Haaland 1984: 30).
7. The radiocarbon date of 6310 ± 170 BC (7455 ± 180 bp; Hv-2775) on a bone sample found at Enneri Bardagué in the Tibesti mountains (Chad), often mentioned in reviews of early African cattle husbandry (e.g., Marshall and Hildebrand 2002: 110, Fig. 2) is not considered here because the dated animal bone was not identified (Gabriel 2002: 387). Besides cattle, the site yielded remains of elephant, giraffe, buffalo (*Bubalus*), antelope, gazelle, sheep and ostrich (Gabriel 1977: 52). A review of the radiocarbon dates associated with early domestic cattle in Africa seems necessary since uncritical acceptance of dubious contexts is frequent.
8. Gazelle (*Gazella dorcas*), addax (*Addax nasomaculatus*) and hare (*Lepus capensis*) were identified by Hubert Berke, University of Cologne.
9. In a refuse pit at Ennedi Erg S99/1, cattle bones bearing butchering marks were even found in association with mega-segments (Berke 2001: 245).
10. For studies of recent pastoral societies in relation to this matter, see Cribb (1991: 82) and Hodder (1982).
11. B. Zach (personal communication).
12. For instance, at Nabta Playa in the Egyptian Western Desert (Applegate, Achilles and Duncan 2001: 486–87; Wendorf and Schild 2001: 668) and at Adrar Bous Site 1 in Niger (Paris 1997: 115). For a review, see di Lernia (2006).
13. At site Conical Hill S02/3, cattle bones make up nearly 100 per cent, whereas at site Conical Hill S02/4, small livestock herding is also attested, sheep and goat representing 1.4 per cent of the faunal remains (Godhoff 2005: 25–26). Among the faunal remains of site Abu Tabari

S02/28, cattle are largely dominant, accounting for about 95 per cent of the identified bones. Fishing and hunting only played a minor role (Pöllath and Peters 2007: 65).

14. Both regions have undergone intensive research by ACACIA projects, but, more importantly, the Ennedi represents a core area of prehistoric art production in the Eastern Sahara.
15. Reports and photographs include Newbold (1924, 1928); the information contained there has been supplemented as a result of the Cologne University BOS and ACACIA projects.
16. Pachur and Altmann (2006: 564) have reconstructed an annual precipitation rate of 400 to 700 mm for the period around 7000 BC.
17. Research into the archaeology and rock art of the region has covered four more or less representative regions along a west–east transect (Lenssen-Erz and von Czerniewicz 2005; Keding et al. 2007) where archaeological sites, paintings and engravings can be found in different relative quantities.
18. For a discussion of so-called 'Bovidien' art, see Heckendorf (2008).
19. Due to research history, the two terms 'Leiterband' and 'Hohou' are used for the same type of pottery, the former in the Wadi Howar region and the latter in the Ennedi (Keding 1998).
20. See also Coote (1994: 250) for further references on this matter.
21. Compare the regionalisation which Bailloud (1997) grasped by recognising eight, mainly locality-bound, styles of the Bovine period within a very restricted area of about 70 by 60 kilometres.
22. This view is supported by linguistic evidence which not only indicates a linguistic fragmentation in the area but also contacts between east and west: see, e.g., Blench (1999) and Dimmendaal (2007).

References

Andigue, J., S. Bonnaventure and A. Adoum. 2006. 'Climat', in *Atlas du Tchad.* Paris: Les Editions J.A., pp.12–13.

Applegate, A., G. Achilles and S. Duncan. 2001. 'The North Tumuli of the Nabta Late Neolithic Ceremonial Complex', in F. Wendorf et al. (eds), *Holocene Settlement of the Egyptian Sahara: The Archaeology of Nabta Playa*, New York: Kluwer Academic/Plenum, pp.468–88.

Arnold, J.E. 2000. 'Revisiting Power, Labour Rights, and Kinship: Archaeology and Social Theory', in M.B. Schiffer (ed.), *Social Theory in Archaeology.* Salt Lake City: University of Utah Press, pp.14–30.

Bailloud, G. 1969. 'L'Evolution des Styles Céramiques en Ennedi (République du Tchad)', in *Mémoires I, Actes du Premier Colloque International d'Archéologie Africaine, Fort Lamy (Rép. du Tchad) 1966.* Fort Lamy: Institut National Tchadien pour les Sciences Humaines, pp.31–45.

——— 1997. *Art Rupestre en Ennedi.* Saint-Maur: Editions Sépia.

Baxter, P.T.W. 1996. 'Cattle Complex', in A. Barnard and J. Spencer (eds), *Encyclopedia of Social and Cultural Anthropology.* London: Routledge, pp.91–92.

Bechhaus-Gerst, M., and B. Keding. 2007. 'Meat or Milk? Archaeological and Linguistic Evidence for the Peridynastic Use of Cattle in the Eastern Sahara', *SUGIA* 18: 17–35.

Behrens, H. 1963. 'Neolithisch-frühmetallzeitliche Tierskelettfunde aus dem Niltal und ihre religionsgeschichtliche Bedeutung', *Zeitschrift für Ägyptische Sprache* 88: 75–83.

Berke, H. 2001. 'Gunsträume und Grenzbereiche: Archäozoologische Beobachtungen in der Libyschen Wüste, Sudan und Ägypten', in B. Gehlen, M. Heinen and A. Tillman (eds), *Zeit-Räume: Gedenkschrift für Wolfgang Taute*. Bonn: Habelt, pp.237–56.

Binford, L.R. 1980. 'Willow Smoke and Dogs' Tails: Hunter-gatherer Settlement Systems and Archaeological Site Formation', *American Antiquity* 45: 4–20.

Blench, R. 1999. 'The Westward Wanderings of Cushitic Pastoralists. Explorations in the Prehistory of Central Africa', in C. Baroin and J. Boutrais (eds), *L'Homme et l'Animal dans le Bassin du Lac Tchad*. Paris: Editions IRD, pp.39–80.

Blench, R.M. and K.C. MacDonald (eds). 2000. *The Origins and Development of African Livestock: Archaeology, Genetics, Linguistics and Ethnography*. London: University College London Press.

Bonnet, C. 1996. 'C-Gruppe', in D. Wildung (ed.), *Sudan: Antike Königreiche am Nil*. Tübingen: Ernst Wasmuth Verlag, pp.51–70.

Brass, M. 2007. 'Reconsidering the Emergence of Social Complexity in Early Saharan Pastoral Societies, 5000–2500 BC', *Sahara* 18: 7–22.

Chaix, L., and A. Grant. 1992. 'Cattle in Ancient Nubia', *Anthropozoologica* 16: 61–66.

Clanet, J.-C., J. Lemoalle and L. Dara. 2006. 'Elevage et Pêche', in *Atlas du Tchad*. Paris: Les Éditions J.A., pp.34–37.

Clutton-Brock, J. (ed.). 1989. *The Walking Larder: Patterns of Domestication, Pastoralism, and Predation*. London: Unwin Hyman.

Comaroff, J.L., and J. Comaroff. 1990. 'Goodly Beasts, Beastly Goods: Cattle and Commodities in a South African Context', *American Ethnologist* 17(2): 195–216.

Coote, J. 1994. '"Marvels of Everyday Vision": The Anthropology of Aesthetics and the Cattle-keeping Nilotes', in J. Coote and A. Shelton (eds), *Anthropology, Art and Aesthetics*. Oxford: Clarendon Press, pp.245–73.

Cribb, R. 1991. *Nomads in Archaeology*. Cambridge: Cambridge University Press.

Cziesla, E. 1986. 'Excavations at Wadi Sahal', in M. Krause (ed.), *Nubische Studien. Tagungsakten der 5. Internationalen Konferenz der International Society for Nubian Studies. Heidelberg, 22–25 September 1982*. Mainz: Philipp von Zabern, pp.143–49.

Dahl, G., and A. Hjort. 1976. 'Having Herds: Pastoral Herd Growth and Household Economy,' *Stockholm Studies in Social Anthropology* 2. Stockholm: University of Stockholm.

Davis, S.J.M. 1987. *The Archaeology of Animals*. New Haven, CT: Yale University Press.

De Garine, I., and G.J.A. Koppert. 1988. 'Coping with Seasonal Fluctuations in Food Supply among Savanna Populations: The Massa and Mussey of Chad and Cameroon', in I. De Garine and G.A. Harrison (eds), *Coping with Uncertainty in Food Supply*. Oxford: Clarendon Press, pp.210–59.

De Waal, A. 1989. *Famine that Kills: Darfur, Sudan, 1984–1985*. Oxford: Oxford University Press.

di Lernia, S. 2006. 'Building Monuments, Creating Identity: Cattle Cult as a Social Response to Rapid Environmental Changes in the Holocene Sahara', *Quaternary International* 151: 50–62.

Dimmendaal, G. 2007. 'The Wadi Howar Diaspora: Linking Linguistic Diffusion to Palaeoclimatological and Archaeological Findings', in O. Bubenzer, A. Bolten and F. Darius (eds), *Atlas of Cultural and Environmental Change in Arid Africa*. Cologne: Heinrich-Barth-Institut, pp.148–49.

Dupuy, C. 1998. 'Réflexions sur l'Identité des Guerriers Représentés dans les Gravures Rupestres de l'Adrar des Iforas et de l'Aïr', *Sahara* 10: 31–54.

Dyson-Hudson, R., and N. Dyson-Hudson. 1970. 'The Food Production System of a Semi-nomadic Society: The Karimojong, Uganda', in P.F. McLoughlin (ed.), *African Food Production Systems: Cases and Theory*. Baltimore, MD: Johns Hopkins University Press, pp.91–123.

Eckl, A.E. 2000. 'Language, Culture and Environment: The Conceptualization of Herero Cattle Terms', in M. Bollig and J.-B. Gewald (eds), *People, Cattle and Land: Transformations of a Pastoral Society in Southwestern Africa*. Cologne: Rüdiger Köppe, pp.401–31.

El-Mahi, A.T., and R. Haaland. 1984. 'Archaeological Research in the Area of Rabak and Atbara, Sudan, 1983–84', *Nyame Akuma* 24/25: 28–32.

Ferguson, J. 1985. 'The Bovine Mystique: Power, Property and Livestock in Rural Lesotho', *Man* 20(4): 647–74.

Ferhat, N. 2003. 'Mankhor : Une Nécropole de Bovidés au Tassili N'ajjer (Sahara Algérien)', *Dossiers d'Archéologie* 282: 79.

Ferhat, N., K.-H. Striedter and M. Tauveron. 1996. 'Un Cimetière de Boeufs dans le Sahara Central: La Nécropole de Mankhor', in G. Aumassip (ed.), *La Préhistoire de l'Afrique de l'Ouest*. Saint-Maur: Éditions Sépia, pp.103–7.

Francke, U. 1986a. 'Camp 49 Re-examined', in M. Krause (ed.), *Nubische Studien*. Mainz: Philipp von Zabern, pp.137–42.

——— 1986b. 'Untersuchungen zur Keramik des dritten und zweiten Jahrtausends v. Chr. aus dem Wadi Shaw, Nord-Sudan', MA dissertation. Cologne: University of Cologne.

Gabriel, B. 1977. *Zum Ökologischen Wandel im Neolithikum der Östlichen Zentralsahara*. Berlin: Institut für Physische Geographie der Freien Universität Berlin.

——— 2002. 'Alter und Ursprung des Rinderhirtennomadismus in Afrika', *Erdkunde* 56: 385–400.

Galaty, J.G. 1990. 'Cattle and Cognition: Aspects of Maasai Practical Reasoning', in J. Clutton-Brock (ed), *The Walking Larder: Patterns of Domestication, Pastoralism, and Predation*. London: Unwyn Hyman, pp.215–30.

Garcea, E.A.A. 2004. 'An Alternative Way Towards Food Production: The Perspective from the Libyan Sahara', *Journal of World Prehistory* 18: 107–53.

Gautier, A. 1987. 'Prehistoric Men and Cattle in North Africa: A Dearth of Data and a Surfeit of Models', in A.E. Close (ed.), *Prehistory of Arid North Africa*. Dallas, TX: Southern Methodist University Press, pp.163–87.

——— 1989. 'A General Review of the Known Prehistoric Fauna of the Central Sudanese Nile Valley', in L. Krzyżaniak and M. Kobusiewicz (eds), *Late Prehistory of the Nile Basin and the Sahara*. Poznań: Poznań Archaeological Museum, pp.353–57.

——— 2001. 'The Early to Late Neolithic Archaeofaunas from Nabta and Bir Kiseiba', in F. Wendorf et al. (eds), *Holocene Settlement of the Egyptian Sahara: The Archaeology of Nabta Playa I*. New York: Kluwer Academic Press, pp.609–35.

Germer, R. 1985. *Flora des Pharaonischen Ägypten*. Mainz: Philipp von Zabern.

Ghoneim, W. 1977. *Die Ökonomische Bedeutung des Rindes im Alten Ägypten*. Bonn: Habelt.

Gifford-Gonzalez, D. (ed.). 2008. *Adrar Bous: Archaeology of a Central Saharan Granitic Ring Complex in Niger*. Tervuren: Royal Museum for Central Africa.

Godhoff, H.F. 2005. 'Von den Ufern des Nils zu den Weiden der Wüste. Interkulturelle Beziehungen im 4. und 3. Jahrtausend v.Chr. am Beispiel des Fundplatzes Conical Hill 02/4 im Unteren Wadi Howar/Sudan', MA dissertation. Cologne : University of Cologne.

Gratien, B. 1985. *Sai, I: La Nécropole Kerma*. Paris: Centre National de la Recherche Scientifique.

Günther, S. 1995. 'Djabarona 84/19: Keramik und Befunde des Vierten bis Zweiten Jahrtausends v. Chr. aus dem Wadi Howar (Nord-Sudan)', MA dissertation. Cologne: University of Cologne.

Haberland, E. 1983. 'Mensch-Tier-Beziehungen: das Beispiel der nordostafrikanischen Hirten', in H. Müller-Karpe (ed.), *Zur Frühen Mensch-Tier-Symbiose*. München: C.H. Beck, pp.175–89.

Hachi, S., and C. Roubet. 2003. 'Les Inhumations en Nécropoles de l'Homme de Mechta-Afalou', *Préhistoire d'Algérie Dossiers d'Archéologie* 282: 16–21.

Hassan, F.A. (ed.). 2002. *Droughts, Food and Culture: Ecological Change and Food Security in Africa's Later Prehistory*. New York: Kluwer Academic/Plenum.

Heckendorf, R. 2008. *'Bubalin' und 'Bovidien' in Südmarokko*. Wiesbaden: Reichert.

Herskovits, M.J. 1926. 'The Cattle Complex in East Africa.' *American Anthropologist* N.S. 28: 230–72; 361–88; 494–528; 633–64.

Hodder, I. 1982. *Symbols in Action: Ethnoarchaeological Studies of Material Culture*. Cambridge: Cambridge University Press.

Hoelzmann, P., B. Keding, H. Berke, S. Kröpelin and H.J. Kruse. 2001. 'Environmental Change and Archaeology: Lake Evolution and Human Occupation in the Eastern Sahara during the Holocene', *Palaeogeography, Palaeoclimatology, Palaeoecology* 169: 193–217.

Holl, A. 1998. 'The Dawn of African Pastoralisms: An Introductory Note', *Journal of Anthropological Archaeology* 17: 81–96.

Holst, D. 2003. 'Das Steininventar des Fundplatzes S 99/1 (NW-Sudan). Rohmaterialversorgung und Artefaktproduktion bei Neolithischen Pastoralnomaden, *Archäologische Information* 26(1): 207–13.

Honegger, M. 2005. 'Kerma et les Débuts du Néolithique Africain', *Genava* N.S. 53: 239–49.

Jesse, F. 2003. *Rahib 80/87. Ein Wavy-Line Fundplatz in Wadi Howar und die früheste Keramik in Nordafrika*. Cologne: Heinrich-Barth-Institut.

_______ 2005. 'Rock Art in Lower Wadi Howar, Northwest Sudan', *Sahara* 16: 27–38.

_______ 2006. 'Cattle, Sherds and Mighty Walls – The Wadi Howar from Neolithic to Kushite Times', *Sudan and Nubia* 10: 43–54.

——— 2007. 'Pastoral Groups in the Southern Libyan Desert: The Handessi Horizon (c. 2400–1100 BC)', in K. Kroeper, M. Chlodnicki and M. Kobusiewicz (eds), *Archaeology of Early Northeastern Africa*. Poznań: Poznań Archaeological Museum, pp.987–1004.

Jesse, F., and B. Keding. 2002. 'Death in the Desert: Burials in the Wadi Howar Region (Eastern Sahara)', in Jennerstr. 8. (eds), *Tides of the Desert/Gezeiten der Wüste*. Cologne: Heinrich-Barth-Institut, pp.277–93.

——— 2007. 'Holocene Settlement Dynamics in the Wadi Howar Region (Northern Sudan) and the Ennedi Mountains (Chad)', in O. Bubenzer, A. Bolten and F. Darius (eds), *Atlas of Cultural and Environmental Change in Arid Africa*. Cologne: Heinrich-Barth-Institut, pp.42–43.

Jesse, F., B. Keding, N. Pöllath, M. Bechhaus-Gerst and T. Lenssen-Erz. (2007) 'Cattle Herding in the Southern Libyan Desert', in O. Bubenzer, A. Bolten and F. Darius (eds), *Atlas of Environmental Change and Human Adaptation in Arid Africa*. Cologne: Heinrich-Barth-Institut, pp.44–47.

Jesse, F., S. Kröpelin, M. Lange, N. Pöllath and H. Berke. 2004. 'On the Periphery of Kerma – The Handessi Horizon in Wadi Hariq, Northwestern Sudan', *Journal of African Archaeology* 2: 123–64.

Johnson, A.L. 2002. 'Cross-cultural Analysis of Pastoral Adaptations and Organizational States: A Preliminary Study', *Cross-Cultural Research* 36: 151–80.

Keding, B. 1997a. *Djabarona 84/13. Untersuchungen zur Besiedlungsgeschichte des Wadi Howar Anhand der Keramik des 3. und 2. Jahrtausends v.Chr.* Cologne: Heinrich-Barth-Institut.

_______ 1997b. 'Prehistoric Investigations in the Wadi Howar Region: A Preliminary Report on the 1995/96 Season', *Kush* 17: 33–46.

_______ 1998. 'The Yellow Nile: New Data on Settlement and the Environment in the Sudanese Eastern Sahara', *Sudan and Nubia* 2: 2–12.

_______ 2000. 'New Data on the Holocene Occupation of the Wadi Howar Region (Eastern Sahara/Sudan)', in L. Krzyżaniak, K. Kroeper and M. Kobusiewicz (eds), *Recent Research into the Stone Age of Northeastern Africa.* Poznań: Poznań Archaeological Museum, pp.89–104.

——— 2004. 'The Yellow Nile: Settlement Shifts in the Wadi Howar Region (Sudanese Eastern Sahara) and Adjacent Areas from between the Sixth to the First Millennium BC', in T. Kendall (ed.), *Nubian Studies 1998: Proceedings of the Ninth Conference of the International Society of Nubian Studies.* Boston: Department of African-American Studies, Northeastern University, pp.95–108.

_______ 2006. 'Pottery of the Wadi Howar – Traditions, Transformations and Their Cultural Implications', in K. Kroeper, M. Chlodnicki and M. Kobusiewicz (eds), *Archaeology of Early Northeastern Africa. In Memoriam of Lech Krzyżaniak.* Poznań: Poznań Archaeological Museum, pp.235–59.

Keding, B., T. Lenssen-Erz and A. Pastoors. 2007. 'Pictures and Pots from Pastoralists. Investigations into the Prehistory of the Ennedi Highlands in NE Chad', *Sahara* 18: 23–46.

Kröpelin, S. 1993. *Zur Rekonstruktion der Spätquartären Umwelt am unteren Wadi Howar (Südöstliche Sahara/NW-Sudan).* Berlin: Institut für Physische Geographie der Freien Universität Berlin.

——— 2004. 'New Petroglyph Sites in the Southern Libyan Desert (Sudan-Chad)', *Sahara* 15: 111–17.

——— 2007. 'The Wadi Howar', in O. Bubenzer, A. Bolten and F. Darius (eds), *Atlas of Cultural and Environmental Change in Arid Africa.* Cologne: Heinrich-Barth-Institut, pp.38–39.

Kröpelin, S., et al. 2008. 'Climate Driven Ecosystem Succession in the Sahara: The Past 6000 Years', *Science* 320: 765–68.

Kuper, R. 1981. 'Untersuchungen zur Besiedlungsgeschichte der östlichen Sahara: Vorbericht über die Expedition 1980', *Beiträge zur Allgemeinen und Vergleichenden Archäologie* 3: 215–75.

——— 1995. 'Prehistoric Research in the Southern Libyan Desert: A Brief Account and Some Conclusions of the B.O.S. Project', Actes de la VIIIe Conférence Internationale d'Etudes des Nubiennes Lille 11–17 septembre 1994, *Cahiers de recherché de l'Institut de Papyrologie et Egyptologie de Lille* 17(1): 123–40.

Kuper, R. and S. Kröpelin. 2006. 'Climate-controlled Holocene Occupation in the Sahara: Motor of Africa's Evolution', *Science* 313: 803–7.

Laudien, A. 2000. 'Osteometrische Untersuchungen zum altägyptischen Langhornrind', PhD dissertation. Munich: Ludwig-Maximilians-Universität.

Lenssen-Erz, T. 2007. 'Rock Art in African Highlands. Ennedi Highlands, Chad: Artists and Herders in a Lifeworld on the Margins', in O. Bubenzer, A. Bolten and F. Darius (eds), *Atlas of Cultural and Environmental Change in Arid Africa.* Cologne: Heinrich-Barth-Institut, pp.48–51.

Lenssen-Erz, T., and M. von Czerniewicz. 2005. 'Résultats Préliminaires des Recherches Archéologiques dans l'Ennedi', *Revue Scientifique du Tchad* 7(2): 5–18.

Lévi-Strauss, C. 1962. *Totemism*. Boston: Beacon Press.
Mair, L. 1985. 'The Cattle Complex', *Man* N.S. 20(4): 743.
Marshall, F., and E. Hildebrand. 2002. 'Cattle Before Crops: The Beginnings of Food Production in Africa', *Journal of World Prehistory* 16: 99–143.
Mtetwa, J. 1982. *Man and Cattle in Africa: A Study of Social Change in Pastoral Societies in Botswana*. Saarbrücken: Breitenbach.
Murray, G.W. 1926. 'Graves of Oxen in the Eastern Desert of Egypt', *Journal of Egyptian Archaeology* 12(3/4): 248–49.
Muzzolini, A. 2001. 'Saharan Africa', in D.S. Whitley (ed.), *Handbook of Rock Art Research*. Walnut Creek, CA: Altamira Press, pp.605–36.
Neumann, K. 1989. 'Zur Vegetationsgeschichte der Ostsahara im Holozän: Holzkohlen aus prähistorischen Fundstellen (mit einem Exkurs über die Holzkohlen von Fachi-Dogonboulo/Niger)', in R. Kuper (ed.), *Forschungen zur Umweltgeschichte der Ostsahara*. Cologne: Heinrich-Barth-Institut, pp.13–181.
_______ 2005. 'The Romance of Farming: Plant Cultivation and Domestication in Africa', in A. Brower Stahl (ed.), *African Archaeology. A Critical Introduction*. Malden: Blackwell, pp.249–75.
Newbold, D. 1924. 'A Desert Odyssey of a Thousand Miles', *Sudan Notes and Records* 7: 43–92.
——— 1928. 'Rock-pictures and Archaeology in the Libyan Desert', *Antiquity* 2: 261–91.
Nussbaum, S., S. Kröpelin and F. Darius. 2007. 'The Flora and Vegetation of Wadi Howar', in O. Bubenzer, A. Bolten and F. Darius (eds), *Atlas of Cultural and Environmental Change in Arid Africa*. Cologne: Heinrich-Barth-Institut, pp.40–41.
Pachur, H.-J. 1999. 'Paläo-Environment und Drainagesysteme der Ostsahara im Spätpleistozän und Holozän', in E. Klitzsch and U. Thorweihe (eds), *Nordost-Afrika: Strukturen und Ressourcen. Ergebnisse aus dem Sonderforschungsbereich 'Geowissenschaftliche Probleme in ariden und semiariden Gebieten'.* Weinheim: Wiley-VCH, pp.366–445.
Pachur, H.-J. and N. Altmann. 2006. *Die Ostsahara im Spätquartär. Ökosystemwandel im Größten Hyperariden Raum der Erde*. Berlin/Heidelberg: Springer.
Pachur, H.-J. and S. Kröpelin. 1987. 'Wadi Howar: Paleoclimatic Evidence from an Extinct River System in the Southeastern Sahara', *Science* 237: 298–300.
Pachur, H.-J. and B. Wünnemann. 1996. 'Reconstruction of the Palaeoclimate along 30°E in the Eastern Sahara during the Pleistocene/Holocene Transition', *Palaeoecology of Africa and the Surrounding Islands* 24: 1–32.
Paris, F. 1997. 'Les Inhumations de Bos au Sahara Méridional au Néolithique', *Archaeozoologia* 9: 113–22.
_______. 2000. 'African Livestock Remains from Saharan Mortuary Contexts', in F.A. Hassan (ed.), *Droughts, Food and Culture. Ecological Change and Food Security in Africa's Later Prehistory*. New York: Kluwer Academic/Plenum, pp.111–26.
Peters, J. and N. Pöllath. 2004. 'Holocene Faunas of the Eastern Sahara: Zoogeographical and Palaeoecological Aspects', in R.C.G. Lauwerier and I. Plug (eds), *The Future from the Past: Archaeozoology in Wildlife Conservation and Heritage Management. Proceedings of the 9th Conference of the International Council of Archeozoology, Durham 2002*. Oxford: Oxbow Books, pp.34–50.
Pöllath, N. and J. Peters. 2005. 'On the Possible Use of the LSI Scaling Technique for Stature Analysis in Cattle', *Revue de Paléobiologie*, Vol. Spéc. 10: 225–35.
_______. 2007. 'Holocene Climatic Change, Human Adaptation and Landscape Degradation in Arid Africa as Evidenced by the Faunal Record', in O. Bubenzer, A. Bolten and

F. Darius (eds), *Atlas of Cultural and Environmental Change in Arid Africa.* Cologne: Heinrich-Barth-Institut, pp.64–7

Rhotert, H. 1952. *Libysche Felsbilder.* Darmstadt: L.C. Wittich.

Robertshaw, P. 1990. 'The Development of Pastoralism in East Africa', in J. Clutton-Brock (ed.), *The Walking Larder: Patterns of Domestication, Pastoralism, and Predation.* London: Unwyn Hyman, pp.207–14.

Sadr, K. 1991. *The Development of Nomadism in Ancient Northeast Africa.* Philadelphia: University of Philadelphia Press.

Sauvet, G., R.H. Layton, T. Lenssen-Erz, P. Taçon and A. Wlodarczyk. 2006. 'La Structure Iconographique d'un Art Rupestre : Est-elle une Clef pour Son Interprétation?' *Zephyrus* 59: 97–110.

Schön, W. 2002. 'Survey before GPS: Some Remarks on Archaeological Sites on the Northern Bank of the Middle Wadi Howar (Sudan)', in Jennerstrasse 8 (eds), *Tides of the Desert/ Gezeiten der Wüste.* Cologne: Heinrich-Barth-Institut, pp.313–9.

Shipton, P. 1990. 'African Famines and Food Security: Anthropological Perspectives', *Annual Review of Anthropology* 19: 353–94.

Smith, A.B. 2004. 'A Prehistory of Modern Saharan Pastoralists', *Sahara* 15: 43–58.

_______ 2005. *African Herders. Emergence of Pastoral Traditions.* Walnut Creek: Altamira Press.

Smith, S.E. 1980. 'The Environmental Adaptation of Nomads in the West African Sahel: A Key to Understanding Prehistoric Pastoralists', in M.A.J. Williams and H. Faure (eds), *The Sahara and the Nile: Quarternary Environments and Prehistoric Occupation in Northern Africa.* Rotterdam: Balkema, pp.467–87.

Tilley, C. 1994. *A Phenomenology of Landscape.* Oxford/Providence: Berg.

Tubiana, M.-J. and J. Tubiana. 1977. *The Zaghawa from an Ecological Perspective.* Rotterdam: A. A. Balkema.

Van Neer, W. 1989. 'Holocene Fish Remains from the Sahara', *Sahara* 2: 61–68.

Van Neer, W. and H.-P. Uerpmann. 1989. 'Palaeoecological Significance of the Holocene Faunal Remains of the B.O.S.-Missions', in R. Kuper (ed.), *Forschungen zur Umweltgeschichte der Ostsahara.* Cologne: Heinrich-Barth-Institut, pp.307–41.

von Maydell, H.-J. 1990. *Trees and Shrubs of the Sahel: Their Characteristics and Uses.* Weikersheim: Markgraf.

Wendorf, F. 1968. 'Late Palaeolithic Sites in Egyptian Nubia', in F. Wendorf (ed.), *The Prehistory of Nubia.* Vol. 2. Dallas: Fort Burgwin Research Center and Southern Methodist University Press, pp.791–953.

Wendorf, F. and R. Schild. 2001. 'Conclusions', in F. Wendorf, R. Schild and Associates (eds), *Holocene Settlement of the Egyptian Sahara. Vol. 1 The Archaeology of Nabta Playa.* New York: Kluwer Academic/Plenum, pp.648–75.

CHAPTER 3

Trajectories of Pastoralism in Northern and Central Kenya

An Overview of the Archaeological and Environmental Evidence

Paul Lane

Introduction

This chapter offers a review of the archaeological, environmental and related evidence concerning the introduction of and transition to pastoralism in northern and central Kenya (see Figure 3.1), from the earliest evidence for livestock husbandry in the north of the region to the recent past. The chapter also aims to examine how these landscapes were transformed over millennia in terms of their material and ecological characteristics. In order to situate these events and processes, the chapter begins with an outline sketch of the origins and expansion of cattle, camel and small stock pastoralism in the region, drawing on a combination of archaeological, genetic and linguistic data. This is followed by a discussion of the changing nature of archaeological landscapes on the Laikipia and Leroghi plateaus in north-central Kenya from the middle Holocene (circa 5000 bp)[1] to around 100 years ago, as reconstructed from multiple sources of information.

The history of pastoralism and stock keeping in East Africa – defined here as encompassing Kenya, Uganda and Tanzania and the border areas of southern Sudan, Ethiopia and Somalia – extends over 4,500 to 5,000 years. Over the millennia a diverse range of herd management strategies and settlement systems has been followed by an equally diverse range of ethnic groups and pastoralist societies. In terms of their mobility practices these have encompassed full-blown nomadism, semi-nomadism, seasonal transhumance and settled agro-pastoralism.[2] The documentary record for the region includes several examples of mobile stock herders becoming sedentary farmers and agro-pastoralists (e.g., Spear and

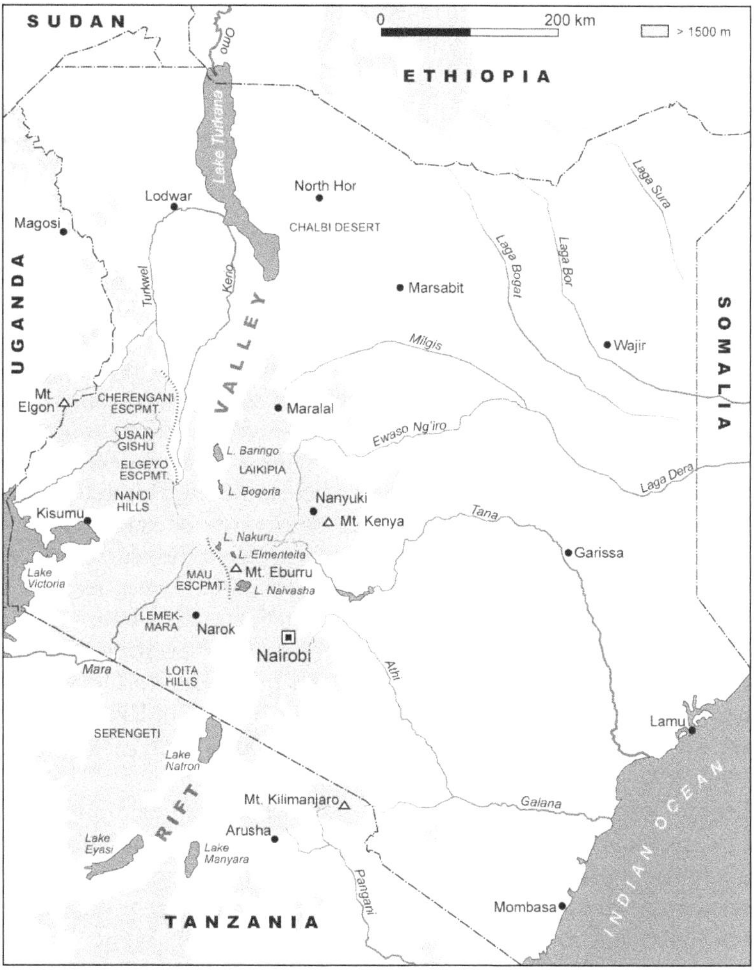

Figure 3.1: Map showing the key areas and places mentioned in the text.

Nurse 1992); hunter-gatherers acquiring livestock and adopting a pastoralist lifestyle and the cultural traditions of their pastoralist neighbours (e.g., Cronk 1989); and impoverished pastoralists becoming either temporary or more long-term foragers (e.g., Spencer 1973). Other attested subsistence strategies include the integration of herding and fishing (e.g., Barthelme 1985; Sobania 1988). There is also widespread evidence for 'ethnic shifting' (e.g., Galaty 1982; Hodder 1982: 87–104), the absorption of neighbouring pastoralist and non-pastoralist societies by more dominant pastoralist groups during one or other expansionary phases (e.g., Lamphear 1988), and even the formation of new ethnicities comprised of speakers of two distinct language families, as in the case of the Ariaal of northern Kenya, whose origins can be traced back to Cushitic-speaking Rendille and Nilotic-speaking Samburu (Falkenstein 1995). All of this indicates that there has been no single trajectory towards either becoming or being a 'pastoralist'. This is clearly reflected in the archaeological record for the region. It also poses many challenges for the interpretation of that record, especially since the identification of a particular society as 'pastoral' rests more on their beliefs and ideology than on simply the presence or absence of livestock (see Robertshaw 1990: 299–301).

Origins of Livestock Domestication

Of the five principal domestic animals (cattle, sheep, goats, camels and donkeys), there is no evidence that a possible wild progenitor of sheep or goats once lived on the African continent, implying that both must have been domesticated elsewhere and introduced into Africa at different times in the past. On current evidence, sheep (*Ovis aries*) and goats (*Capra hircus*) were first domesticated in south-western Asia (often referred to in older literature as the Fertile Crescent) around 9,500 to 10,500 years ago (Zeder and Hesse 2000). Domestic ovicaprines may have spread into Egypt along two separate routes: one along the northern coastal zone of Sinai into the Nile Delta, the other from the southern Sinai Peninsula across the Gulf of Suez. Domestic ovicaprines have been recovered from the sites of El Awag and Abu Gedar 2 in southern Sinai. These have been dated to around 6300 to 5200 cal BC and 5600 to 4900 cal BC respectively (Close 2002). In Egypt, some of the earliest remains from sites in the Eastern Desert have been dated to around 6000 to 5300 cal BC, while in the Nabta-Kiseiba area of the Western Desert ovicaprines first appear around 6000 to 5500 cal BC (see Riemer 2007: 107–13).

The exact wild progenitor of the one-humped Arabian or dromedary camel (*Camelus dromedarius*) is unknown but its range may have extended from western Asia to north-east Africa. Currently available genetic and biometric evidence suggests that the dromedary was first domesticated in central or southern Arabia between 6,000 and 3,400 years ago (Jianlin et al. 2000). However, the existence of other possible centres of independent domestication cannot be ruled out entirely (Mburu et al. 2002). Camel bones have been recovered from mid- to late-Holocene levels at Gobedra in Ethiopia (Phillipson 1977a) and Ele Bor in northern Kenya (Gifford-Gonzalez 2003). The dating of these finds is problematic, however (Marshall 2000: 211), and the earliest securely dated finds are probably those from Napatan levels at Qasr Ibrim in the Nile Valley, about 140 km south of Aswan, Egypt (Rowley-Conwy 1988).

With regard to cattle domestication, it is now generally accepted that there was more than one centre of domestication. These include the middle Euphrates Basin, where indigenous populations of the wild progenitor *Bos primigenius* were domesticated around 10,500 years ago (Götherström et al. 2005), giving rise to a new species, *B. taurus*; and the Indian subcontinent, where humped-back cattle (*B. indicus* or zebu), probably descended from *B. primigenius namadicus* or possibly *B. planiforms*, were already present in Baluchistan at sites such as Mehrgarh by about 9,000 to 4,500 years ago (Meadow and Patel 2003). Based on a combination of recent genetic, osteological and archaeological evidence, some scholars have argued that North African populations of *B. primigenius* were domesticated independently by groups occupying areas of the Sahara around 9,000 years ago during a period of climatic amelioration and higher rainfall at the start of the Holocene (e.g., Wendorf, Schild and Close 1984; Hanotte et al. 2002). Other researchers, however, have argued that the large bovid remains from sites in the Nabta and Kiseiba regions in the Egyptian Eastern Sahara dated to the period between 9,500 and 8,400 years ago which have formed the basis of claims for early, independent domestication, have not been definitively identified on osteometric grounds as being derived from domesticated cattle (Riemer 2007: 106; cf. Gautier 2007: 77–82). Moreover, far from indicating a drawn-out process of the gradual spread of domesticated cattle from the ninth millennium cal BC, the pattern of radiocarbon dates for clearly attested evidence of domestication instead points to 'a rapid expansion from the end of the seventh through the sixth millennium calBC' (Riemer 2007: 107).[3] These early populations of *B. taurus* cf. *africanus* most probably formed the ancestors of the first cattle in eastern Africa, while humped zebu cattle (*B. indicus*) are attested in the regional archaeological record by 1,800 to 800 years ago (Marshall 2000).[4] The crossbreeding of taurine and zebu cattle, probably initially in the northern part of eastern Africa, also gave rise to a hybrid breed, the Sanga, now widely distributed across many parts of sub-Saharan Africa.

Donkeys (*Equus asinus*), often overlooked in many studies of modern pastoralist societies partly because they are used for transport and porterage rather than food, and possibly partly because their management is typically undertaken by women rather than men,[5] were, on current evidence, domesticated around 6,000 years ago, probably in either Egypt or the Horn of Africa (Blench 2000). The wild ancestor of the donkey is the African wild ass (*E. africanus*) and two subspecies are known. These are the Nubian wild ass (*E. africanus africanus*), long believed to have been the original progenitor of the domestic donkey, and the Somali wild ass (*E. africanus somaliensis*). Recent genetic and biometric research on modern donkeys and African wild asses, coupled with evidence from the dating and distribution of domesticated donkeys in the archaeological record, has raised the possibility that the domestic donkey may in fact have separate maternal origins derived from both of these subspecies (Rossel et al. 2008).

Whereas the currently reconstructed centres of domestication for all five species lie outside the area under discussion here, their simple presence in eastern Africa does not necessarily imply that domestic livestock were introduced by human population movements. This is because other processes, including livestock exchanges and knowledge transfers, could as easily account for the gradual southward spread of these domesticates. However, this said, the case for population migration is given independent support by other data, including details of the linguistic histories of speakers of three (Afro-Asiatic, Nilo-Saharan and Niger-Congo) out of the four main language phyla of eastern Africa (e.g., Bender 2000; Hayward 2000; Ehret 2003); the genetic histories of modern eastern African populations as reconstructed from mitochondrial DNA and Y-chromosome markers, and the distribution of genetic tracers of lactose tolerance (e.g., Salas et al. 2002; Tishkoff et al. 2007; Henn et al. 2008); and archaeological evidence. The correlation of these different strands of evidence remains problematic, however (e.g., Krings et al. 1999), and it is increasingly clear that several different trajectories were followed. These were also shaped, to greater or lesser degrees, by changing regional climatic conditions and more localised environmental parameters.

Climate Change and Vegetation Mosaics: Macro-scale Models

In East Africa, the late Pleistocene and early Holocene (circa 12500 to 5000 bp) were considerably wetter than today,[6] as indicated by the widespread evidence for higher lake levels across the Rift Valley lakes and the Ethiopian crater lakes (Butzer et al. 1972; Gasse and Street 1978; Stager and Mayewski 1997; Dühnforth, Bergner and Trauth 2006) during an early Holocene humid optimum.[7] Although there was a drier phase centred around circa 10,000 bp, lake levels at Lake Turkana were still around 80 metres higher than they are today (Butzer et al. 1972). Here, Late Stone Age (hereafter LSA) hunter-gatherer groups began to focus activities around the lake and on the exploitation of fish and other lacustrine resources, possibly becoming more sedentary in the process (see below). Following the onset of drier conditions toward the end of the early Holocene, reduced rainfall resulted in the Rift Valley lakes (Naivasha, Nakuru and Elmenteita) at times drying out altogether circa 5500 to 3000 bp (Richardson and Richardson 1972; Stager and Johnson 2000). While this increased aridity may have made the area less attractive for settlement, climate and vegetation changes during the mid Holocene in East Africa may have actually facilitated pastoral immigration through the narrow Central Rift Valley of Kenya. This was a major avenue of entry from Lake Turkana to the plains of southern Kenya and northern Tanzania (Ambrose 1984a, 1998). In the warmer and wetter conditions of the early Holocene, the latter area would probably have been more wooded than today and grasslands suitable for grazing were likely restricted to areas close to the lake shores (Richardson and Richardson 1972). It has been suggested that tsetse fly (*Glossina* genus), the carrier of parasitic protozoa belonging to the *Trypanosoma* genus which cause sleeping sickness, would have thrived in these moister habitats, and so may have provided an effective barrier to the southward movement of both people and livestock (Gifford-Gonzalez 2000). However, with the end of the humid optimum, which from available palaeoenvironmental records appears to have occurred within the space of perhaps a couple of centuries in some areas, though not in others,[8] it has been argued that the reduction in rainfall and fall in lake levels would have encouraged the retreat of forest margins to higher altitudes allowing the expansion of grasslands more suited to grazing. During this dry period, the Central Rift Valley lakes reduced in size as lake levels fell, drying out completely between around 3400 to 3000 bp (Richardson and Richardson 1972). Moreover, pollen records from Lake Naivasha indicate there was a decline in forest and lowland trees beginning around 6000 bp and a corresponding shift toward open grassland habitats (Maitima 1991), while changes in soil carbon isotope ratios indicate that the savannah/forest boundary rose to above 2,600 metres from its previous level around 1,940 metres during the early Holocene humid phase (Ambrose and Sikes 1991). As a result, the tsetse belt may have shrunk, allowing the southward movement of people and livestock (Marean 1992).

Regional Archaeological Sequences

Archaeologists generally use the term Pastoral Neolithic (hereafter PN) to refer to the era in eastern Africa during which domesticated livestock were first integrated into subsistence economies. However, the term has also been employed by different archaeologists to convey rather different meanings,[9] and the very use of the word 'Neolithic' has been contested on the grounds that it encourages undue emphasis on herding as a subsistence strategy relative to the utilisation of other food resources, and a reliance on a few key (and supposedly diagnostic) artefact types (or *fossiles directeurs*) as both chronological and cultural markers (Karega-Mũnene 2002; 2003: 19–20). The definition of the PN as 'cultures of eastern Africa which (1) relied substantially on domesticated stock for their livelihood; (2) used pottery, and (3) employed typical Late Stone Age technologies for the manufacture of edged tools' (Bower and Nelson 1978: 562), is still perhaps the most concise, but masks considerable diversity in terms of the relative importance of livestock in subsistence strategies, preferred ceramic styles, the range of formal stone tools and associated manufacturing techniques and debris, site distributions and placement within the landscape, and associated burial practices and forms. Hence, as Wright has observed recently, 'labelling an East African Neolithic site as "Eburran" or "PN" is not a simple endeavour because the grey area between a purely foraging and a purely pastoral existence is large' (Wright 2007: 29). These points notwithstanding, and despite increasing recognition by archaeologists working in the region of the shortcomings of the term, it is still widely used as a shorthand in the literature to refer to the kind of archaeological traces covered by Bower and Nelson's definition. It is in this sense that the term is used here, but with recognition of its limitations.

Lake Turkana Basin

Overall, current evidence suggests that the transition to generalised, mixed cattle-and-sheep/goat pastoralism, had occurred in some parts of the northern lowlands bordering Lake Turkana by around 4200 bp (Barthelme 1985), around 3800 to 3400 bp in highland central, southern and western Kenya (Marshall 2000; Karega-Mũnene 2002; Marshall and Hildebrand 2002), and by around 3800 bp in the Tsavo and middle Sabaki River areas of south-eastern Kenya (Kusimba and Kusimba 2005; Wright 2005, 2007), although in all localities the dated evidence for the first presence of domesticated small stock is somewhat earlier. As discussed above, the initial introduction of livestock was due in part to the southward expansion of small groups of herders from Sudan, Ethiopia and possibly Somalia. The historical linguistic evidence suggests that these early pastoralists were speakers of proto-Southern Cushitic languages (Ehret 1998). Nevertheless, as stressed by Marshall (2000) and Marshall and Hildebrand (2002), the transition to food production in East Africa was a very uneven process characterised by often quite localised uptake and stretched over several millennia. The available evidence certainly suggests that initially the southward spread of herding economies was more of a 'trickle' than a full-blown 'wave' or 'splash' (Bower 1991), involving alternating moving and static 'frontiers' between herders and foragers (Lane 2004; Prendergast 2011), with the latter possibly obtaining stock for themselves as often through theft or subterfuge (Ambrose 1984a) as by exchange. Equally, there were times when, and places where, herders combined their activities with the cultivation of crops, and/or the hunting of wild animals, fishing and gathering of plant foods, and others when they focused their productive activities more exclusively on keeping and raising livestock. This frequent and recurrent 'identity' shifting also seems to have a long ancestry since, even during the initial phase of livestock domestication in the region, pottery and stone tool styles exhibit considerable variation (Table 3.1). This would suggest that far from there being a single 'unified cultural group', the PN landscape was comprised instead of numerous 'autonomous groups … in contact with one another who contemporaneously occupied a contiguous territory' (Wright 2007: 29).

Recently, Marshall and colleagues have suggested that the spread of domestic stock and the changes in material culture, settlement strategies, burial practices and inferred social relations that accompanied this southward movement can be divided into two phases: an initial exploratory period dating roughly to between 5,000 and 3,500 years ago, and a later period of consolidation, settlement and social fragmentation between about 3,500 and 2,000 years ago (Marshall, Grillo and Arco 2011). During the former, communities with livestock, who were probably already experienced herders, moved into the area of the Lake Turkana Basin in response to the increasing desiccation of the Sahara, which had already begun, in the central parts of the Egyptian Western Desert, by around 5300 cal BC and was to continue for several millennia (Kuper and Kröpelin 2006: 805–7). These pioneer communities interacted with local fisher-hunter-gatherers whose economic strategy was largely based on the hunting of terrestrial animals (typically medium to small bovids such as oryx, wildebeest, Grant's gazelle and warthog), fishing and, perhaps less frequently, hunting aquatic mammals such as crocodile and hippopotamus (Stewart 1989: 160–69). Stewart (ibid.: 227–35) has argued that fishing was largely opportunistic in nature and conducted as part of a seasonal round with an emphasis on Nile perch (*Lates*) and the larger cichlids (tilapias), and to a much lesser extent mudfish (*Clarias*). These early Holocene fisher-hunters used harpoons and spears as part of their fishing technology, as

Table 3.1: Approximate dates of Pastoral Neolithic and Pastoral Iron Age pottery traditions (based on Gifford-Gonzalez 1998a, 2000; Karega-Mũene 2002).

Age bp (kyr)	Archaeological periodisation	Selected LSA stone tool industries	LSA, PN and PIA ceramic traditions – i.e. associated with the hunter-gatherer-fisher and pastoral populations discussed in the text	Iron Age ceramic traditions – i.e. associated with iron-using and farming populations
0	Historically documented ethnic groups			
1	Later IA and Pastoral IA		Kisima c. 750–200 bp Lanet/Sirikwa c. 1200–300 bp (PIA)	MIA and LIA regional variants
2	Early Iron Age – overlaps with PN and LSA	Elmenteitan	Turkwel c. 1800–950 bp Akira c. 1900–1200 bp Maringishu ?? to c. 1700 bp Elmenteitan c. 3300–1300 bp Narosura c. 2800–1400 bp	Kwale Urewe
4.5	Pastoral Neolithic – overlaps with later LSA	Eburran 5 LSA 'Bone harpoon' sites	Nderit and Ileret c. 4500–1500 bp (PN)	
6	Late LSA	Eburran 4		
8			Kansyore c. 8250–1200 bp (LSA)	

attested by the recovery of bone points and barbed harpoons from several early Holocene sites around Lake Turkana, such as Lothagam (Robbins 1980) and Lowasera (Phillipson 1977b) (Figure 3.2). It is perhaps significant that these artefact types are not found in secure contexts at the mid-Holocene sites associated with the earliest evidence for domestic livestock, and Stewart (1989: 235), among others, lists this as one of the indicators that the early herding communities may have been intrusive. In addition to these bone tools, the early-Holocene occupants of the lake edge also produced stone tools (notably backed microliths, *outils écaillés* and core tools) on local chert and lava, with rare use of quartz and obsidian (Barthelme 1985: 273–75), and made pottery decorated with 'wavy-line' motifs similar, but not identical, to pottery associated with hunter-fishing sites found along the Nile in the Sudan and in parts of the Sahara in southern Egypt (cf. Kuper and Kröpelin 2006).

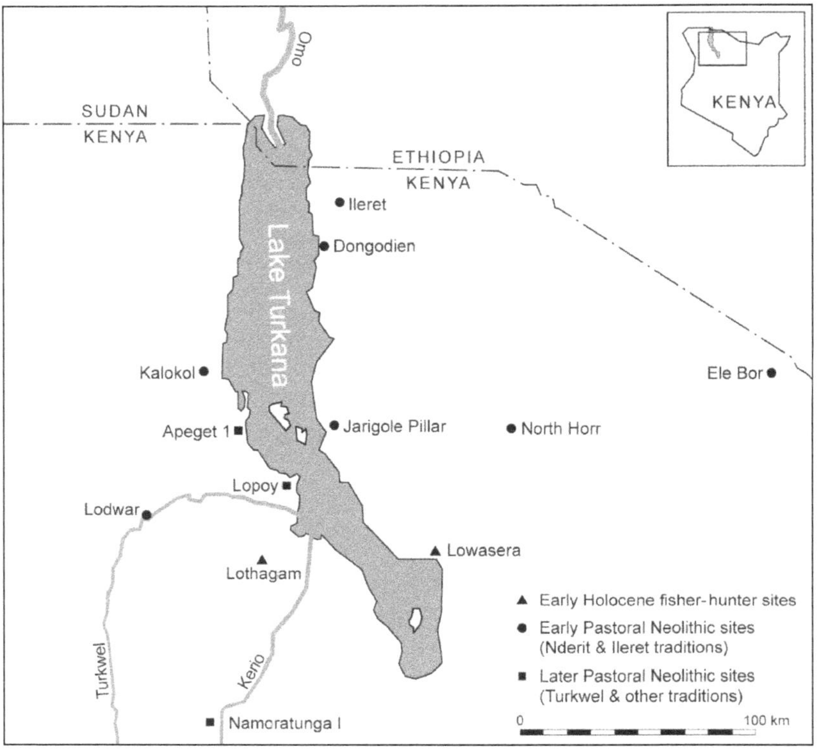

Figure 3.2: Location of Holocene sites in the Lake Turkana Basin mentioned in the text.

Around 4500 to 4100 bp, at least two new pottery traditions appear for the first time in the archaeological record of the Lake Turkana Basin. Termed 'Nderit' and 'Ileret' pottery, these are distinguished in terms of the manner in which they were decorated, the range of decorative motifs employed and the preferred shape(s) of the vessels produced (Figure 3.3.1 and 3.3.2). The earliest dated remains of domestic stock, both cattle and ovicaprines, are found at sites associated with these new pottery traditions, such as North Horr, Dongodien and Ileret. These early herders continued to exploit fish, but in contrast to the LSA inhabitants of the basin, a more diverse range of taxa, typically at least eight, were exploited. The use of stone tools continued, with obsidian, probably obtained from local sources (Merrick and Brown 1984: 148), becoming the preferred raw material. Another development is the first appearance of stone bowls, made by a combination of pecking and grinding pieces of lava, pumice and similar 'soft' stones. Taken together, these changes in the archaeological record lend support to the hypothesis that there was population migration into the Lake Turkana Basin and that these newcomers brought with them domestic stock and new ways of exploiting the landscape.

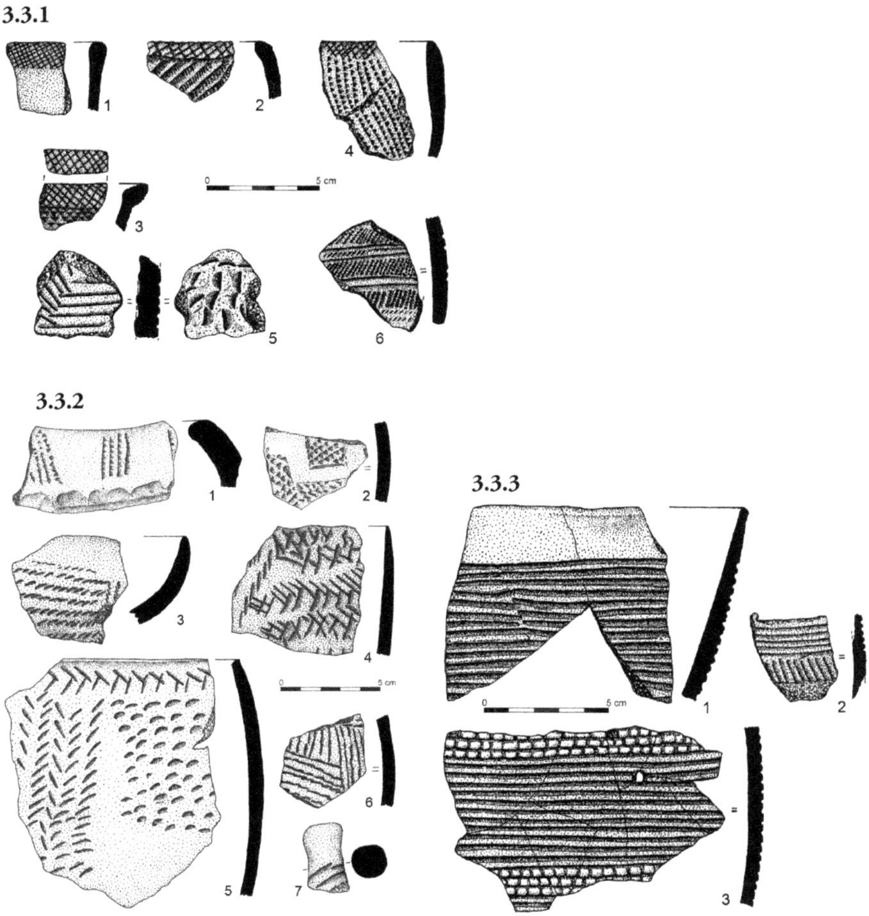

Figure 3.3: Examples of Pastoral Neolithic pottery from the Lake Turkana Basin.

3.3.1: Northern Nderit pottery: 1–4 – decorated rims; 5–6 – decorated body sherds. 1–3,5 from Dongodien, 4,6 from East Turkana. After Barthelme (1985); 1: 160, Fig. 61/d; 2: 158, Fig. 59/b; 3: 160, Fig. 61/c; 4: 268, Fig. 98/e; 5: 159, Fig. 60/a; 6: 268, Fig. 98/f. © John Barthelme, reproduced with permission.

3.3.2: Ileret pottery: 1,3–5 – decorated rims; 2,6 – decorated body sherds; 7 – decorated lug. 1–2,7 from Ileret; 3–6 from East Turkana; 7 from Dongodien. After Barthelme (1985); 1: 201, Fig. 76/a; 2: 201, Fig. 76/e; 3: 271, Fig. 101/c; 4: 271, Fig. 101/a; 5: 270, Fig. 100/a; 6–7: 201, Fig. 76/g. © John Barthelme, reproduced with permission.

3.3.3: Turkwel pottery from Lopoy. 1 – decorated rim; 2–3 – decorated body sherds. After Lynch and Robbins (1979: 326, Fig. 3/a.b.d). © Cambridge University Press, reproduced with permission.

The appearance of domestic livestock and a new artefactual repertoire from around 4500 to 4100 bp in the Turkana Basin also seems to correlate well with a phase of fluctuating lake levels between higher and intermediate positions relative to today as attested by the date of various fossil shorelines (Owen et al. 1982: 524–25), the diatom record of sediment cores from the southern basin (Halfman et al. 1992), and changes in carbon and oxygen isotope ratios (specifically $\delta^{13}C$ and $\delta^{18}O$) in the sediment record (Johnson, Halfman and Showers 1991). These trends were probably associated with greater humidity on the Ethiopian Plateau, since lake levels at Lake Turkana are strongly influenced by the rate of inflow from the Omo River at its northern end. Environmental studies also indicate that there was a marked drop in lake levels from around 3900 bp, although unlike the Rift Valley lakes further south, Lake Turkana was never fully desiccated. Nevertheless, the lake level dropped by about 30 metres over this period, probably contributing to the abandonment of the hunter-fisher encampment sites (such as those at Lowasera and Lothagam) near the 80 metre contour. The fall in lake level 'was followed by a 2000 y[ea]r long shift in atmospheric circulation over the Turkana watershed that has been sustained to the present' (Ricketts and Johnson 1996: 15), as well as a change from an open to a closed basin hydrology and a corresponding increase in lake salinity. The establishment of a bimodal seasonal distribution of rainfall across the region is believed to be associated with this shift in atmospheric circulation. One consequence may have been an increase in alternating El Niño related events of higher rainfall and La Niña related periods of extreme drought. These are both related to variations in sea surface temperatures (SSTs) in the western Indian Ocean (higher for El Niño, lower for La Niña events), which in turn result in either a strengthening or weakening of the winter monsoons (Nicholson and Kim 1997). Wright (2007: 27–28) has argued, with particular reference to the Tsavo area of south-west Kenya, that because the higher frequency of El Niño related events during the mid to later Holocene would have reduced the predictability of seasonally available resources to hunter-gatherers, this may have encouraged changes in food procurement strategies.

Further changes in subsistence strategies and ceramic styles are evident in the archaeological record of the Lake Turkana Basin from around 1800 bp as indicated by the appearance of a new pottery tradition (Figure 3.3.3), mostly on sites in the Turkwel area in the southwest of the Basin, and hence the designation 'Turkwel tradition' (Lynch and Robbins 1979; Robbins 1984). However, possible Turkwel ceramics have also been reported from around Lake Baringo at the site of Ngenyn (Hivernel 1976), and parallels can be drawn between this ceramic and pottery recovered from Jebel Kathangor in southern Sudan (Robertshaw and Mawson 1981) and North Horr on the east side of Lake Turkana (Phillipson 1979). The site of Lopoy, located about 7 kilometres from the current shoreline of Lake Turkana, is one of the key sites (Robbins 1980). This was occupied around 1000 to 950 bp and has two main components distinguished by differences in the ceramics, lithics and faunal material. It covers a very large area (about 650 by 150 metres) that includes several middens, hearths, butchery areas and ashy patches that may represent burned livestock enclosures. Domesticated ovicaprines and bovids are present in the faunal assemblage. However, several bone concentrations point to the intensive exploitation of lacustrine resources, such as crocodile and Nile perch. Whether the site was utilised by a single community practising a mixed pastoral, hunting and fishing economy or by separate hunting-fishing and pastoral communities is a matter of debate. Other Turkwel sites in this area include Apeget 1, dated to about 1800 bp.

Central Rift Valley

The final phases of the LSA in central Kenya date to around 7000 to 1700 bp and are associated with 'Eburran' stone-tool-flaking technologies (Ambrose 1985, 2003).[10] In its initial phases at least, the Eburran is associated with a purely hunting-gathering subsistence economy. Evidence from several excavated rock-shelters and cave sites indicates that these LSA hunter-gatherer groups were highly mobile populations who lacked pottery, exploited a range of habitats and a variety of medium- to large-sized animals, and participated in wide-ranging exchange networks that gave them access to different lithic raw materials, such as obsidian and chert (Ambrose 1998; Kusimba and Kusimba 2005). The evidence also suggests that there were different adaptations to highland and lowland environments, with the groups utilising the former tending to select caves and rock-shelters situated close to the boundary between montane forest and grassland habitats, while the latter generally occupied open-air sites on the savannahs. Faunal assemblages from excavated sites situated in these different ecological zones may also reflect a preference for species that are typically found in these different habitats. For example, although the faunal assemblage throughout the entire LSA sequence at the site of Enkapune ya Muto rock-shelter, located on a steep slope about 2,400 metres above sea level on the Mau Escarpment to the southeast of Lake Naivasha (Figure 3.4), is dominated by the remains of smaller bovids (Marean 1992: 102), during the earliest phase of Eburran occupation commencing around 6350 bp most of the species were derived from a montane forest environment. In contrast, and consistent with independent evidence for drier climatic conditions and upward shift of the forest/savannah ecotone after the mid-Holocene humid phase (see above), the Eburran phase 4 and early phase 5 levels contain a higher proportion of bush and open habitat species (Ambrose and DeNiro 1989; Marean 1992). There also appears to have been a hiatus in occupation[11] of rock-shelters near the floor of the Rift Valley (for example, Lion Hill Cave, Gamble's Cave, Naivasha Railway Shelter, Salasun, Masai Gorge Rock-shelter) between about 6000 to 3000 bp, which may well be indicative of a marked decrease in the exploitation of lower elevations (Ambrose 1985: 62).

Honey, although not yet detectable in the East African archaeological record, may have been an important source of carbohydrates in the diet of these hunter-gatherers, much as it was among Okiek groups in the more recent past and to an extent also today (Blackburn 1971). Honey may also have been exchanged

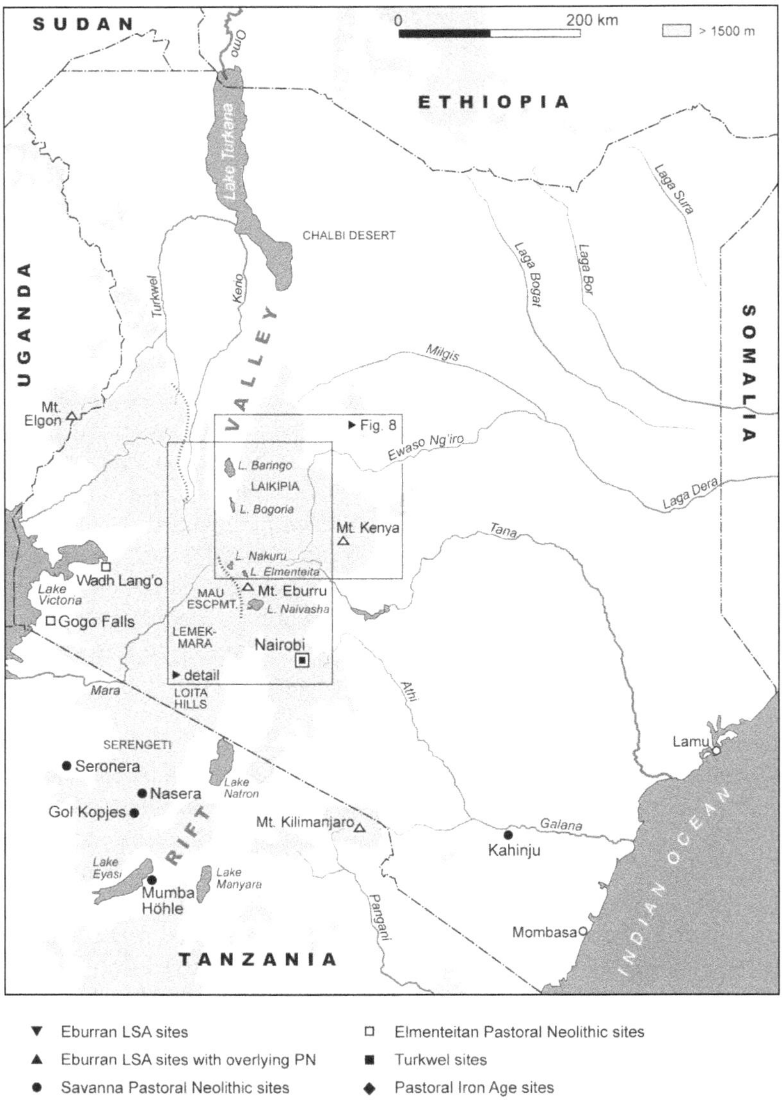

Figure 3.4.1: Location of mid- to late-Holocene sites in the Kenyan Central Rift Valley and adjacent areas mentioned in the text.

and used in various rituals, again based on analogies with historically documented hunter-gatherer groups such as the Okiek and Mukogodo. Additionally, honey is known to have medicinal effects – for instance, it has good antibacterial properties and has been used in more recent times by Maasai and others as a wound-healing salve (Bodeker, Ryan and Ong 1999). Studies of recent hunter-gatherers occupying high woodland and forest areas of eastern Africa indicate

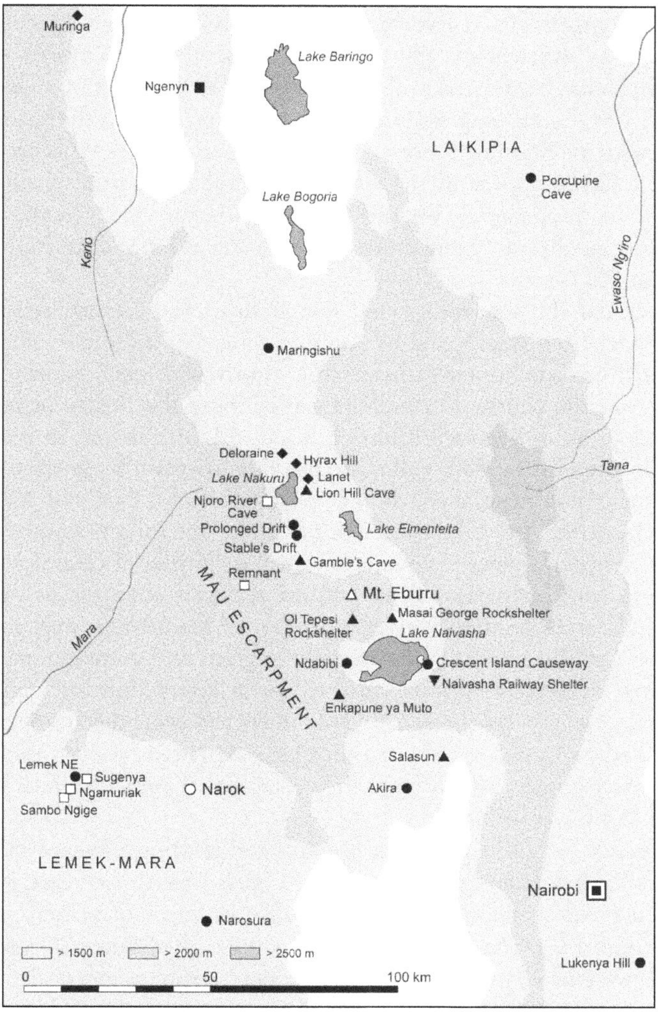

Figure 3.4.2: Detailed map of site locations in the Central Rift Valley.

that honey extraction is often organised on an individual basis, with private ownership of hives and exclusive rights over territory (Mutundu 1999: 69). While there is no way of establishing whether similar principles applied during the later LSA, these observations raise the possibility that the resource tenure system of later Eburran hunter-gatherers may have meant that there was less resistance to the adoption of domestic stock and an associated system of individual or family ownership of subsistence resources than has been noted among hunter-gatherer societies that place a strong emphasis on communal ownership (ibid.: 67–77).

In this regard, it is notable that the Enkapune ya Muto rock-shelter has produced some of the earliest dates for the presence of domestic stock in southern

Kenya. This comprises just a few ovicaprine teeth associated with typical Eburran 5 lithics and an overwhelmingly wild faunal assemblage recovered from level RBL2.1, which has been dated to about 4500 to 4300 bp. Pottery, including samples of the Nderit and Ileret traditions,[12] first occurs lower in the sequence, also associated with an Eburran 5 stone tool industry, at around 4860 bp (Ambrose 1998: 381). Cattle appear in the sequence at Enkapune after circa 3400 bp, although domestic stock only become dominant after around 3280 bp, by which time the associated pottery and stone tool industry is more characteristic of the Elmenteitan PN (see below).

The particular characteristics of the faunal, lithic and ceramic assemblages at Enkapune have been interpreted by Ambrose (ibid.) as indicative of interaction by an autochthonous hunter-gatherer community with neighbouring pastoralist groups over the course of a millennium or more that led ultimately to the full-scale adoption of livestock. Other Eburran 5 LSA horizons in rock-shelters elsewhere in the Central Rift Valley are similarly overlain by horizons containing PN Elmenteitan material, such as at Gamble's Cave, Lion Hill Cave, Masai Gorge Rock-shelter, and Ol Tepesi Rock Shelter. The long period between the initial occurrence of domesticates and/or small quantities of ceramics associated further north with early pastoralists, and the full-scale adoption of a pastoralist economy and corresponding changes in artefactual assemblages may point to the existence of a cultural 'frontier' between immigrant and autochthonous groups that involved degrees of various kinds of social interaction but neither complete assimilation nor integration.[13] Over time, however, the steady expansion of herders had an impact on the resource base exploited by the autochthonous hunter-gatherer groups, which ultimately restricted them to habitats less suited to herding (Ambrose 1984a).

The evidence for a distinct stock keeping and herding presence in the Central Rift Valley landscapes after about 3000 bp is much more extensive, albeit such evidence exhibits considerable variability in terms of pottery traditions, stone-tool industries, the ratio of wild to domesticated fauna in assemblages, the overall diversity of artefact assemblages, site locations, and evidence for the exploitation and exchange of different obsidian sources.[14] The evidence from the central Kenya Rift Valley and adjacent highlands, as well as from the Serengeti and the foothills of Mount Kilimanjaro, northern Tanzania, is conventionally divided into three broad groupings.[15] These are the Savannah Pastoral Neolithic (SPN), Elmenteitan and Eburran 5 (Ambrose 2003). Within these categories there is considerable inherent diversity and 'fuzziness' in the material traditions of broadly contemporary sites in adjacent areas. Thus, for instance, the Narosura, Prolonged Drift and Crescent Island Causeway sites have all been assigned to the SPN, yet in terms of their faunal assemblages there are some marked differences between them in the proportion of wild to domesticated fauna, and in the relative proportions of cattle to ovicaprines among the domesticated component. Moreover, although Elmenteitan and SPN material tends to occur in 'adjacent, and mutually exclusive geographical ranges' (Gifford-Gonzalez 1998b: 127), it remains far from clear which aspects (if any) of the three groupings represent the material expression of different 'cultures', 'ethnicities' or language clusters,[16] and which are due to variations in the preferred economic and subsistence strategy of the occupants of individual sites, the opportunities provided by the specific environmental setting of each site (Robertshaw and Collett 1983a, 1983b), and/or even the changing fortunes of different communities (see, e.g., Gifford, Isaac and Nelson 1980: 86–99; Mutundu 1999: 78–85). Some of the key aspects of the SPN and Elmenteitan groupings are summarised below.

Savannah Pastoral Neolithic

This has the widest spatial distribution and has been subdivided into Lowland SPN (found in northern Kenya at elevations below 1,100 metres) and Highland SPN (found in central and southern Kenya and northern Tanzania, above 1,500 metres) (Ambrose 1984a). More recently, Ambrose (2003: 106) has proposed an alternative division of the SPN into two macro traditions, one comprising Nderit and Ileret ceramic traditions, and the other comprising Narosura, Akira and Maringishu ceramic traditions. The former might represent an exploratory phase, while the latter could be associated with a period of consolidation (cf. Marshall, Grillo and Arco 2011).

The SPN in the Central Rift Valley spans roughly 3400 to 1300 bp; however, dating of the various components is poorly resolved. SPN sites are generally found in vegetation zones associated with open habitats, such as along the Central Rift Valley floor, the Loita Plains of south-western Kenya, and the Serengeti and western Kilimanjaro, northern Tanzania, between about 1,500 and 2,050 metres above sea level. Settlement sites are typically large and there seems to have been a preference for locating these in ecological and topographical settings similar to those preferred by modern Maasai (see Western and Dunne 1979), such as on gently sloping ridges and hills, and on well-drained soils. While SPN ceramics occur at sites with a high percentage of wild fauna (for example, Prolonged Drift, see below), in general terms SPN faunal assemblages tend to have high proportions of cattle and ovicaprines relative to wild game. Inhumation burial under stone cairns, in rock-shelters and in rock-filled crevices formed the dominant burial tradition. Accompanying grave goods can include stone bowls, pestles, ochre palettes and large obsidian blades. The obsidian typically comes from sources in the Naivasha and Nakuru-Elmenteita basins in the Central Rift Valley (Merrick and Brown 1984: 145–48). Stone bowls, various polished stone artefacts and rubbing stones also occur on settlement sites. Polished and ground stone 'axes', some of which may have been used for horn shaping and straightening (Brown 1990) while others may have been used as hoes (Robertshaw and Collett 1983b: 71–73), also occur. SPN sites are associated with a range of different ceramic traditions, some of which may be contemporary while others may represent later developments (Figure 3.5). SPN ceramics also occur at several sites (such as Enkapune ya Muto, Hyrax Hill, Masai Gorge) in association with Eburran 5 lithics.

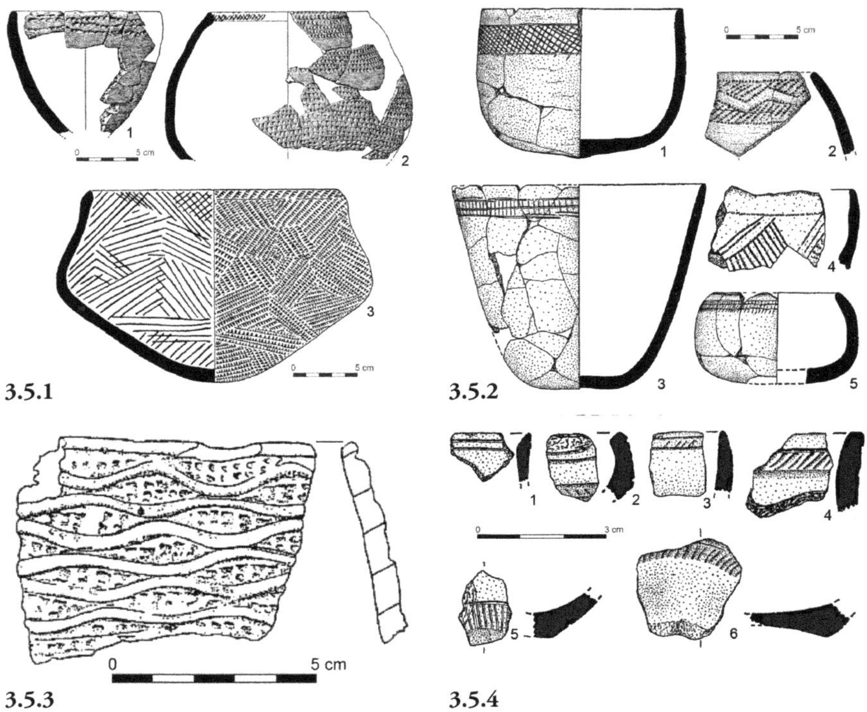

Figure 3.5: Examples of Savannah Pastoral Neolithic pottery from the Central Rift Valley.

3.5.1: Nderit pottery. 1 – Wide-mouthed, tapered beaker from Hyrax Hill 'Neolithic' site (Leakey 1945: 309, fig. 9/3); 2 – highly decorated bowl with incurving rim from the same site (ibid.: fig. 9/4); 3 – schematic drawing of a decorated bowl from Stable's Drift (Sutton 1964: 29, fig. 3/1). © Taylor and Francis (nos. 1 and 2) and John Sutton (no. 3), reproduced with permission.

3.5.2: Narosura pottery. 1 – decorated hemispherical bowl; 2 – rim with comb-stamp decoration; 3 – decorated beaker; 4 – rim with comb-stamp decoration; 5 – decorated narrow-mouthed bowl; 6 – decorated bowl with slightly everted rim. 1–3, 5 and 6 from Narosura; 4 from Tsavo East. Sources: Odner (1972); 1: 65, Fig. 24/i; 2: 61, Fig. 21/b; 3: 61, Fig. 21/e; 5: 61, Fig. 21/d; 6: 61, Fig. 21/c; 4: Wright (2003: 185, Fig. 1). © BIEA & individual authors, reproduced with permission.

3.5.3: Decorated Maringishu rim sherd from Maringishu (Bower et al. 1977: 130, fig. 2/f). © BIEA & individual authors, reproduced with permission.

3.5.4: Akira pottery from Seronera. (1–4 – decorated rims; 5–6) – bases. After Bower (1973); 1: 96, Fig. 11/g; 2: 96, Fig. 11/h; 3: 96, Fig. 11/f; 4: 96, Fig. 11/k; 5: 96, Fig. 11/c; 6: 96, Fig. 11/d. © BIEA & individual authors, reproduced with permission.

The earliest ceramics belong to the Nderit tradition (Figure 3.5.1).[17] Sites with Nderit pottery in southern Kenya include Stable's Drift, Hyrax Hill, Enkapune ya Muto, Lukenya Hill GvJm3 and GvJm44, and Salasun; and in northern Tanzania, Seronera, Nasera, and Gol Kopjes. At both Enkapune ya Muto and Gol Kopjes, wild fauna dominate the assemblages (and at Gol Kopjes no domesticates were encountered). Gifford-Gonzalez (1998a: 186–90) has noted that whereas Nderit pottery occurs around Lake Turkana in association with elaborate funerary sites and an accompanying range of other artefactual material, this is not the case in the Central Rift Valley. Instead, this material is often found on sites with scanty evidence of domestic stock and in association with autochthonous hunter-gatherer stone-tool industries. She argues that this may indicate that in the Central Rift Valley area Nderit ceramics could have been used as prestige items acquired by hunter-gatherer groups through exchange.

Narosura tradition ceramics (Figure 3.5.2)[18] have been dated to between about 3000 to 1600 bp. Other than the type site, sites with Narosura pottery include Prolonged Drift, Salasun, Ndabibi, Crescent Island Causeway, Maringishu and possibly Lemek North-East (all Kenya). Maringishu tradition ceramics (Figure 3.5.3)[19] are found in the Lake Nakuru basin and immediately adjacent areas. Only a few sites are known; they include Maringishu and Hyrax Hill (Bower et al. 1977; Karega-Mũnene 2003), and Kahinju on the Galana River in Tsavo (Wright 2005). The only dated site is Maringishu itself, which was occupied around 1700 bp, where it occurred in association with Narosura pottery and a largely cattle and ovicaprine faunal assemblage. On the Laikipia Plateau, Maringishu ceramics have been found in association with Akira material (Siiriäinen 1984). Pottery assigned to the Akira ceramic tradition (Figure 3.5.4) was first identified at Seronera in the Serengeti and site GvJM 44 at Lukenya Hill, south-east of Nairobi, Kenya. It is also known from Nasera, Mumba Höhle and other sites around Lake Eyasi, northern Tanzania. Available radiocarbon dates suggest it was produced between about 2050 to 1000 bp (Wandibba 1980), and it may well represent the final stage of PN expansion (Bower 1991). Examples of this material have also been found in the Lemek-Mara region of south-western Kenya (where the material occurs in the same contexts as Elmenteitan lithics), and on the Laikipia Plateau (Siiriäinen 1984; Robertshaw 1990: 198–200). As with Nderit pottery, it has been found on sites with a predominantly wild faunal assemblage (especially at sites in the Serengeti ecosystem), and in view of its quite wide geographical distribution it may well have been a highly valued exchange item that was perhaps made by hunter-gatherers (Robertshaw 1990: 200, 294), and possibly even had ceremonial uses (Soper 1989).

Elmenteitan

Most sites with Elmenteitan material are found in a band stretching from the Naivasha-Nakuru area across the Loita Hills and the Mara Plains as far as the south-eastern margins of Lake Victoria. However, Elmenteitan material has also been reported from the Laikipia Plateau on the eastern side of the Kenya Rift Valley (see below). A range of locations were utilised; these include open sites situated in grassland and bushland habitats, and rock-shelters. The open sites are often large (in the Lemek-Mara area, for instance, they are significantly bigger than modern Maasai settlements), and are typically single component sites suggesting relatively short occupancies. Along the Mau Escarpment open sites occur at higher elevations while rock-shelters are found at lower altitudes (Robertshaw 1988: 62–64), whereas around Lake Naivasha they are typically found above 1,900 metres near the savannah/forest ecotone (Ambrose 1984a, 1985). Most sites date to between about 3300 and 1200 bp.

The faunal assemblages from sites in the Central Rift Valley (such as Masai Gorge) and especially the neighbouring Loita-Mara region (for instance, Ngamuriak, Sugenya, Sambo Ngige), are indicative of subsistence economies heavily reliant on the intensive use of domestic stock, even in areas where wild game was probably abundant (Robertshaw 1988; Marshall 1990a, 1990b). Assemblages from the lower elevation rock-shelters along the Mau Escarpment (such as Enkapune ya Muto), on the other hand, are dominated by ovicaprines, and their occupation may well have been on a seasonal basis.

Analysis of the faunal material from Ngamuriak indicates a herd management and culling strategy focusing on adults and young adults. The bones of both cattle and small stock were highly fragmented, suggesting intensive use to extract bone nutrients. The combined evidence from the site also implies an absence of stress, with the inhabitants able to feed surplus animals until they reached full meat weight (Marshall 1990b). The large number of mature animals in the assemblage has also been interpreted as evidence for considerable dietary reliance on milk. This development may have been related to the evolution of a bimodal rainfall regime, which would have allowed reduced mobility and so reduced the stress experienced by cattle under more nomadic conditions which prevents them from lactating (Marshall 1990a; 1994: 34).

In terms of their size, Elmenteitan sites in the Lemek-Mara region exhibit a bimodal, or possibly tripartite, distribution, ranging in extent between about 10 to 230 metres in diameter. The majority are relatively small (about 30 to 120 metres in diameter), with a few being at least twice this size. The smaller sites have only localised dung accumulations, or lack these altogether, while the largest sites (Sugenya and Oldorotua 3) have extensive dung deposits. Artefact density is variable, but appears not to correlate with site size (Robertshaw et al. 1990: 41–43). On the basis of these patterns, Robertshaw (1990: 293–99) suggested that while the differences may simply relate to site function or economic specialisation (with the smaller sites being more associated with agricultural production, as seems to be implied by the nature of their faunal assemblages), they may also reflect differences in the wealth and status of the site occupants. Thus, for instance, at Sugenya there is evidence of comparatively high slaughter rates of juvenile cattle, which could be indicative of conspicuous consumption on the part of wealthier settlements or individual herders (ibid.). Moreover, Sugenya and Oldorotua 3 occupy rather different ecological and topographic settings, yet they are the only sites in this systematically surveyed area to have evidence of extensive livestock enclosures, suggesting that other, more socially determined factors allowing for selective accumulation of livestock might account for their size. In particular, Robertshaw argued that differential access to exchange networks, and particularly supplies of obsidian for stone tool manufacture from Central Rift Valley sources, may have given the occupants of Sugenya and Oldorotua a competitive edge over

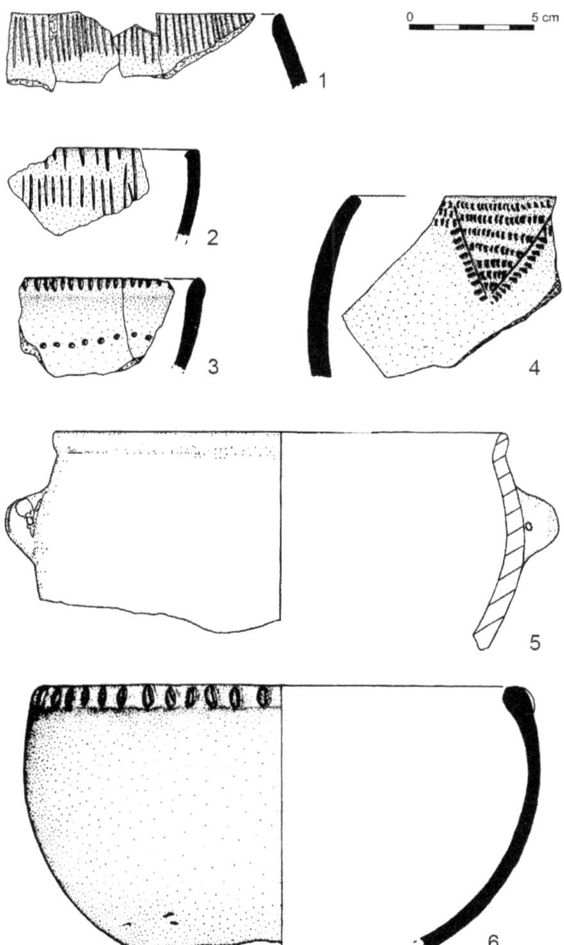

Figure 3.6: Pastoral Neolithic pottery of Elmenteitan tradition. 1–3: decorated rims from Ngamuriak; 4: decorated rim (late variant) from Deloraine; 5: wide-mouthed pot with pierced lug-handles from Mau Narok; 6: hemispherical bowl with nicked rim from Ngamuriak. *Sources*: 1–3: Robertshaw (1990: 191, fig. 9.5/c, e & g); 4: Collett (1984: 89, fig. 3/9); 5: Bower et al. (1977: 139, fig. 6); 6: Robertshaw (1990: 188, fig. 9.2/m). © BIEA & individual authors, reproduced with permission.

those of the smaller sites and helped lay the basis for a nascent political hierarchy (ibid.: 296–97).

Simons (2005: 38–40) has also argued that inter-site variation in culling patterns and relative proportions of cattle to ovicaprines may be indicative of wealth differences, although her data contradict aspects of Robertshaw's model. Thus, for example, she suggests that the occupants of Ngamuriak (which belongs to the 'small' class of sites) and Sugenya (one of only two 'large' sites), may have been wealthier than occupants of other sites since both have a higher ratio of cattle to ovicaprines, whereas faunal assemblages from other 'small' sites such as Oldorotua 1 (see also Siiriäinen 1990) exhibit higher ratios of ovicaprines to cattle. However, such differences are complicated once other variables are considered. Thus, for instance, at both Ngamuriak and Oldorotua 1, kill-off patterns suggest that livestock were raised to maximum meat weight before slaughter, implying that in neither case were the site economies under particular stress. Moreover, although the butchery practices at all three sites (Sugenya, Ngamuriak and Oldorotua 1) were similar, more juvenile cattle were culled at Sugenya (about 46% under one year old) than at Ngamuriak (about 12%). As Simons (2005: 39) notes, various scenarios might account for these differences, and far more research on inter-assemblage variations is needed before differences in status or other socio-economic characteristics can be asserted with confidence.

Sites closer to Lake Victoria, however, such as Gogo Falls (Robertshaw 1991; Marshall and Stewart 1995) and Wadh Lang'o (Lane et al. 2007; Prendergast 2008), where Elmenteitan levels directly overlie LSA Kansyore horizons, contain more mixed wild and domestic assemblages, as well as evidence for fishing and shellfish collection. It has been suggested that the degree of diversity at Gogo Falls may have been a necessary adaptation to an environment where tsetse fly was locally prevalent (Marshall and Stewart 1995); alternatively, the evidence could be taken as indicative of '"poor" pastoralists in the process of re-establishing their herds after some sort of calamity' (Robertshaw 1990: 300). The gathering of wild plant foods is also attested at Gogo Falls, and may well become more widely documented as sampling for botanical remains becomes more regularly practised. A great many sites are located in areas where the cultivation of domestic crops would have been viable, and grinding and rubbing stones have been recovered from several sites (Robertshaw and Collett 1983b). The presence of grinding and rubbing stones in burial contexts may be indicative of the value attached to the activities associated with the use of these objects. Limited stable isotope analyses of human skeletal remains from possible Elmenteitan burials also point to an even mix of plant and animal foods in local diets (Ambrose and DeNiro 1986) although crop remains have still to be recovered from archaeological contexts. By around 1800 bp, iron-using cultivators of Bantu stock would certainly have been present within the general area, and both crops and iron may have been obtained from them in exchange for livestock and possibly other products.

The Elmenteitan stone-tool technology differs from the Eburran 5 industry in terms of form, flaking technique and typology. The production of long, broad, thin blades using the punch technique on prepared cores with small, coarsely abraded striking platforms is the main diagnostic feature of the industry (Ambrose 2003). Ground stone material includes stone bowls, rubbing stones and stone palettes. Compared with SPN pottery, Elmenteitan ceramics (Figure 3.6)[20] exhibit far less spatial and temporal diversity. A further difference between Elmenteitan and SPN is in the preference for cremation rather than inhumation burial. Like SPN inhumations, these burials are sometimes accompanied by stone bowls (typically, these are broken, however), utilised obsidian blades, and stone pestles and grinding stones. Pottery and powdered red ochre have also been recovered from some cremation burials. The richest assemblage, comprising seventy-eight burials, comes from Njoro River Cave. The excellent preservation conditions allowed a detailed reconstruction of burial practices, which seem to have involved several stages. Prior to burial the corpse was wrapped in skins and bound in a contracted position. Another feature of these burials, and which seems to be an Elementeitan trait, is that the central incisors of the lower jaws had been removed (presumably during life). The body was then placed in a shallow pit along with a range of grave goods. The recovered material included eight hundred drilled stone beads, over four thousand sedge-seed beads, pendants made from quartz, chalcedony, agate and other types of stone, and obsidian blades. A carved wooden vessel, gourds, pieces of basketry and leather, mats and cordage were also recovered during excavation. The burial may well have then been covered by a thin layer of soil before being burnt. At some stage in the process, red ochre was sprinkled over the burial. The site was clearly used repeatedly over a period of time as several of the grave pits have been cut by later additions (Leakey and Leakey 1950). The four radiocarbon dates available for the site all cluster around 3000 to 2900 bp (Merrick and Monaghan 1984).

Pastoral Iron Age

From around 1900 to 1300 bp, there was yet another shift in the Central Rift Valley and adjacent areas back to more highly mobile settlement strategies and an increased reliance on foraged wild resources, even among groups of previously specialised pastoralists (Bower 1991). Further changes probably occurred around 800 to 900 years ago following the adoption of iron manufacturing technologies among pastoralist communities of the Central Rift Valley (Collett and Robertshaw 1983). The processes by which this technological change occurred remain poorly understood, but it is possible that among other changes it triggered (or accompanied, or even followed) changes in weaponry, military methods and tactics, settlement design, and even the consolidation of a warrior age-set in at least some pastoralist societies (Sutton 1993a).[21] Since the period of transition overlaps that of the consolidation of settlement in the region of iron-using farming communities, who spoke a variety of Bantu languages (see Lane 2004), it is possible that new relationships developed between farmers and stock keepers involving exchange of agricultural produce and other commodities, which in turn permitted the development of more specialised livestock management and production systems among pastoralist communities (Robertshaw 1990: 297).

Archaeologists term this phase the Pastoral Iron Age (PIA). One of the earliest dated occurrences of this transition comes from Deloraine 'main site' (about AD 700 to 1000), near Rongai on the edge of the Mau Escarpment to the north-west of Lake Nakuru (Ambrose 1984b) (see Figure 3.4.2). This site contains abundant remains of cattle and evidence of cereal cultivation, as well as evidence for the manufacture and use of iron implements, and may well represent a development of earlier PN Elmenteitan traditions (Sutton 1993b: 123), as implied by continuities in the layout of ceramic decoration between earlier Elmenteitan materials and pottery from sites such as Deloraine and Lanet (Collett 1984: 87). From around AD 500, there is increasing evidence of the adoption of iron by other PN groups, which seems to have precipitated a marked qualitative decline in stone-tool production, especially in the use of obsidian. The ensuing centuries witnessed the consolidation of PIA economies along the Central Rift Valley and adjacent highlands, out of which many of the ethnic identities and linguistic clusters that characterise these areas today were formed. Around the same time, there was a greater shift towards the type of specialised pastoralism which has often been held to characterise the economies of Nilotic speakers in more recent times. This is best exemplified by the evidence from the site of Deloraine.

A characteristic of PIA sites in the Central Rift Valley, such as Lanet (Posnansky 1967) and Hyrax Hill (Leakey 1945) on the edge of Nakuru, and Muringa and other sites across the Western Highlands (Sutton 1973), is the presence

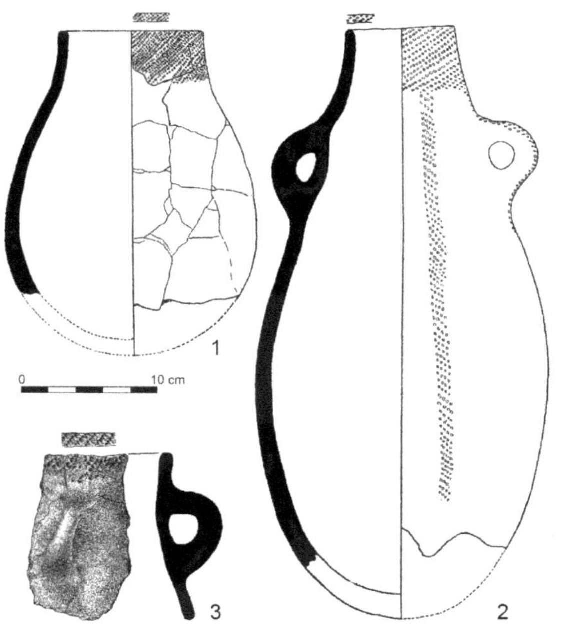

Figure 3.7: Pastoral Iron Age pottery of Lanet tradition. 1–2: reconstructed elongated, gourd-shaped pots with cord-roulette decoration from Hut A; 3: rim with cord-roulette decoration and handle, from below earthwork bank. After Posnansky (1967); 1–2: 99, fig. 6; 3: 101, fig. 8/1. © BIEA & individual authors, reproduced with permission.

of Lanet ware (Figure 3.7), a PIA ceramic tradition that dates to between 800 and 200 bp. This pottery type has a wide distribution, encompassing most of the areas previously occupied by Elmenteitan and SPN groups. Lanet ware is also associated with the 'Sirikwa hole' type of earthwork sites that are especially common along the Elgeyo Escarpment, Uasin Gishu and in the Nandi area. These earthworks appear as saucer-shaped depressions, roughly 10 metres in diameter and between 0.5 and 1.0 metre deep. They are found on hillsides in a broad band running roughly north-west from the Central Rift Valley from Lake Elmenteita across the Western Highlands as far as the foothills of Mount Elgon and the Cherangani Escarpment. They are often in groups, from five to fifty individual hollows, and many are associated with a low mound. Excavations of these indicate that the earthworks are the remains of livestock enclosures that had been deliberately hollowed out. Each has a narrow entranceway facing downhill, which from the alignment of surviving postholes appears to have been protected by some form of gate. Up to three circular houses were attached to the outer edge of the enclosure.[22] Those opposite the entrance were used for dwelling and sleeping, those to one side were kitchens, and those by the entrance were probably guard houses. The mounds, which occur just beyond the entrance, are the remains of dumps of dung and household rubbish (Sutton 1973; 1987: 41–51). The faunal remains from the majority of sites indicate a subsistence economy with an emphasis on specialised pastoralism that combined dairy cattle with small stock raised for their meat. Another characteristic of these sites is the presence of humped cattle (*Bos indicus* or *B. indicus/taurus* crossbreed), as at Hyrax Hill and Lanet, both dated to the sixteenth century AD. Evidence from Hyrax Hill indicates that donkeys were also kept (Kyule 1997). Overall, these sites indicate a considerable degree of sedentism, and Sirikwa are probably best envisaged as stock keepers who practised seasonal transhumance with their livestock, rather than as nomadic pastoralists (Sutton 1993a: 46).

Sirikwa sites range in date from about the twelfth to the eighteenth century AD. Surviving oral traditions suggest that the Sirikwa were an early Kalenjin people. Not all scholars agree with this interpretation, however, arguing that the appearance of Lanet ware correlates far better with the expansion of Ongamo-Maa (Eastern Nilotic) speakers, who, to judge from their oral traditions, probably reached the northern Rift Valley during the ninth century AD, and the southern Rift by the seventeenth century AD, perhaps earlier (Sommer and Vossen 1993: 27–29).

Further to the south, in the north-eastern Mara Plains, the transition from the PN Elmenteitan to later cultural traditions is mainly represented by changes in ceramics, a diminishing reliance on stone tools, and changes in obsidian sources and associated exchange networks (Robertshaw 1990: 294; Siiriäinen, Seitsonen and Laurén 2009). These later sites, dating to about 1100 to 1400 AD, have been tentatively attributed to the 'Oldorotua' tradition (Siiriäinen, Seitsonen and Laurén 2009). Although similar to Elmenteitan material, the stone-tool industry at these sites is more impoverished, and contains higher proportions of grey obsidian relative to other types (ibid.: 168), which could be indicative of changed mobility patterns and perhaps the disruption of pre-existing exchange networks linking the Lemek-Mara region with the Central Rift Valley sources around Lake Naivasha and Mount Eburru (Merrick and Brown 1984). Evidence from site GuJe 12 on the Mara River indicates that iron technology had been introduced by AD 560 to 980, which correlates quite well with the evidence from Deloraine (Siiriäinen, Seitsonen and Laurén 2009: 168–70; Robertshaw 1990: 194–96). Direct evidence for iron production from this area is scarce, however, and mostly limited to a single excavated furnace near Aitong (site GuJe 32). This had a rounded base and was roughly triangular in plan, with tuyère ports at each corner (Siiriäinen, Seitsonen and Laurén 2009: 180–81). Although undated, it bears some similarity to examples excavated on Laikipia (see below), and may well be broadly contemporary with these latter sites and the PIA settlement site GuJf 70.

From Hunting and Gathering to Herding on the Laikipia Plateau

The evidence currently available from the Laikipia Plateau concerning the adoption and spread of pastoralism is less extensive, but is broadly consistent with that from the better researched Central Rift Valley. For the most part, this evidence is derived from the four rock-shelter sites on the Laikipia Plateau that have been investigated in some detail (Figure 3.8), coupled with information from surface discoveries and smaller exploratory investigations of a few other rock-shelters. No open LSA sites have been excavated yet. Of the four key sites, those at Shulumai and Kakwa Lelash close to Dol Dol in the Mukogodo Hills provide long, dated sequences that span terminal phases of the Middle Stone Age (MSA), the subsequent development of the LSA and the appearance of PN ceramics and stone-tool industries. Detailed analysis of the faunal remains from the upper levels at Shulumai indicates that the assemblage was dominated by remains of wild animals, with a preference for small species such as hyrax, dik-dik and bush pig, with only small numbers of domestic cattle and sheep and goats (Mutundu 1999: 50–56). Although this pattern of resource use may have been influenced by the ecological conditions in the immediate vicinity of these rock-shelters, Mutundu argues that the characteristics of the assemblage are fairly typical of specialised hunter-gatherers with relatively restricted settlement mobility in the process of adopting livestock herding. This specific shift in economic strategy can be linked directly with ethnohistorical data from the same area (Cronk 1989). In the Mukogodo case, this may have been as late as the 1920s and 1930s (Cronk 1989; Mutundu 1999). The evidence from the sites of Porcupine Cave (KFR-A5) and Ol Ngoroi also relates to the end of the LSA, the first appearance of domestic livestock on the Plateau, and subsequent utilisation of these rock-shelters by stock keeping communities.

Excavations at Porcupine Cave (about 35 kilometres north-east of Rumuruti) and a limited number of other rock-shelters on the Laikipia Plateau by Siiriäinen (1977, 1984) suggest that the first appearance of domestic stock in the Ewaso Basin was between 4000 and 3000 bp. These dates are broadly consistent with the

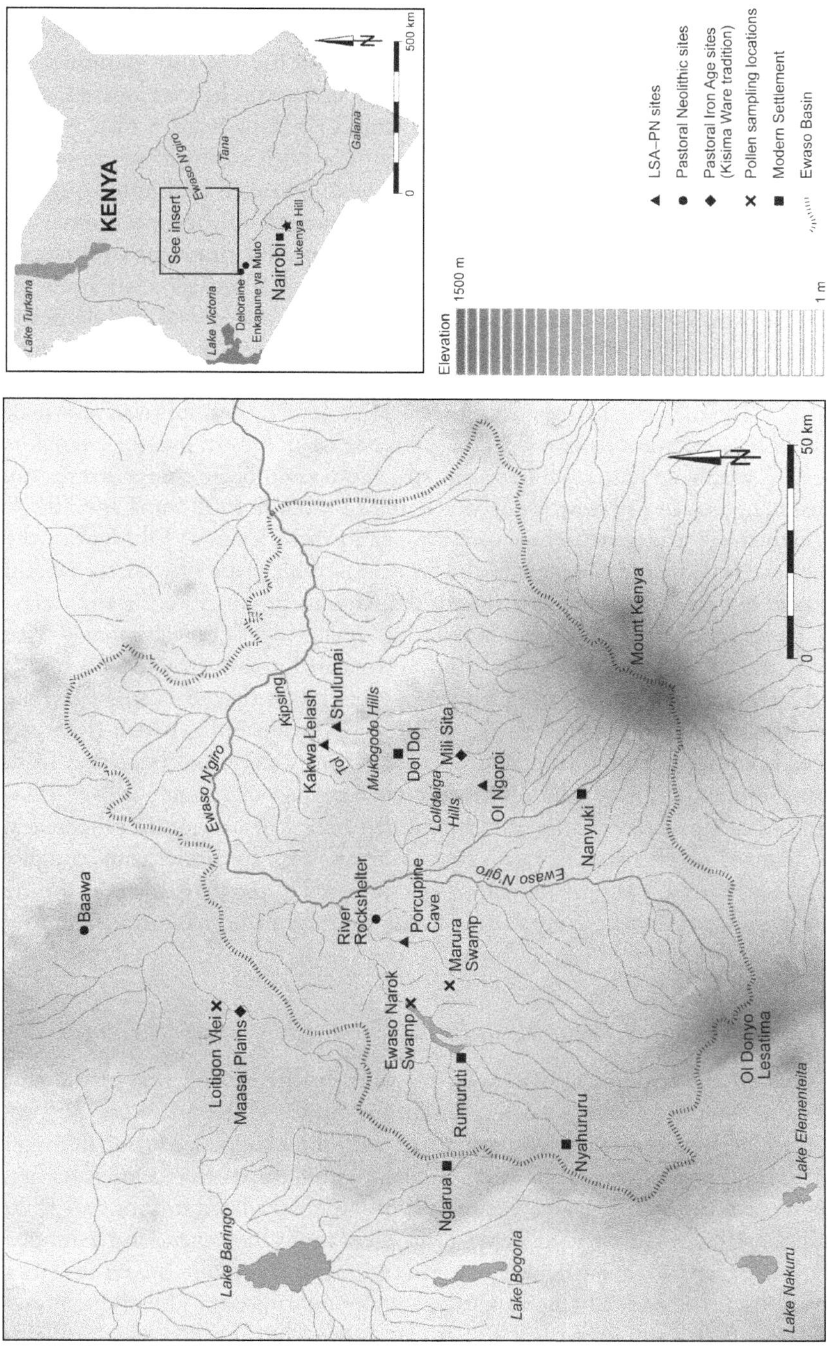

Figure 3.8: Location of main archaeological sites on the Laikipia and Leroghi plateaux mentioned in the text.

current evidence from other parts of the region discussed above. Also, Siiriäinen (1984) found considerable typological continuity in the stone-tool assemblages from different levels at the sites he excavated, spanning levels dated between about 2000 to 800 bp, and perhaps even up to 600 bp. He also found a wide range of different PN pottery types represented at the sites he investigated, with no single type dominating until the appearance of Kisima ware, a PIA ceramic tradition (see below), during the last millennium. Both observations would be consistent with a view that the rock-shelters were occupied by hunter-gatherer communities that either gradually made the transition to food production or were able to gain access to livestock through exchange relationships with neighbouring pastoralist communities. This is borne out by the limited information on the faunal assemblages of these sites, which contain a mix of wild and domestic taxa (Siiriäinen 1984: 88), implying that hunting remained an important aspect of the subsistence strategies of these early herding communities.

At Ol Ngoroi rock-shelter on Lolldaiga Hills Ranch, despite relatively shallow deposits, excavations revealed the presence of domestic livestock remains associated with wild fauna, a microlithic stone tool assemblage comprised mostly of obsidian, and *in situ* hearth deposits dated to around 4000 bp (Lane 2011). Although sheep and goat bones were present in these levels at Ol Ngoroi, the bulk of the identifiable elements in the faunal assemblage are of wild fauna, and the range of taxa reflects the structure of the habitats proximal to the site today, with hyraxes and small bovids predominating. A charcoal sample associated with pottery and other material above these levels yielded a date around 730 bp. In broad terms, the assemblage from these levels resembles that recovered from the upper levels at Shulumai rock-shelter in the Mukogodo Hills, which is known to have been occupied by Mukogodo groups during a period of transition from hunting and gathering to a herding subsistence strategy (Mutundu 1999). Elsewhere on Laikipia, recent surveys have located over 250 previously unrecorded sites, many of which are stone cairns or rock-art sites. Of the settlements, roughly six times as many are of probable PIA date compared to those associated with the earlier Pastoral Neolithic suggesting a steady increase in human activity during the last millennium.[23]

The Consolidation of Pastoralism in the Eastern Highlands of Kenya

PIA sites on the Laikipia and Leroghi plateaus on the eastern side of the Rift Valley are associated with a pottery tradition known as 'Kisima ware' (Figure 3.9),[24] which has been tentatively associated with the Laikipiak Maasai although quite possibly made by neighbouring 'forager' communities (see below). Its first appearance in the archaeological record was provisionally dated to about AD 1500 (Siiriäinen 1984). More recent surveys and excavations, however, have extended its chronological range; particularly important in this regard are the results from excavations at the Mili Sita open settlement site on Lolldaiga Hills Ranch, and the Maasai Plains site on Mugie Ranch. Both the Mili Sita and Maasai Plains sites are large, open-air pastoralist settlements associated with Kisima ware. The Maasai Plains site on Mugie is the older, and based on available radiocarbon dates was occupied around AD 1400 to 1480. The site is situated in an area of open grassland surrounded by woody vegetation, about 1.5 kilometres to the west of Loitigon Vlei, which has been sampled for pollen and other environmental remains (Taylor et al. 2005). It consists of three concentric and roughly circular arrangements of low ash mounds (123 in all) between some 0.35 and 1.0 metre high, covering an area approximately 750 metres in diameter, with a pair of mud-wallows or shallow waterholes in the centre. The mounds, although not especially distinctive on the ground, show up as darker patches on vertical aerial photographs and even on moderately low resolution satellite images. Two of the mounds have been examined by excavation, as have some of the level areas in between (Lane 2011).

In terms of their form, composition, stratigraphic structure and contents, the mounds at the Maasai Plains site closely resemble the dung-and-trash heaps commonly observed outside modern Samburu homesteads (*nkan'g*; Maasai, *enkang*), formed by regular cleaning of dung from animal kraals and the discard of household waste.[25] No house remains have yet been located in the areas between the mounds to confirm this, although the presence of low density mixed artefact and faunal scatters in these areas would be consistent with the kind of debris generated around a living area (Causey 2008). The regular patterning of the mounds and the obvious clear area at the centre of the site suggest that, minimally, the mounds making up at least one 'ring' were created at the same time. Given the layering in the mounds, it seems likely that each mound was built up progressively. The extent of the site is much larger than any other recorded archaeological pastoralist site in the region, and even covers an area larger than most ethnographically documented Maasai *enkang*. It is entirely possible, however, that other 'rings' were created at different times, and that not all mounds relate to separate houses. It is also possible that the site was only occupied for short periods at a time, and was perhaps reoccupied on several occasions.

The apparent date of occupation of the Maasai Plains site roughly coincides with evidence from Lake Naivasha for a brief high-stand, possibly indicative of a period of increased precipitation (Verschuren, Laird and Cumming 2000; Lamb, Darbyshire and Verschuren 2003), which would have made the Laikipia Plateau more attractive to pastoralists, especially for seasonal grazing, given the relative scarcity of permanent water sources. This climatic amelioration could also account for why PIA sites are more numerous on the Laikipia Plateau compared with PN sites. High-resolution environmental records from various locations throughout the region indicate that the following period, from the late sixteenth to the late eighteenth century AD, appears to have been characterised by long intervals when levels of effective precipitation in eastern Africa were much reduced relative to the present (Verschuren, Laird and Cumming 2000). As the records from Lake Naivasha and elsewhere indicate, against this general drying trend there were periodic wet episodes interspersed with periods of very low rain-

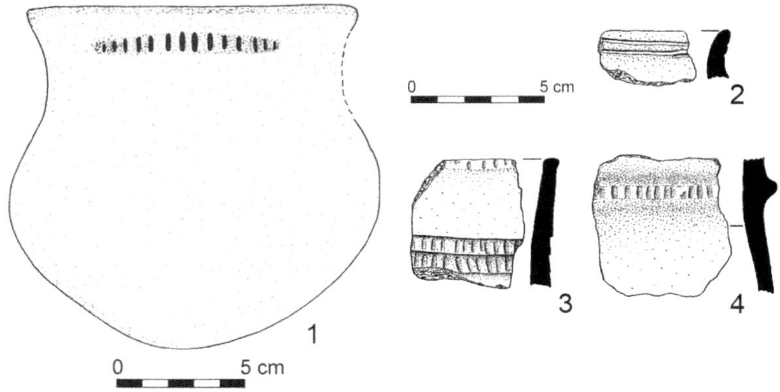

Figure 3.9: Examples of Pastoral Iron Age Kisima ware from Laikipia. 1: reconstructed pot from Maasai Plains; 2–3: decorated rim sherds; 4: nicked, beaded body sherd. (2 and 4 from KFR-A4; 3 from KFR-A12). *Sources*: 1: Causey (2010: 116, fig. 3); 2–4: Siiriäinen (1977: 171, figs. 8/a. b and e). © BIEA & individual authors, reproduced with permission.

fall and ensuing drought, including one centred around AD 1780 (Bessems et al. 2008). In view of the relative scarcity of water sources on Laikipia, it might be expected that the utilisation of this landscape by pastoralists would have exhibited a cycle of expansion and contraction directly correlated with the 'wet' and 'dry' phases.

One site which could well have been occupied during one of the later phases of higher rainfall is that at Mili Sita (see Figure 3.8). This is located on a low col towards the end of a northward-trending ridge in the Lolldaiga Hills, and like the Maasai Plains site appears to have been used as an area of pastoralist settlement. Nearby are numerous distinct scatters of iron slag mixed with tuyère fragments and the remains of several shallow bowl furnaces forming two distinct clusters. All of the furnaces range between 30 and 40 centimetres in diameter and have a maximum depth of 30 centimetres. Those in Cluster A were circular in plan with steeply sloping walls tapering downwards, whereas the Cluster B furnaces were roughly square in plan with tuyère ports in each corner and near vertical walls; charcoal samples suggest a date falling between around AD 1660 to 1810. Traces of iron smelting activity occur elsewhere in the Lolldaiga Hills, including at a locality designated Cattle Dip, about 3 kilometres to the north-west of Mili Sita, dated to around AD 1640 to 1670 (Iles and Martinón-Torres 2009).

On the basis of the available radiocarbon dates, the Mili Sita site appears to have been occupied around AD 1640 to 1730. If one accepts Maasai Purko oral traditions that the Laikipiak arrived from the north-east sometime prior to AD 1600 (Jacobs 1972: 82), then this would imply that the site was occupied by Laikipiak Maasai. If this is correct, it might mean that the Laikipiak were the makers of Kisima ware. In which case, the presence of Kisima pottery at the Maasai Plains site dated to about AD 1400 to 1480 would imply that Laikipiak were present on the Laikipia Plateau up to two centuries earlier than is suggested by available oral histories. However, it is possible that Kisima pottery, although abundant on both these sites, was made by neighbouring communities (such as Mukogodo, who are known historically and ethnographically to have supplied Samburu with pottery) and exchanged with Laikipiak for other products. Also, it should be noted that there are differences in scholarly opinion over when the first Maa speakers in general arrived in the Rift Valley and adjacent highlands, which complicate matters. Arguing from the position of historical linguistics, Sommer and Vossen suggest that the ancestors of modern Maa speakers reached the Rift Valley 'by the end of the ninth century' (Sommer and Vossen 1993: 25), whereas Galaty (1993) has argued that these represent 'Early Maa-speakers', who can be distinguished from 'Later Maasai' who represent another expansionary phase of pastoral genesis in the area stretching from Lake Baringo to Lake Natron from which they spread in a spiral fashion (see also Sutton 1993a). In this model the Laikipiak only attained dominance over the Ewaso Basin around the early nineteenth century.[26]

Conclusion

The overwhelming conclusion to be drawn from the archaeological, environmental and historical evidence reviewed above is that multiple trajectories towards a pastoralist existence have been followed in eastern Africa since the initial introduction of domesticated livestock into the region. There is good evidence to suggest that at times these involved demic migration of groups from further north, who not only brought with them domesticated livestock but also new languages, artefactual traditions and cultural practices.[27] Some elements of these were long-lasting, while others underwent successive transformations and mutations. It is also apparent that there have been several episodes of migration up to quite recent times. It would be mistaken, however, to envisage the majority of such movements as if they constituted aggressive invasions of territory, not least because there is a general lack of evidence of physical violence or the burning of settlements dating to the period when domestic stock and/or new material culture traditions first appear. Also, since the lithic assemblages at many sites exhibit considerable continuity between levels where domestic stock and/or PN pottery are absent and those in which they first appear, it can be reasonably inferred that these sites continued to be occupied, whether wholly or partially, by descendants of the autochthonous hunter-gatherer populations. If encounters with immigrant stock keepers had tended to be violent, on the other hand, one would expect widespread evidence for abandonment of earlier hunter-gatherer sites and more marked stylistic changes in artefact assemblages and in the technologies of their manufacture.

There are a few exceptions, however, such as the burial of an adolescent male in one of the rock-shelters on Kisima Farm (KFR-A5) on the Laikipia Plateau excavated by Siiriäinen (1984: 37–39), which revealed a crescent microlith embedded in the man's spine. Moreover, since there have been times more recently when pastoralist groups used force for the purposes of territorial expansion – as happened during the Loikop Wars of the later nineteenth century, for instance – the possibility of similar conflict happening further back in the past should not be ruled out entirely. On balance, though, the initial immigration of stock keepers seems to have been on a fairly small scale,[28] and characterised by initial, exploratory, 'moving frontiers' (Lane 2004) during which cultural, linguistic and ethnic boundaries may have been quite fluid – especially on the part of immigrant groups. In the Central Rift Valley and adjacent Loita-Mara area this phase may have lasted a thousand years or more, and eventually culminated in the formation of more 'static frontiers' and the formation of more bounded social groups that employed a repertoire of material culture differences and stylistic attributes to signal group identity, with interaction being mediated through mechanisms of long-distance exchange possibly involving some prestige items. As these early PN communities became more established and population densities increased there was also a shift towards greater reliance on livestock, reduced mobility and the emergence of 'specialised' pastoralism, especially in some Elmenteitan areas where herds were clearly being managed for milk production. On the Laikipia Plateau, this phase of consolidation and, possibly, settlement expansion may well have been accompanied by an increase in the use of burning to control and modify local vegetation and so might mark a full-scale 'domestication' of these landscapes (cf. Terrell et al. 2003) with profound consequences for the local ecology. Contrary to the common and well-entrenched environmental narrative that all pastoralist systems result in overgrazing and rangeland degradation,[29] these ecological consequences were not necessarily detrimental. In particular, the regular, repetitive creation of nutrient-enriched patches as a result of the occupation and abandonment of pastoralist settlements, which as recent work on Laikipia has shown can survive within savannah environments for upwards of six hundred years (Causey 2008; Lane 2011), may have been of considerable importance to the well-being of wild ungulate populations and enhanced local biodiversity more generally.[30] There is thus good reason to believe that the ecologically complex, spatially heterogeneous and temporally dynamic vegetation mosaics that characterise modern East African savannahs have been significantly shaped by the history of livestock grazing and the human occupation and utilisation of these landscapes (Homewood and Rodgers 1991).

Nevertheless, the precise details of how these landscapes have been domesticated, the range of ecological feedback mechanisms generated in different habitats, the influence of climate change and other abiotic factors on these processes, and the nature and changing pattern of pastoralist activity in many parts of the region, still remain poorly understood. This is partly because large swathes of eastern Africa remain virtually unknown archaeologically, and even where there has been a long tradition of archaeological research, such as in the Kenyan Central Rift Valley, the necessary palaeoenvironmental records are either lacking or exhibit poor temporal resolution for many of the critical time periods. The general lack of archaeobotanical data from PN and PIA sites, despite enhanced recovery techniques and analytical methods, is a further stumbling block. Differentiating the effects (at all spatial scales) of climate change, other natural processes and anthropogenic activities also continues to be problematic. Equally, whereas the functioning of these landscapes has been informed by recent theoretical advances concerning 'shifting patch dynamics' and 'ecological resilience', such studies as have been carried out (e.g., Gillson 2004) have largely failed to take the evidence for the long-term human manipulation of these landscapes into account. Coordinated, cross-disciplinary and collaborative research on such issues, employing the central precepts of long-term landscape historical ecology (Crumely 1994) would clearly pay dividends and in this author's view should be one of the key priorities for future research.

Notes

The research on which this chapter is based was conducted as part of a wider British Institute in Eastern Africa project 'Settlement and Land-use Change in Semi-arid Areas of Eastern and Southern Africa', generously funded by the British Academy under the auspices of its BASIS New Initiatives Scheme, launched in 2002. Additional fieldwork was conducted as part of an EU-funded Marie Curie Excellence Grant, 'Historical Ecologies of East African Landscapes' (MEXT-CT-2006-042704) between 2007 and 2011. Both funding bodies are warmly thanked. The research was undertaken under the auspices of Kenya Research Permit No. MOEST 13/014 issued by the Ministry of Education. The following institutions also provided additional funding, support in kind, use of laboratory and computing facilities, and/or logistical assistance: Borana Ranch, Laikipia; Department of Archaeology, University of Oxford; Department of Archaeology and Rock Art Research Institute, Witwatersrand University; Department of Environmental Studies, Kenyatta University; Department of Geography, Trinity College Dublin; Department of Geography, University of Cambridge; Department of Palynology, National Museums of Kenya; Lolldaiga Hills Ltd., Laikipia; OnSite Archaeology, York; Mugie Ranch, Laikipia; Malilangwe Trust; Mpala Ranch, Laikipia; School of Agriculture, Food and Rural Development, University of Newcastle upon Tyne, all of whom are thanked. I would also like to thank the following for their assistance and input: Robert Wells, Claus Mortenson, John Sutton, Fiona Marshall, John Giblin, Michael Causey, Louise Iles, Benjamin Smith, Robert Payton, Kennedy Mutundu, Bilinda Straight, Charles Hilton, Carolyn Thorp, Fumi Mizutani, Ann Powys, Michael and Nicky Dyer, Max Graham, Bill Adams, Richard Waller, David Taylor, Rob Marchant, Kathleen Ryan, Nick Pearson, Philip Owiti, Stephen Manoa and all the students who have assisted on the field projects. Particular thanks are also due to the editors of this volume for their patience and many thoughtful suggestions, and Tim Evans and Lutz Hermsdorf-Knauth for help with preparing the illustrations. Any remaining errors are solely my responsibility.

1. All dates in uncalibrated radiocarbon years (bp) unless stated in calibrated form (cal BC/AD).
2. For overlapping definitions and distinctions of these terms, see, e.g., Dyson-Hudson and Dyson-Hudson (1980) and Smith (1992).
3. See Riemer (2007: 107–13) for further discussion of this point.
4. This is somewhat later than proposed initially by Marshall (1989).
5. F. Marshall (personal communication, June 2006).

6. Rainfall in south-central Kenya, for instance, could have been 16 to 32 per cent higher than today (Bergner, Trauth and Bookhagen 2003). Note, however, that estimates for different parts of the region and by different scholars are quite variable. For instance, Dühnforth, Bergner and Trauth (2006: 291) have suggested that in the Lake Naivasha catchment rainfall was even higher, at around 45 per cent above today's levels, whereas Hastenrath and Kutzbach (1983) argued that for the Turkana, Victoria and Naivasha basins precipitation levels were only 10 to 20 per cent higher than today.
7. The period between about 9500 and 6000 bp is often referred to as the African Humid Period (for a synopsis of the most recent dating evidence for the Eastern Sahara, see Bubenzer and Riemer 2007: 609–11), and has long been recognised as being characterised by a greater frequency of higher lake levels than in either the previous or subsequent periods (Street and Grove 1979). Some recent studies (e.g., Kröpelin et al. 2008), however, suggest that there is less evidence to support the idea of continent-wide synchronicity of climatic amelioration as inferred from marine sediment records from offshore Mauritania (deMenocal et al. 2000), and that, far from ending abruptly, the transition to hyper-arid conditions in the Eastern Sahara was quite gradual.
8. See Marchant and Hooghiemstra (2004) for a review of possible causes.
9. For a short history, see Karega-Mũnene (2003).
10. Ambrose (1984a) has divided the LSA Eburran stone-tool industry (formerly Kenya Capsian) into five phases. Overall, geometric and non-geometric microliths, produced from narrow blades, are the main diagnostic characteristics of the Eburran, often accompanied by the use of the microburin technique. Phases 1 to 3, dated to circa 12000 to 7000 bp, are characterised by large backed blades, whereas Phases 4 and 5 have smaller backed tools, notably microliths and scrapers. Platform preparation is highly diagnostic, with most being relatively wide and heavily microfacetted.
11. As indicated by archaeologically sterile aeolian sediments and unconformities.
12. Salasun pottery was found in association with these PN ceramics. The Salasun tradition is poorly defined, but is principally found in rock-shelters in the Central Rift Valley and around Lukenya Hill, and may well have been made by autochthonous hunter-gatherers (Ambrose 2003: 101).
13. See Lane (2004) for a review of the core issues. It should be pointed out that securely dated sites with a more definitively 'pastoralist' faunal signature older than circa 3400 bp representing immigrant 'pastoralist' populations have yet to be encountered in the archaeological record of the Central Rift Valley. This does not mean that they do not await discovery.
14. See especially Gifford-Gonzalez (1998a: 173–90) for a review of this variability.
15. Additionally, domestic stock are now also securely attested from sites around eastern Lake Victoria in association with LSA Kansyore ceramics from circa 3400 to 3200 bp (circa 1700 to 1450 cal BC) (Robertshaw 1991; Karega-Mũnene 2002; Lane et al. 2007; Prendergast 2011), and around Lake Eyasi, northern Tanzania (Prendergast 2008, 2011).
16. For instance, Ehret (1998: 98, fig. 3) has suggested that the two SPN pottery groupings may have been associated with, respectively, Ma'a Southern Cushites and Rift Southern Cushites. The makers of Elmenteitan material have often been identified as Southern Nilotes (e.g., Ambrose 1984a).
17. First reported from the site of Stable's Drift on the Nderit River and originally labelled Gumban A (Leakey 1931: 198–204).
18. After the type site Narosura, south of Narok in south-western Kenya (Odner 1972).
19. Originally known as the 'Hyrax Hill variant of the Stone Bowl Culture' (Leakey 1945: 302).

20. Originally known as Remnant ware, after the site where these ceramics were first documented.
21. However, Ehret (2003: 173) seems to be of the opinion that the tradition of warrior age-sets has been fundamental to the expansion of Nilotic and Proto-Nilotic speakers over the entire roughly 5,000 year history of this language family.
22. Interestingly, the layout of Sirikwa holes bears some resemblance to recent Kuria homesteads (compare Sutton 1993a: 45, fig. 2.1 with Ruel 2000: 65, fig. 1). The Kuria occupy areas that straddle the western Kenya–Tanzania border. Linguistically, Kuria belongs to the Eastern Bantu group of languages in the East Nyanza grouping of Nurse and Philippson (1980). However, their history has been closely linked to their Luo (Western Nilotic) and Maasai (Eastern Nilotic) neighbours; they also have a long tradition of cattle keeping and stock raiding, and are almost certainly of diverse origins (Abuso 1980).
23. See Lane (2011) for further discussion.
24. Named after Kisima Ranch on Laikipia, where it was first identified archaeologically.
25. In recent Maasai homesteads, the practice is for household rubbish to be dumped either on the inside of the perimeter fence, or externally between 3 and 5 metres from the entrance (Mbae 1990: 281–83).
26. See Lane (2011) for additional details.
27. Although demic migration – i.e. actual physical movements of people – has been long suspected and inferred from a range of possible signatures, recent evidence concerning the distribution of genetic markers associated with lactose tolerance (which appears to have developed in eastern Africa between 3,000 and 7,000 years ago) among East African pastoralists offers, currently, the most convincing proof (Tishkoff et al. 2007). From other preliminary studies (Henn et al. 2008; Gomes et al. 2010) it appears that while there is considerable genetic diversity among modern pastoralists, this is only weakly correlated with linguistic affinity. It might also be anticipated that the distribution of mtDNA and Y-chromosome haplotypes relative to linguistic affinity may exhibit different patterns in view of their different, sex-related lines of transmission. Overall, however, the genetic histories of East Africa's pastoralist communities remain poorly understood.
28. If the results of a recent study that examined aspects of the later expansion of pastoralists into southern Africa (Henn et al. 2008) provide a suitable analogy, then the actual demographic contribution of immigrants may have been relatively minor.
29. For one of the earliest critiques of this narrative, see Homewood and Rodgers (1987); see also Collett (1987).
30. For studies of the ecological dynamics in more recent nutrient-enriched patches in savannah environments, see, e.g., Reid and Ellis (1995), Young, Partridge and Macrae (1995), Augustine (2003) and Augustine and McNaughton (2004).

References

Abuso, A.P. 1980. *A Traditional History of the Abakuria: c. AD 1400–1914*. Nairobi: Kenya Literature Bureau.

Ambrose, S.H. 1984a. 'The Introduction of Pastoral Adaptations to the Highlands of East Africa', in J.D. Clark and S.A. Brandt (eds), *From Hunters to Farmers: The Causes and Consequences of Food Production in Africa*. Berkeley: University of California Press, pp.212–39.

——— 1984b. 'Excavations at Deloraine, Rongai, 1978', *Azania* 19: 79–104.

——— 1985. 'Excavations at Masai Gorge Rock-shelter, Naivasha', *Azania* 20: 29–67.

——— 1998. 'Chronology of the Later Stone Age and Food Production in East Africa', *Journal of Archaeological Science* 25: 377–92.

——— 2003. 'East African Neolithic', in P.N. Peregrine and M. Ember (eds), *Encyclopedia of Prehistory*, Vol. 1: *Africa*. London: Kluwer Academic/Plenum, pp.97–109.

Ambrose, S.H., and M.J. DeNiro. 1986. 'Reconstruction of African Human Diet Using Bone Collagen Carbon and Nitrogen Isotope Ratios', *Nature* 319: 321–24.

——— 1989. 'Climate and Habitat Reconstruction Using Stable Carbon and Nitrogen Isotope Ratios of Collagen in Prehistoric Herbivore Teeth from Kenya', *Quaternary Research* 31: 407–22.

Ambrose, S.H. and N.E. Sikes. 1991. 'Soil Carbon Isotope Evidence for Holocene Habitat Change in the Kenya Rift Valley', *Science* 253: 1402–5.

Augustine, D.J. 2003. 'Long-term, Livestock Mediated Redistribution of Nitrogen and Phosphorus in an East African Savannah', *Journal of Applied Ecology* 40: 137–49.

Augustine, D.J., and S.J. McNaughton. 2004. 'Regulation of Shrub Dynamics by Native Browsing Ungulates on East African Rangeland', *Journal of Applied Ecology* 41: 45–58.

Barthelme, J. 1985. *Fisher-hunters and Neolithic Pastoralists in East Turkana, Kenya*. Oxford: British Archaeological Reports.

Bender, L.M. 2000. 'Nilo-Saharan', in B. Heine and D. Nurse (eds), *African Languages: An Introduction*. Cambridge: Cambridge University Press, pp.43–73.

Bergner, A.G.N., M.H. Trauth and B. Bookhagen. 2003. 'Paleoprecipitation Estimates for the Lake Naivasha Basin (Kenya) During the Last 175 k.r. Using a Lake-balance Model', *Global Planetary Change* 36: 117–36.

Bessems, I., et al. 2008. 'Paleolimnological Evidence for Widespread Late 18th Century Drought across Equatorial East Africa', *Palaeogeography, Palaeoclimatology, Palaeoecology* 259: 107–20.

Blackburn, R.H. 1971. 'Honey in Okiek Personality, Culture and Society', PhD dissertation. East Lansing: Michigan State University.

Blench, R.M. 2000. 'A History of Donkeys and Mules in Africa', in R.M. Blench and K.C. MacDonald (eds), *The Origin and Development of African Livestock: Archaeology, Genetics, Linguistics and Ethnography*. London: University College Press, pp.339–54.

Bodeker, G.C., T.J. Ryan and C.-K. Ong. 1999. 'Traditional Approaches to Wound Healing', *Clinics in Dermatology* 17: 93–98.

Bower, J.R.F. 1973. 'Seronera: Excavations at a Stone Bowl Site in the Serengeti National Park, Tanzania', *Azania* 8: 71–104.

——— 1991. 'The Pastoral Neolithic of East Africa', *Journal of World Prehistory* 5: 49–82.

Bower, J.R.F., and C.M. Nelson. 1978. 'Early Pottery and Pastoral Cultures of the Central Rift Valley, Kenya', *Man* 13: 554–66.

Bower, J.R.F., et al. 1977. 'The University of Massachusetts' Later Stone Age/Pastoral "Neolithic" Comparative Study in Central Kenya: An Overview', *Azania* 12: 119–46.

Brown, J. 1990. 'Horn-shaping Ground Stone Axe Hammers', *Azania* 25: 57–67.

Bubenzer, O., and H. Riemer. 2007. 'Holocene Climatic Change and Human Settlement Between the Central Sahara and the Nile Valley: Archaeological and Geomorphological Results', *Geoarchaeology* 22: 607–20.

Butzer, K.W., et al. 1972. 'Radiocarbon Dating of East African Lake Levels', *Science* 175: 1069–76.

Causey, M. 2008. 'Delineating Pastoralist Behaviour and Long-term Environmental Change: A GIS Landscape Approach on the Laikipia Plateau, Kenya', PhD dissertation. Oxford: University of Oxford.

——— 2010. 'New Archaeological Discoveries from the Laikipia Plateau, Kenya', *Azania* 45: 112–36.

Close, A.E. 2002. 'Sinai, Sahara, Sahel: The Introduction of Domestic Caprines to Africa', in Jennerstr. 8 (eds), *Tides of the Desert/Gezeiten der Wüste.* Cologne: Heinrich-Barth-Institut, pp.471–83.

Collett, D. 1984. 'The Finds', in S.H. Ambrose (ed.), 'Excavations at Deloraine, Rongai. 1978', *Azania* 19: 83–98.

——— 1987. 'Pastoralists and Wildlife: Images and Reality in Kenya Maasailand', in D.M. Anderson and R.H. Grove (eds), *Conservation in Africa: People, Policies and Practice.* Cambridge: Cambridge University Press, pp.129–48.

Collett, D., and P.T. Robertshaw. 1983. 'Problems in the Interpretation of Radiocarbon Dates: The Pastoral Neolithic of East Africa', *African Archaeological Review* 1: 57–74.

Cronk, L. 1989. 'From Hunters to Herders: Subsistence Change as a Reproductive Strategy among the Mukogodo', *Current Anthropology* 30: 224–34.

Crumley, C.L. (ed.). 1994. *Historical Ecology: Cultural Knowledge and Changing Landscapes.* Albuquerque, NM: School of American Research Press.

deMenocal, P., et al. 2000. 'Coherent High and Low-latitude Climate Variability during the Holocene Warm Period', *Science* 288: 2198–202.

Dühnforth, M., A.G.N. Bergner and M.H. Trauth. 2006. 'Early Holocene Water Budget of the Nakuru-Elmenteita Basin, Central Kenya Rift', *Journal of Paleolimnology* 36: 281–94.

Dyson-Hudson, R., and N. Dyson-Hudson. 1980. 'Nomadic Pastoralism', *Annual Review of Anthropology* 9: 15–61.

Ehret, C. 1998. *An African Classical Age: Eastern and Southern Africa in World History, 1000 BC to AD 400.* Oxford: James Currey.

——— 2003. 'Language Family Expansions: Broadening Our Understandings of Cause from an African Perspective', in P. Bellwood and C. Renfrew (eds), *Examining the Farming/Language Dispersal Hypothesis.* Cambridge: McDonald Institute Monographs, pp.163–76.

Falkenstein, M. 1995. 'Concepts of Ethnicity and Inter-ethnic Migration among the Ariaal of Kenya', *Zeitschrift für Ethnologie* 120: 201–25.

Galaty, J. 1982. 'Being "Maasai", Being "People of Cattle": Ethnic Shifters in East Africa', *American Ethnologist* 9: 1–20.

——— 1993. 'Maasai Expansion and the New East African Pastoralism', in T. Spear and R. Waller (eds), *Being Maasai: Ethnicity and Identity in East Africa.* London: James Currey, pp.61–86.

Gasse, F., and F.A. Street. 1978. 'Late Quaternary Lake-level Fluctuations and Environments of the Northern Rift Valley and Afar Region (Ethiopia and Djibouti)', *Palaeogeography, Palaeoclimatology, Palaeoecology* 24: 279–325.

Gautier A. 2007. 'Animal Domestication in North Africa', in M. Bollig et al. (eds), *Aridity, Change and Conflict in Africa.* Cologne: Heinrich-Barth Institut, pp.75–89.

Gifford, D.P., G.Ll. Isaac and C.M. Nelson. 1980. 'Evidence for Predation and Pastoralism at Prolonged Drift: A Pastoral Neolithic Site in Kenya', *Azania* 15: 57–108.

Gifford-Gonzalez, D. 1998a. 'Early Pastoralists in East Africa: Ecological and Social Dimensions', *Journal of Anthropological Archaeology* 17: 166–200.

——— 1998b. 'Gender and Early Pastoralists in East Africa', in S. Kent (ed.), *Gender in African Prehistory.* Walnut Creek, CA: Altamira Press, pp.115–37.

——— 2000. 'Animal Disease Challenges to the Emergence of Pastoralism in Sub-Saharan Africa', *African Archaeological Review* 17: 95–139.

——— 2003. 'The Fauna from Ele Bor: Evidence for the Persistence of Foragers into the Later Holocene of Arid North Kenya', *African Archaeological Review* 20: 81–119.

Gillson, L. 2004. 'Evidence of Hierarchical Patch Dynamics in an East African Savannah?' *Landscape Ecology* 19: 883–94.

Götherström, A.C., et al. 2005. 'Cattle Domestication in the Near East was Followed by Hybridization with Auroch Bulls in Europe', *Proceedings of the Royal Society B* 272: 2345–50.

Gomes, V., et al. 2010. 'Digging Deeper into East African Human Y Chromosome Lineages', *Human Genetics* 127: 603–13.

Halfman, J.D., et al. 1992. 'Fossil Diatoms and the Mid to Late Holocene Paleolimnology of Lake Turkana, Kenya: A Reconnaissance Study', *Journal of Paleolimnology* 7: 23–35.

Hanotte, O., et al. 2002. 'African Pastoralism: Genetic Imprints of Origins and Migrations', *Science* 296: 336–39.

Hastenrath, S., and J.E. Kutzbach. 1983. 'Paleoclimatic Estimates from Water and Energy Budgets of East African Lakes', *Quaternary Research* 19: 141–53.

Hayward, R.J. 2000. 'Afroasiatic', in B. Heine and D. Nurse (eds), *African Languages: An Introduction.* Cambridge: Cambridge University Press, pp.74–98.

Henn, B.N., et al. 2008. 'Y-chromosomal Evidence of a Pastoralist Migration through Tanzania to Southern Africa', *Proceedings of the National Academy of Sciences* 105: 10693–98.

Hivernel, F. 1976. 'Preliminary Report on Excavations at Ngenyn, Kenya', *Azania* 10: 140–44.

Hodder, I.R. 1982. *Symbols in Action.* Cambridge: Cambridge University Press.

Homewood, K.M., and W.A. Rodgers. 1987. 'Pastoralism, Conservation and the Overgrazing Controversy', in D.M. Anderson and R.H. Grove (eds), *Conservation in Africa: People, Policies and Practice.* Cambridge: Cambridge University Press, pp.111–28.

——— 1991. *Maasailand Ecology: Pastoralist Development and Wildlife Conservation in Ngorongoro, Tanzania.* Cambridge: Cambridge University Press.

Iles, L., and M. Martinón-Torres. 2009. 'Pastoralist Iron Production on the Laikipia Plateau, Kenya: Implications for Wider Archaeometallurgical Studies', *Journal of Archaeological Science* 36: 2314–26.

Jacobs, A. 1972. 'The Discovery and Oral History of the Narosura Site', *Azania* 7: 79–87.

Jianlin, H., et al. 2000. 'Application of New World Camelidae Microsatellite Primers for Amplification of Polymorphic Loci in Old World Camelids', *Animal Genetics* 31: 404–19.

Johnson, T.C., J.D. Halfman and W.J. Showers. 1991. 'Paleoclimate of the Past 4000 Years at Lake Turkana, Kenya, Based on the Isotopic Composition of Authigenic Calcite', *Palaeogeography, Palaeoclimatology, Palaeoecology* 85: 189–98.

Karega-Mũnene. 2002. *Holocene Foragers, Fishers and Herders of Western Kenya.* Oxford: Archaeopress.

——— 2003. 'The East African Neolithic: A Historical Perspective', in C.M. Kusimba and S.B. Kusimba (eds), *East African Archaeology: Foragers, Potters, Smiths and Traders.* Philadelphia: University of Pennsylvania Museum of Archaeology and Anthropology, pp.17–32.

Krings, M., et al. 1999. 'mtDNA Analysis of Nile River Valley Populations: A Genetic Corridor or a Barrier to Migration?' *American Journal of Human Genetics* 64: 1166–76.

Kröpelin, S., et al. 2008. 'Climate-driven Ecosystem Succession in the Sahara: The Past 6000 Years', *Science* 320: 765–68.

Kuper, R., and S. Kröpelin. 2006. 'Climate-controlled Holocene Occupation in the Sahara: Motor of Africa's Evolution', *Science* 313: 803–7.

Kusimba, C.M., and S.B. Kusimba. 2005. 'Mosaics and Interactions: East Africa, 2000 b.p. to the Present', in A.B. Stahl (ed.), *African Archaeology: A Critical Introduction.* Oxford: Blackwell, pp.392–419.

Kyule, D. 1997. 'The Sirikwa Economy: Further Work at Site II on Hyrax Hill', *Azania* 32: 21–30.

Lamb, H., I. Darbyshire and D. Verschuren. 2003. 'Vegetation Response to Rainfall Variation and Human Impact in Central Kenya during the Past 1100 Years', *Holocene* 13: 285–92.

Lamphear, J. 1988. 'The People of the Grey Bull: The Origin and Expansion of the Turkana', *Journal of African History* 29: 27–39.

Lane, P.J. 2004. 'The "Moving Frontier" and the Transition to Food Production in Kenya', *Azania* 39: 243–64.

——— 2011. 'An Outline of the Later Holocene Archaeology and Precolonial History of the Ewaso Basin, Kenya', *Smithsonian Contributions to Zoology* 632: 11–30.

Lane, P., et al. 2007. 'The Transition to Farming in Eastern Africa: New Faunal and Dating Evidence from Wadh Lang'o and Usenge, Kenya', *Antiquity* 81: 62–81.

Leakey, L.S.B. 1931. *The Stone Age Cultures of Kenya Colony*. London: Frank Cass.

Leakey, M.D. 1945. 'Report on the Excavations at Hyrax Hill, Nakuru, Kenya Colony', *Transactions of the Royal Society of South Africa* 30: 271–409.

Leakey, M.D., and L.S.B. Leakey. 1950. *Excavations at Njoro River Cave*. Oxford: Clarendon Press.

Lynch, B.M., and L.H. Robbins. 1979. 'Cushitic and Nilotic Prehistory: New Archaeological Evidence from North-west Kenya', *Journal of African History* 20: 319–28.

Maitima, J.M. 1991. 'Vegetation Response to Climatic Change in Central Rift Valley, Kenya', *Quaternary Research* 35: 234–45.

Marchant, R.A., and H. Hooghiemstra. 2004. 'Rapid Environmental Change in Tropical Africa and Latin America about 4000 Years before Present: A Review', *Earth Science Reviews* 66: 217–60.

Marean, C.W. 1992. 'Hunter to Herder: Large Mammal Remains from the Hunter-gatherer Occupation at Enkapune ya Muto Rock-shelter', *African Archaeological Review* 10: 65–127.

Marshall, F.B. 1989. 'Rethinking the Role of *Bos indicus* in Sub-Saharan Africa', *Current Anthropology* 30: 235–40.

——— 1990a. 'Origins of Specialized Pastoral Production in East Africa', *American Anthropologist* 92: 873–94.

——— 1990b. 'Cattle Herds and Caprine Flocks', in P.T. Robertshaw (ed.), *Early Pastoralists of South-western Kenya*. Nairobi: British Institute in Eastern Africa, pp.205–46.

——— 1994. 'Archaeological Perspectives on East African Pastoralism', in E. Fratkin, K.A. Galvin and E.A. Roth (eds), *African Pastoralist Systems*. Boulder, CO: Lynne Reinner, pp.17–43.

——— 2000. 'The Origins and Spread of Domestic Animals in East Africa', in R.M. Blench and K.C. MacDonald (eds), *The Origins and Development of African Livestock: Archaeology, Genetics, Linguistics and Ethnography*. London: University College Press, pp.191–221.

Marshall, F., K. Grillo and L. Arco. 2011. 'Prehistoric Pastoralists and Social Responses to Climatic Risk in East Africa', in N.F. Miller, K.M. Moore and K. Ryan (eds), *Sustainable Lifeways: Cultural Persistence in an Ever-changing Environment*. Philadelphia: University of Pennsylvania Museum Publications, pp.39–74.

Marshall, F.B., and E. Hildebrand. 2002. 'Cattle before Crops: The Beginnings of Food Production in Africa', *Journal of World Prehistory* 16: 99–143.

Marshall, F.B., and K.M. Stewart. 1995. 'Hunting, Fishing and Herding Pastoralists of Western Kenya: The Fauna from Gogo Falls', *Archaeozoologia* 7: 7–27.

Mbae, N.B. 1990. 'The Ethnoarchaeology of Maasai Settlements and Refuse Disposal Patterns in the Lemek Area', in P.T. Robertshaw (ed.), *Early Pastoralists of South-western Kenya*. Nairobi: British Institute in Eastern Africa, pp.279–92.

Mburu, D.N., et al. 2002. 'Genetic Diversity and Relationships of Indigenous Kenyan Camel (*Camelus dromedarius*) Populations: Implications for their Classification', *Animal Genetics* 34: 26–32.
Meadow, R., and A.K. Patel. 2003. 'From Mehrgarh to Harappa and Dholavira: Prehistoric Pastoralism in Northwestern South Asia from the Neolithic Through the Harappan Period', in S.A. Weber and W.R. Belcher (eds), *Indus Ethnobiology: New Perspectives from the Field.* Lanham, MD: Lexington Books, pp.65–94.
Merrick, H.V., and F.H. Brown. 1984. 'Obsidian Sources and Patterns of Source Utilization in Kenya and Northern Tanzania: Some Initial Findings', *African Archaeological Review* 2: 129–52.
Merrick, H.V., and M.C. Monaghan. 1984. 'The Date of the Cremated Burials in Njoro River Cave', *Azania* 19: 7–11.
Mutundu, K.K. 1999. *Ethnohistoric Archaeology of the Mukogodo in North Central Kenya.* Oxford: Archaeopress.
Nicholson, S.E., and E. Kim. 1997. 'The Relationship of the El Niño Southern Oscillation to African Rainfall', *International Journal of Climatology* 17: 117–35.
Nurse D., and G. Philippson. 1980. 'Bantu Languages of East Africa: A Lexicostatistical Survey', in E.C. Polome and C.P. Hill (eds), *Language in Tanzania.* London: Oxford University Press, pp.26–67.
Odner, K. 1972. 'Excavations at Narosura, a Stone Bowl Site in the Southern Kenya Highlands', *Azania* 7: 25–92.
Owen, R.B., et al. 1982. 'Palaeolimnology and Archaeology of Holocene Deposits North-east of Lake Turkana, Kenya', *Nature* 298: 523–29.
Phillipson, D.W. 1977a. 'The Excavation of Gobedra Rock-shelter, Axum', *Azania* 12: 53–82.
——— 1977b. 'Lowasera', *Azania* 12: 1–32.
——— 1979. 'The Origins of Prehistoric Farming in East Africa', *Hadith* 7: 41–63.
Posnansky, M. 1967. 'Excavations at Lanet, 1957', *Azania* 2: 89–114.
Prendergast, M.E. 2008. 'Forager Variability and Transitions to Food Production in Secondary Settings: Kansyore and Pastoral Neolithic Economies in East Africa', PhD dissertation. Cambridge, MA: Harvard University.
——— 2011. 'Hunters and Herders at the Periphery: The Spread of Herding in Eastern Africa', in H. Jousse and J. Lesur (eds), *People and Animals in Holocene Africa: Recent Advances in Archaeozoology.* Frankfurt am Main: Africa Magna Verlag, pp.43–58.
Reid, R.S., and J.E. Ellis. 1995. 'Impacts of Pastoralists on Woodlands in South Turkana, Kenya: Livestock-mediated Tree Recruitment', *Ecological Applications* 5: 978–92.
Richardson, J.L., and A.E. Richardson. 1972. 'History of an African Rift Lake and its Climatic Implications', *Ecological Monographs* 42: 499–534.
Ricketts, R.D., and T.C. Johnson. 1996. 'Climate Change in the Turkana Basin as Deduced from a 4000 Year Long δO^{18} Record', *Earth and Planetary Science Letters* 142: 7–17.
Riemer, H. 2007. 'When Hunters Started Hunting: Pastro-foragers and the Complexity of Holocene Economic Change in the Western Desert of Egypt', in M. Bollig et al. (eds), *Aridity, Change and Conflict in Africa.* Cologne: Heinrich-Barth Institut, pp.105–44.
Robbins, L.H. 1980. *Lopoy: A Late Stone-age Fishing and Pastoralist Settlement in the Late Turkana Basin, Kenya.* East Lansing: Michigan State University Museum.
——— 1984. 'Late Prehistoric Aquatic and Pastoral Adaptations West of Lake Turkana, Kenya', in J.D. Clark and S.A. Brandt (eds), *From Hunters to Farmers: The Causes and Consequences of Food Production in Africa.* Berkeley: University of California Press, pp.206–12.
Robertshaw, P.T. 1988. 'The Elmenteitan: An Early Food-producing Culture in East Africa', *World Archaeology* 20: 57–69.

——— 1991. 'Gogo Falls: Excavations at a Complex Archaeological Site East of Lake Victoria', *Azania* 26: 63–195.

——— (ed.). 1990. *Early Pastoralists of South-western Kenya.* Nairobi: British Institute in Eastern Africa.

Robertshaw, P.T., and D. Collett. 1983a. 'A New Framework for the Study of Early Pastoral Communities in East Africa', *Journal of African History* 24: 289–301.

——— 1983b. 'The Identification of Pastoral Peoples in the Archaeological Record: An Example from East Africa', *World Archaeology* 15: 67–78.

Robertshaw, P.T., and A. Mawson. 1981. 'Excavations in Eastern Equatoria, Southern Sudan, 1980', *Azania* 16: 55–95.

Robertshaw, P.T., T. Pilgrim, A. Siiriainen and F. Marshall. 1990. 'Archaeological Surveys and Prehistoric Settlement Patterns', in P.T. Robertshaw (ed.), *Early Pastoralists of South-western Kenya.* Nairobi: British Institute in Eastern Africa, pp.36–51.

Rossel, S., et al. 2008. 'Domestication of the Donkey: Timing, Processes, and Indicators', *Proceedings of the National Academy of Sciences* 105: 3715–20.

Rowley-Conwy, P. 1988. 'The Camel in the Nile Valley: New Radiocarbon Accelerator (AMS) Dates from Qasr Ibrim', *Journal of Egyptian Archaeology* 74: 245–48.

Ruel, M. 2000. 'The Kuria Homestead in Space and Time', *Journal of Religion in Africa* 30: 62–85.

Salas, A., et al. 2002. 'The Making of the African mtDNA Landscape', *American Journal of Human Genetics* 71: 1082–111.

Siiriäinen, A. 1977. 'Later Stone Age Investigation in the Laikipia Highlands, Kenya: A Preliminary Report', *Azania* 12: 161–86.

——— 1984. *Excavations in Laikipia: An Archaeological Study of the Recent Prehistory in the Eastern Highlands of Kenya.* Helsinki: Suomen Muinais-muistoyhdistksen Aikakauskirja Finska Fornminnesförenengens Tidskrift.

——— 1990. 'Excavations of Elmenteitan Sites in the Oldorotua Valley', in P.T. Robertshaw (ed.), *Early Pastoralists of South-western Kenya.* Nairobi: British Institute in Eastern Africa, pp.261–66.

Siiriäinen, A., O. Seitsonen and J. Laurén. 2009. 'Pastoralists in the Northeastern Mara Plains, Kenya: Archaeological Investigations of the Pastoral Neolithic and Pastoral Iron Age', *Azania* 44: 163–93.

Simons, A. 2005. 'Exchange Networks, Socio-political Hierarchies and the Archaeological Evidence for Differential Wealth amongst Pastoralists in Southwestern Kenya', *Nyame Akuma* 64: 36–40.

Smith, A.B. 1992. *Pastoralism in Africa: Origins and Development Ecology.* London: Hurst.

Sobania, N. 1988. 'Fishermen Herders: Subsistence, Survival and Cultural Change in Northern Kenya', *Journal of African History* 29: 41–56.

Sommer, G., and R. Vossen. 1993. 'Dialects, Sectiolects, or Simply Lects? The Maa Language in Time Perspective', in T. Spear and R. Waller (eds), *Being Maasai: Ethnicity and Identity in East Africa.* London: James Currey, pp.25–37.

Soper, R.C. 1989. 'History of Pottery in Kenya', in J. Barbour and S. Wandibba (eds), *Kenyan Pots and Potters.* Nairobi: Oxford University Press, pp.6–26.

Spear, T., and D. Nurse. 1992. 'Maasai Farmers: The Evolution of Arusha Agriculture', *International Journal of African Historical Studies* 25: 481–503.

Spencer, P. 1973. *Nomads in Alliance: Symbiosis and Growth among the Rendille and Samburu of Kenya.* Oxford: Oxford University Press.

Stager, J.C., and T.C. Johnson. 2000. 'A 12,400 C-14 yr Offshore Diatom Record from East Central Lake Victoria, East Africa', *Journal of Paleolimnology* 23: 373–83.

Stager, J.C., and P.A. Mayewski. 1997. 'Abrupt Early to Mid-Holocene Climatic Transition Registered at the Equator and the Poles', *Science* 276: 1834–36.

Stewart, K.M. 1989. *Fishing Sites of North and East Africa in the Late Pleistocene and Holocene: Environmental Change and Human Adaptation*. Oxford: British Archaeological Reports.

Street, F.A., and A.T. Grove. 1979. 'Global Maps of Lake-level Fluctuations Since 30,000 Years BP', *Quaternary Research* 12: 83–118.

Sutton J.E.G. 1964. 'A Review of Pottery from the Kenya Highlands', *South African Archaeological Bulletin* 19: 27–35.

——— 1973. *The Archaeology of the Western Highlands of Kenya*. Nairobi: British Institute in Eastern Africa.

——— 1987. *A Thousand Years in East Africa*. Nairobi: British Institute in Eastern Africa.

——— 1993a. 'Becoming Maasailand', in T. Spear and R. Waller (eds), *Being Maasai: Ethnicity and Identity in East Africa*. London: James Currey, pp.38–60.

——— 1993b. 'Deloraine: Further Excavation, and the Iron Age Sequence of the Central Rift', *Azania* 28: 103–25.

Taylor, D., et al. 2005. 'Holocene Vegetation Dynamics on the Laikipia Plateau, Kenya', *Holocene* 15: 837–46.

Terrell, J.E., et al. 2003. 'Domesticated Landscapes: The Subsistence Ecology of Plant and Animal Domestication', *Journal of Archaeological Method and Theory* 10: 323–68.

Tishkoff, S.A., et al. 2007. 'Convergent Adaptation of Human Lactase Persistence in Africa and Europe', *Nature Genetics* 39: 31–40.

Verschuren, D., K.R. Laird and B.R. Cumming. 2000. 'Rainfall and Drought in Equatorial East Africa During the Past 1,100 Years', *Nature* 403: 410–14.

Wandibba, S. 1980. 'The Application of Attribute Analysis to the Study of Later Stone Age/Neolithic Pottery Ceramics in Kenya (Summary)', in R. Leakey and B. Ogot (eds), *Proceedings of the 8th Panafrican Congress of Prehistory and Quaternary Studies, Nairobi, 5 to 10 September 1977*. Nairobi: TILLMIAP, pp.283–85.

Wendorf, F., R. Schild and A. Close. 1984. *Cattle-keepers of the Eastern Sahara: The Neolithic of Bir Kiseiba*. Dallas, TX: Department of Anthropology, Southern Methodist University.

Western, D., and T. Dunne. 1979. 'Environmental Aspects of Settlement Site Decisions among Pastoral Maasai', *Human Ecology* 7: 75–98.

Wright, D.K. 2003. 'Archaeological Investigations of Three Pastoral Neolithic Sites in Tsavo National Park, Kenya', *Azania* 38: 183–88.

——— 2005. 'New Perspectives on Early Regional Interaction Networks of East African Trade: A View from Tsavo National Park, Kenya', *African Archaeological Review* 22: 111–40.

——— 2007. 'Tethered Mobility and Riparian Resource Exploitation Among Neolithic Hunters and Herders in the Galana River Basin, Kenyan Coastal Lowlands', *Environmental Archaeology* 12: 25–47.

Young, T.P., N. Partridge and A. Macrae. 1995. 'Long-term Glades in Acacia Bushland and their Edge Effects in Laikipia, Kenya', *Ecological Applications* 5: 97–108.

Zeder, M.A., and B. Hesse. 2000. 'The Initial Domestication of Goats (*Capra hircus*) in the Zagros Mountains 10,000 Years Ago', *Science* 287: 2254–57.

Chapter 4
From the First Stock Keepers to Specialised Pastoralists in the West African Savannah

Veerle Linseele

Introduction

The present chapter deals with the emergence and spread of pastoralism in the western parts of Africa (see Figure 4.1). In its simplest and most general sense, 'pastoralism' is synonymous with domestic stock keeping. Except perhaps for guineafowl, and for very recent experiments with the domestication of local wild species, none of the domestic animals kept in western Africa nowadays are indigenous to the region. A major part in any study of the history of stock keeping in the West African savannah therefore concerns the date and mode of introduction of each of the domesticated animal species, and its adaptation to this new environment. At the other end of the extreme, 'pastoralism' refers to specialised pastoralists, who have stock keeping as their principle economic activity, are usually highly mobile, and obtain crops through exchange relations with farmers. How and when this mode of livelihood developed in the savannah of West Africa is also investigated.

Figure 4.1: Map of West Africa with eco-climatic zones (after Le Houérou 1989: 7).

The discussion of the history of pastoralism in the West African savannah starts from archaeozoological data obtained from sites in northern Burkina Faso and in north-eastern Nigeria, near the shores of Lake Chad (Figures 4.2 to 4.4) (Linseele 2007). The data set comes from a few dozen sites, covering the entire chronological sequence between about 2000 BC and the present.[1] The fauna of these sites was excavated and studied with similar methods, which allows for easy comparisons. Moreover, archaeobotanical data are also available from the same sites. Other regions of the West African savannah are only patchily investigated archaeologically and the available archaeozoological information is of variable quality, but is included where relevant.

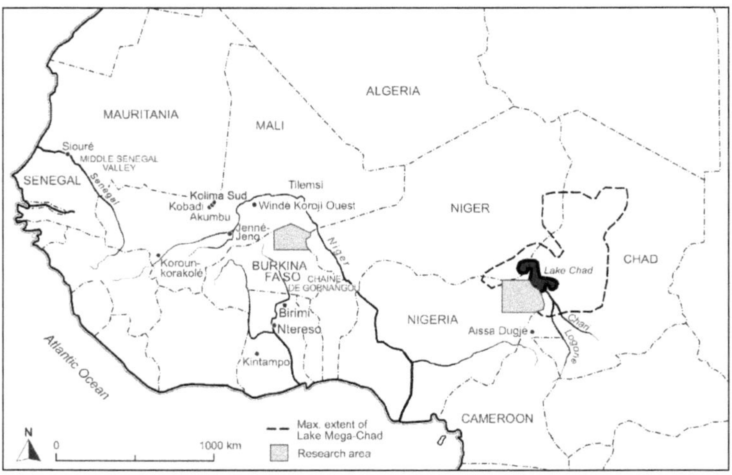

Figure 4.2: Map of West Africa with locations mentioned in the text. See Figures 4.3 and 4.4 for detailed maps of northern Burkina Faso and north-eastern Nigeria.

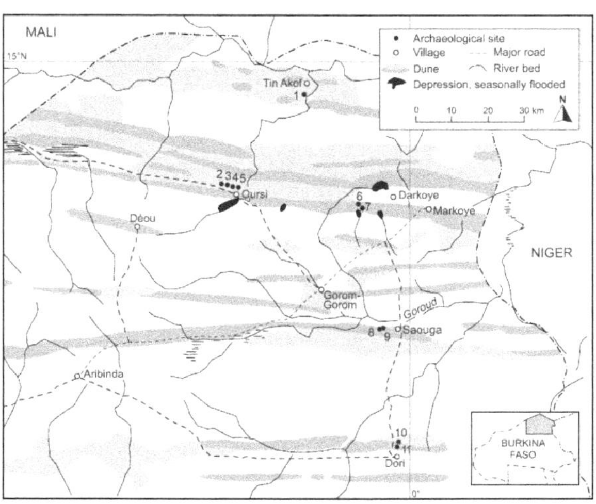

Figure 4.3: Detailed map of northern Burkina Faso showing the locations of archaeological sites mentioned in Tables 4.2 and 4.3.
1: BF 94/133 (Tin Akof), 2: BF 94/45 (Oursi), 3: BF 97/5 (Corcoba), 4: BF 97/30 (Oursi hu-beero), 5: BF 97/13 (Oursi village), 6: BF 96/22 (Kissi 22), 7: BF 97/31 (Kissi 40), 8: BF 94/120 (Saouga 94/120), 9: BF 95/7 (Saouga 95/7), 10: BF 94/40 (Dori 94/40), 11: BF 94/96 (Dori 94/96).

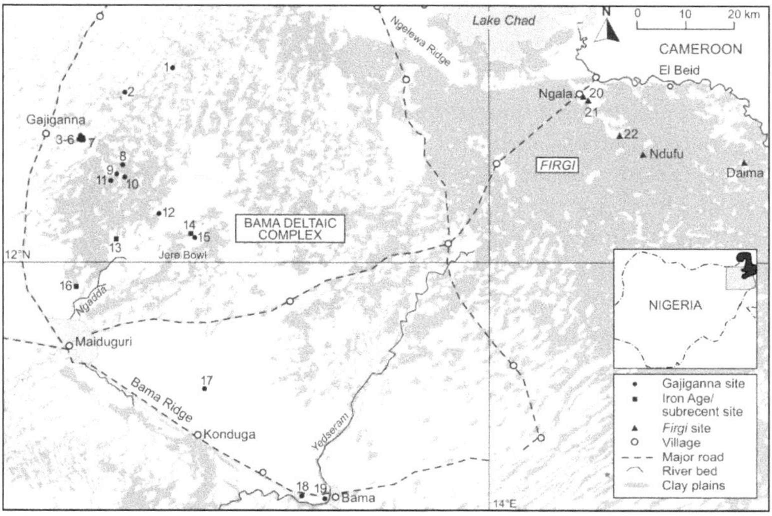

Figure 4.4: Detailed map of north-eastern Nigeria showing the locations of archaeological sites mentioned in Tables 4.2 and 4.3.
1: NA 97/37 (Zilum), 2: NA 92/2C (Galaga), 3: NA 90/5A (Gajiganna A), 4: NA 90/5BI–II (Gajiganna BI and BII), 5: NA 90/5C (Gajiganna C), 6: NA 90/5D (Gajiganna D), 7: NA 99/65 (Gilgila), 8: NA 93/36, 9: NA 93/10, 10: NA 93/42, 11: NA 91/1A, 12: NA 97/18 (Dumge A), 13: NA 99/75 (Elkido North), 14: NA 97/24 (Labe), 15: NA 97/26 (Labe Kanuri), 16: NA 97/13 (Dorota), 17: NA 97/33 (Bukarkurari), 18: NA 95/1 (Kariari C), 19: NA 96/45 (Kelumeri), 20: Ngala (NA 93/45), 21: Kursakata (NA 93/46), 22: Mege (NA 94/7).

The West African Savannah and Pastoralism at Present

For a good understanding of the history of pastoralism in the West African savannah, a description is first necessary of the particularities of the region today, as well as of the modes of livelihood of its specialised pastoralists. In addition, the main domestic animal species of the area are discussed.

Geography

The savannah of West Africa, between the Atlantic Coast and Lake Chad, is an almost 1,500 kilometre wide zone reaching from the Sahara Desert in the north to the rainforest in the south. The climate in the area is characterised by high temperatures and by markedly seasonal rainfall. The rainy season is shorter and rainfall becomes more restricted and more irregular from south to north. The rainfall isohyets allow the subdivision of the savannah into three different zones: the Guinean, Sudan, and Sahel zone (see Figure 4.1). Besides its unstable climate, another ecological constraint of sub-Saharan West Africa, mainly south of the Sahel zone, is the occurrence of several endemic, humidity-related diseases, such as malaria and sleeping sickness or trypanosomiasis (Connah 1987).

The 400 millimetre isohyet is approximately the northern border of the region where farming communities of the West African savannah can settle (Gallais 1975: 167–68; Krings 1980: 15). At lower rainfall levels, farming-based subsistence is impossible because farmers need a minimum amount of rainfall for crop growing, unless compensated by the presence of flood waters. Moreover, enough drinking water has to be present year round near to their permanent settlements. In areas which do not receive enough rainfall to practise agriculture, livestock keeping is the main economic activity. In the West African savannah of today, the economic importance of domestic stock within a household generally rises with declining rainfall, culminating in an almost total dependence on stock keeping in (sub-)desert zones (Amanor 1995; FAO 2001).

(Specialised) Pastoralists

The stock keepers of the arid zones of West Africa usually need to have a mobile lifestyle in order to be able to provide their animals with the necessary food and water the whole year round (see Figure 4.5). Simplified, their movements can be either 'transhumant' – that is, movement between fixed points – in order to exploit the seasonal availability of pastures (FAO 2001), or 'nomadic' – that is, more irregular (Khazanov 1994: 16). In the case of the former, there is a permanent homestead and base where the older members of the community stay all year round (FAO 2001), while in the latter the whole community is involved in movements (Khazanov 1994: 16).

Figure 4.5: Shuwa Arabs in the Dikwa area (Borno State, Nigeria), migrating with their cattle herds. Photograph by Peter Breunig.

Grain crops are always an important part of the pastoral diet (e.g., Smith 1978; Froment 1988: 47–50). Pastoralists obtain them either from farming communities, often in exchange for dairy products, or grow crops themselves seasonally (Dahl and Hjort 1976: 155; McIntosh 1997; FAO 2001). In the latter case they usually make transhumant types of movements. In contrast to transhumant pastoralists, nomadic pastoralists need exchange relations with their settled neighbours for survival (Khazanov 1994). Because of these exchange relations, nomadic pastoralists are also called specialised pastoralists.[2] What holds true for nomadic pastoralists equally counts for other specialised groups of the West African savannah. Specialised fishers also have a high component of cultivated plants in their diet, which they often obtain through trade or exchange with farmers (Sundström 1972: 134–36). Specialised fully sedentary farmers and specialised nomadic pastoralists are the extreme ideal types. Many intermediate lifestyles exist in the present West African savannah, which can be classified according to the degree of mobility versus sedentism of the human groups on the one hand, and their degree of dependence on stock keeping versus crop cultivation, on the other (e.g., Gallais 1975: 188–89).

Pastoralists generally obtain more calories from milk than from meat (FAO 2001). Cattle are the main milk providers in Africa and their productivity rises with the number of adult females in a cattle herd, their fecundity, and the quality of the available forage (Dahl and Hjort 1976: 142, 200). Fecundity and forage availability are largely related to the quantity and periodicity of rainfall. Pastoralists from West Africa, which has only one rainy season, can therefore rely less on milk than their counterparts in East Africa, where the bimodal rainfall regime allows for more constant milk production throughout the year.

Domestic Livestock

Cattle and ovicaprines are the main livestock species of the West African savannah. Pastoral nomads keep mainly zebu-type or humped cattle, although nowhere in West Africa do pure types occur (see below). Zebu need less food and water and are better suited to long distance migration than taurine or humpless cattle.[3] Moreover, their milk production is less affected by food shortages, and they are less sensitive to ticks, gastrointestinal parasites and rinderpest (cattle plague) (Marshall 1989; FAO 2001). On the other hand, taurines are mainly kept in transhumant systems or by farmers (FAO 2001). They are more resistant to trypanosomiasis, the main disease threatening livestock in areas where agriculture can be practised. Sanga, usually humpless but sometimes with a small, muscular hump, are sometimes considered as a third type of African cattle (Grigson 1991). They are less resistant to rinderpest and produce less milk than zebu, but they are trypanotolerant (Porter 1991: 193). At the end of the nineteenth century AD, a rinderpest epidemic eradicated taurine and Sanga cattle from most of the African continent (Payne 1964).

The great majority of the ovicaprines in the West African savannah are sedentary village livestock (e.g., Bourn et al. 1994). The main types of sheep and goats have been described by Epstein (1971). Savannah type ovicaprines, kept by both farmers and pastoral nomads, are usually rather large. They are well adapted to arid conditions but susceptible to trypanosomiasis. In forested zones, on the other hand, dwarf forms of both sheep and goat occur, that is, animals with a withers height below 50 centimetres. They can occasionally also be found further north. Dwarf sheep and goat are extremely hardy and trypanotolerant and are solely associated with sedentary farmers.

Dromedaries are the only livestock animals which enable the exploitation of areas receiving less than 200 millimetres of annual precipitation (Köhler-Rollefson 1996), and they are therefore the prevalent species in desert areas. Their southern distribution limit is situated at the southern end of the Sahel zone, because of diseases further south (Wilson 1984: 17). Pigs are kept all over the forest savannah belt of West Africa, usually by settled farmers (Blench 2000c), but their status as taboo animals according to Islam restricts their spread (Simoons 1994: 40–43). Moreover, pigs have high water requirements and are unable to feed on cellulose-rich plants, requiring that they are kept close to settlements (Zeder 1991: 30–31). Other domestic mammals that can be found in the West African savannah are dogs, horses, including large (Dongolowi and Barb) as well as small, pony types (Epstein 1971: 436–80), and donkeys. Domestic poultry, especially chicken and guineafowl, are also important in West African economies (Bourn et al. 1994).

First Domestic Animals in West Africa

The earliest domestic animals were to be found in the north-eastern parts of Africa, whence they gradually propagated over the rest of the continent (see Table 4.1).[4] During the early and middle Holocene climatic optimum (Gasse 2000), groups of stock keepers made a living in several regions of the Sahara. The available archaeological evidence indicates that they moved southwards with the onset of the mid- to late-Holocene dry phase, thus introducing the first livestock into sub-Saharan West Africa (Kuper and Kröpelin 2006).

Table 4.1: Earliest evidence for the main domestic livestock animals in Africa, and West Africa in particular. For more details on genetics see Gifford-Gonzalez and Hanotte (2011).

Domestic fowl (*Gallus gallus* f. domestica) Earliest evidence on African continent: multiple depictions from middle 2nd millennium BC, Egypt (MacDonald 1992). Earliest finds in sub-Saharan West Africa: late 6th century AD, Kirikongo (Burkina Faso) (Dueppen 2011), AD 500-850, Jenné-Jeno (Mali) (MacDonald 1992; 1995; MacDonald and MacDonald 2000).
Helmeted guineafowl (*Numida meleagris*) Distinction between wild and domestic form impossible from bone remains. Earliest finds in sub-Saharan West Africa: first half 2nd millennium BC, Kintampo (Ghana) (Stahl 1985); Gajiganna phase IIa (1500–1200 BC), northeastern Nigeria (Linseele 2007).
Dog (*Canis lupus* f. familiaris) By 6th millennium BC in Northeast Africa, arriving together with domestic sheep and goat from the Levant as herding dogs (Gallant 2002: 51; Gautier 2002)? Earliest finds in sub-Saharan West Africa: Gajiganna phase IIb (1200–1000 BC), northeastern Nigeria (Linseele 2007).

Donkey (*Equus africanus* f. asinus) Earliest finds in Africa: second half 4th millennium BC, several sites in Egypt (Marshall 2007). Earliest finds in sub-Saharan West Africa: AD 0–250, Siouré (Senegal) (MacDonald and MacDonald 2000).
Horse (*Equus ferus* f. caballus) Earliest finds in Africa: Second Intermediate Period (1786–1552 BC), Tell Heboua, Sinai (Egypt) (Chaix 2000). Earliest finds in sub-Saharan West Africa: 8th century AD, Aissa Dugjé (northern Cameroon) (MacEachern et al. 2001); AD 600–1000, Akumbu (Mali) (MacDonald and MacDonald 2000).
Pig (*Sus scrofa* f. domestica) Earliest finds in Africa: 5th millennium BC, Merimde (Egypt) (von den Driesch and Boessneck 1985: 23–29). Arrives in West Africa with coming Europeans or in early 2nd millennium AD (Blench 2000c)?
Dromedary (*Camelus thomasi* f. dromedarius) Earliest finds in Africa: 1st millennium BC, Qasr Ibrim (Egypt) (Rowley-Conwy 1988). Earliest finds in sub-Saharan West Africa: AD 250–400, Siouré (Senegal) (MacDonald and MacDonald 2000).
Sheep (*Ovis ammon* f. aries) and **goat** (*Capra aegagrus* f. hircus) Earliest finds in Africa: 6th millennium BC, in Eastern and Western Desert of Egypt (Gautier 2002). Earliest finds in sub-Saharan West Africa: 2200–850 BC, Windé Koroji Ouest (Mali) (MacDonald 1996); c. 1600 BC, Kobadi (Mali) (Jousse and Chenal-Vélardé 2001–2002); 1600/1500 BC, Bukarkurari, northeastern Nigeria (Linseele 2007).
Taurine cattle (*Bos primigenius* f. taurus) and **Sanga cattle** Earliest taurine/Sanga cattle finds in Africa: 9th millennium BC, Western Desert of Egypt (Gautier 1987; 2002; 2007), but disputed (e.g., Riemer 2007). Secure evidence in Africa from around 6000 BC, Western Desert of Egypt (Gautier 2011; Riemer 2007). Earliest taurine/Sanga cattle finds in sub-Saharan West Africa: 2200–850 BC, Windé Koroji Ouest (Mali) (MacDonald 1996); ca. 1600 BC, Kobadi (Mali) (Jousse and Chenal-Vélardé 2001–2002); 1600/1500 BC Bukarkurari, northeastern Nigeria (Linseele 2007).
Zebu cattle (*Bos primigenius* f. indicus) Known in New Kingdom Egypt (1569–1076 BC) through feudal tributes and spoils of war (Nicolotti and Guérin 1992). On African continent in considerable numbers from 4th century AD (Epstein 1971: 519). Distinction between zebu and other cattle types difficult from bone remains. Earliest finds in sub-Saharan West Africa: possibly AD 800–1400, Jenné Jeno (Mali) and Siouré (Senegal) (MacDonald 1995; MacDonald and MacDonald 2000).

First Domestic Animals in Northern Burkina Faso

The oldest sites discovered in the north of Burkina Faso date to between 2200 and 1000 BC. The sites are extremely small and shallow, and are presumably the remnants of highly mobile groups originating from the Sahara (Vogelsang, Albert and Kahlheber 1999; Vogelsang 2000). These Late Stone Age (LSA) sites have not yielded a single bone from domestic animals (see Tables 4.2 and 4.3). However, small numbers of bones from domestic stock have been reported from Windé Koroji Ouest (2200 to 950 BC), a broadly contemporary settlement mound in the neighbouring eastern Inland Niger Delta of Mali (MacDonald 1996;

MacDonald and MacDonald 2000). However, measurements of these bones fit better with those of wild than those of domestic bovids, and the alleged domestic cattle and ovicaprines from Windé Koroji could instead be buffalo (*Syncerus caffer*) and bush duiker (*Sylvicapra grimmia*) respectively (Linseele 2007: 132, 137). On the other hand, the cultural material found at the site has affinities with that from the Tilemsi Valley, in the Malian Sahara, where domestic livestock are known to have existed from at least the second millennium BC (Smith 1974, 1975). Looking at this evidence from Mali, it seems possible that there were domestic animals in LSA northern Burkina Faso but that their bones are missing because of the small size of the available samples, increasing the chances that rare species are missing. The recovered plant remains from LSA Burkina Faso consist mainly of fruits of wild trees and shrubs, but some sparse cultivated pearl millet (*Pennisetum americanum*) was also present (Neumann, Breunig and Kahlheber 2000). AMS-dating run on grains found at Ti-n-Akof yielded a date of around 1000 BC (Vogelsang, Albert and Kahlheber 1999; Vogelsang 2000), but the crop was already present in the oldest of the site's occupational phases (Kahlheber and Neumann 2007), which began around 1800 BC according to the archaeological inventory (Kühltrunk 2000).

Virtually no archaeological traces are known in the north-east of Burkina Faso from the entire first millennium BC, but around the beginning of the first millennium AD, the Iron Age sets in and large settlement mounds appear, marking a shift to a sedentary way of life (Breunig and Neumann 2002a). With the beginning of the Iron Age, the earliest archaeozoological evidence for domestic livestock, cattle, sheep, goat and dog appears in the area (Tables 4.2 and 4.3). The period also seems to have brought a shift to full agriculture, the archaeobotanical evidence showing a diversification of the crops grown. Pearl millet remained the staple crop, but farming became more productive thanks to the introduction of mixed cropping and agroforestry systems (Kahlheber 2004).

First Domestic Animals in North-eastern Nigeria

Geographically, north-eastern Nigeria is largely dominated by the presence of Lake Chad. During the early and middle Holocene climatic optimum, it was about as large as the Caspian Sea is today, but because of increasing aridity after the mid Holocene the lake gradually retreated. The new ground thereby becoming available for human use, around 2000 BC, was the geographical unit known as the Bama Deltaic Complex. About a millennium later, the so-called *firgi* area also emerged from the regression of the lake. Large parts of this area are still inundated annually (Bawden 1972).

The Gajiganna Culture represent the first colonists of the Bama Deltaic Complex. The origins of these colonists should probably be looked for in the Central Sahara (Breunig and Neumann 2002a). The Gajiganna Culture has been subdivided into three major phases, phase I (1800 to 1400 BC), phase IIa–c (1500 to 800 BC) and phase III (600 to 400 BC) (Breunig et al. 2006; Wendt 2007). Domestic animals are present from Gajiganna phase I (Tables 4.2 and 4.3). At that time there

Table 4.2: Numbers of osteologically identified bone remains by major economic strategy and phase at archaeological sites in northern Burkina Faso and north-eastern Nigeria.

				Fishing	Hunting and fowling*	Domestic stock keeping*
Burkina Faso	BF 94/133	LSA	2200–1000 BC	1	1	—
	BF 94/96	LSA	2200–1000 BC	—	—	—
	BF 97/5	LSA	2200–1000 BC	1410	27	—
	BF 97/5	EIA	AD 0–500	892	—	—
	BF 94/45	LSA	2200–1000 BC	30	10	—
	BF 94/45	EIA	AD 0–500	1474	196	310
	BF 97/13	EIA	AD 0–500	55	4	36
	BF 97/13	MIA	AD 500–1000	45	19	274
	BF 97/13	LIA	AD 1000–1400	55	35	349
	BF 97/30	LIA	AD 1000–1400	79	27	863
	BF 96/22	MIA	AD 500–1000	90	96	312
	BF 97/31	LIA	AD 1000–1400	129	98	280
	BF 94/120	LIA	AD 1000–1400	403	313	1285
	BF 95/7	LIA	AD 1000–1400	178	169	701
Bama Deltaic Complex (northeastern Nigeria)	NA 90/5C	Gaj IIa	1500–1200 BC	—	8	16
	NA 90/5A	Gaj IIb	1200–800 BC	415	121	1349
	NA 90/5BI	Gaj IIa/b	1500–800 BC	821	41	487
	NA 90/5BII	Gaj IIc	1000–800 BC	8	115	241
	NA 91/1A	Gaj I	1800–1400 BC	—	—	6
	NA 93/42	Gaj I	1800–1400 BC	71	6	33
	NA 97/18	Gaj I	1800–1400 BC	32	—	1
	NA 97/24	Gaj I	1800–1400 BC	—	—	14
	NA 93/36	Gaj IIa	1500–1200 BC	9	4	8
	NA 99/65	Gaj IIa	1500–1200 BC	—	—	1
	NA 99/65	Gaj IIb	1200–800 BC	61	11	55
	NA 99/65	Gaj IIc	1000–800 BC	5	18	56
	NA 93/10	Gaj IIc	1000–800 BC	—	—	29
	NA 97/37	Gaj III	600–400 BC	262	55	150
	NA 97/33	Gaj I	1500–1200 BC	4	11	599
	NA 97/33	Gaj IIa/b	1500–1000 BC	7	21	244
	NA 96/45	Gaj IIa/b	1500–1000 BC	40	16	320
	NA 95/1	Gaj IIa/b	1500–1000 BC	55	51	559
	NA 97/26	IA	AD 0–700	58	32	54
	NA 99/75	IA	AD 0–700	90	28	54
	NA 97/13	IA	AD 0–700	—	12	49
	NA 92/2C	subrecent	19th century AD	124	125	922
Firgi (northeastern Nigeria)	NA 93/46	LSA	1300/800–800/500 BC	928	—	36
	NA 93/46	EIA	800/500 BC–AD 400/600	24942	94	342
	NA 94/7	I	550 BC–AD 50	2261	198	593
	NA 94/7	II	AD 50–700	854	17	55
	NA 94/7	III	AD 700–1150	706	28	40
	NA 94/7	IV	AD 1150–1983	468	7	27
	NA 93/45	I	Mixed/?–7th century AD	2088	36	182
	NA 93/45	II	8–10/11th century AD	4488	67	207
	NA 93/45	IIIa	10–12th century AD	1841	23	116
	NA 93/45	IIIb	10–12th century AD	1261	23	119
	NA 93/45	IV	19–20th century AD	1611	14	132

Table 4.2 Notes:
EIA – Early Iron Age; Gaj – Gajiganna phase; IA – Iron Age; LIA – Late Iron Age; LSA – Late Stone Age; MIA – Middle Iron Age.

*Remains impossible to identify more precisely than large galliform, medium-sized carnivore, small bovid, large bovid, and cattle or buffalo (*Syncerus caffer*), could pertain to either domestic or hunted animals. Their numbers were added to the more precisely identified taxa in the same proportions as they occurred in the accurately identified taxa of these groups. For example, when a site yields 100 bones identified as small bovid, 186 as domestic ovicaprines, and 14 as antelopes of the same size class, then 93 bones of the small bovids are put with the domestic mammals, and 7 with the wild mammals. In cases where there were no bones within a certain group, which could be identified either as wild or as domestic, the bones of 'wild or domestic' animals were excluded from the counts.

are as yet no indications of agriculture (Klee and Zach 1999; Klee, Zach and Stika 2004). Gajiganna phase I sites have been interpreted as short-term camps (Breunig and Neumann 2002a). Only one of these, Bukarkurari (NA 97/33), yielded a large archaeofaunal sample. It is dominated by domestic cattle, but sheep and goat are also present. Dated to around 1600 to 1500 BC, these domestics belong to the oldest known examples in sub-Saharan West Africa (Table 4.1) (MacDonald and MacDonald 2000; Jousse 2004). Domestic dog is attested in the southern Lake Chad area from Gajiganna phase IIb (1200 to 1000 BC) onwards (Table 4.3). No older dog finds are known from the rest of the West African savannah, but the animals were probably present from earlier in the second millennium BC. It is assumed that they spread from North-east Africa, together with domestic cattle and ovicaprines, as shepherding dogs (Gautier 2000: 203; 2001: 620; Gallant 2002: 51), and that they did not arrive as a separate introduction. During Gajiganna phase II, people apparently grew pearl millet in the Bama Deltaic Complex (Klee and Zach 1999; Klee, Zach and Stika 2004). The degree of sedentism of the settlements increased concomitantly, reflected by the larger size of the sites (Breunig and Neumann 2002a).

In the *firgi*, traces of human occupation appear between 1000 and 800 BC (Gronenborn 1998). The first occupants of the area probably originated from areas north of Lake Chad (Wiesmüller 2001). Domestic animals (Tables 4.2 and 4.3), as well as cultivated crops (Klee, Zach and Neumann 2000), are present in the area from then on. Farming does not seem to have been of great economic importance yet, and the existence of sedentary farming communities need not be supposed. Mobile lifestyles are also confirmed by the poor bone productivity of the LSA phases at the *firgi* site of Kursakata (NA 93/46), compared to its later occupation periods (Linseele 2007: 89).

Stimuli and Constraints for the Propagation of the Earliest Domestic Animals

Saharan stock keepers brought the earliest domestic animals to western Africa in the first half of the second millennium BC, but it may have taken until the beginning of our era before they arrived in northern Burkina Faso. In the south of the

Table 4.3: Numbers of osteologically identified bone remains from domestic animals by phase at archaeological sites in northern Burkina Faso and north-eastern Nigeria.

				Domestic fowl	Helmeted guineafowl	Domestic dog	Horse	Donkey/horse	Sheep/goat	Sheep	Goat	Cattle
Burkina Faso	BF 94/133	LSA	2200–1000 BC	—	—	—	—	—	—	—	—	—
	BF 94/96	LSA	2200–1000 BC	—	—	—	—	—	—	—	—	—
	BF 97/5	LSA	2200–1000 BC	—	—	—	—	—	—	—	—	—
	BF 97/5	EIA	AD 0–500	—	—	—	—	—	—	—	—	—
	BF 94/45	LSA	2200–1000 BC	—	—	—	—	—	—	—	—	—
	BF 94/45	EIA	AD 0–500	—	—	1	—	—	201	7	23	108
	BF 97/13	EIA	AD 0–500	—	—	1	—	—	27	—	1	8
	BF 97/13	MIA	AD 500–1000	—	1	39	—	1	159	2	11	76
	BF 97/13	LIA	AD 1000–1400	—	—	34	—	2	212	9	9	103
	BF 97/30	LIA	AD 1000–1400	28	—	65	—	2	656	67	52	114
	BF 96/22	MIA	AD 500–1000	—	—	18	—	—	275	3	15	19
	BF 97/31	LIA	AD 1000–1400	—	—	31	—	—	217	2	11	32
	BF 94/120	LIA	AD 1000–1400	81	1	276	1	1	879	29	74	149
	BF 95/7	LIA	AD 1000–1400	46	1	253	—	2	356	7	11	46
Bama Deltaic Complex (northeastern Nigeria)	NA 90/5C	Gaj IIa	1500–1200 BC	—	1	—	—	—	6	—	1	10
	NA 90/5A	Gaj IIb	1200–800 BC	—	3	6	—	—	500	1	73	849
	NA 90/5BI	Gaj IIa/b	1500–800 BC	—	1	3	—	—	61	1	13	426
	NA 90/5BII	Gaj IIc	1000–800 BC	—	34	—	—	—	84	—	8	155
	NA 91/1A	Gaj I	1800–1400 BC	—	—	—	—	—	—	—	—	6
	NA 93/42	Gaj I	1800–1400 BC	—	—	—	—	—	2	—	1	31
	NA 97/18	Gaj I	1800–1400 BC	—	—	—	—	—	—	—	—	1
	NA 97/24	Gaj I	1800–1400 BC	—	—	—	—	—	—	—	—	14
	NA 93/36	Gaj IIa	1500–1200 BC	—	—	—	—	—	—	—	—	8
	NA 99/65	Gaj IIa	1500–1200 BC	—	1	1	—	—	—	—	—	1
	NA 99/65	Gaj IIb	1200–800 BC	—	—	—	—	—	45	—	4	10
	NA 99/65	Gaj IIc	1000–800 BC	—	—	—	—	—	27	2	6	23
	NA 93/10	Gaj IIc	1000–800 BC	—	—	—	—	—	4	—	—	25
	NA 97/37	Gaj III	600–400 BC	—	—	3	—	—	31	1	2	119
	NA 97/33	Gaj I	1500–1200 BC	—	—	—	—	—	78	—	1	521
	NA 97/33	Gaj IIa/b	1500–1000 BC	—	—	—	—	—	43	—	1	202
	NA 96/45	Gaj IIa/b	1500–1000 BC	—	—	—	—	—	127	1	3	193
	NA 95/1	Gaj IIa/b	1500–1000 BC	—	—	—	—	—	113	—	12	446
	NA 97/26	IA	AD 0–700	—	—	1	—	—	39	—	—	15
	NA 99/75	IA	AD 0–700	—	—	1	—	—	37	3	1	17
	NA 97/13	IA	AD 0–700	—	1	2	—	—	43	—	1	6
	NA 92/2C	subrec.	19th century AD	18	2	2	—	—	622	65	40	282
Firgi (northeastern Nigeria)	NA 93/46	LSA	1300/800–800/500 BC	—	—	—	—	—	16	—	1	20
	NA 93/46	EIA	800/500 BC–AD 400/600	—	—	—	—	—	242	6	10	100
	NA 94/7	I	550 BC–AD 50	—	—	1	—	—	46	1	5	547
	NA 94/7	II	AD 50–700	—	—	—	—	—	8	—	1	47
	NA 94/7	III	AD 700–1150	—	—	—	—	—	22	—	—	16
	NA 94/7	IV	AD 1150–1983	—	—	—	—	—	13	—	1	14
	NA 93/45	I	Mixed/?–7th century AD	—	—	3	—	—	117	3	11	45
	NA 93/45	II	8–10/11th century AD	—	—	—	—	—	114	3	8	53
	NA 93/45	IIIa	10–12th century AD	—	—	—	—	—	64	2	2	47
	NA 93/45	IIIb	14/16–18th century AD	—	—	—	—	—	31	3	1	85
	NA 93/45	IV	19–20th century AD	—	—	—	—	—	59	1	5	64

country, in the Chaîne de Gobnangou, stock keeping started even later. Until AD 1000, only wild plants and animals, with the possible exception of domestic dogs, were apparently exploited there (Frank et al. 2001). Similarly, a hunter-gatherer lifestyle possibly persisted at Korounkorokalé in Mali until as recently as the mid to late first millennium AD (MacDonald 1997). MacDonald (ibid.) proposed a model for the peopling of West Africa featuring a long-term autochthonous human presence and a continuation of pure hunter-gatherer economies south of the Sahara. The archaeological evidence in support of his model is scarce and it also relies on local oral history. In the southern Lake Chad area, domestic stock keeping could perhaps have spread at an early date and on a relatively large scale because of the absence of autochthonous populations, as the area was uninhabitable until shortly before Saharan stock keepers arrived. Inversely, in areas such as northern Burkina Faso, the presence of hunter-gatherer groups, making it more difficult for immigrating groups to settle there than in unpopulated areas, may have retarded the introduction of domestic livestock. However, this remains speculative since former hunter-gatherer populations of West Africa are poorly documented and archaeological evidence in the area from the period before 2000 BC is scarce. It seems, moreover, that the presence of local hunter-gatherers did not necessarily retard the introduction of domestic stock. For example, domestic cattle and ovicaprines appear at an early date, during the first half of the second millennium BC, at Kintampo Culture sites on the forest margin of Ghana where the evidence available indicates that they were incorporated into a local tradition (Anquandah 1993; Stahl 1993; Gautier and Van Neer 2005; Gifford-Gonzalez 2005; Watson 2005). In fact, the complex is the only known example from the West African savannah of interaction between Saharan herders and indigenous populations.

In the southern Lake Chad area, the absence of an autochthonous population would not have been the only reason for the relatively early presence of domesticated animals. Its attractive environmental setting, offering plentiful drinking water and pasture, would certainly have been a factor in its favour. Comparable conditions must have been available in the two other large floodplains of arid West Africa, the Inland Niger Delta and the Middle Senegal Valley (McIntosh 1999). In the Inland Niger Delta, early livestock have been uncovered at Kobadi (about 1600 BC), where cattle, but no ovicaprines, have been identified (Jousse and Chenal-Vélardé 2001). The presence of this favourable area north of northern Burkina Faso may also have slowed down the introduction of domestic animals there, because there was probably no ecological pressure on these stock keepers to move further south. The available archaeozoological evidence indicates that domestic livestock may not have been present as early in the Middle Senegal Valley as in the two other large floodplains of the West African Savannah. The first records of domestic livestock there only date back to the beginning of our era (MacDonald and MacDonald 2000).

The finds from the Kintampo sites of Ntereso and Kintampo in Ghana, dated to the first half of the second millennium BC, indicate that domestics spread very rapidly over the West African savannah once south of the Sahara (Carter and Flight

1972; Stahl 1985). This suggests that diseases may not have slowed down the spread of livestock as markedly as in eastern or southern Africa (Gifford-Gonzalez 2000; Marshall 2000), perhaps because the first livestock entering sub-Saharan West Africa was already trypanotolerant. However, the question of when such trypanotolerant stock developed in west Africa is still a matter of debate. There is a possible, but unproven link between trypanotolerance and dwarfism (Epstein 1971: 276; Clutton-Brock 2000; MacDonald and MacDonald 2000). If there is indeed such a link, then the first cattle in sub-Saharan West Africa may already have been trypanotolerant, since they were apparently of rather small size (MacDonald and MacDonald 2000; Jousse 2004; Linseele 2007; see below).

Introduction Waves and Local Developments of Livestock

Multiple, later introductions took place after the very first appearance of domestic animals in sub-Saharan West Africa. The arrival of domestic livestock in the area seems to have happened in four main 'waves', labelled the Late Stone Age (LSA), Iron Age, and the Arab and European waves (Linseele 2007). Once propagated across West Africa, livestock locally underwent changes and new types developed.

Late Stone Age Wave (Second Millennium BC to the Mid First Millennium BC)

The first wave of introduction during the LSA has been described above. It included all the earliest introductions, notably the first cattle, ovicaprines and dogs. The animals arrived with Saharan stock keepers, moving southwards due to increasing aridification. It is not clear whether the cattle propagated in the West African savannah during the LSA were of either taurine or Sanga type, or both, since the cattle types are very difficult to differentiate from osteological remains (Grigson 1991). While domestic animals were well known in some parts of the West African savannah by the LSA, they were still missing or of very limited importance in others. The beginnings of pearl millet cultivation also occurred in the LSA, but not always simultaneously with stock keeping.

Iron Age Wave (Mid First Millennium BC to the Early First Millennium AD)

Although (good) archaeozoological evidence is missing, a small type of horse, ancestral to modern pony breeds, was possibly already present in West Africa by the Early Iron Age (Law 1980: 3; Blench 2000b). Clearer than the introduction of new domestic animal species during the Iron Age is the appearance of new cultivated crops, of which sorghum was the most important as a staple (Kahlheber and Neumann 2007). The innovations of the Iron Age seem to have been associated with population movements, and for some areas, such as the *firgi*, renewed influence from the south-western Sahara and adjacent Sahelian regions has been postulated (Wiesmüller 2001: 220, 233).

Arab Wave (Late First to Early Second Millennium AD)

The Arab wave included domestic fowl, the ancestors of the larger horse breeds, donkeys, dromedary camels, and finally, zebu cattle. Arabs, mostly traders, gradually began to penetrate West Africa after their invasion of northern Africa during the seventh century AD (Mones 1992). The animals of the Arab wave were not (necessarily) trade products, but their introduction is probably associated with increased inter-regional contacts during this period. Nevertheless, the new species seem to have arrived earlier at trade centres, such as Jenné-Jeno in Mali (MacDonald 1995), than in the hinterland. Besides domestic animals, some new plant species – such as wheat (*Triticum aestivum*, *T. turgidum* conv. *durum*), dates (*Phoenix dactylifera*), cotton (*Gossypium* sp.) and watermelon (*Citrullus lanatus*) – also appeared during the Arab wave, either as traded items or as locally grown plants (Kahlheber and Neumann 2007).

Dromedaries and zebu are particularly important introductions for pastoral economies. The former were the main transport animal used in trans-Saharan trade. In fact, dromedary remains from AD 250 to 400 at Siouré in the Middle Senegal Valley (MacDonald and MacDonald 2000) are adduced as proof of the existence of such trade relations already before the arrival of Arabs (e.g., Magnavita et al. 2002). As mentioned above, zebu have several characteristics that make them better suited for keeping by pastoral nomads than taurine and Sanga cattle. Introduced from Asia via the Horn of Africa, zebu must have dispersed rapidly over Africa between AD 700 and 1400 (Grigson 1991; Blench 1993; Meghen, MacHugh and Bradley 1994). Genetic research on living African cattle breeds has shown that they all have taurine origins (Loftus and Cunningham 2000). Zeboid characters, both morphological and genetic, are the result of varying degrees of zebu introgression. The zebu introgression in African cattle declines from east to west and from north to south (Meghen, MacHugh and Bradley 1994; MacHugh et al. 1997), corresponding with the routes along which zebu spread over the continent. The presence of zebu-type cattle is very difficult to prove from osteological remains, since only few features allow distinction between them and taurine or Sanga cattle (Epstein 1971: 388; Grigson 1980, 1991, 2000). Moreover, there is a lack of data on osteological characters displayed by crossbreeds. Zebu are possibly present in AD 800 to 1400 contexts at Jenné-Jeno (Mali) and Siouré (Senegal) (MacDonald 1995; MacDonald and MacDonald 2000). However, with the exception of a putative zebu thoracic vertebra from Galaga (NA 92/2C), a sub-recent site in Nigeria, the studied assemblages in Burkina Faso and north-eastern Nigeria did not yield any evidence of zebu cattle (Linseele 2007).

European Wave (Second Half of Second Millennium AD)

The biological characteristics of pigs (Zeder 1991: 30–31) prevented them from spreading together with cattle and ovicaprines from North to West Africa by way of the Sahara. Its absence from archaeological bone inventories suggests that the pig was only known in West Africa by the time of European contact. Nevertheless, linguists have claimed that a pre-European introduction of the pig is possible (Blench 2000c).

Local Developments

Apart from very recent domestication experiments with local wild species, guineafowl is the only possibly local West African domesticate, being the only species with an indigenous wild ancestor. However, the living conditions of 'domestic' guineafowl in the area are so close to those of the wild form that the former can hardly be considered domesticated (Blench 2000a). Moreover, criteria for distinguishing between bone remains of the wild and the domestic form are unknown. Domestication must have happened sometime before the fifth century BC, when domestic guineafowl appear in Europe (Mongin and Plouzeau 1984). It is not very clear which subspecies was domesticated, but it need not (only) have been the West African one (Gautier 1990: 4–5; Donkin 1991: 16). Evidence for guineafowl domestication from West African sites is missing, and, in fact, bones from this bird are usually not very frequent (Table 4.3) (MacDonald and MacDonald 2000).

Once cattle and ovicaprines had reached the savannah of West Africa, dwarf types presumably evolved locally. Dwarfism is a consequence of adaptation to (unfavourable) local circumstances and would especially be associated with trypanotolerance, although this has not yet been proven (Epstein 1971: 276; Clutton-Brock 2000; MacDonald and MacDonald 2000). The first cattle in West Africa, including those from Gajiganna sites, were already small, but it is not clear whether these should be considered as dwarf types (MacDonald and MacDonald 2000; Jousse 2004; Linseele 2007). They equally fall in the size range of modern N'Dama, which are also relatively small. Later sites have equally failed to produce unequivocal evidence of dwarf cattle. The oldest West African examples of dwarf goats have been reported for the second millennium BC Kintampo Culture of Ghana (Carter and Flight 1972), but the only available measurement is on a mandible. Relations between cranial measurements and body size are unclear (Uerpmann and Uerpmann 1994), and the Kintampo data therefore do not prove that the goats were of a dwarf type. Moreover, Gautier and Van Neer (2005) have questioned the differentiation between goats and sheep in the Kintampo assemblages. Second in date are the early first millennium BC dwarf goat remains recorded at Kolima Sud (Mali) (MacDonald 1994: 104; MacDonald and Van Neer 1994). In the southern Lake Chad area, dwarf goats were identified at Zilum (NA 97/37), the type site of Gajiganna phase III (600 to 400 BC) (Linseele 2007).

Diachronic and Geographic Variation in the Relative Importance of Domestic Stock

In northern Burkina Faso as well as in north-eastern Nigeria, domestic mammals were predominant over wild ones from their first appearance in the archaeozoological record. However, in both regions hunting and fowling supplemented the diet throughout the entire investigated sequence (Table 4.2). The continued exploitation of wild game was probably part of strategies to deal with the risks involved in living in the instable environment of the West African savannah (cf. Zvelebil 1992). The low milk productivity of cattle in West Africa, due to the region's single rainy season, may have been another reason why wild resources continued to be a necessary supply of animal-derived protein. The complete abandonment of hunting seems to be rare over the whole of Africa, although in parts of East Africa an almost exclusive dependence on domesticates for animal proteins is in evidence from around the end of the second millennium BC (Gifford-Gonzalez 2005: 207). This shift has been connected to the beginning of the bimodal rainfall regime of the region, resulting in higher milk productivity (Marshall 1990). In the West African savannahs, fishing also continued after the introduction of domestic animals (Table 4.2). Its importance seems to be related to the type of water body in the vicinity of settlements. For example, in the *firgi*, fishing must have been a rewarding and economically important activity, judging from the large number of fish remains and the diversity of species found there (Table 4.2) (Linseele 2007). This may have compensated for the constraints on stock keeping imposed in this area by the limited available grazing during yearly inundations.

The high economic importance of domesticates at archaeological sites in northern Burkina Faso and in north-eastern Nigeria differentiates these sites from others in more southerly parts of the West African savannah, and especially from those in wooded areas where domesticates were only a secondary source of food and where meat was mainly obtained through hunting (Van Neer 2002; Gautier and Van Neer 2005). The greater importance of hunting at West African forest sites compared to savannah environments stands in apparent contradiction to the higher density of game in the latter (Bourlière 1963). Therefore the economic importance of game is probably related more to difficulties with keeping domestic stock – such as humidity-related diseases – than to the potential of hunting.

From Generalised to Specialised Economies

General Economies of the Second Millennium BC

For early Saharan African stock keepers, stock keeping was only one way of acquiring food, besides wild plant gathering, fishing and hunting (e.g., Marshall 1998). The main advantage of keeping domestic animals over pure hunting and gathering must have been the predictable availability of domestic stock (Marshall and Hildebrand 2002). In West Africa, the appearance of pearl millet as the first domestic crop during the early second millennium BC did not substantially change established lifestyles (Kahlheber and Neumann 2007). The survival strategies of human groups living in the West African savannah during the second millennium BC show many similarities with those of Saharan hunter-gatherers and livestock keepers. They exploited diverse food resources, including wild and, depending on the area, also domesticated plants and/or animals. As a rule, in the

West African savannah zone, agriculture was not yet of great importance during that period (ibid.). Only at Birimi in Ghana, one of the northernmost sites of the Kintampo Culture, might pearl millet cultivation already have been the dominant economic activity (D'Andrea, Klee and Casey 2001). LSA lifestyles in sub-Saharan West Africa during the second millennium BC were usually not sedentary, although the degree of mobility varied. The main determining factor seems to have been the importance of agriculture. In the Bama Deltaic Complex, for example, people became more sedentary between Gajiganna phases I and II, parallel with the appearance of the first domestic crops. Only at Birimi in Ghana, where pearl millet cultivation was probably the predominant activity, does a sedentary village lifestyle seem to have existed already (ibid.).

The Transitional First Millennium BC

In the north of Burkina Faso, the first millennium BC is almost invisible archaeologically. In contrast, the Bama Deltaic Complex of north-eastern Nigeria saw some distinct new developments during Gajiganna phase III (600 to 400 BC): the appearance of a new crops such as cowpea (*Vigna unguiculata*), new cropping systems based on intercropping, and early urban features (Magnavita, Kahlheber and Eichorn 2004). After 400 BC, the Gajiganna phase III sites were abandoned, and only at the beginning of the first millennium AD is there evidence for renewed settlement. In the *firgi* there is no long gap in the archaeological evidence for the first millennium BC, although most sites there also show interruptions in their stratigraphy (Gronenborn 1998). Not only Burkina Faso and Nigeria, but also other parts of West Africa, such as Ghana after the Kintampo Culture (Stahl 1993), show evidence for discontinuity during the first millennium BC. According to Breunig and Neumann (2002b) this was a consequence of severe aridity at the time, inducing major changes that became visible around the beginning of our era.

Specialisation from the Beginning of Our Era

The appearance of sedentary communities of fully fledged agriculturalists, now growing more than just pearl millet, was one of the major changes of the Early Iron Age in sub-Saharan West Africa (Kahlheber and Neumann 2007). Since specialised herders and fishers living in the area today cannot survive without obtaining agricultural products from farmers (see above), it is likely that it was the turn to advanced crop growing that allowed the development of economic specialisation (Linseele 2010). With the advent of the Iron Age, herders were able to intensify their pastoral activities, with an increase in mobility, because they could acquire agricultural produce from farmers instead of having to grow it themselves.[5] Increasing specialisation led to higher interdependence between different groups, and inversely, more cooperation between groups allowed further specialisation. From the second half of the first millennium AD, the introduction of zebu cattle, which are better suited for pastoral nomadism and are more milk productive than other cattle types, must have further stimulated the development of specialist herders in sub-Saharan Western Africa. Moreover, the introduction of dromedaries now also allowed the exploitation of very arid areas. Long-distance trade between West and North Africa would also have contributed to economic specialisation. Arab traders left historical sources documenting the co-occurrence of specialised farmers and nomadic pastoralists in West Africa. For instance, both groups are mentioned for the Middle Senegal Valley from about the twelfth century AD (Levtzion and Hopkins 1981: 184, 400).

Large amounts of cultural remains and results from archaeobotanical research point to permanent settlement by sedentary farmers at all Iron Age sites in northern Burkina Faso as well as in the Bama Deltaic Complex and the *firgi* area of north-eastern Nigeria (Linseele 2007). Ovicaprines, and especially goats, predominate in the Iron Age archaeofaunal samples of the former two, while cattle come second (Table 4.3). In northern Nigeria nowadays, such a herd composition is typical of village livestock (Bourn et al. 1994). Cattle are usually more difficult to keep by settled farmers than ovicaprines because they need more and better pasture, and larger amounts of drinking water (Dahl and Hjort 1976: 249, 253). In the *firgi*, ovicaprines predominate in the Early Iron Age, but cattle are more numerous in the later phases (Table 4.3). It may be concluded that inundations of the *firgi* plains became less severe through time, resulting in larger grazing areas near the sites which enabled larger cattle herds to be kept (Linseele 2007).

Figure 4.6: Late Iron Age (circa AD 1000) house of fully fledged, settled farmers at Oursi hu-beero (BF 97/30) in northern Burkina Faso. Photograph by Maya von Czerniewicz.

Although it was postulated that economic specialisation, including specialised pastoralism, in West Africa began during the Iron Age, archaeological sites of specialised pastoralists are missing for the period. This could probably be explained by the poor archaeological visibility of the latter. Nomadic pastoralists leave little archaeological traces because of their low population density, the temporary nature of their settlements and the small amounts of cultural material they carry with them. Furthermore, recent pastoralists in the West African Sahel do not have their own ceramic traditions (MacDonald 1999), which would make them even harder to recognise in the archaeological record. Even though no Iron Age archaeological traces left by nomadic pastoralists were excavated in Burkina Faso and Nigeria, the presence of such groups can be inferred from indirect archaeozoological evidence. More precisely, at some sites, osteometrical data suggesting two or more size groups of ovicaprines may be interpreted in terms of stature differences between small animals kept by the inhabitants themselves and larger animals obtained through exchange with pastoral nomads (Linseele 2007: 132). A similar interpretation has been put forward by MacDonald (1995) to explain the co-occurrence of a smaller and a larger cattle type at Jenné-Jeno in Mali, between AD 400 and 800, where the smaller type is likewise supposed to have been kept locally while the larger would have been obtained through relationships with pastoral nomads visiting the area. At Late Iron Age Oursi hu-beero in northern Burkina Faso (Figure 4.6), there were also indications of two differently sized groups of cattle, which could probably be similarly explained (Linseele 2007: 137). This kind of evidence for specialised pastoralists is admittedly rather poor, but as yet no systematic archaeological research programmes have focused on the identification of such groups in the West African savannah. The absence of archaeological traces from certain periods in a region is often interpreted as habitation by nomadic pastoralists. Thus Albert (2000) postulated a return to nomadic lifestyles after the abandonment of the Iron Age settlement mounds of Burkina Faso marking the end of the regional archaeological sequence in the middle of the second millennium AD. If this was the case, then, judging from modern parallels, the nomadic pastoral groups of the area must have maintained exchange relations with sedentary farmers in order to survive.

Summary and Conclusions

All true domestic animals of the West African savannah were introduced from outside at some point in the past. The first species – cattle, sheep, goats and dogs – had reached the area by the second millennium BC. They arrived together with Saharan stock keepers who were being pushed southwards due to increasing climatic aridification. Once beyond the Sahara, domestic stock propagated quickly to the southern margins of the West African savannah. However, before the beginning of our era, it had spread only patchily over the area, possibly depending on local geographic circumstances and on the presence of indigenous hunter-gatherer populations. In some areas local hunter-gatherer lifestyles may have persisted until late in the first millennium AD. After the arrival of the first domestic animals, later introductions followed, of which the most important ones for pastoral nomadic economies – that is, zebu and dromedary camels – seem to be associated with increased inter-regional contacts after the Arab conquest of North Africa in the seventh century AD.

Once domestic stock reached sub-Saharan West Africa, it became the main source of animal protein in most areas, but the exploitation of wild animal resources was never abandoned. The use of diverse food resources must have been a way of managing the risks involved in living in the unstable environments of the West African savannah. Although richer in game, hunting in the West African savannah was less important than in the forest zone. Apparently, the importance of hunting depended more on the difficulties with keeping livestock than on hunting opportunities. Fishing usually also remained a source of animal protein in sub-Saharan West Africa after the introduction of food production. In some cases, probably depending on local environmental circumstances, its economic importance was substantial, such as in the Nigerian *firgi*.

There seem to have been two main phases in the development of food producing economies in the West African savannah, one with generalised and one with more specialised economies, separated by a transitional phase. The beginning of true agriculture after the turn of our era triggered the emergence of nomadic pastoralists and other specialist groups. On the other hand, many intermediate forms between the different types of specialists must have remained, as is the case today. In the course of the Iron Age the development of specialised pastoralism was further stimulated by the introduction of zebu cattle. However, communities with this mode of livelihood are almost entirely absent from the archaeological record where remains of settled farmers prevail on account of their higher visibility.

Notes

The study of faunal remains from northern Burkina Faso and north-eastern Nigeria was accomplished thanks to a grant by the Belgian Federal Science Policy Office. The material was excavated under the direction of Peter Breunig of Frankfurt University (Germany), in the framework of the multidisciplinary research project 'Cultural Development and Language History in the West African Savannah' (SFB 268). I am most grateful to Professor Breunig and all his collaborators. Wim Van Neer (Royal Belgian Institute for Natural Sciences) is acknowledged for supervising the faunal studies and for having taken the time to critically read through an earlier draft of the present chapter, and many other manuscripts. Sincere thanks are also due to the editors for inviting me to contribute to this book.

1. Unless otherwise specified, all dates in this text are calendar dates or radiocarbon dates calibrated to calendar dates. Details on radiocarbon dates of the sites mentioned can be found in Linseele (2007).
2. Compare the definition of economic specialisation of Alchian and Allen (1969: 197–217).
3. For a summary of the issues, see Marshall (1989) and FAO (2001).
4. See also the models of Payne (1964), Gautier (1987), Chenal-Vélardé (1997) and Jousse (2004).
5. Compare Robertshaw and Collett (1983) for East Africa.

References

Albert, K.-D. 2000. 'Montée et Abandon des Collines d'Occupation de l'Age de Fer au Nord du Burkina Faso', *Berichte des Sonderforschungsbereichs* 268(14): 335–51.

Alchian, A., and W. Allan. 1969. *Exchange and Production Theory in Use*. Belmont, CA: Wadsworth Publishing.

Amanor, K.S. 1995. 'Dynamics of Herd Structures and Herding Strategies in West Africa: A Study of Market Integration and Ecological Adaptation', *Africa* 65: 351–94.

Anquandah, J. 1993. 'The Kintampo Complex: A Case Study of Early Sedentism and Food Production in Sub-Sahelian West Africa', in T. Shaw et al. (eds), *The Archaeology of Africa: Food, Metals and Towns*. London: Routledge, pp.255–60.

Bawden, M.G. 1972. 'Geomorphology', in P.J. Aitchison et al. (eds), *The Land Resources of North East Nigeria*, Vol. 1: *The Environment*. Surbiton: Land Resource Division, pp.60–70.

Blench, R.M. 1993. 'Ethnographic and Linguistic Evidence for the Prehistory of African Ruminant Livestock, Horses and Ponies', in T. Shaw, P. Sinclair, B. Andah and A. Okpoko (eds), *The Archaeology of Africa: Food, Metals and Towns*. London: Routledge, pp.71–103.

——— 2000a. 'African Minor Livestock Species', in R.M. Blench and K.C. MacDonald (eds), *The Origins and Development of African Livestock: Archaeology, Genetics, Linguistics and Ethnography*. London: UCL Press, pp.314–38.

——— 2000b. 'A History of Donkeys, Wild Asses and Mules in Africa', in R.M. Blench and K.C. MacDonald (eds), *The Origins and Development of African Livestock: Archaeology, Genetics, Linguistics and Ethnography*. London: UCL Press, pp.339–54.

——— 2000c. 'A History of Pigs in Africa', in R.M. Blench and K.C. MacDonald (eds), *The Origins and Development of African Livestock: Archaeology, Genetics, Linguistics and Ethnography*. London: UCL Press, pp.355–67.

Bourlière, F. 1963. 'Observations on the Ecology of Some Large African Mammals', in F.C. Howel and F. Bourlière (eds), *African Ecology and Human Evolution*. New York: Wenner-Gren Foundation, pp.43–54.

Bourn, D.M., et al. 1994. 'Nigerian Livestock Resources Survey', *World Animal Review* 78: 49–58.

Breunig, P., and K. Neumann. 2002a. 'From Hunters and Gatherers to Food Producers: New Archaeological and Archaeobotanical Evidence from the West African Sahel', in F.A. Hassan (ed.), *Droughts, Food and Culture: Ecological Change and Food Security in Africa's Later Prehistory*. New York: Kluwer Academic/Plenum, pp.123–55.

——— 2002b. 'Continuity or Discontinuity? The 1st Millennium BC Crisis in West African Prehistory', in Jennerstr. 8 (eds), *Tides of the Desert/Gezeiten der Wüste*. Cologne: Heinrich-Barth-Institut, pp.491–505.

Breunig, P., et al. 2006. 'G(l)anz ohne Eisen: Große Siedlungen aus der Mitte des 1. Jahrtausends v. Chr. im Tschadbecken von Nordost-Nigeria', in H.-P. Wotzka (ed.), *Grundlegungen: Beiträge zur Europäischen und Afrikanischen Archäologie für Manfred K. H. Eggert*. Tübingen: Francke, pp.255–70.

Carter, P.L., and C. Flight. 1972. 'A Report on the Fauna from the Sites of Ntereso and Kintampo Rock Shelter Six in Ghana: With Evidence for the Practice of Animal Husbandry during the Second Millennium BC', *Man* 7(2): 277–82.

Chenal-Vélardé, I. 1997. 'Les Premières Traces de Boeuf Domestique en Afrique du Nord: Etat de la Recherche Centré sur les Données Archéozoologiques', *Archaeozoologia* 9: 11–40.

Clutton-Brock, J. 2000. 'Cattle, Sheep, and Goats South of the Sahara: An Archaeozoological Perspective', in R.M. Blench and K.C. MacDonald (eds), *The Origins and Development of African Livestock: Archaeology, Genetics, Linguistics and Ethnography*. London: UCL Press, pp.30–37.
Connah, G. 1987. 'An Optimal Zone: The West African Savanna', in G. Connah (ed.), *African Civilizations*. Cambridge: Cambridge University Press, pp.97–120.
Dahl, G., and A. Hjort. 1976. *Having Herds: Pastoral Herd Growth and Household Economy*. Stockholm: Liber Tryck.
D'Andrea, C., M. Klee and J. Casey. 2001. 'Archaeobotanical Evidence for Pearl Millet (*Pennisetum Glaucum*) in Sub-Saharan West Africa', *Antiquity* 75: 341–48.
Donkin, R.A. 1991. *Meleagrides: An Historical and Ethnographical Study of the Guinea Fowl.* London: Ethnographica.
Dueppen, S.A. 2011. 'Evidence for Chickens at Iron Age Kirikongo (c. AD 100–1450), Burkina Faso', *Antiquity* 85: 142–57.
Epstein, H. 1971. *The Origin of the Domestic Animals of Africa,* 2 vols. New York: Africana Publishing.
FAO. 2001. 'Pastoralism in the New Millennium'. FAO Animal Production and Health Paper 150. Rome: Food and Agriculture Organization.
Frank, T., et al. 2001. 'The Chaîne de Gobnangou, South-east Burkina Faso: Archaeological, Archaeobotanical, Archaeozoological and Geomorphological Studies', *Beiträge zur Allgemeinen und Vergleichenden Archäologie* 21: 127–90.
Froment, A. 1988. *Le Peuplement Humain de la Boucle du Niger*. Paris: Editions de l'ORSTOM.
Gallais, J. 1975. *Pasteurs et Paysans du Gourma: La Condition Sahélienne.* Paris: Centre National de Recherche Scientifique.
Gallant, J. 2002. *The Story of the African Dog.* Pietermaritzburg: University of Natal Press.
Gasse, F. 2000. 'Hydrological Changes in the African Tropics since the Last Glacial Maximum', *Quaternary Science Reviews* 19: 189–211.
Gautier, A. 1987. 'Prehistoric Men and Cattle in North Africa: A Dearth of Data and a Surfeit of Models', in F.Wendorf, R. Schild and A.E. Close (eds), *Prehistory of Arid North Africa.* Dallas, TX: Southern Methodist University Press, pp.163–87.
——— 1990. *La Domestication: Et l'Homme Créa ses Animaux.* Paris: Editons Errance.
——— 2000. 'The Evidence for the Earliest Livestock in North Africa: Or Adventures with Large Bovids, Ovicaprids, Dogs and Pigs', in F.A. Hassan (ed.), *Droughts, Food and Culture: Ecological Change and Food Security in Africa's Later Prehistory.* New York: Kluwer Academic/Plenum, pp.195–207.
——— 2001. 'The Early to Late Neolithic Archaeofaunas from Nabta and Bir Kiseiba', in F. Wendorf et al., *Holocene Settlement of the Egyptian Sahara,* Vol. 1: *The Archaeology of Nabta Playa.* New York: Kluwer Academic/Plenum, pp.609–35.
——— 2007. 'Animal Domestication in North Africa', in M. Bollig et al., *Aridity, Change and Conflict in Africa.* Cologne: Heinrich-Barth-Institut, pp.75–89.
Gautier, A., and W. Van Neer. 2005. 'The Continuous Exploitation of Wild Animal Resources in the Archaeozoological Record of Ghana', *Journal of African Archaeology* 3: 195–212.
Gifford-Gonzalez, D. 2000. 'Animal Disease Challenges to the Emergence of Pastoralism in Sub-Saharan Africa', *African Archaeological Review* 17(3): 95–139.
——— 2005. 'Pastoralism and its Consequences', in A.B. Stahl (ed.), *African Archaeology: A Critical Introduction.* Malden, MA: Blackwell, pp.187–224.
Gifford-Gonzalez, D., and O. Hanotte. 2011. 'Domesticating Animals in Africa: Implications of Genetic and Archaeological Findings', *Journal of World Prehistory* 24: 1–23.

Grigson, C. 1980. 'The Craniology and Relationships of Four Species of Bos. V. *Bos indicus* L', *Journal of Archaeological Science* 7: 3–32.

——— 1991. 'An African Origin for African Cattle? Some Archaeological Evidence', *African Archaeological Review* 9: 119–44.

——— 2000. '*Bos Africanus* (Brehm)? Notes on the Archaeozoology of the Native Cattle of Africa', in R.M. Blench and K.C. MacDonald (eds), *The Origins and Development of African Livestock: Archaeology, Genetics, Linguistics and Ethnography*. London: UCL Press, pp.38–60.

Gronenborn, D. 1998. 'Archaeological and Ethnohistorical Investigations along the Southern Fringes of Lake Chad, 1993–1996' *African Archaeological Review* 15: 225–57.

Jousse, H. 2004. 'A New Contribution to the History of Pastoralism in West Africa', *Journal of African Archaeology* 2(2): 187–201.

Jousse, H. and I. Chenal-Vélardé. 2001. 'Nouvelles Données sur la Faune Mammalienne de Kobadi (Mali) au Néolithique: Implications Paléoéconomiques et Paléoenvironnementales', *Préhistoire Anthropologie Méditerranéennes* 10/11: 145–58.

Kahlheber, S. 2004. 'Perlhirse und Baobab: Archäobotanische Untersuchungen im Norden Burkina Fasos', PhD dissertation. Frankfurt am Main: Johann Wolfgang Goethe-University.

Kahlheber, S., and K. Neumann. 2007. 'The Development of Plant Cultivation in Semi-arid West Africa', in T.P. Denham, J. Iriarte and L. Vrydaghs (eds), *Rethinking Agriculture: Archaeological and Ethnoarchaeological Perspectives*. Walnut Creek, CA: Left Coast Press, pp.320–45.

Khazanov, A.M. 1994. *Nomads and the Outside World*, trans. J. Crookenden. Madison: University of Wisconsin Press.

Klee, M., and B. Zach. 1999. 'The Exploitation of Wild and Domestic Food Plants at Settlement Mounds in North-east Nigeria (1800 cal BC to Today)', in M. Van der Veen (ed.), *The Exploitation of Plant Resources in Ancient Africa*. New York: Kluwer Academic/Plenum, pp.81–88.

Klee, M., B. Zach and K. Neumann. 2000. 'Four Thousand Years of Plant Exploitation in the Chad Basin of North-east Nigeria, Part I: The Archaeobotany of Kursakata', *Vegetation History and Archaeobotany* 9: 223–37.

Klee, M., B. Zach and H.-P. Stika. 2004. 'Four Thousand Years of Plant Exploitation in the Lake Chad Basin (Nigeria), Part III: Plant Impressions in Potsherds from the Final Stone Age Gajiganna Culture', *Vegetation History and Archaeobotany* 13: 131–42.

Köhler-Rollefson, I. 1996. 'The One-humped Camel in Asia: Origin, Utilization and Mechanisms of Dispersal', in D. Harris (ed.), *The Origins and Spread of Agriculture and Pastoralism in Eurasia*. London: UCL Press, pp.282–95.

Krings, T.F. 1980. *Kulturgeographischer Wandel in der Kontaktzone von Nomaden und Bauern im Sahel von Obervolta: Am Beispiel des Oudalan (Nordost-Obervolta)*. Hamburg: Ferdinand Hirt.

Kühltrunk, P. 2000. 'Taphonomische und Typologische Keramikanalyse des Steinzeitlichen Fundplatzes Tin-Akoff im Norden Burkina Fasos', MA thesis. Frankfurt am Main: Johann Wolfgang Goethe University.

Kuper, R., and S. Kröpelin. 2006. 'Climate-controlled Holocene Occupation in the Sahara: Motor of Africa's Evolution', *Science* 313: 803–7.

Law, R. 1980. *The Horse in West African History: The Role of the Horse in the Societies of Pre-colonial West Africa*. London: Oxford University Press.

Le Houérou, H.N. 1989. *The Grazing Land Ecosystems of the African Sahel.* Berlin/Heidelberg/New York: Springer.

Levtzion, N. and Hopkins, J. 1981. *Corpus of Early Arabic Sources for West African History.* Cambridge: Cambridge University Press.

Linseele, V. 2007. *Archaeofaunal Remains from the Past 4,000 Years in Sahelian West Africa: Domestic Livestock, Subsistence Strategies and Environmental Change.* Oxford: Archaeopress.

——— 2010. 'Did Specialized Pastoralism Develop Differently in Africa than in the Near East? An Example from the West African Sahel', *Journal of World Prehistory* 23: 43–77.

Loftus, R., and P. Cunningham. 2000. 'Molecular Genetic Analysis of African Zeboid Populations', in R.M. Blench and K.C. MacDonald (eds), *The Origins and Development of African Livestock: Archaeology, Genetics, Linguistics and Ethnography.* London: UCL Press, pp.251–58.

MacDonald, K.C. 1994. 'Socio-economic Diversity and the Origins of Cultural Complexity along the Middle Niger (2000 BC to AD 300)', PhD dissertation. Cambridge: University of Cambridge.

——— 1995. 'Analysis of the Mammalian, Avian and Reptilian Remains', in S.K. McIntosh (ed.), *Excavations at Jenné-Jeno, Hambarketolo, and Kaniana (Inland Niger Delta), the 1981 Season.* Berkeley: University of California Press, pp.291–318.

——— 1996. 'The Windé Koroji Complex: Evidence for the Peopling of the Eastern Inland Niger Delta (2100–500 BC)', *Préhistoire Anthropologie Méditerranéennes* 5: 147–65.

——— 1997. 'Korounkorokalé Revisited: The Pays Mande and the West African Microlithic Technocomplex', *African Archaeological Review* 14: 161–200.

——— 1999. 'Invisible Pastoralists: An Inquiry into the Origins of Nomadic Pastoralism in the West African Sahel', in C. Gosden and J. Hather (eds), *The Prehistory of Food: Appetites for Change.* London: Routledge, pp.333–49.

MacDonald, K.C., and R.H. MacDonald. 2000. 'The Origins and Development of Domesticated Animals in Arid West Africa', in R.M. Blench and K.C. MacDonald (eds), *The Origins and Development of African Livestock: Archaeology, Genetics, Linguistics and Ethnography.* London: UCL Press, pp.127–62.

MacDonald, K.C., and W. Van Neer. 1994. 'Specialised Fishing Peoples in the Later Holocene of the Méma Region (Mali)', in W. Van Neer (ed.), *Fish Exploitation in the Past.* Tervuren: Annales du Musée Royal de l'Afrique Centrale, pp.243–51.

MacHugh, D., et al. 1997. 'Microsatellite DNA Variation and the Evolution, Domestication and Phylogeography of Taurine and Zebu Cattle (*Bos Taurus* and *B. Indicus*)', *Genetics* 146: 1071–86.

McIntosh, R.J. 1997. 'Agricultural Beginnings in Sub-Saharan Africa', in J.O. Vogel (ed.), *Encyclopedia of Precolonial Africa: Archaeology, History, Languages, Cultures and Environments.* Walnut Creek, CA: Altamira Press, pp.409–18.

McIntosh, S.K. 1999. 'A Tale from Two Floodplains: Comparative Perspectives on the Emergence of Complex Societies and Urbanism in the Middle Niger and Senegal Valleys', in P. Sinclair (ed.), *Proceedings of the Second World Archaeological Congress Intercongress, Mombasa.* Retrieved 4 June 2008 from: http://www.arkeologi.uu.se/afr/projects/BOOK/Mcintosh/mcintosh.htm.

Magnavita, C., S. Kahlheber and B. Eichhorn. 2004. 'The Rise of Organisational Complexity in Mid-first Millennium BC Chad Basin', *Antiquity* 78(301). http://antiquity.ac.uk/projgall/magnavita/index.html

Magnavita, S., et al. 2002. 'Nobles, Guerriers, Paysans: Une Nécropole de l'Age de Fer et son Emplacement dans l'Oudalan Pré- et Protohistorique', *Beiträge zur Allgemeinen und Vergleichenden Archäologie* 22: 21–64.
Marshall, F. 1989. 'Rethinking the Role of *Bos Indicus* in Sub-Saharan Africa', *Current Anthropology* 30(2): 235–39.
——— 1990. 'Origins of Specialized Pastoral Production in East Africa', *American Anthropologist* 92: 873–94.
——— 1998. 'Early Food Production in Africa', *Review of Archaeology* 19(2): 47–58.
——— 2000. 'The Origins and Spread of Domestic Animals in East Africa', in R.M. Blench and K.C. MacDonald (eds), *The Origins and Development of African Livestock: Archaeology, Genetics, Linguistics and Ethnography*. London: UCL Press, pp.191–221.
Marshall, F., and E. Hildebrand. 2002. 'Cattle before Crops: The Beginnings of Food Production in Africa', *Journal of World Prehistory* 16(2): 99–143.
Meghen, C., D.E. MacHugh and D.G. Bradley. 1994. 'Genetic Characterization and West African Cattle', *World Animal Review* 78: 59–66.
Mones, H. 1992. 'The Conquest of North Africa and Berber Resistance', in I. Hrbek (ed.), *General History of Africa*, Vol. III: *Africa from the Seventh to the Eleventh Century*, abridged edn. Paris: Unesco Publishing, pp.118–29.
Mongin, P., and M. Plouzeau. 1984. 'Guineafowl', in I.L. Mason (ed.), *Evolution of Domesticated Animals*. London: Longman, pp.322–25.
Neumann, K., P. Breunig and S. Kahlheber. 2000. 'Early Food Production in the Sahel of Burkina Faso', *Berichte des Sonderforschungsbereichs* 268(14): 327–34.
Payne, W.J.A. 1964. 'The Origin of Domestic Cattle in Africa', *Empire Journal of Experimental Agriculture* 32: 97–113.
Porter, V. 1991. *Cattle: A Handbook to the Breeds of the World.* London: Christopher Helm/A and C Black.
Robertshaw, P. and D. Collett. 1983. 'The Identification of Pastoral Peoples in the Archaeological Record: An Example from East Africa', *World Archaeology* 15: 67–78.
Simoons, F.J. 1994. *Eat Not this Flesh: Food Avoidances from Prehistory to the Present.* Madison: University of Wisconsin Press.
Smith, A.B. 1974. 'Preliminary Report of Excavations at Karkarichinkat Nord and Karkarichinkat Sud, Tilemsi Valley, Republic of Mali, Spring 1972', *West African Journal of Archaeology* 4: 33–55.
——— 1975. 'A Note on the Flora and Fauna from the Post-Palaeolithic Sites of Karkarichinkat Nord and Sud', *West African Journal of Archaeology* 5: 201–4.
Stahl, A. 1985. 'Reinvestigation of Kintampo 6 Rock Shelter, Ghana: Implications for the Nature of Culture Change', *African Archaeological Review* 3: 117–50.
——— 1993. 'Intensification in the West African Late Stone Age: A View from Central Ghana', in T. Shaw et al. (eds), *The Archaeology of Africa: Food, Metals and Towns*. London: Routledge, pp.261–73.
Smith, S.E. 1978. 'The Environmental Adaptation of Nomads in the West African Sahel: A Key to Understanding Prehistoric Pastoralists', in W. Weissleder (ed.), *The Nomadic Alternative: Modes and Models of Interaction in the African-Asian Desert and Steppes*. Paris: Mouton Publishers, pp.75–96.
Sundström, L. 1972. *Ecology and Symbiosis: Niger Water Folk.* Uppsala: Almqvist and Wiksell
Uerpmann, M., and H.-P. Uerpmann. 1994. 'Animal Bones', in F. Højlund and H. Andersen (eds), *Qala'at al Bahrein*, Vol. 1: *The Northern City Wall and the Islamic Fortress*. Aarhus: Aarhus University Press, pp.417–54.

Van Neer, W. 2002. 'Food Security in Western and Central Africa during the Late Holocene: The Role of Domestic Stock Keeping, Hunting and Fishing', in F.A. Hassan (ed.), *Droughts, Food and Culture: Ecological Change and Food Security in Africa's Later Prehistory.* New York: Kluwer Academic/Plenum, pp.251–74.

Vogelsang, R. 2000. 'Archäologische Forschungen in der Sahel-Region Burkina Fasos: Ergebnisse der Grabungskampagnen 1994, 1995, 1996', *Beiträge zur Allgemeinen und Vergleichenden Archäologie* 20: 173–203.

Vogelsang, R., K.D. Albert and S. Kahlheber. 1999. 'Le Sable Savant: Les Cordons Dunaires Sahéliens au Burkina Faso Comme Archive Archéologique et Paléoécologique du Holocène', *Sahara* 11: 51–68.

Von den Driesch, A., and J. Boessneck. 1985. *Die Knochenfunde aus der Neolitischen Siedlung von Merimde-Benisalâme am Westlichen Nildelta.* Munich: Institut für Paläoanotomie, Domestikationsforschung und Geschichte der Tiermedizin.

Watson, D.J. 2005. 'Under the Rocks: Reconsidering the Origin of the Kintampo Tradition and the Development of Food Production in the Savanna-Forest/Forest of West Africa', *Journal of African Archaeology* 3: 3–55.

Wendt, P. 2007. *Gajiganna: Analysis of Stratigraphies and Pottery of a Final Stone Age Culture of North-east Nigeria.* Frankfurt am Main: Africa Magna Verlag/Journal of African Archaeology.

Wiesmüller, B. 2001. 'Die Entwicklung der Keramik von 3000 BP bis zu Gegenwart in den Tonebenen Südlich des Tschadsees', PhD dissertation. Frankfurt am Main: Johann Wolfgang Goethe-University.

Wilson, T.R. 1984. *The Camel.* London: Longman.

Zeder, M.A. 1991. *Feeding Cities: Specialised Animal Economies in the Ancient Near East.* Washington: Smithsonian Institution Press.

Zvelebil, M. 1992. 'Hunting in Farming Societies: The Prehistoric Perspective', *Anthropozoologica* 16: 7–18.

Chapter 5
A Short History of Early Herding in Southern Africa

Karim Sadr

Introduction

Over five hundred years ago, Portuguese seafarers encountered domesticated livestock on the coast of southernmost Africa (Raven-Hart 1967). Cattle, sheep and goats, which are not native to sub-Saharan Africa, must have been brought from farther north (Mitchell 2002: 231; Phillipson 2005: 272). One episode of cattle domestication may have taken place some 10,000 to 11,000 calendar years ago in what is now the Eastern Sahara (Wendorf and Schild 1998; cf. Riemer 2007), but sheep and goats are of West Asian origin (Zeder and Hesse 2000; MacHugh and Bradley 2001). Key questions in the history of Southern African herding thus have been: How, when and why did livestock come to be herded in the subcontinent? The questions matter because the early exploitation of domesticated livestock (or plants, for that matter) heralds a transition from hunting and gathering – where food is extracted from what nature freely provides – to a strategy where food is grown or nurtured, quasi produced. Producing food is, in effect, the accumulation of private property. To make this economic transition, egalitarian hunting and gathering societies with a strict ethos of sharing would need to adopt new social rules to accommodate the concept and practice of private ownership. Egalitarianism itself would be threatened since the surplus from food production ultimately facilitates the appropriation of others' labour. Many thus see food production, be it herding or farming, as the first step towards the rise of economically, socially and politically more complex societies, and ultimately of civilisation (e.g., Diamond 1998).

For the past few decades, conventional wisdom has stated that it was Khoekhoe pastoralists from northern Botswana who introduced the first sheep and ceramic vessels to southernmost Africa about two thousand years ago (Elphick 1985: 14–22; Smith 1992: 96–98; Ehret 1998: 240). It is said that the Khoekhoe (the people who were referred to as Hottentots by the first European settlers at the Cape of South Africa) had learnt herding and pottery from Early Iron Age (EIA),[1] Bantu-speaking farmers and herders who were themselves immigrants on their way south from East Africa (Elphick 1985: 10–14; Smith 1990: 65; 1992: 83; Boonzaier et al. 1996: 25, 27). In a variation on this theme, Ehret (1998: 216–18, 240) proposed that Eastern Sahelian-speakers preceded the Bantu, and introduced the first livestock and ceramics to the ancestors of the Khoekhoe in the middle Zambezi Basin.

Much of the evidence of these migrations and transfers of skills and technologies comes from linguistic studies (e.g., Bleek 1929; Westphal 1963; Ehret 1967, 1982; Elphick 1985; Güldemann 2008). Archaeology, meanwhile, has provided absolute dates for the earliest livestock bones and ceramics in Southern Africa (e.g., Klein 1986; Sealy and Yates 1994; Pleurdeau et al. 2012), and it has more or less arbitrarily assigned Khoekhoe agency to selected material remains. Thus, Khoekhoe potters, for example, are seen as the makers of Southern African fine, thin-walled ceramic vessels, which are clearly different from coarse, thick-walled EIA pots (Schofield 1948: 56–70; Sydow 1967: 63; Rudner 1968: 442; Sadr and Smith 1991; Sadr and Sampson 1999). Fine-line rock paintings of sheep and their herders are said to depict immigrant Khoekhoe (Cooke 1969: 108). The Khoekhoe in turn painted geometric shapes and left handprints on rocks all along their route from Central Africa to the Western Cape (Smith and Ouzman 2004; Eastwood and Smith 2005). Similarly, the immigrant Khoekhoe are said to have used relatively large ostrich eggshell beads (Jacobson 1987; Smith et al. 1991: 87–88), while some changes in stone tool assemblages around two thousand years ago are attributed to their influence as well (Smith et al. 1991: 87; Sadr 1997: 109, fig. 4; cf. Sampson 1974: 401; Deacon 1984: 315; Schrire and Deacon 1989: 111–12).

Recent finds refute many of these attributions, and the archaeological evidence fails to implicate Khoekhoe pastoralists as the source of livestock and ceramic technology in southernmost Africa. In fact, evidence for pastoralism – that is to say a way of life that was economically, socially and symbolically focused on livestock management – is very rare and arguably absent from the Southern African Late Stone Age (LSA) in the first millennium AD. Below, this archaeological evidence, remains of livestock, stone tools and ceramic pots found in association with Southern African LSA sites, are discussed in more detail.

Livestock

There is no doubt that livestock, especially sheep, are associated with two-millennia-old LSA sites (Sealy and Yates 1994; Henshilwood 1996; Vogel, Plug and Webley 1997; Robbins et al. 2005; Pleurdeau et al. 2012). But, with a very few notable exceptions the bones of domesticated animals found in LSA sites represent less than 10 per cent of the mammalian faunal remains (Table 5.1). Do such low proportions of livestock suggest pastoralism? To put these numbers in perspective, Southern African Iron Age pastoralists who revered their cattle, placed them at the centre of their symbolic universe, and rarely ate them except on special occasions,[2] have left behind kitchen middens where livestock bones

Table 5.1: Frequency of faunal remains from some Late Stone Age sites in Southern Africa dating to the last 2,000 years. Note the low proportions of domesticated livestock remains.

Site	Location	Layer	Mammals	Ovicaprids	Cattle	Livestock %	References
Kasteelberg De	West Coast	all	1501	963	57	67.95	Kasteelberg Archaeological Project (KAP), unpublished
Jakkalsberg	Namaqualand	area B	456	253	0	55.48	Webley 1997; Brink and Webley 1996
Kabeljous River Mouth	Eastern Cape	SM1A	11	6	0	54.55	Binneman 1995
Kabeljous River Mouth	Eastern Cape	SM1B	7	3	0	42.86	Binneman 1995
Jakkalsberg	Namaqualand	area A	1197	446	0	37.26	Webley 1997; Brink and Webley 1996
Makwe RS	Zambia	6, 5	107	2	36	35.51	Phillipson 1976
Kasteelberg A	West Coast	all	390	122	8	33.33	KAP
MAS North Mound	Eastern Cape	all	15	1	4	33.33	Derricourt 1977
Kasteelberg M	West Coast	all	24	7	0	29.17	KAP
Kasteelberg G	West Coast	5	4	1	0	25.00	KAP
Makwe RS	Zambia	4	78	2	12	17.95	Phillipson 1976
Kasteelberg E	West Coast	all	229	35	0	15.28	KAP
Kasteelberg G	West Coast	1	17	2	0	11.76	KAP
Boomplaas	Little Karoo	DGL & BLD	157	17	0	10.83	Klein 1978
Kasteelberg N	West Coast	all	210	18	1	9.05	KAP
Atlantic Beach	Table Bay	AB1/3	463	35	0	7.56	Sealy et al. 2004
Kasteelberg A	West Coast	all	3671	268	6	7.46	Klein and Cruz-Uribe 1989
Die Kelders	Walker Bay	1 to 4	469	29	2	6.61	Schweitzer 1979
Atlantic Beach	Table Bay	AB1/2	782	42	0	5.37	Sealy et al. 2004
Mzinyashana	KwaZulu-Natal	1, 2, 3	188	5	5	5.32	Plug 2002
Kreeftebaai	Saldanha Bay	all	38	0	2	5.26	Smith et al. 1991
Big Elephant Shelter	Namibia	top 4	38	1	1	5.26	Wadley 1979
Kasteelberg B	West Coast	Lower	4985	257	4	5.24	Klein and Cruz-Uribe 1989

Table 5.1: *continued*

Kasteelberg G	West Coast	4	44	2	0	4.55	KAP
Spoegrivier	Namaqualand	2	842	36	0	4.28	Webley 1992
Drie Susters Main	Saldanha Bay	all	127	3	2	3.94	Smith et al. 1991
Tortoise Cave	Elands Bay	2b	326	12	0	3.68	Klein and Cruz-Uribe 1987
Kasteelberg B	West Coast	Middle	9324	309	13	3.45	Klein and Cruz-Uribe 1989
Spoegrivier	Namaqualand	1	818	28	0	3.42	Webley 1992
Klein Kliphuis	Oliphants River W.Cape	Upper	101	3	0	2.97	Van Rijssen 1992
Kasteelberg G	West Coast	3	35	1	0	2.86	KAP
Geduld	Namibia	ALL	215	6	0	2.79	Smith and Jacobson 1995
Kasteelberg G	West Coast	1	37	1	0	2.70	KAP
Tortoise Cave	Elands Bay	2a	500	10	0	2.00	Klein and Cruz-Uribe 1987
Mzinyashana	KwaZulu-Natal	4	270	2	3	1.85	Plug 2002
Kasteelberg B	West Coast	Upper	3563	58	7	1.82	Klein and Cruz-Uribe 1989
Tortoise Cave	Elands Bay	1b	137	2	0	1.46	Klein and Cruz-Uribe 1987
Mhlawizini Cave	KwaZulu-Natal	1 & 2	71	1	0	1.41	Mazel 1990
Spoegrivier	Namaqualand	3	268	3	0	1.12	Webley 1992
Tortoise Cave	Elands Bay	3a,b	197	2	0	1.02	Klein and Cruz-Uribe 1987
Steenbokfontein	Elands Bay	2	115	1	0	0.87	Jerardino and Yates 1996
Elands Bay Cave	Elands Bay	2	443	3	0	0.68	Klein and Cruz-Uribe 1987
Atlantic Beach	Table Bay	AB3/3	587	3	0	0.51	Sealy et al. 2004
Tortoise Rock Shelter	Elands Bay	1a	635	3	0	0.47	Klein and Cruz-Uribe 1987
Elands Bay Cave	Elands Bay	4	230	1	0	0.43	Klein and Cruz-Uribe 1987
iNkolimahashi shelter	KwaZulu-Natal	2 & 3	301	1	0	0.33	Badenhorst 2003; Mazel 1999
Steenbokfontein	Elands Bay	1	355	1	0	0.28	Jerardino and Yates 1996
Die Kelders	Walker Bay	7 to 12	1502	3	0	0.20	Schweitzer 1979
Elands Bay Cave	Elands Bay	1	566	1	0	0.18	Klein and Cruz-Uribe 1987
Voelvlei	Berg River	1 & 2	67	0	0	–	Smith et al. 1991
Pancho's Kitchen Midden	Elands Bay	1	189	0	0	–	Jerardino 1998
Elands Bay Open	Elands Bay	1	166	0	0	–	Klein and Cruz-Uribe 1987

Faraoskop	Oliphants River W.Cape	1	88	0	0	–	Manhire 1993
Pancho's Kitchen Midden	Elands Bay	2	48	0	0	–	Jerardino 1998
Kasteelberg G	West Coast	Upper	8	0	0	–	KAP
Renbaan	Oliphants River W.Cape	Bottom	104	0	0	–	Kaplan 1987
Duiker Eiland	West Coast	1	30	0	0	–	Robertshaw 1979
Elands Bay Cave	Elands Bay	6	150	0	0	–	Klein and Cruz-Uribe 1987
Stofbergsfontein	Saldanha Bay	Middle	32	0	0	–	Robertshaw 1978a
Kasteelberg G	West Coast	2	31	0	0	–	KAP
Witklip	West Coast	3	275	0	0	–	Smith et al. 1991
Klein Kliphuis	Oliphants River W.Cape	Lower	290	0	0	–	Van Rijssen 1992
Steenbokfontein	Elands Bay	3	440	0	0	–	Jerardino and Yates 1996
Pancho's Kitchen Midden	Elands Bay	3	42	0	0	–	Jerardino 1998
Pancho's Kitchen Midden	Elands Bay	4	117	0	0	–	Jerardino 1998
Witklip	West Coast	4	58	0	0	–	Smith et al. 1991
Pancho's Kitchen Midden	Elands Bay	6	108	0	0	–	Jerardino 1998
Elands Bay Cave	Elands Bay	3	171	0	0	–	Klein and Cruz-Uribe 1987
Witklip	West Coast	2	30	0	0	–	Smith et al. 1991
Pancho's Kitchen Midden	Elands Bay	5	48	0	0	–	Jerardino 1998
Pancho's Kitchen Midden	Elands Bay	7	249	0	0	–	Jerardino 1998
Faraoskop	Oliphants River W.Cape	2	76	0	0	–	Manhire 1993
Falls Rock Shelter	Namibia	Pottery	96	0	0	–	Kinahan 2001
Snake Rock Shelter	Namibia	Pottery	41	0	0	–	Kinahan 2001
Fackeltrager	Namibia	I & II	67	0	0	–	Richter 1991
Etemba 2	Namibia	all	340	0	0	–	Richter 1991
Austerlitz	Namibia	Upper	206	0	0	–	Richter 1991
iNkolimahashi shelter	KwaZulu-Natal	4	228	0	0	–	Badenhorst 2003; Mazel 1999

Table 5.1: *continued*

iNkolimahashi shelter	KwaZulu-Natal	5	604	0	0	–	Badenhorst 2003; Mazel 1999
iNkolimahashi shelter	KwaZulu-Natal	6	262	0	0	–	Badenhorst 2003; Mazel 1999
iNkolimahashi shelter	KwaZulu-Natal	7	160	0	0	–	Badenhorst 2003; Mazel 1999
iNkolimahashi shelter	KwaZulu-Natal	8	182	0	0	–	Badenhorst 2003; Mazel 1999
Mbabane	KwaZulu-Natal	3	69	0	0	–	Mazel 1986
Mbabane	KwaZulu-Natal	4&5	371	0	0	–	Mazel 1986
KwaThwaleyakhe Shelter	KwaZulu-Natal	2	460	0	0	–	Mazel 1993
Collingham Shelter	KwaZulu-Natal	Upper	826	0	0	–	Mazel 1992
Collingham Shelter	KwaZulu-Natal	Lower	1466	0	0	–	Mazel 1992
Clarke's shelter	KwaZulu-Natal	2	180	0	0	–	Mazel 1984
Clarke's shelter	KwaZulu-Natal	3	100	0	0	–	Mazel 1984
Mhlawizini Cave	KwaZulu-Natal	4	110	0	0	–	Mazel 1990
Mhlawizini Cave	KwaZulu-Natal	5	336	0	0	–	Mazel 1990
Mhlawizini Cave	KwaZulu-Natal	6	395	0	0	–	Mazel 1990
Mzinyashana	KwaZulu-Natal	5	219	0	0	–	Plug 2002
Mzinyashana	KwaZulu-Natal	6,7,8	382	0	0	–	Plug 2002
Thandwe Rock Shelter	Zambia	all	55	0	0	–	Phillipson 1976
Makwe RS	Zambia	3	55	0	0	–	Phillipson 1976
Kalemba RS	Zambia	S	8	0	0	–	Phillipson 1976
Kalemba RS	Zambia	Q	6	0	0	–	Phillipson 1976

Table 5.2: Faunal remains from some 'Iron Age' sites in Southern Africa during the last 2,000 years. Note the high proportions of domesticated livestock remains.

Site	Location	Layer	Mammals	Ovicaprids	Cattle	Livestock %	Reference
uMgungundlovu	KZN	all	2599	8	2576	99.42	Plug & Roodt 1990
Tavhatshena	Soutpansberg	T3/3	70	21	39	85.71	De Wet-Bronner 1994
Stayt	Limpopo Valley	all	230	85	103	81.74	Plug 2000

Tavhatshena	Soutpansberg	T1/4-5	304	46	202	81.58	De Wet-Bronner 1994
Tshirululuni	Soutpansberg	T1/5-6	249	14	188	81.12	De Wet-Bronner 1995b
Tshitheme	Soutpansberg	T1/5-6	53	13	29	79.25	De Wet-Bronner 1995a
Tshitheme	Soutpansberg	T1/1-4	175	24	114	78.86	De Wet-Bronner 1995a
Mapungubwe	Limpopo Valley	all	441	200	142	77.55	Plug 2000
K2	Limpopo Valley	all	709	290	226	72.78	Plug 2000
2229AD5	Limpopo Valley	all	420	158	117	65.48	Plug 2000
Schroda	Limpopo Valley	all	710	263	201	65.35	Plug 2000
Mmatshetshele	Rustenburg	all	133	37	46	62.41	Pistorius 2001
Millbank	Makgabeng	all	328	87	116	61.89	Van Schalkwyk 2000
Thulamela	Limpopo Valley	all	2479	594	916	60.91	Plug 2000
Wildebeestfontein	Mpumalanga Province	IA midden	31	7	10	54.84	Plug 1979
Commando Kop	Limpopo Valley	all	143	46	29	52.45	Plug 2000
Ondini	KZN	all	235	24	97	51.49	Watson & Watson 1990
Sentinel Ranch	Limpopo Valley	all	23	8	3	47.83	Plug 2000
Dzata	Soutpansberg	T2/1-6	212	15	85	47.17	De Wet-Bronner 1995a
Pont Drift	Limpopo Valley	all	380	119	59	46.84	Plug 2000
Icon	Limpopo Valley	all	86	20	6	30.23	Voigt 1979
Pa8.1	Limpopo Valley	all	53	16	0	30.19	Plug 2000
Diamant	Limpopo Valley	all	412	93	24	28.40	Plug 2000
Harmony	Limpopo Province	all	307	32	7	12.70	Welbourne 1979
Hapi Pan	Limpopo Valley	all	38	0	4	10.53	Plug 2000

rarely account for less than half the mammalian faunal remains (Table 5.2) (see also Reid, Sadr and Hanson-James 1998: 88).

Although pastoralism cannot be defined simply by the number of livestock remains in archaeological sites, the discrepancy between the faunal composition in Iron Age and LSA middens must point to a significant difference in the people's relations with livestock. From the archaeological evidence, we cannot judge how symbolically important those sheep and cattle were to the LSA people of first-millennium AD Southern Africa. Kraals were not central features of their settlements, as they were in Iron Age sites. Paintings of livestock are rare and do not suggest great symbolic importance. But from the numbers of their bones, we can tell that livestock contributed very little to the LSA human diet. The dearth of LSA livestock remains might indicate instead of pastoralism a subsistence strategy similar to one recorded among mid- to late-twentieth-century Kalahari hunter-gatherers who kept small flocks of goats (Ikeya 1993; Kent 1993). The Kalahari goat-herders perhaps represent a transitional stage from hunting and gathering to true pastoralism, or maybe they represent a distinct subsistence strategy more common in the past. In either case, as a potential analogy for the LSA hunter-herders, the Kalahari example not only shows how such small-scale herding can be made viable with few livestock (contra Smith 2006: 70), but also how livestock can pass down the line from one group of hunter-gatherers to the next, without the intermediary of a wave of migrating pastoralists.

Having addressed the majority of the LSA sites with few livestock, mention must also be made of the few sites in Table 5.1 with exceptionally high proportions of livestock. Do they represent sites of true pastoralists? Makwe rock-shelter in eastern Zambia has a sizeable bone sample and is located close to an Iron Age site (Phillipson 1976), so the high proportions of livestock may tell us more about contact with farmers and herders (cf. Sadr and Plug 2001) than about pre-contact pastoralism in the LSA. The Eastern Cape sites, on the other hand, are all coastal shell middens with very small samples of mammalian remains. The high proportion of livestock at these sites is puzzling. Proximity to Eastern Cape Iron Age settlements may have been a factor here. Also, the collection strategy for bones on some of the sites was not unbiased, so livestock numbers may be inflated (Binneman 1995). Or, these may indeed be sites of pre-contact LSA pastoralists, but larger, more controlled samples of bones and more dates would be needed to clinch the argument.

Of the other LSA sites rich in livestock, a few are neighbouring localities on the hill Kasteelberg, some 140 kilometres north of Cape Town. The other example is at Jakkalsberg near the mouth of the Orange River (Brink and Webley 1996). Both have large faunal samples, dated to the second half of the first millennium AD, and contain ceramics of the same tradition. Some consider Kasteelberg and Jakkalsberg to be pastoralist Khoekhoe sites (Webley 1997; Smith 2006: 71). An alternative reading of the evidence is that they represent special-purpose locations for feasting events (Sadr 2004).

The relevant faunal evidence thus can be summed us as follows: the very low proportions of livestock bones on the majority of Southern African LSA sites of the last two millennia do not immediately bring pastoralism to mind. At face value, the evidence shows some livestock appearing in some hunter-gatherer sites in widely separated parts of the subcontinent around two thousand years ago (Sealy and Yates 1994; Henshilwood 1996; Robbins et al. 2005; Pleurdeau et al. 2012). For the most part, these domesticated animals apparently did not cause a significant change in local subsistence and settlement strategies: their appearance does not correlate with significant changes in other domains of material culture. The exceptionally sheep-rich localities in the Eastern Cape are an enigma, but on the west coast of South Africa they can be interpreted as special purpose locations for feasts where fat-rich foods were conspicuously consumed. They may have been sites of pastoralists, or they may have more to say about the local development of a 'Big Man' economy in the LSA of the west coast (Sadr 2004). This issue remains to be settled.

Lithics

Stone-tool assemblages in the Holocene LSA sites of Southern Africa are diverse. Some interpret this as a sign of diverse cultural groups (e.g., Goodwin and Van Riet Lowe 1929; Sampson 1972: 205). Others consider different 'toolkits' in different places as signals of different activities rather than emblems of different cultures (Parkington 1980). This is an old debate and not confined to the subcontinent.[3] Besides diversity across space, chronological change in lithic assemblages also requires explanation. Around two thousand years ago, for example, scraping tools became generally more common as backed elements became rare (Deacon 1984: 311, figs. 13–15). The backed elements such as bladelets, points, segments and so on are often interpreted as points and barbs for arrows, and their proportional decline can be seen as a drop in the importance of bow-and-arrow hunting. The increase in the number of scrapers in turn can be seen as evidence for the increased importance of hide preparation, perhaps for trade and exchange. Coeval with these typological changes in the stone toolkit, in some parts fine-grained stone raw materials gave way to coarser grained assemblages (Smith et al. 1991).

Some consider these material changes as a sign of the arrival of Khoekhoe pastoralists (Smith et al. 1991; Sadr 1997). Others suggest that different factors could have influenced such changes in stone-tool assemblages (Humphreys and Thackeray 1983: 292; Deacon 1984: 304–15; Schrire and Deacon 1989). Timing is important in this debate. If the changes occurred at around the time that ceramics and livestock bones appeared in the archaeological sites, the lithic changes could be attributed more convincingly to incoming pastoralists. If the timing was off, or if the same changes occurred without livestock and ceramics, then the link with incoming pastoralists is broken.

The matter can be settled with a simple comparison of lithics in pre- and post-ceramic phases at several sites across the subcontinent. For example, we can turn to the west coast of South Africa, where stone tools recorded in an archaeological survey of the northern half of the Vredenburg Peninsula during late 1991 and early 1992 (Sadr, Gribble and Euston-Brown 1992) provide a useful database. The survey was carried out to test the material distinction proposed by Smith et al. (1991) between Bushman hunter-gatherer and Khoekhoe herder sites. It covered a cross section of the different types of landscape, and systematic surface collections of artefacts were made at 129 sites (Sadr, Gribble and Euston-Brown 1992: 41, fig. 2). An important point here is that surface collection strategies were fully standardised, making the results quantitatively comparable between sites. Being near the coast, the inhabitants of all these sites exploited marine resources, and samples of their discarded limpet and mussel shells were sent for radiocarbon dating (Sadr 2009). In this way, for about half of the surveyed sites we have a date associated with an analysed sample of stone tools, and this helps to test whether the changes in stone tools were coincident with the arrival of livestock and ceramic technology in this landscape.

Given that the shell and lithic samples were collected from the surface of exposed sites, where overprinting of different occupations and other disturbances often lead to spurious associations between dates and stone tools, the noise-to-signal ratio in the patterns presented below is quite high. This is seen in the spikiness of the charts in Figure 5.1. Despite the noise, however, the signal comes through, and it shows some interesting trends (Sadr and Gribble 2010). First, fine-grained raw materials like silcrete did indeed give way to more quartz dominated assemblages through time (Figure 5.1.1 and 5.1.2). Second, the numbers of different types of formally retouched tools did in fact drop (Figure 5.1.3), with backed elements and associated bits of lithic technology (such as the production of bladelets) becoming numerically less important over time (Figure 5.1.4 and 5.1.5). But these changes, as the date lines in the diagrams show, started gradually and very early, some before 1000 BC or even 2500 BC. And the changes went through several stages. Thus, backed elements (arrowheads perhaps, but also cutting implements) were more frequent before 2500 BC, adzes (woodworking implements) were common from 2500 to 500 BC (Figure 5.1.6), and (hide-working) scrapers became the ubiquitous tools thereafter.

The full details of the survey, associated finds and dates will be published separately, but even this cursory examination of the data should suffice to show the absence of evidence for the kind of abrupt change in stone-tool assemblages around two thousand years ago that could signal the sudden arrival of pastoralist immigrants in the Vredenburg Peninsula. The observed changes suggest more an evolution in lithic manufacturing and utilisation habits, perhaps just the natural 'life cycle' of an industry (Deacon 1972: 37–38). Deacon had already remarked that, 'sites where both pre- and post-pottery/domestic stock assemblages occur show no significant difference in the stone artefacts through this sequence' (Deacon 1984: 323). Several other recent and current studies of change in lithic technology and typology across this boundary repeatedly confirmed this pattern (Rivat 2006; Byrne 2007; Modikwa 2008). More such studies, from other parts of Southern Africa, are planned for the near future.

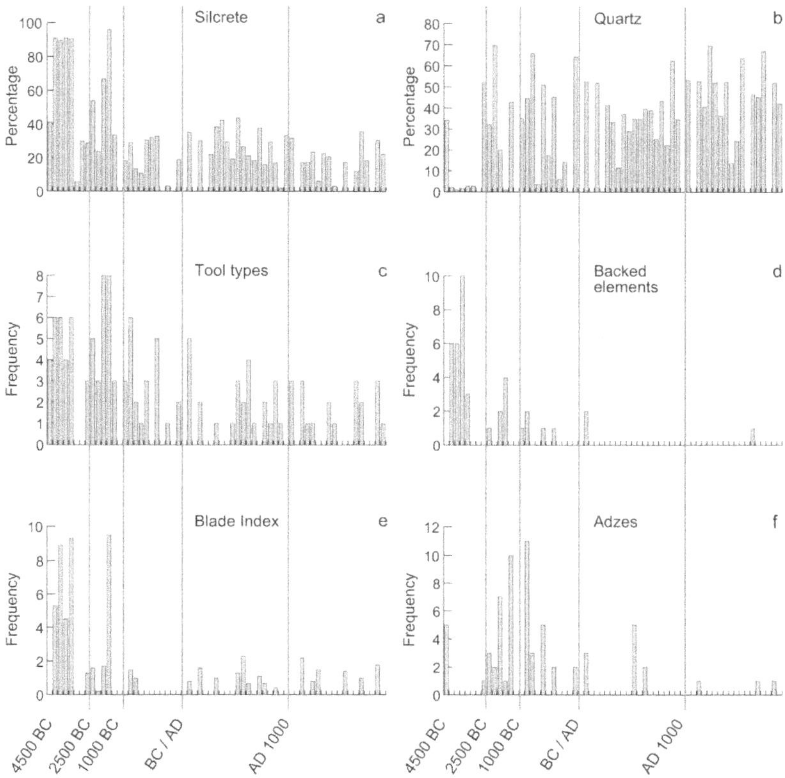

Figure 5.1: Stone-tool changes through time on the Vredenburg Peninsula. Each bar represents one of the sixty-three dated sites from the survey. The oldest are on the left.

Ceramics

Ceramics are often used to document migrations in later prehistory. Well-known examples include the Lapita stamped ware, which spread across 5,000 kilometres of island chains in Oceania from 1600 to 1000 BC (Spriggs 1995; Sand 1997; Bedford, Spriggs and Regenvanu 2006). Lapita pots are found in association with the remains of domestic animals, distinctive shell ornaments and large villages, all of which together formed a clearly intrusive package of traits in Oceania (Spriggs 1995). Similarly, Linearbandkeramik or LBK documents demic diffusion in the Danubian Neolithic between 5300 and 4900 BC (Quitta 1960; Gronenborn 1999; Dolukhanov et al. 2005). As in the case of Lapita ware, LBK appeared abruptly in new lands, and always in the context of clearly intrusive traits such as new village and house styles, and new subsistence economies involving agriculture and herding. In Africa, the Bantu migrations have also been documented archaeologically with the help of ceramics (Maggs 1984; Huffman 1989; Phillipson 2005).[4] Here, the intrusive package includes villages with huts arranged around a central kraal, evidence of cereal farming and cattle herding, use and manufacture of iron tools, and most visibly, distinct globular, thick-walled, coarse, mineral-tempered clay pots known as Chifumbaze pottery.

In these ceramic examples, two lines of evidence are considered important for documenting migration and demic diffusion. First, the abrupt appearance of an intrusive suite of material remains with no local antecedents. Second, similarities in ceramics along a 'stylistic chain', which traces the route and rate of migration. Southern African ceramic LSA sites from the first few centuries AD do not exhibit either of these lines of evidence. The ceramics and sheep bones do appear abruptly, but always within the context of LSA sites whose nature, location and artefacts indicate much cultural continuity (e.g., Deacon 1984: 323; Sadr 2003). In fact, thin-walled LSA ceramics significantly predate EIA thick-walled ceramics in the subcontinent, and show no technological links with ceramics north of the Zambezi (Sadr and Sampson 2006). Whether or not pottery was invented locally, as seems possible (ibid.), LSA ceramic styles show a great deal of regional diversity (Sadr 1998, 2008; Motloung 2007).

The marked regional diversity in LSA ceramics presents no clear stylistic chain. But does this necessarily refute migration? It can be argued that the ceramic and associated material 'chains' evident in Oceania, the Danube Valley and the eastern seaboard of sub-Saharan Africa represent the demic diffusion of sedentary farming populations, technologically and economically quite distinct from the hunting-and-gathering inhabitants of the new lands they colonised. The Khoekhoe, on the other hand, would have been an immigrant herder-hunter population with a culture, technology, settlement system and subsistence economy not very different from those of autochthonous hunter-gatherers.

In this light, rather than looking to Lapita, LBK and Chifumbaze as sources of analogy, the archaeology of the expansion of 'pastro-foragers' out of the Sahara and into the Nile and Niger River valleys at the end of the early- to mid-Holocene humid phase (Smith 1992, 2005; Phillipson 2005; Kuper and Kröpelin 2006) may provide a more appropriate comparison for the Khoekhoe migration. Here, the material traces of migration are much harder to find. In West Africa, the evidence is vague and often goes no further than a few shared stone-tool types or similar techniques of decorating pottery (MacDonald 1996; Raimbault and Commelin 2002; Watson 2005). In the east, the archaeological evidence of a population influx from the Sahara to the Nile in the fourth millennium BC is even more difficult to identify in the ceramics (e.g., Warfe 2003).

Another Look at Regional Diversity in LSA Ceramic Styles

Although a clear stylistic chain and package of new traits eludes the archaeologist looking for evidence of immigrant LSA pastoralists in Southern Africa, there are other changes evident in the ceramic assemblages that hint at some population infiltration and hybridisation. To adequately describe these changes, some background information is necessary. A key date is the sixth century AD, which represents a time of significant stylistic changes in ceramics in many parts of the subcontinent. Prior to this watershed, in the first half of the first millennium AD, there appear to have been two supra-regional ceramic stylistic 'families' among the thin-walled LSA wares of Southern Africa: a northern style with decorated and sometimes spouted pots, and a southern style with undecorated burnished pots and no sign of spouts (Sadr 2008).

The northern family consist of the so-called Bambata pottery of Zimbabwe and northern Botswana, as well as the diverse northern Namibian traditions (Albrecht et al. 2001; Vogelsang and Eichhorn 2011).[5] These northern pottery assemblages were decorated with a mix of simple techniques such as comb stamping and incisions, sometimes found alongside more complex ones, such as broad incisions subsequently modified by burnishing, the so-called rippled rims (Schofield 1941; Walker 1983; Huffman 1994, 2007; Smith and Jacobson 1995; Albrecht et al. 2001; Kinahan 2001).

The contemporary southern stylistic family, on the other hand, had few if any decorated pots. The southern style is best exemplified by the large sample from Die Kelders, Layer 12, which includes eight reconstructed vessels (Schweitzer 1979). There are also smaller contemporary assemblages from the southern and western Cape coasts, which are devoid of any decorated sherds (e.g., Poggenpoel and Robertshaw 1981; Leslie 1989: 27; Henshilwood 1996; Webley 1992: 5, 7). In some cases, such as at Byneskranskop, Witklip and Boomplaas, where a few decorated sherds are found, it is not clear whether these are associated with the early or mid-first-millennium dates, and the probability that the decorated potsherds filtered down the profile cannot be discounted (Deacon et al. 1978: 44, 54 table 7; Schweitzer and Wilson 1982: 92; Smith et al. 1991: 73–74.; Sadr 2007). Farther east and north, the earliest pottery in the central interior (Humphreys 1974; Sampson et al. 1989: 13),[6] and in KwaZulu-Natal Province (Maggs and Ward 1980; Mazel 1984, 1986, 1990, 1992, 1993; Maggs and Whitelaw 1991) were also undecorated. An important characteristic that distinguishes the two supra-regional ceramic families is spouted vessels in the northern sites. No indisputable spouts have been reported from early-first-millennium AD southern LSA sites.

We are, of course, still a long way from understanding what these ceramic stylistic families or even their constituent regional styles represent in social, cultural and economic terms. Perhaps they tell us something about the extent, direction and intensity of interaction networks of the potters. Whatever its exact significance, this supra-regional dichotomy disappeared around the mid first millennium AD. Although the Namibian branch of the northern family seems to have changed at its own pace and in its own particular direction (Kinahan 2001; Vogelsang and Eichhorn 2011), the thin-walled Bambata pots of Botswana and Zimbabwe were replaced by EIA thick-walled ware, including Happy Rest, Gokomere, Dambwa, Kumadzulo and other types (Sadr 2008). These EIA ceramic types share some interesting similarities in techniques of decoration with the Bambata pottery, a link that remains to be investigated further.

Significantly, at this time around the mid first millennium AD, the southern thin-walled pots took on many of the traits seen earlier in the north. It is these changes in the south that may provide proxy evidence for an influx of people. They require closer examination. To do so, it is necessary first to describe in a little more detail so-called Bambata pottery. Although it remains poorly defined, most authors have emphasised tight comb-stamped patterns on thin-walled sherds as a defining characteristic of Bambata pottery (Sadr 2008: 107, table 2). Geographically, Bambata pottery is found in sites from Lake Ngami (Campbell 1992; Huffman 1994) along the Boteti (Denbow and Campbell 1980; Huffman 2005) to the Makgadikgadi Pans (Denbow 1984) and on into the Limpopo Basin (Walker 1983, 1994; Robbins 1985; Wadley 1996; Hall and Smith 2000; Van Doornum 2005; Van der Ryst 2007) (Figure 5.2).

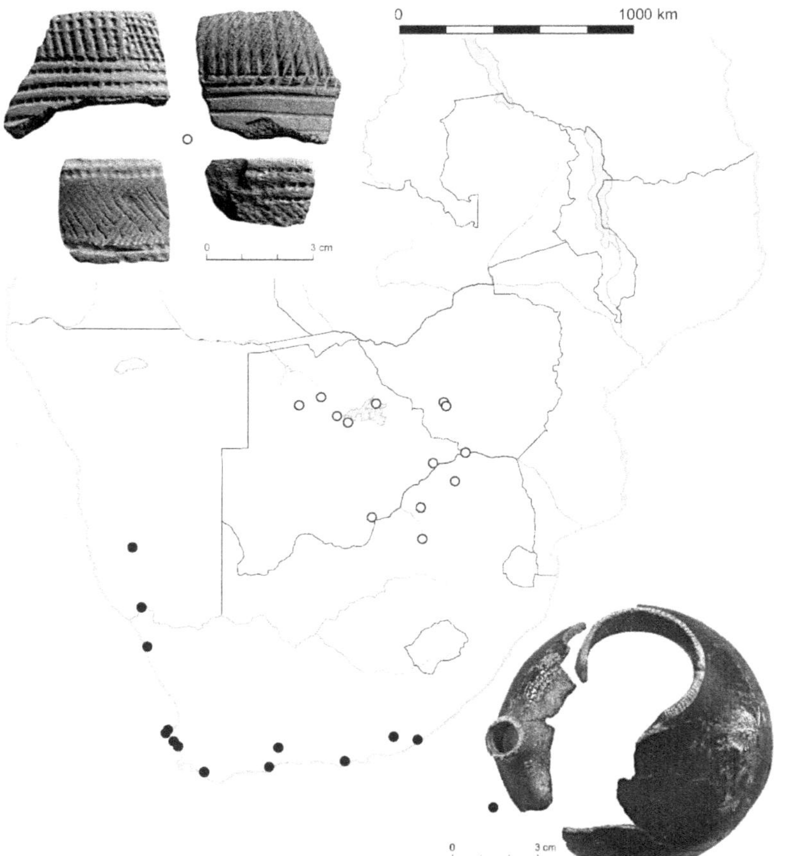

Figure 5.2: Map of Southern Africa showing the distribution of sites with Bambata pottery (first half of first millennium AD, open circles) and Cape coastal spouted wares (second half of first millennium AD, filled circles). Sherds from Bambata Cave, Zimbabwe (photograph by R. Burrett) and fragment of a spouted pot from Die Kelders, South Africa (photograph by K. Sadr).

Beyond this clearly delimited area, no indisputable Bambata sherds have been found. Although there are thin-walled pots in northern Namibia, they lack the characteristic decorations of Bambata pottery. No Bambata sherds have been reported from the Tsodilo Hills. Even just west of Lake Ngami, contemporary thin-walled pots do not show the classic Bambata traits.[7] Likewise, no Bambata pottery is known in the lands north of the Boteti River, or in the Kalahari to the south. Possible outlier sherds have been reported at Great Zimbabwe and Mabveni east of the Matopos (Huffman 2005), but not everyone agrees with these identifications (Walker 1983).

Over the years, various authors have attributed the manufacture of Bambata pottery to Bushmen hunter-gatherers, Khoekhoe pastoralists, and Bantu farmers (Sadr 2008: 107, table 1). By now everyone agrees they are principally associated with LSA sites, but we are still far from understanding how this (or any other) LSA ceramic type correlates with economic, social, cultural or linguistic entities. We know that there are eastern and western varieties not only in Bambata ceramic decoration, vessel shapes and manufacturing techniques (Motloung 2007), but also in the proportions of different types of stone tools associated with the LSA sites of these areas (Walker 1989, 1994). As for subsistence activities, large and small livestock are associated with Bambata pottery at some of the sites and perhaps even with the pre-Bambata layers at Toteng 1 near Lake Ngami (Walker 1983; Campbell 1992; Robbins et al. 2005). In the Makgabeng, the sites with Bambata pottery may be associated with rock paintings of sheep or even with the red geometric finger-painting tradition, but the matter cannot be settled without dating the art. Most importantly for the present argument, none of the sites with Bambata pottery seem to contain significant proportions of livestock bones, and considering that most Bambata pottery comes from rock-shelters it is difficult to see evidence of pastoralism here. If not pastoralists, however, it is worth noting that several LSA sites in the Matopos Hills contain storage pits that suggest a delayed-return or accumulation-based economy (Walker 1995; Mitchell 2002: 161), indicating a relatively complex social, economic and perhaps political organisation among these (trans-egalitarian?) hunter-gatherers.

Thin-walled Bambata pottery is not very accurately dated (Walker 1983; Campbell 1992; Huffman 2005; Sadr and Sampson 2006). Most agree that it belongs in the first half of the first millennium AD, some think earlier. After the mid millennium, thick-walled vessels replaced the thin-walled Bambata pots. We cannot tell precisely what became of the hunter-gatherers and hunters-with-livestock who made and used Bambata pottery. Some no doubt carried on as foragers for a while, but without thin-walled pottery (Van Doornum 2005). At least some were eventually absorbed into the newly arrived farming communities (Hall and Smith 2000), while others may have retreated from the expanding farmer frontier, which in this case would have to mean either to the south or the west.

Having described Bambata pottery and its associated archaeology from the first half of the first millennium AD in the north, I now turn to the south and west coasts of South Africa after the mid first millennium AD. In the first half of the first millennium, southern ceramics were rarely, if ever, decorated. Starting around the sixth century AD, however, a remarkably uniform style of pottery appeared along coastal areas all the way from southern Namibia to the region of today's East London, and even partly into the central Karoo (see Figure 5.2). At first glance, these mainly coastal pots, with a narrow mouth, short neck and somewhat pointed base, hardly resemble Bambata pottery. In some aspects, however, the two pottery assemblages share distinct similarities. It goes without saying that they are both thin-walled wares, but that is a characteristic of nearly all LSA pottery in Southern Africa. More interestingly, both assemblages include spouts. In the case of Bambata and other early-first-millennium northern assemblages, only a few pots were spouted (Schofield 1941; Walker 1983; Campbell 1992; Smith and Jacobson 1995; Albrecht et al. 2001), whereas in the coastal assemblage most of the pots in the late first millennium seem to have had spouts (Rudner 1968; Sadr and Smith 1991: 108). The spouts were decorated in the same manner as the vessel's main opening. Given that there is no clear evidence of spouted vessels in the southern assemblages before the mid millennium, their appearance in the Cape coastal sites at the same time that Bambata disappeared in the north almost certainly represents a link.

The link extends to ceramic decoration techniques. The two main decorative techniques of Bambata pottery were fine comb stamping and broad incisions. The Bambata comb may sometimes have been a notched bone, an example of which was recently excavated on the Makgabeng Plateau. The coastal spouted pots also exhibit two principal decoration techniques: comb stamping and broad incisions. But here, the 'comb' was the edge of a white mussel shell, *Donax serra* (Wilson and Halkett 1980). Not surprisingly, inland examples suggest the use of a different implement for a comb. Sampson et al. (1989: 4) suggest that the edge of a freshwater mussel or a fragment of ostrich eggshell was used for the comb-stamped decoration found in the Karoo. Despite the variations in the nature of the 'comb' itself, it seems more than coincidental that around the mid first millennium not only spouts but also the two main decoration techniques of Bambata pottery appeared in southern areas, just as Bambata pottery disappeared in the north. It is tempting to see in this particular constellation of ceramic evidence a population movement out of Botswana and western Zimbabwe, perhaps directly in response to the arrival of EIA farmers.

However, whether and to what extent there was a cultural, linguistic or genetic link between the earlier first-millennium makers and users of Bambata ceramics in the north and those of the later first-millennium coastal spouted wares in the south remains an open question. The degree of similarity in other material remains – for example, stone-tool technology – remains to be thoroughly investigated. Certainly there are significant differences to bear in mind. The northern ceramic family was associated with an inland adaptation while the spouted southern pots are often associated with coastal sites and marine faunal remains. And there are lesser differences evident in, for example, the range of ceramic vessel shapes, average ostrich eggshell bead diameters and so on. Given these similarities and differences, it is difficult to argue for a mid-first-millennium AD demographic and cultural replacement due to mass migration from north to south. An assimilation of new elements by southerners, which hints at infiltration and hybridisation rather than migration and displacement, seems more probable.

Late Migration?

Before jumping to the conclusion that the ceramic evidence shows a mid-first-millennium AD infiltration by the Khoekhoe, it is worth pointing out other potential episodes of population movement from north to south in later periods. I have suggested elsewhere (Sadr 1998) that a certain aspect of early-second-millennium AD thin-walled ceramics, pierced lugs, might be considered evidence of Khoekhoe migration. At the time of first contact with Europeans, such lugged vessels were seen in the possession of Cape herders (Rudner 1979: 9, fig. 4). But does this suffice to label all lugged pots Khoekhoe pottery? In the archaeological record there is much diversity in the morphology of lugged vessels, and it is by no means sure that they can be linked in a stylistic chain from Botswana to the Cape (Sadr 1998). Nor can one assume that all makers of lugged pots were necessarily speakers of a Khoe language.

Further indications of a late migration include the hand prints found in caves and rock-shelters in the Western Cape, which most likely date to the early second millennium AD (Yates, Manhire and Parkington 1994). They are related to the red geometric art found in Central Africa and along the Limpopo and Orange River valleys as well (Smith and Ouzman 2004). Smith and Ouzman have cogently argued that this style of art provides strong evidence for a migration of people from north to south, but precisely when this happened and whether they spoke a Khoe language are questions that are not easily answered.

There are probably other strands of evidence from different periods that will eventually come to light, and which will indicate other possible episodes of migration in the subcontinent. That is not unexpected. Historians have noted – uncharitably, in the case of Trevor-Roper (1965: 5) – what African oral traditions amply demonstrate, that people are constantly on the move. Comparative linguistic studies point in the same direction. Migrations happened, but they would have varied in the number of people involved, the rate and the direction of the movement, as well as in their cause and effect. Archaeological remains indicate several potential episodes of migration and infiltration, but which of them actually involved Khoe speakers cannot be answered with the bare bones of archaeological evidence. Languages leave no archaeological traces except in writing, and there is no simple, universal, static, predictable link between linguistic groups and artefact style (Hodder 1982; Lemonnier 1986; David et al. 1991; MacEachern 1994). What is more, none of the potential episodes of migration listed above show any faunal evidence to indicate that true pastoralists were involved, nor do any of them seem to be coeval with the first appearance of sheep and pottery in southernmost Africa.

When did true pastoralism emerge in Southern Africa? It is now clear that in the EIA cattle were crucial for obtaining wives, and settlements were organised according to the Central Cattle Pattern (Huffman 2007: 341), so one can assume that large livestock already occupied a central position in the ideological universe of EIA people. In the LSA, true pastoralism seems to be a later development. The evidence elaborated indicates the absence of evidence for South African pastoralism in the first millennium AD. The evidence from Namibia suggests quite intensive pastoralism in the early second millennium AD, with transhumance and cattle posts identified in the material record (Kinahan 2001). Physical evidence for livestock kraals is also found in the early-second-millennium AD Karoo of the central interior (Sampson 2010), but these were not located in the centre of settlements and the ideological and economic importance of livestock at that time is not easy to assess (see also Jerardino and Maggs 2007). If not in the early second millennium, certainly by the seventeenth and eighteenth centuries AD, there seem to have been true pastoralists in the Cape region whose settlements were organised around central kraals and whose seasonal movements were dictated by the requirements of their livestock (Jopp 1979). The site KFS5 on the Vredenburg Peninsula suggests that such pastoralists may have operated there even a few centuries earlier (Fauvelle-Aymar et al. 2006). But the question of precisely how and why herder-hunters became true pastoralists in these settings, although it has received some attention (Kinahan 1996, 2001; Sadr 2002: 478–79), requires considerably more work.

Conclusion

The absence of evidence for a two-thousand-year-old migration of pastoralists from northern Botswana to the south-western Cape has sometimes been attributed to the archaeological invisibility of pastoralists (Smith 2005: 44–48). Nomadic herders, it is said, move often and leave little behind. Ethnographic examples of barely visible, recently abandoned herder camps have supported this argument (e.g., Robertshaw 1978b; Smith 1992). But such ethnoarchaeological observations are from the plastic and metal age, where pots do not break so easily and stone flakes and chips are not produced by the dozen every time a cutting edge is required. If Stone Age sites of highly mobile hunter-gatherers can be found without difficulty even in the open landscape (e.g., Sampson 1985; Manhire 1987; Wilmsen 1989; Sadr, Gribble and Euston-Brown 1992), there is no reason why it should be more difficult to find sites of mobile pastoralists.

In fact, ephemeral sites of fifteenth-century herders have been recorded in the sandy areas north of Kasteelberg on the Vredenburg Peninsula, in the form of extensive but low-density surface scatters of artefacts and vitrified dung (Fauvelle-Aymar et al. 2006). Similar late-herder sites are now also known from the Breede and Berg river valleys (Arthur 2008), and examples have been known from Namibia for some time (Kinahan 1996, 2001). If we are able to find such late-herder sites, archaeological invisibility can hardly explain their earlier absence. Certainly in the intensively surveyed coastal landscapes of the Vredenburg Peninsula, where site visibility is ideal and where late-herder camps have been identified, their absence in the first millennium AD may simply mean that pastoralism in these parts did not appear until later.

This would mean that the origin of pastoralism at the Cape need not be linked to the origin of livestock herding or pottery manufacture. We have tended to bundle these variables in order to explain their appearance with one slash of Occam's razor, in a single migration of the Khoekhoe two thousand years ago. But the principle of parsimony may not be the best guide in this case. That the first livestock and pottery in southernmost Africa arrived long before the emergence of true pastoralist communities in this area seems archaeologically evident. There is certainly not a shred of material evidence for a mass migration with livestock and ceramic technology from the north to the Cape around two thousand years ago. Several episodes of infiltration or small-scale demic diffusion, with subsequent hybridisation, may account for the vague similarities between Bambata and Cape coastal spouted pottery, and perhaps the later similarities in the use of lugged wares as well as the red geometric art style (for the latest genetic evidence see Schlebusch et al. 2012). The faunal evidence, in any case, fails to document pastoralists in these potential movements; it shows only hunters with livestock.

Notes

I would like to thank Michael Bollig for the invitation to write this chapter, and ACACIA for having considered me a co-researcher in their project and supporting my research over the last few years. An earlier version of this chapter was presented at a colloquium in Paarl in 2006. Alan Barnard, Chris Ehret, François-Xavier Fauvelle-Aymar, Detlef Gronenborn, Tom Güldemann, Wilfried Haacke, Bernd Heine, John Kinahan, Christa König, Andy Smith, Ben Smith and Himla Soodyall offered several helpful comments and much constructive criticism on that occasion. Hans-Peter Wotzka suggested other improvements to this new version, for which I am extremely grateful.

1. Since the EIA overlaps with the most recent parts of the Later Stone Age (LSA), they cannot represent two separate 'Ages'. A more appropriate terminology has not yet evolved, so we stick to the classic labels for now. To lighten the terminological load, only the abbreviated initials 'LSA' and 'EIA' are used in this chapter, and then only sparingly. See Sadr and Sampson (2006) for a short description of the salient material differences between the two.
2. This applies to the commoners. At elite residences and royal capitals, cattle were consumed more regularly (Tom Huffman, personal communication, 2008).
3. See, e.g., the 1960s debate between Binford (e.g., Binford and Binford 1966) and Bordes (e.g., Bordes and De Sonneville-Bordes 1970).
4. See also Eggert (2005: 319–23) for a critique of the arguments.
5. Thanks also to R. Vogelsang (personal communication, 2007) on this issue.
6. Walker (personal communication, 2005).

References

Albrecht, M., H. Berke, B. Eichhorn, T. Frank, R. Kuper, P. Prill, R. Vogelsang, R., and S. Wenzel. 2001. 'Oruwanje 95/1: A Late Holocene Stratigraphy in Northwestern Namibia', *Cimbebasia* 17: 1–22.

Arthur, C.I. 2008. 'The Khoekhoen of the Breede River Swellendam: An Archaeological and Historical Landscape Study', MSc thesis. Cape Town: University of Cape Town.

Bedford, S., M. Spriggs and R. Regenvanu. 2006. 'The Teouma Lapita Site and the Early Human Settlement of the Pacific Islands', *Antiquity* 80: 812–28.

Binford L.R., and S. Binford. 1966. 'A Preliminary Analysis of Functional Variability in the Mousterian of Levallois Facies', *American Anthropologist* 68: 238–95.

Binneman, J.N.F. 1995. 'Symbolic Construction of Communities During the Holocene Later Stone Age in the South-eastern Cape', PhD dissertation. Johannesburg: University of the Witwatersrand.

Bleek, D.F. 1929. *Comparative Vocabularies of Bushman Languages*. Cambridge: Cambridge University Press.

Boonzaier, E., et al. 1996. *The Cape Herders: A History of the Khoikhoi of Southern Africa*. Cape Town: David Philip.

Bordes, F., and D. de Sonneville-Bordes. 1970. 'The Significance of Variability in Palaeolithic Assemblages', *World Archaeology* 2(1): 61–73.

Brink, J., and L. Webley. 1996. 'Faunal Evidence for Pastoralist Settlements at Jakkalsberg, Richtersveld, Northern Cape Province', *Southern African Field Archaeology* 5: 70–79.

Byrne, P.J. 2007. 'Kasteelberg G: The Introduction of Livestock and its Effects on the Archaeological Sequence', BSc honours thesis. Johannesburg: University of the Witwatersrand.

Campbell, A.C. 1992. *Southern Okavango Integrated Water Development Study: Archaeological Survey of the Proposed Maun Reservoir*. Gaborone: Department of Water Affairs.

Cooke, C.K. 1969. *Rock Art of Southern Africa*. Cape Town: Books of Africa.

David, N., K.B. Gavua, A.S. MacEachern and J.A. Sterner. 1991. 'Ethnicity and Material Culture in North Cameroon', *Canadian Journal of Archaeology* 15: 171–77.

Deacon, H.J., J. Deacon, M. Brooker and M. Wilson. 1978. 'The Evidence for Herding at Boomplaas Cave in the Southern Cape, South Africa', *South African Archaeological Bulletin* 33: 39–65.

Deacon, J. 1972. 'Wilton: An Assessment After 50 Years', *South African Archaeological Bulletin* 27: 10–45.

——— 1984. 'Later Stone Age People and Their Descendants in Southern Africa', in R.G. Klein (ed.), *Southern African Prehistory and Palaeoenvironments*. Rotterdam: Balkema, pp.221–329.

Denbow, J.R. 1984. 'Report on Preliminary Archaeological Reconnaissance of BP Soda Ash Project, Sowa Pan, Botswana', unpublished report commissioned by BP Botswana.

Denbow, J.R., and A.C. Campbell. 1980. 'National Museum of Botswana: Archaeological Research Programme', *Nyame Akuma* 17: 3–9.

De Wet-Bronner, E. 1994. 'The Faunal Remains from Four Late Iron Age Sites in the Soutpansberg Region: Part I: Tavhatshena', *Southern African Field Archaeology* 3: 33–43.

——— 1995a. 'The Faunal Remains from Four Late Iron Age Sites in the Soutpansberg Region: Part II: Tshitheme and Dzata', *Southern African Field Archaeology* 4: 18–29.

——— 1995b. 'The Faunal Remains from Four Late Iron Age Sites in the Soutpansberg Region: Part III: Tshirululuni', *Southern African Field Archaeology* 4: 109–19.

Derricourt, R. 1977. *Prehistoric Man in the Ciskei and Transkei*. Cape Town and Johannesburg: Struik.

Diamond, J. 1998. *Guns, Germs and Steel: A Short History of Everybody for the Last 13,000 Years*. London: Vintage.

Dolukhanov, P., A. Shukorov, D. Gronenborn, D. Sokoloff, V. Timofeev, and G. Zaitseva. 2005. 'The Chronology of Neolithic Dispersal in Central and Eastern Europe', *Journal of Archaeological Science* 32: 1441–58.

Eastwood, E.B., and B.W. Smith. 2005. 'Fingerprints of the Khoekhoen: Geometric and Handprinted Rock Art in the Central Limpopo Basin, Southern Africa', *South African Archaeological Society Goodwin Series* 9: 63–76.

Eggert, M. 2005. 'The Bantu Problem and African Archaeology', in A.B. Stahl (ed.), *African Archaeology: A Critical Introduction*. Malden, MA: Blackwell, pp.301–26.

Ehret, C. 1967. 'Cattle-keeping and Milking in Eastern and Southern African History: The Linguistic Evidence', *Journal of African History* 8(1): 1–17.

——— 1982. 'The First Spread of Food Production to Southern Africa', in C. Ehret and M. Posnansky (eds), *The Archaeological and Linguistic Reconstruction of African History*. Berkeley: University of California Press, pp.158–81.

——— 1998. *An African Classical Age: Eastern and Southern Africa in World History, 1000 BC to AD 400*. Oxford: James Currey.

Elphick, R. 1985. *Khoikhoi and the Founding of White South Africa*. Johannesburg: Ravan Press.

Fauvelle-Aymar, F.-X., K. Sadr, F. Bon and D. Gronenborn. 2006. 'The Visibility and Invisibility of Herders' Kraals in South Africa, with Reference to a Possible Early Contact Period Khoekhoe Kraal at KFS 5 (Western Cape)', *Journal of African Archaeology* 4: 253–71.

Goodwin, A.J.H., and C. Van Riet Lowe. 1929. 'The Stone Age Cultures of South Africa', *Annals of the South African Museum* 27: 1–276.

Gronenborn, D. 1999. 'A Variation on a Basic Theme: The Transition to Farming in Southern Central Europe', *Journal of World Prehistory* 13: 123–210.

Güldemann, T. 2008. 'A Linguist's View: Khoe-Kwadi Speakers as the Earliest Food-producers of Southern Africa', in K. Sadr and F.-X. Fauvelle-Aymar (eds), 'Khoekhoe and the Origins of Herding in Southern Africa', *Southern African Humanities* 20(1): 93–132.

Hall, S.L., and B.W. Smith 2000. 'Empowering Places: Rock Shelters and Ritual Control in Farmer–Forager Interactions in the Northern Province', *South African Archaeological Society Goodwin Series* 8: 30–46.

Henshilwood, C. 1996. 'A Revised Chronology for Pastoralism in Southernmost Africa: New Evidence of Sheep at c. 2000 b.p. from Blombos Cave, South Africa', *Antiquity* 70: 945–49.

Hodder, I. 1982. *Symbols in Action*. Cambridge: Cambridge University Press.

Huffman, T.N. 1989. *Iron Age Migrations*. Johannesburg: Wits University Press.

——— 1994. 'Toteng Pottery and the Origins of Bambata', *Southern African Field Archaeology* 3: 3–9.

——— 2005. 'The Stylistic Origin of Bambata and the Spread of Mixed Farming in Southern Africa', *Southern African Humanities* 17: 57–79.

——— 2007. *Handbook to the Iron Age: The Archaeology of Pre-colonial Farming Societies in Southern Africa*. Pietermaritzburg: KZN Press.

Humphreys, A.J.B. 1974. 'A Preliminary Report on Test Excavations at Dikbosch Shelter I, Herbert District, Northern Cape', *South African Archaeological Bulletin* 29: 115–19.

Humphreys, A.J.B., and A.I. Thackeray. 1983. *Ghaap and Gariep: Later Stone Age Studies in the Northern Cape*. Cape Town: South African Archaeological Society.

Ikeya, K. 1993. 'Goat Raising Among the San in the Central Kalahari', Kyoto University Center for African Area Studies, *African Study Monographs* 14(1): 39–52.

Jacobson, L. 1987. 'The Size Variability of Ostrich Eggshell Beads from Central Namibia and its Relevance as a Stylistic and Temporal Marker', *South African Archaeological Bulletin* 42: 55–59.

Jerardino, A. 1998. 'Excavations at Pancho's Kitchen Midden, Western Cape Coast, South Africa: Further Observations into the Megamidden Period', *South African Archaeological Bulletin* 53:16–25.

Jerardino, A., and T. Maggs. 2007.' Simon se Klip at Steenbokfontein: The Settlement Pattern of a Built Pastoralist Encampment on the West Coast of South Africa', *South African Archaeological Bulletin* 62(186): 104–14.

Jopp, W. (ed.). 1979. *Unter Hottentoten 1705–1713: Die Aufzeichnungen von Peter Kolb. Alte abenteuerliche Reiseberichte.* Basel: Erdmann.

Kaplan, J. 1987. 'Settlement and Ss]ubsistence at Renbaan Cave', in J. Parkington and M. Hall (eds), *Papers in the Prehistory of the Western Cape, South Africa. British Archaeological Reports*, International Series 332,ii. Oxford: B.A.R., pp.350–76.

Kent, S. 1993. 'Sharing in an Egalitarian Kalahari Community', *Man* 28: 479–514.

Kinahan, J. 1996. 'A New Archaeological Perspective on Nomadic Pastoralist Expansion in South-western Africa', in J.E.G. Sutton (ed.), *The Growth of Farming Communities in Africa from the Equator Southwards*. Nairobi: British Institute in Eastern Africa, pp.211–26.

——— 2001. *Pastoral Nomads of the Namib Desert: The People History Forgot*. Windhoek: Namibia Archaeological Trust.

Klein, R.G. 1978. 'A Preliminary Report on the Larger Mammals from the Boomplaas Stone Age Cave Site, Cango Valley, Oudtshoorn District, South Africa', *South African Archaeological Bulletin* 33: 66–75.

——— 1986. 'The Prehistory of Stone Age Herders in the Cape Province of South Africa', *South African Archaeological Society Goodwin Series* 5: 5–12.

——— 1989. 'Faunal Evidence for Prehistoric Herder-forager Activities at Kasteelberg, Western Cape Province, South Africa', *South African Archaeological Bulletin* 44: 82–97.

Klein, R.G. and K. Cruz-Uribe. 1987. 'Large Mammal and Tortoise Bones from Eland's Bay Cave and Nearby Sites, Western Cape Province, South Africa', in J. Parkington and M. Hall (eds), *Papers in the Prehistory of the Western Cape, South Africa*. Oxford: British Archaeological Reports International Series 332(i), pp. 132–64.

——— 1989. 'Faunal Evidence for Prehistoric Herder-Forager Activities at Kasteelberg, Western Cape Province, South Africa', *South African Archaeological Bulletin* 44: 82–97.

Kuper, R., and S. Kröpelin. 2006. 'Climate-controlled Holocene occupation in the Sahara: Motor of Africa's Evolution', *Science* 313: 803–7.

Lemonnier, P. 1986. 'The Study of Material Culture Today: Toward an Anthropology of Technical Systems', *Journal of Anthropological Archaeology* 5: 147–86.

Leslie, M. 1989. 'The Holocene Sequence from Uniondale Rock Shelter in the Eastern Cape', *South African Archaeology Society Goodwin Series* 6: 17–32.

MacDonald, K.C. 1996. 'The Windé Koroji Complex: Evidence for the Peopling of the Eastern Inland Niger Delta (2100–500 BC)', *Préhistoire Anthropologie Méditerranéennes* 5: 147–66.

MacEachern, A.S. 1994. '"Symbolic Reservoirs" and Cultural Relations between Ethnic Groups: West African Examples', *African Archaeological Review* 12: 203–22.

MacHugh, D.E., and D.G. Bradley. 2001. 'Livestock Genetic Origins: Goats Buck the Trend', *Proceedings of the National Academy of Sciences* 98: 5382–84.

Maggs, T. 1984. 'The Iron Age South of the Zambezi', in R.G. Klein (ed.), *Southern African Prehistory and Palaeoenvironments*. Rotterdam: Balkema, pp.329–60.
Maggs, T., and V. Ward. 1980. 'Driel Shelter: Rescue at a Late Stone Age Site on the Tugela River', *Annals of the Natal Museum* 24(1): 35–70.
Maggs, T., and G. Whitelaw. 1991. 'A Review of Recent Archaeological Research on Food Producing Communities in Southern Africa', *Journal of African History* 21: 3–24.
Manhire, A. 1987. *Later Stone Age Settlement Patterns in the Sandveld of the South-western Cape Province*. Oxford: British Archaeological Reports.
——— 1993. 'A Report on the Excavations at Faraoskop Rock Shelter in the Graafwater District of the South-Western Cape', *Southern African Field Archaeology* 2: 3–23.
Mazel, A.D. 1984. 'Diamond 1 and Clarke's Shelter: Report on Excavations in the Northern Drakensberg, Natal, South Africa', *Annals of the Natal Museum* 26(1): 25–70.
——— 1986. 'Mbabane Shelter and Esinhlonhlweni Shelter: The Last Two Thousand Years of Hunter-gatherer Settlement in the Central Thukela Basin, Natal, South Africa', *Annals of the Natal Museum* 27: 389–453.
——— 1990. 'Mhlwazini Cave: The Excavation of Late Holocene Deposits in the Northern Natal Drakensberg, Natal, South Africa', *Natal Museum Journal of Humanities* 2: 95–133.
——— 1992. 'Collingham Shelter: The Excavation of Late Holocene Deposits, Natal, South Africa', *Natal Museum Journal of Humanities* 4: 1–51.
——— 1993. 'KwaThwaleyakhe Shelter: The Excavation of Mid and Late Holocene Deposits in the Central Thukela Basin, Natal, South Africa', *Natal Museum Journal of Humanities* 5: 1–36.
——— 1999. 'iNkolimahashi Shelter: the Excavation of Later Stone Age Rock Shelter Deposits in the Central Thukela Basin, KwaZulu-Natal, South Africa', *Natal Museum Journal of Humanities* 11: 1–21.
Mitchell, P. 2002. *The Archaeology of Southern Africa*. Cambridge: Cambridge University Press.
Modikwa, B. 2008. 'Lithic Technology and the Introduction of Pottery in Southern Africa', MSc thesis. Johannesburg: University of the Witwatersrand.
Motloung, A. 2007. 'Morphological Analysis of First Millennium Thin Walled Pottery', MSc thesis. Johannesburg: University of the Witwatersrand.
Parkington, J. 1980. 'Time and Place: Some Observations on Spatial and Temporal Patterning in the Later Stone Age Sequence in Southern Africa', *South African Archaeological Bulletin* 35: 75–83.
Phillipson, D.W. 1976. *The Prehistory of Eastern Zambia*. Nairobi: British Institute in East Africa.
——— 2005. *African Archaeology*. Cambridge: Cambridge University Press.
Pistorius, J.C.C. 2001. 'Late Iron Age Sites on Mmatshetshele Mountain in the Central Bankeveld of the North West Province, South Africa, with Appendix by S. Badenhorst & I. Plug', *South African Archaeological Bulletin* 56: 46–56.
Pleurdeau, D., E. Imalwa, F. Détroit, J. Lesur, A. Veldman, J.-J. Bahain, and E. Marais. 2012. '"Of Sheep and Men": Earliest Direct Evidence of Caprine Domestication in Southern Africa at Leopard Cave (Erongo, Namibia)', *PLoS ONE* 7(7): e40340. doi:10.1371/journal.pone.0040340.
Plug, I. 1979. 'The Faunal Remains from Wildebeestfontein', *South African Archaeological Society Goodwin Series* 3: 130–32.
——— 2000. 'Overview of the Iron Age Fauna from the Limpopo Valley', *South African Archaeological Society Goodwin Series* 8: 117–26.

_______ 2002. 'Faunal Remains from Mzinyashana, a Later Stone Age Site in KwaZulu-Natal, South Africa', *Southern African Humanities* 14: 51–63.
Plug, I. and F. Roodt. 1990. 'The Faunal Remains from Recent Excavations at uMgungundlovu', *South African Archaeological Bulletin* 45: 47–52.
Poggenpoel, C., and P.T. Robertshaw. 1981. 'The Excavation of Smitswinkelbaai Cave, Cape Peninsula', *South African Archaeological Bulletin* 36: 29–35.
Quitta, H. 1960. 'Zur Frage der ältesten Bandkeramik in Mitteleuropa', *Prähistorische Zeitschrift* 38: 153–89.
Raimbault, M., and D. Commelin. 2002. 'La Poterie du Site Néolithique de Kobadi dans le Sahel Malien', *Préhistoire Anthropologie Méditerranéennes* 10/11: 107–16.
Raven-Hart, R. 1967. *Before Van Riebeek*. Cape Town: Struik.
Reid, D.A.M., K. Sadr and N. Hanson-James. 1998. 'Herding Traditions', in P. Lane, A. Reid and A. Segobye (eds), *Ditswa Mmung: The Archaeology of Botswana*. Gaborone: Pula Press and Botswana Society, pp.81–100.
Richter, J. 1991. *Studien zur Urgeschichte Namibias*. Cologne: Heinrich-Barth-Institut.
Riemer, H. 2007. 'When Hunters Started Herding: Pastro-foragers and the Complexity of Holocene Economic Change in the Western Desert of Egypt', in M. Bollig, O. Bubenzer, R. Vogelsang and H.-P. Wotzka (eds), *Aridity, Change and Conflict in Africa*. Cologne: Heinrich-Barth-Institut, pp.105–44.
Rivat, F. 2006. 'Les Derniers Tailleurs de Pierre d'Afrique Australe: Apport de la Technologie Lithique pour la Caractérisation de l'Identité Socio-économique des Groupes de l'Extrême Fin du Late Stone Age', Mémoire de Maîtrise I. Toulouse: Université de Toulouse-Le Mirail.
Robbins, L.H. 1985. 'The Manyana Rock Paining Site', *Botswana Notes and Records* 17: 1–14.
Robbins, L.H., A.C. Campbell, M.L. Murphy, G.A. Brook, P. Srivastava, and S. Badenhorst. 2005. 'The Advent of Herding in Southern Africa: Early AMS Dates on Domestic Livestock from the Kalahari Desert', *Current Anthropology* 46: 671–77.
Robertshaw, P.T. 1975–1977. 'Archaeological Investigations at Langebaan Lagoon, Cape Province', *Palaeoecology of Africa and the Surrounding Islands* 10/11: 139–48.
——— 1978a. 'The Archaeology of an Abandoned Pastoralist Camp-site', *South African Journal of Science* 74: 29–31.
——— 1978b. 'Archaeological Investigations at Langebaan Lagoon, Cape Province', *Palaeoecology of Africa* 10: 139–48.
——— 1979. 'Excavations at Duiker Eiland, Vredenburg District, Cape Province', *Annals of the Cape Provincial Museums (Human Sciences)* 1(1): 1–26.
Rudner, J. 1968. 'Strandloper Pottery from South and South West Africa', *Annals of the South African Museum* 49: 441–663.
——— 1979. 'The Use of Stone Artefacts and Pottery among the Khoisan Peoples in Historic and Protohistoric Times', *South African Archaeological Bulletin* 34: 3–17.
Sadr, K. 1997. 'Kalahari Archaeology and the Bushman Debate', *Current Anthropology* 38: 104–11.
——— 1998. 'The First Herders at the Cape of Good Hope', *African Archaeological Review* 15: 101–32.
——— 2002. 'Ancient Pastoralists in the Sudan and in South Africa', in Jennerstr. 8 (ed.), *Tides of the Desert*. Cologne: Heinrich-Barth-Institut, pp.471–84.
——— 2003. 'The Neolithic of Southern Africa', *Journal of African History* 44: 195–209.
——— 2004. 'Feasting on Kasteelberg? Early Herders on the West Coast of South Africa', *Before Farming* 2004/3, article 2.

——— 2007. 'Early First Millennium Pastoralists on Kasteelberg? The UB/UCT Excavation at KBA', *South African Archaeological Bulletin* 62(186): 154–61.

——— 2008. 'An Ageless View of First Millennium AD Southern African Ceramics', *Journal of African Archaeology* 6(1): 103–29.

——— 2009. 'Marine Shell Dates and Surface Lithic Assemblages on the West Coast of South Africa', *Journal of Archaeological Science* 36: 2713–29.

Sadr, K., and J. Gribble. 2010. 'The Stone Artefacts from the Vredenburg Peninsula Archaeological Survey, West Coast of South Africa', *Southern African Humanities* 22: 19–88.

Sadr, K., J. Gribble and G. Euston-Brown. 1992. 'The Vredenburg Peninsula Survey, 1991/1992 Season', in A.B. Smith and B. Mütti (eds), *Guide to the Archaeological Sites in the Southwestern Cape*. Cape Town: University of Cape Town, pp.41–43.

Sadr, K., and I. Plug. 2001. 'Faunal Remains in the Transition from Hunting to Herding in Southeastern Botswana', *South African Archaeological Bulletin* 56: 76–82.

Sadr, K., and C.G. Sampson. 1999. 'Khoekhoe Ceramics of the Upper Seacow River Valley', *South African Archaeological Bulletin* 54: 3–15.

——— 2006. 'Through Thick and Thin: Early Pottery in Southern Africa', *Journal of African Archaeology* 4: 235–52.

Sadr, K., and A. Smith. 1991. 'On Ceramic Variation in the Southwestern Cape, South Africa', *South African Archaeological Bulletin* 46: 107–15.

Sampson, C.G. 1972. *The Stone Age Industries of the Orange River Scheme and South Africa*. Bloemfontein: National Museum of Bloemfontein.

——— 1974. *The Stone Age Archaeology of Southern Africa*. New York: Academic Press.

——— 1985. *Atlas of Stone Age Settlement in the Central and Upper Seacow Valley*. Bloemfontein: National Museum of Bloemfontein.

——— 2010. 'Chronology and Dynamics of Later Stone Age Herders in the Upper Seacow River Valley, South Africa', *Journal of Arid Environments* 74(7): 842–48.

Sampson, C.G., T.J.G. Hart, D. Wallsmith and J.D. Blagg. 1989. 'The Ceramic Sequence in the Upper Seacow Valley: Problems and Implications', *South African Archaeological Bulletin* 44: 3–16.

Sand, C. 1997. 'The Chronology of Lapita Ware in New Caledonia', *Antiquity* 71: 539–47.

Schlebusch, C.M., P. Skoglund, P. Sjödin, L.M. Gattepaille, D. Hernandez, F. Jay, S. Li, M. De Jongh, A. Singleton, M.G.B. Blum, H. Soodyall, and M. Jakobsson. 2012. 'Genomic Variation in Seven Khoe-San Groups Reveals Adaptation and Complex African History'. *Science*. doi: 10.1126/science.1227721.

Schofield, J.F. 1941. 'A Report on the Pottery from Bambata Cave', *South African Journal of Science* 37: 361–72.

——— 1948. *Primitive Pottery: An Introduction to South African Ceramics, Prehistoric and Protohistoric*. Cape Town: South African Archaeological Society.

Schrire, C., and J. Deacon. 1989. 'The Indigenous Artefacts from Oudepost I, a Colonial Outpost of the VOC at Saldanha Bay, Cape', *South African Archaeological Bulletin* 44: 105–13.

Schweitzer, F. 1979. 'Excavations at Die Kelders, Cape Province, South Africa', *Annals of the South African Museum* 78: 102–233.

Schweitzer, F.R., and M.L. Wilson. 1982. 'Byneskranskop1a Late Quaternary Living Site in the Southern Cape Province, South Africa', *Annals of the South African Museum* 88(1): 1–203.

Sealy, J., T. Maggs, A. Jerardino and J. Kaplan. 2004. 'Excavations at Melkbosstrand: Variability among Herder Sites on Table Bay, South Africa', *South African Archaeological Bulletin* 59: 17–28.
Sealy, J., and R. Yates. 1994. 'The Chronology of the Introduction of Pastoralism to the Cape,
Smith, A.B. 1990. 'On Becoming Herders: Khoikhoi and San Ethnicity in Southern Africa', *African Studies* 49: 51–73.
——— 1992. *Pastoralism in Africa.* Johannesburg: Wits University Press.
——— 2005. *African Herders: Emergence of Pastoral Traditions.* Walnut Creek, CA: Alta Mira Press.
——— 2006. *Excavations at Kasteelberg, and the Origins of the Khoekhoen in the Western Cape, South Africa.* Oxford: British Archaeological Reports.
Smith, A., and L. Jacobson. 1995. 'Excavations at Geduld and the Appearance of Early Domestic Stock in Namibia', *South African Archaeological Bulletin* 50: 3–14.
Smith, A.B., K. Sadr, J. Gribble and R. Yates. 1991. 'Excavations in the South-western Cape, South Africa, and the Archaeological Identity of Prehistoric Hunter-gatherers within the Last 2000 Years', *South African Archaeological Bulletin* 46: 71–91.
Smith, B., and S. Ouzman. 2004. 'Taking Stock: Identifying Khoekhoen Herder Rock Art in Southern Africa', *Current Anthropology* 45: 499–526.
Spriggs, M. 1995. 'The Lapita Culture and Austronesian Prehistory in Oceania', in P. Bellwood, J.J. Fox and D. Tryon (eds), *The Austronesians: Historical and Comparative Perspectives.* Canberra: Australian National University, pp.112–33.
Sydow, W. 1967. *The Pre-European Pottery of South West Africa.* Windhoek: State Museum.
Trevor-Roper, H. 1965. *The Rise of Christian Europe.* London: Thames and Hudson.
Van der Ryst, M. 2007. 'Seeking Shelter: Later Stone Age Hunters, Gatherers and Fishers of Olieboomspoort in the Western Waterberg, South of the Limpopo', PhD dissertation. Johannesburg: University of the Witwatersrand.
Van Doornum, B. 2005. 'Changing Places, Spaces and Identity in the Shashe-Limpopo Region of Limpopo Province, South Africa', PhD dissertation. Johannesburg: University of the Witwatersrand.
Van Rijssen, W.J. 1992. 'The Late Holocene Deposits at Klein Kliphuis Shelter, Cedarberg, Western Cape Province', *South African Archaeological Bulletin* 47: 34–39.
Van Schalkwyk, J.A. 2000. 'Excavation of a Late Iron Age site in the Makgabeng, Northern Province', *Southern African Field Archaeology* 9: 75–82.
Vogel, J., I. Plug and L. Webley. 1997. 'New Dates for the Introduction of Sheep into South Africa: The Evidence form Spoegrivier Cave in Namaqualand', *South African Journal of Science* 93: 246–8.
Vogelsang, R., and B. Eichhorn. 2011. *Under the Mopane Tree: Holocene Settlement in Northern Namibia.* Africa Praehistorica 24. Köln: Heinrich Barth Institut.
Voigt, E. 1979. 'The Faunal Remains from Icon', *South African Archaeological Society Goodwin Series* 3: 80–85.
Wadley, L. 1979. 'Big Elephant Shelter and Its Role in the Holocene Prehistory of Central South West Africa', *Cimbebasia Series B* 3(1): 1–76.
——— 1996. 'Changes in the Social Relations of Precolonial Hunter-gatherers after Agropastoralist Contact: An Example from the Magaliesberg, South Africa', *Journal of Anthropological Archaeology* 15: 205–17.
Walker, N.J. 1983. 'The Significance of an Early Date for Pottery and Sheep in Zimbabwe', *South African Archaeological Bulletin* 38: 88–92.

——— 1989. 'The Late Stone Age', in P. Lane, A. Reid and A. Segobye (eds), *Ditswa Mmung: The Archaeology of Botswana*. Gaborone: Pula Press and Botswana Society, pp.65–80.

——— 1994. 'The Late Stone Age of Botswana: Some Recent Excavations', *Botswana Notes and Records* 26: 1–35.

——— 1995. *Late Pleistocene and Holocene Hunter-gatherers of the Matopos: An Archaeological Study of Change and Continuity in Zimbabwe*. Uppsala: Societas Archaeologica Upsaliensis.

Warfe, A.R. 2003. 'Cultural Origins of the Egyptian Neolithic and Predynastic: An Evaluation of the Evidence from the Dakhleh Oasis (South Central Egypt)', *African Archaeological Review* 20: 175–202.

Watson, D.J. 2005. 'Under the Rocks: Reconsidering the Origin of the Kintampo Tradition and the Development of Food Production in the Savanna-Forest/Forest of West Africa', *Journal of African Archaeology* 3(1): 3–56.

Watson, E.J. and V. Watson. 1990. 'Of Commoners and Kings: Faunal Remains from Ondini', *South African Archaeological Bulletin* 45: 33–46.

Webley, L. 1992. 'Early Evidence for Sheep from Spoegrivier Cave, Namaqualand', *Southern African Field Archaeology* 1: 3–13.

——— 1997. 'Jakkalsberg A and B: The Cultural Material from Two Pastoralist Sites in the Richtersveld, Northern Cape', *Southern African Field Archaeology* 6: 3–20.

Welbourne, R.G. 1979. 'Animal Remains from the Harmony Salt Factory, Northeast Transvaal', *South African Archaeological Society Goodwin Series* 3: 108.

Wendorf, F., and R. Schild. 1998. 'Nabta Playa and its Role in Northeastern African Prehistory', *Journal of Anthropological Archaeology* 17(2): 97–123.

Westphal, E.O.J. 1963. 'The Linguistic Prehistory of Southern Africa: Bush, Kwadi, Hottentot, and Bantu Linguistic Relationships', *Africa* 33: 237–64.

Wilmsen, E.N. 1989. 'Antecedents of Contemporary Pastoralism in Western Ngamiland', *Botswana Notes and Records* 20: 29–40.

Wilson, M.L., and D. Halkett. 1980. 'The Use of Marine Shell for Decorating Cape Coastal (Khoisan) Pottery', *South African Archaeological Bulletin* 36: 43–44.

Yates, R., A. Manhire and J. Parkington. 1994. 'Rock Painting and History in the Southwestern Cape', in T.A. Dowson and D. Lewis-Williams (eds), *Contested Images: Diversity in Southern African Rock Art Research*. Johannesburg: Wits University Press, pp.29–60.

Zeder, M.A., and B. Hesse. 2000. 'The Initial Domestication of Goats (*Capra hircus*) in the Zagros Mountains 10,000 Years Ago', *Science* 287(5461): 2254–57.

Part II

Historical and Contemporary Dynamics of Pastoralism

CHAPTER 6

Establishing a Precolonial 'Modern' Cattle-and-Gun Society

(Re-)Pastoralisation, Mercantile Capitalism and Power amongst Herero in Nineteenth-century Central Namibia

Dag Henrichsen

Introduction

The history of the Herero is generally perceived as being the history of cattle pastoralists. In the Southern African context, Herero have been considered archetypal pastoralists and historians and anthropologists have in general focussed their interest on a people 'with cattle' and living in a specific 'cattle culture' (see Hagolani 1967; Vivelo 1977).[1] The literature attests to cattle in Herero society being what Jean and John Comaroff have called 'the medium of transformation, in a *total* economy of signs and practices, between a material economy of things and a moral economy of persons' (Comaroff and Comaroff 1992: 128). Such a view of Herero society and history, and its interpretation, has been influenced by perceptions and interpretations held amongst Herero themselves. A Herero informant of the missionary Heinrich Vedder in the early twentieth century stated the following:

> Have not the Hereros been cattle-breeders ever since God created them? As a cattle-breeder, does not one always live in the selfsame way? One treks with the herd wherever water and grazing are to be found and, in the meantime, the cattle increase. Sometimes they are stolen by our enemies; sometimes no thefts take place for many years at a time. That is the life of a Herero; that was the life my great-grandfather lived, that was the life my grandfather lived, and my father lived it, too. (Vedder 1938: 145)[2]

From research on pastoral societies elsewhere we know that such a view has to be regarded as an elitist view of history – that is, belonging to successful cattle pastoralists. It projects an ideological model of the socio-cultural 'hegemony' and 'homogeneity' of cattle owners, irrespective of space and time (Galaty 1977; Spear and Waller 1993: 13). Such a view, and the academic literature which takes the notion of Herero being people 'with cattle' for granted, emphasises the primordial existence of a Herero cattle-based society. It neglects, on the one hand, other histories of Otjiherero-speaking people, of which the history of people with or without little stock or with small stock only is perhaps the most obvious one. On the other hand, it neglects the question how a consolidated cattle-based pastoral society, which did at times exist amongst Herero, came about. As research elsewhere has shown, the existence of cattle-based societies cannot be taken for granted but needs to be viewed as a historical construction and, indeed, a powerful achievement which has had and continues to be (re-)claimed constantly, both in a materialistic and ideological way (Spear and Waller 1993: 13). As David Cohen and Atieno Odhiambo have suggested, 'an economy and a culture organized around the holding and exchange of cattle was a formed and not a given feature ... Its context may lie not in an ancient "cattle complex" ... but rather in the struggles to rebuild wealth ... with their accompanying intense focus on cattle exchanges and management' (Cohen and Odhiambo 1989: 76).

In this chapter I focus on the history of Otjiherero-speaking people in the nineteenth century and analyse a period of intense reconstruction which brought about the establishment of a consolidated cattle-based pastoral society. The process I analyse in particular is what I call a process of (re-)pastoralisation. Whilst the term (re-)pastoralisation (Almagor 1980: 42–45) implies a general process of consolidation of a society based on cattle ownership and transhumance, my analysis is limited to some socio-economic aspects of it.

The underlying assumption of my interpretation is that precolonial communities in central Namibia were characterised by 'fluidity, competition and movement' (Gordon 1992: 19). They were not dependent exclusively on one specific mode of production such as hunting and gathering or stock breeding. The regional economy that existed can perhaps be best described by what Edwin Wilmsen with reference to the precolonial Kalahari area has termed a 'pastoral/pastroforaging social formation' (Wilmsen 1989: 74–76). People pursued various economic strategies simultaneously and at times emphasised one over the other for specific historical and cultural reasons. With respect to pastoralists it is important to note that they never pursued only one economic strategy but tried to construct a wide range of inter-related economic activities and engage in intra- and inter-regional networks of exchange in order to cope with survival, drought, disaster and historical change (Spear and Waller 1993: 4–9). Consequently, pastoralists are to be viewed as 'multi-resource pastoralists' (Salzmann, quoted in Chang and Koster 1994: 12).

The process of (re-)pastoralisation which I analyse here has to be regarded as a strategic and flexible effort to claim resources and construct an economic specialisation at a specific period in time, the result of which was a new economic, social and political differentiation in central Namibia. It also implies, as Hildi

Hendrickson (1992) has suggested, a process of 'ethnicization', although her argument is in part differently rooted than my own.[3] In a way, I deal with the question of how Otjiherero-speaking people, who during the first half of the nineteenth century were often living as impoverished pastoralists, became what they have generally assumed to have always been: members of a cattle-based pastoral society.

There is an immediate danger of being misunderstood. Firstly, I do not suggest that Herero before the 1860s did not own cattle nor do I claim to explain the origins of cattle culture amongst Herero. I do argue, however, that a majority of Otjiherero-speaking people in central Namibia, who in the decades before the 1860s experienced major stock losses and impoverishment and who were often living as hunter-gatherers, consolidated themselves as stock owners in the 1860s and 1870s with amazing flexibility and determination. They formed a new type of cattle-based pastoral society which, as such, did not exist before. The society that came into existence was that of a 'modern' pastoral society, one which was deeply integrated into the system of mercantile capitalism as it expanded from the Cape Colony into south-western Africa. Whilst aspects of the construction of a 'modern' pastoral society amongst Herero have been analysed by Rudolf Lehmann (1955, 1961), Wolfgang Werner (1980, 1998) and Hildi Hendrickson (1992), and the latter two have emphasised the importance of the mercantile capitalist system in this respect, none of them has analysed the process of (re-) pastoralisation in its own right, nor taken the complexities and the regional dynamics of it seriously.

Secondly, I focus primarily on socio-economic aspects without discussing, for example, the complexities of cultural identifications amongst Herero before and during the process of (re-)pastoralisation. This chapter thus does not claim to provide a comprehensive analysis and explanation of the consolidation of the 'modern' cattle-based pastoral society that came into existence in the nineteenth century (Henrichsen 1997).

Raiding, Impoverishment and Accumulation in the 1850s

Processes of impoverishment and accumulation have been experienced by central Namibian pastoralists throughout history. Histories of Herero families mention periodic experiences of raiding, stock losses and dispersal due to impoverishment, and the conquest of wells and (re-)accumulation of stock between the second half of the eighteenth and the first half of the nineteenth century.[4] The raiding of stock amongst pastoralists was – and is – a central strategy for stock accumulation, incorporation of labour and claiming wells and pastures. Oral traditions amongst Herero also stress another common feature of pastoral societies: stock lending to impoverished pastoralists in order to construct and maintain networks amongst stock owners (see Kaputu 1992b: 72–73).[5] Those oral texts which mention men who lived in the late eighteenth and early nineteenth century and who are referred to only in connection with small stock (goats and sheep) or with hunting as a subsistence strategy, though not with cattle, are noteworthy.[6] An analysis of precolonial Otjiherero place names, which commonly refer to wells, point to horticulture as having been another subsistence strategy.[7] All these references point to a variety of subsistence strategies having been pursued by so-called pastoralists and as such point to Otjiherero-speaking groups marked by economic differentiation.

The earliest written sources available mention the loss of stock amongst central Namibian pastoralists for the first time in the 1790s due to more effective raids by southern Namibian pastoralists as a consequence of the increasing importation of iron from the Cape Colony, and hence increasing availability of iron-pointed arrows and spears (Moritz 1915: 193).[8] The impact of new technologies gained momentum with the immigration of Oorlam groups from the Cape into south-central Namibia from the late 1820s onwards. As Brigitte Lau (1987) and Tilman Dedering (1997) have shown, groups such as the Afrikaner group (//Aixa//ain) under *kaptein* Jonker Afrikaner and the Khauas group (Kai/khauan) under *kaptein* Amraal Lambert established a raiding and tribute economy in south-east and south-central Namibia which was based to a large extent on so-called commando groups of mounted men equipped with guns. Their raids of cattle increasingly plunged central Namibia into a region inhabited mainly by impoverished pastoralists on a scale probably unprecedented in history. In contrast to former times, impoverishment and accumulation in central Namibia now took place in the context of the expanding Cape trade network of which the commando groups with their trading system based on cattle in exchange for guns, ammunition, horses and modern consumer goods were the most important agents.[9]

The written records indicate that a large section, if not the majority, of stock owners in central Namibia had lost much of their stock by the 1850s. Already in 1840 missionary Cook noted that in the Swakop-Khuiseb region, '[a]ll the different tribes we found had been completely stripped of their flocks and herds, and now subsist with a few additions from game and other varieties [sic] upon a small but very agreeable bulb which grows chiefly upon the stony hills'.[10] The same observations were made for the south-eastern region of central Namibia.[11] In 1855 the traveller James Chapman claimed that in the Otjimbingue region, that is in south-central Namibia, 'hundreds, perhaps thousands, of Damaras [Herero] live on roots and berries' (Chapman 1971, I: 172, 166). Increasingly this was true of north-central Namibia as well. In 1844 Jonker Afrikaner still could point out that 'the stock of those [southern] Damras [sic; meaning Herero] is nothing compared to what he saw amongst the northern Damras ... There he found also many Bushmen and Bergdamras, who ... are not poor',[12] the latter reference indicating that San and Damara communities (still) possessed stock. In 1857 the Waterberg region was, according to the few written accounts available, depopulated, and Herero lived 'chiefly upon nuts, beans and guinea-fowls'.[13] Wells in that region, which testified to former stock-owning populations, had fallen into disuse (Hahn 1984/85: 986, 993). In the late 1850s several commando groups operated as far as the southern fringes of the Etosha pan.[14]

European observers at the time were struck by the marked differentiation in central Namibia. According to the traveller and trader Charles Andersson, Otjiherero-speaking groups in the 1850s were divided into

> rich and poor Damaras [Herero], or those who subsist on the produce of their herds; and those who have no cattle, or at least, very few, and who lived chiefly by the chase, and what wild fruit and roots they can pick up. These are called *Ovatjimba*, and are looked upon with the utmost contempt by the prosperous classes. (Andersson 1856: 52)

According to missionary Brincker, 'If a Herero becomes poor, he becomes an omutjimba' (Brincker 1886: 157). The term Ovatjimba implied a social category of impoverished Otjiherero-speaking pastoralists from the perspective of cattle owners – that is, Herero. Herero also spoke of Ovanjake with regard to any groups, including Ovatjimba and San, who 'subsisted on the improbable and impossible of the field'.[15]

Andersson's description indicates that the picture of impoverishment, as it was generally painted by European observers, was much more complex. Firstly, the 'poor' people that Europeans encountered seemed often to have been those who, after having lost their stock, had stayed in the region. Many people in fact had fled the region, taking along any stock which they still possessed. In the late 1840s, for example, wealthy 'chiefs' (*ovahona*) such as Kahitjene and Tjamuaha temporarily fled with followers and what remained of their stock to the Waterberg region. Whilst the Kahitjene group afterwards scattered due to raids by other Herero groups, Tjamuaha and his group (re-)allied themselves with the Afrikaner Oorlam.[16] In the south-eastern part of central Namibia, 'rich Damras [Herero] have fled from the Namaquas and Oerlams [sic], who are hardened in their crimes', whilst 'Bushmen and poor Damaras [Ovatjimba] scattered'.[17] Stock owners in fact fled 'as far north as possible',[18] as far indeed as the central Kaoko region north of Otjitambi (Kaputu 1992a; 1992b: 76). Others, indeed whole families, found refuge in various parts of Ovamboland with what stock they could take along.[19]

Secondly, the general picture of impoverished stock owners living in central Namibia does take into consideration the fact that some Herero groups who remained in central Namibia not only kept their stock but also increased their herds, thereby extending their political influence. This was firstly achieved by what missionaries termed an increasing scale of 'stock theft' amongst Herero themselves,[20] or 'conflicts of inheritance amongst relatives'.[21] Oral traditions in turn paint a picture of a civil war ravaging in central Namibia, thus attesting to violent patterns of appropriation of stock, people and wells amongst various Herero groups (Kaputu 1992a, 1992b).This development testifies to the fact that, as Werner has pointed out, the increasing constraints placed on Herero pastoralists by commando raids and tribute relations resulted in internal conflicts (Werner 1980: 66) of which raiding and/or claiming stock from relatives and others was the most visible one for outsiders. The most well-known winners of this situation were the *ovahona* Tjamuaha and Kamaharero of the Ohorongo 'clan' (*oruzo*). In 1853, missionary Schönberg claimed that, 'the only Captain who is left is Katjimuaha [Tjamuaha] ... he owns, according to the people, more than 1,000 stock ... and he has recently been wreaking havoc amongst the scattered Herero'.[22]

As Brigitte Lau has shown, the success of Tjamuaha and Kamaharero in part has to be explained by the close relations they retained with the Afrikaner Oorlam after 1853 (Lau 1987: 107–9). Jonker Afrikaner allowed them and their followers to keep part of the stock obtained through raiding or tribute collecting and tolerated independent raiding/taxing and the acquisition of horses and guns in return for services to the commando groups. A Herero commando group led by Kamaharero in the late 1850s possessed some hundred guns (Hahn 1984/85: 1129).[23]

Thirdly, it would be wrong to assume that central Namibia was populated only by Ovatjimba and Damara without stock and a few groups such as the Tjamuaha-Kamaharero group possessing larger herds. Whilst most groups did lose stock, those who remained in central Namibia still possessed stock, though it seems this in general was too little to guarantee subsistence based on milk products throughout the year. According to Eduard Karokua, who was born in 1844 and interviewed in the early twentieth century, stock owning families during the 1850s possessed on average three to ten cows.[24] Missionary Hahn calculated in 1856 that a family of five persons needed 'at least five cows, 15 goats and 15 sheep, in order to gain sufficient support for its subsistence' (Hahn 1984/85:

Figure 6.1: The Herero 'chief' Kamaharero (left) photographed during the Palgrave Commission of 1876.

920). Given the impoverishment in stock levels, bridewealth amongst Herero, which 'was formerly from 2 to 10 head of cattle', had dropped to 'one cow for a daughter', if available (Chapman 1971, I: 174). People with more stock continued to engage in trading relations with Ovamboland, though it is significant that trading expeditions included one involving up to eighty women who tried 'to get married and stop in Ovamboland if they could find husbands, many to work there as servants and slaves'.[25] These scant observations by outsiders suggest that the actions of such women came about due to stock impoverishment in central Namibia.

How did impoverished pastoralists construct social relations in the 1850s? Did impoverishment necessarily mean the destruction of social relations and dependence on rich stock owners such as Tjamuaha, Kamaharero or various Nama-Oorlam communities?

Being Ovatjimba and Ovanjake

A close reading of the available sources suggests that the fact that people were designated as Ovatjimba or Ovanjake did not necessarily imply dependence and hunger. Thus missionary Cook remarked of Ovatjimba and Damara without stock:

> [These people] are remarkably provident, taking care of every bit of skin suitable to assist in clothing them, carefully saving up the food which they have to spare, and labouring with their hands, both male and female with their children, in collecting roots, honey, and other provisions to have a store against seasons of scarcity.[26]

Others engaged in an elaborate hunting economy. As another missionary reported, Ovatjimba 'frequently secure' wild animals 'by digging deep pits ... They also make large enclosures in which pits are made, and wild animals driven in and speared to death' (Tindall 1959: 59). In addition, hunting products played a role in inter-regional trade, including Ovamboland, and Ovatjimba participated in this (Henrichsen 1997: 176).

The hunting-and-gathering economy also did not necessarily imply a breakdown of social relations and a change in cultural identification, at least as far as the sources of outsiders can attest to such issues. The sources available allow for a sketchy sociology of Ovatjimba. On the one hand, many groups did scatter or dissolve as a result of raiding, stock losses and flight, as indicated above.[27] It seems that Ovatjimba tended to live in smaller groups than was the case with more wealthy stock owning groups. Missionary Hahn in 1845 remarked of Ovatjimba: 'A tribe is often made up of not more than four, five or six families, like for example these poor Omohereros [Ovatjimba]. In former times they were more numerous ... Of such small tribes there are an exceptionally large number ... Whether they are dependent on larger chiefs, is doubtful' (Hahn 1984/85: 219).

The population of the two mission stations in central Namibia in the 1850s, Otjikango and Otjimbingue, was in part a fugitive one of individuals. Others, however, belonged to 'impoverished families', 'villages', 'tribes' or a group of 'run-away and chased-away women'.[28] Judging from the scant evidence, some of these groups existed as groups prior to their impoverishment and before they moved to the stations.[29] Nor can one subscribe to the statement of Brigitte Lau, that the population on the stations lived 'without the support of the traditional kin network' (Lau 1987: 55). On the stations, Ovatjimba 'who belonged together ... built homesteads [*ozonganda*] together, each homestead [*onganda*] consisting of ca. 10–15 huts'.[30] An Omutjimba could, despite stock poverty, live with several wives, indicating that women were not necessarily 'chased away' in times of stock impoverishment, as quoted above.[31] On the other hand, the new and often rather fluid composition of the mission-station population allowed for new associations and the development of new positions of authority. Kamuzandu, who in 1857 was regarded as the 'captain of the Ovatjimba' in Otjikango, and Zeraua in Otjimbingue, who became a well-known 'chief' (*omuhona*) in the 1860s, are two examples.[32] These new positions could in part be achieved due to the political influence of the missionaries and the political and economic spaces resulting from conflicting interests amongst Nama-Oorlam *kapteins*.[33]

Neither were these groups necessarily dependent on influential 'chiefs'. Thus missionary Rath stated in 1852 that Ovatjimba at times allied themselves to influential 'chiefs' but 'do not actually adhere to them, and leave them again at will'.[34] On the other hand, influential 'chiefs' in the 1850s constantly tried to claim authority over and structure dependency amongst the populations in Otjikango and Otjimbingue (Lau 1987: 107–9). Yet there seems to have been room for manoeuvre, and relations of dependency – if they existed – were dynamic. This is most visible amongst those living on the mission stations and elsewhere who could engage in what I call an incipient process of (re-)pastoralisation.

Ovatjimba and the Incipient Process of (Re-)Pastoralisation

Ovatjimba residing on the mission stations Otjikango and Otjimbingue took advantage of the political and economic spaces the stations provided. From the late 1840s they engaged in a successful economy of horticulture, stock breeding and trading. Horticulture was not unknown to them, although missionaries claimed to have introduced it. Already in 1849 missionary Hahn mentioned that Ovatjimba accumulated 'mainly through tobacco cultivation several hundred heads of [small] stock and own several cattle' (Hahn 1984/85: 437).[35] Tobacco had long since played an important role in intra- and inter-regional trade in south-western Africa and Ovatjimba traded it for cattle and cosmetic products such as red ochre (Hahn 1984/85: 448).

As is common in all impoverished pastoral societies, small stock played a vital role. In 1852 the 'chief' (*omuhona*) of Otjikango, Kamuzandu, was regarded by Hahn as 'one of the wealthiest ... I think that he owns more than 20 heads of cattle and several hundred sheep and goats and some good gardens' (Hahn 1984/85: 453). Such a development made the mission stations new places of accumulation and consequently new places for influential 'chiefs' (*ovahona*) and *kapteins* to claim tribute and authority. Consequently, by 1856 the residents of Otjikango had again lost most of their stock to more powerful chiefs (Hahn 1984/85: 920). However, new options for accumulation appeared. The majority of people working for the newly established Walfish Bay Mining Company (WMBC), which mined copper at the Matchless Mine near Otjimbingue between 1855 and 1860,[36] were Ovatjimba. Missionaries mentioned a constant migration to Otjimbingue 'because they [Ovatjimba] can subsist much better' and 'own more' (Hahn 1984/85: 938, 942). As 'chief' Zeraua explained to the Cape Commissioner Coates Palgrave in 1876: 'We were very much cast down in those days ... I wished to move into the Kaoko veldt [sic], but I met Mr Bassingsthwaite [of the WMBC], who persuaded me, as my people were broken, to go and work with my people in the mines' (Palgrave 1991: 25).

Mine workers and labourers who were engaged by European travellers and traders were often paid in small stock, sometimes cows and calves.[37] Increasingly they bargained for guns and ammunition.[38] Here again, as in the case of the Tjamuaha-Kamaharero group, we note the parallel acquisition of stock and guns which became a decisive feature in the process of (re-)pastoralisation in the 1860s and 1870s.

The accumulation of stock and guns resulted in the establishment of a small elite of relatively wealthy stock owners on the two mission stations. In 1858 missionary Hahn observed that 'the Otjimbingue *poor* [Ovatjimba] have become comparatively rich and they own hundreds of sheep and cattle' (Hahn 1984/85: 1120, original emphasis). A visitor to Otjimbingue in 1859 wrote, 'as they take a great pride and pleasure in the breeding of cattle, they are gradually emerging from that state of utter destitution into which they were thrown' (Anon. 1859: 223). In 1859, Hahn remarked that Ovatjimba in Otjimbingue, 'stick together and have only one resolution, one aim ... to become rich' (Hahn 1984/85: 1120). To become rich implied becoming a stock owner and engaging in stock breeding as a primary mode of existence. Ovatjimba obviously engaged in a strategic process of (re-)pastoralisation, the backdrop being the experiences of raiding and impoverishment of the decades before.

It is difficult to judge from the sources available how far this dynamic development of stock accumulation extended beyond the mission stations. In 1857 Hahn visited a group of Ovatjimba near Okahandja who engaged in a successful garden economy and owned 'some goats, sheep and cattle' (Hahn 1984/85: 977). There were probably other groups like these.

Raiding and Accumulation: The *Omavita* of the 1860s

The following two decades saw a dramatic continuation of the process of (re-)pastoralisation. The political and military manoeuvres of the 1860s, triggered by the deaths of the two major chiefs in south-central Namibia – Jonker Afrikaner in August 1861 and Tjamuaha in December of that year – were articulations of a general effort amongst Otjiherero-speaking people, Ovatjimba and Herero alike, to turn the experiences of impoverishment into a major redistribution of cattle wealth. It brought about the collapse of Afrikaner hegemony and signalled the establishment of new Herero chiefdoms throughout central Namibia.

The end of Afrikaner hegemony and the beginnings of the consolidation of power of several Herero leaders and their respective clans is condensed today in Herero oral historiography in the narrative about the *ovita vyongombongange* ('the raid with the grey, white-brownish spotted cattle'). The narrative revolves around the 'raid' (*ovita*, sing.; *omavita*, pl.) of the 'white sheep' (*otjorondu*) that had belonged to Jonker Afrikaner, the subsequent battle between various Herero and Nama-Oorlam groups, and the victory of the Herero groups together with the cow *ongombongange* on 15 July 1863 in Otjimbingue.[39] The details of this particular incident are corroborated by accounts of contemporary Europeans (Henrichsen 1997: 187–92) and both narratives stress the importance of the role of the new stock-owning elite in Otjimbingue before and during the battle. The *ovita vyongombongange* signalled the beginning of a series of massive though not always successful raids throughout south-central Namibia by Ovatjimba/Herero, which in the end left the Afrikaner and Khauas Oorlam as well as many Damara communities in south-central Namibia virtually stockless. Whilst Lau has claimed that after December 1864 Herero 'did not reappear' until 1870, missionary Brincker recorded that in 1866 Afrikaner and Khauas Oorlam were still raided by Herero and Mbanderu (eastern Herero) 'day by day'.[40] In 1868 Herero commandos extended their raiding activities to southern Namaland, which, according to Manasse Zwartbooi, left communities with little stock and caused many 'to die of hunger'.[41] Missionary Viehe came to the conclusion in 1869 that 'when the Namaquas don't have the strength any more, they will no longer fight, and when the Herero have no prospects any longer to find oxen amongst them, they will no longer send raiding parties'.[42]

The raids of 1863/64 have been interpreted by Brigitte Lau as having been engineered by the Otjimbingue traders to bring about the downfall of the Afrikaner Oorlam hegemony in pursuit of their own interests. This denies the interests and agency of Herero groups, overestimates the traders' position, and neglects the fissions in the Afrikaner polity after 1861.[43] Here it is enough to point out that besides the strategic efforts of all Herero/Ovatjimba groups to raid stock, as the most influential chief, Kamaharero carefully constructed an impressive network of political, military and social alliances in an attempt to control both the raids and the interests of the traders. For his efforts, little appreciated and hardly understood by both contemporary Europeans and recent historiography,[44] he is today remembered as *Muniovita* ('owner of the raid').

The multiple raids by Herero/Ovatjimba facilitated a major redistribution of stock ownership in south-central Namibia. Throughout the decade raiding was the most common and effective means of (re-)building herds. As Charles Andersson already recorded in 1866, the 'large herds of cattle and sheep' of Herero were 'at one time ... possessed more or less' by Nama-Oorlam groups.[45] Missionary

Böhm in 1868 aptly remarked, 'The Herero ... became rich because of war'.[46] The role of the traders has to be regarded as important, given their supplies of guns and ammunition (Henrichsen 1997: 330), but this is secondary in the context of the dynamics of re-stocking.

The consequences of this massive redistribution of cattle wealth were in itself dramatic. Firstly, the redistribution seems to have made it possible for many impoverished groups to enrich themselves and regain stock. Thus, missionary Brincker observed that after 1866, when the numbers of raids decreased, those Ovatjimba who at that time were still living at the mission stations were leaving.[47] This no doubt indicates either new stock ownership amongst them or new roles as herders (*ovaries*) in the newly established cattle economy. Many 'chiefs' – including those who already possessed herds, such as Kamaharero and Zeraua – could claim additional herds, as did Kambazembi, who migrated to central Namibia from the north in the mid 1860s.[48] Mbanderu groups could claim a major share in the raids of 1864 in the south-eastern parts of central Namibia.[49] It should be mentioned that, of those Herero who had migrated to southern Namibia along with their livestock during the first half of the nineteenth century to seek employment or because of having been captured by commandos, most left the region in the late 1860s, thereby taking advantage of the crumbling strength of the commandos and taking along with them the stock of their former masters.[50] All these groups, it should be emphasised, now also possessed (more) guns and horses, and some ox wagons, most of which were captured during the raids. Thus not only stock but other modern means of production changed hands in south-central Namibia in the 1860s.

The consequences of the raids seem even more pronounced when one realises that in the 1870s the new cattle wealth led to a complex spatial, social and political reorganisation amongst Herero. This process, which I analyse in detail elsewhere (Henrichsen 1997: 68–71), basically implied the (re-)establishment of networks of wells and pastures throughout central Namibia and the establishment of new chiefdoms. These networks became the basic networks for production – that is, intense stock breeding over the next ten years. Herero deployed a range of strategies in order to maintain and expand the cattle economy which dramatically accelerated the process of (re-)pastoralisation.

The Consolidation of the Cattle Economy

When missionary Brincker returned in 1875 from an extensive tour through central Namibia, then known as Hereroland,[51] he came to the conclusion that the 'Herero are more numerous and wealthier in cattle than hitherto has been assumed'.[52] The Swakop, Omaruru, Nossob and Omatako river regions he had visited were 'much populated' by people and stock, an observation which stands out in marked contrast to reports of the same areas from the 1850s. In 1879 missionary Viehe stated that 'Hereroland ... is overcrowded with cattle, sheep and goats' to such an extent that pastures were becoming scarce.[53] In fact, conflicts with regard to wells and pastures amongst stock owners, particularly in south-central Namibia, increased.[54] Missionary Irle calculated in 1880 that the stock of Herero 'during the last 10 years … has multiplied tenfold … and the population has increased substantially'.[55] Herero society during the 1870s thus developed into a consolidated stock or rather cattle-based society.

The rapid increase in stock and population is not only to be explained by the *omavita* raids and (return) migration of Herero from the north and south to central Namibia. Favourable climatic and military-political conditions – no large scale raids were recorded until 1878/79 – and the absence of pandemics stimulated the process of stock accumulation as well.[56] In rapid succession, homesteads (*ozonganda*) and stock posts (*ozohambo*) were established throughout the area of Otjimbingue, Epako (Gobabis), Otjozondjupa (Waterberg) and Otjitambi, reaching the area of Otjovisume (Tsumeb) by the mid 1880s.[57] As one missionary already stated in 1869, 'the whole nation [*Volk*] divided itself into small homesteads', implying that Herero had established decentralised networks of social relations and production. This process is well testified by oral traditions. For example, in the case of 'chief' Kukuri (alias Kamaisa, died 1903) who had lived in Otjosazu near Okahandja since 1872, a praise song (*omutango*) about him illustrates the process of constructing such networks:

> His cattle 'are grazing far at the drinking places of our cattle where a person had roamed with a tail of a Kudu and the dog of Kozonde of Murongo. The cattle of the *onganda* [household] of Ketjo, the red ones with white stripes, are grazing there. They were scattered at the settlement of Tjorongo, they scattered to the cattle of Munina of Tjapepa. They went together with the cattle of Hengombe … They are by the cattle of Hanguvi, they are at the well of Hijavangere and the father of Hijavazuvare.[58]

The text vividly reflects a process of structuring – that is, the establishment of decentralised networks of men, places (wells) and cattle. Cattle with a predominately red-and-white marking were a symbol of the Ombandi clans, whose head Kukuri was. It is these special cattle which 'construct' the relations between the men and various places. As such the praise song refers to the structuring of a regional framework of chiefly authority and clan identification. On a general level, Herero created a vast 'land of the Ovaherero' (*ehi rOvaherero*) in central Namibia.[59]

Two aspects of stock management and the establishment of power relations need to be emphasised. Firstly, the process of accumulation rested, as in the case of Ovatjimba in the 1850s, in no small part on the breeding of small stock. Sheep (*ozondu*) and goats (*ozongombo*) reproduce much quicker than cattle.[60] For many new stock owners it was small stock that enabled them to eventually enter into 'cattle networks'. But rich cattle owners also saw to it that they owned large herds of small stock as these provided both a buffer against cattle losses and an additional means of exchange in order to accumulate additional cattle.[61] Small stock were thus not only a vital asset in the cattle economy but in themselves added to the expansion of it.

Secondly, the incorporation of labour, notably of male herders, was of major importance for stock breeding and, indeed, the process of (re-)pastoralisation. A Herero proverb states that in order to start a cattle herd one should not start with cattle but rather with people.[62] Herero regarded stock management as 'hard labour' (Irle 1906), which again indicates that it was not the natural reproduction of stock as such but a strategic and, indeed, cooperative effort by groups of men and households that guaranteed the maintenance and increase of herds (Irle 1906: 121). Social networks between homesteads and stock posts provided the basic framework for accumulation and expansion. As missionary Büttner explained in 1882:

> A rich man in Damaraland [Hereroland] cannot keep all his stock at one place ... Consequently each owner with substantial stock keeps it not only at his main homestead but has several cattle posts, where his younger brothers or close relatives, or, in the absence of these, reliable servants keep supervision. (Büttner 1882: 854)[63]

However, given the rapid expansion of cattle wealth, families and their extended social networks could no longer provide enough labour. In 1876 missionary Irle remarked that 'I get the impression that if the Ovaherero continue as they do with their cattle wealth, in 5 to 6 years not only all boys who are able to walk but all the girls as well will be needed as herders'.[64]

Increasingly, servants (*ovandu ozombue*) were integrated into the stock economy. These comprised of Ovatjimba, who still existed and who, as one source put it, 'disappeared to the cattle posts'.[65] Here, they formed 'the poorer class' amongst the stock holders.[66] This inclusive strategy – culturally speaking Ovatjimba were Herero – resulted in the fact that Herero without stock no longer existed in central Namibia in the 1870s, another indication of the widespread and rapid establishment of a cattle-based pastoral society. In fact, as missionary Brincker put it in 1881, 'every herder ... wants to become a "Herero"; that is, a cattle owner, as soon as possible'.[67] Only in the Omaheke region to the east of the Waterberg, and in the Kaoko region, were there 'Ovatjimba, i.e. poor Herero, who possess no stock [*Hornvieh*]'.[68]

Given the acute shortage of labour, Herero started to incorporate additional labour, mainly by utilising Damara but also San and Ovambo. Damara, who lived throughout central Namibia, were increasingly incorporated by violent means. Herero started to raid Damara villages in order to gain access to more wells, pastures and labour, thereby stripping Damara communities of the stock they possessed. In north-central and west-central Namibia, small-scale wars, which included bitter counter raids by Damara, were indicative of this expansive and violent process.[69] Herero at that time were well equipped with guns and horses, as I will show below, and thus could raid much more effectively than Damara, who were deliberately barred from acquiring modern weapons. The rapid expansion of the Herero economy thus resulted simultaneously in increasing impoverishment amongst Damara communities in central Namibia, creating amongst them an even more pronounced subsistence on hunting and gathering. When droughts afflicted south-central Namibia in the early 1880s, it was these communities which had to cope with severe hunger and even starvation, not Herero communities.[70] The dynamics of the process of (re-)pastoralisation thus in itself structured difference, economically, politically and socially, in central Namibia and as such displayed the ambivalences inherent in the new regional economy.

By the late 1870s several Herero chiefdoms of loosely demarcated territories[71] had evolved. The records of missionary Irle, who drew up a register of chiefs, petty chiefs and their followers in 1876, provides a sketchy overview of the most important clans.[72] Accordingly, Kamaharero (circa 1820 to 1890), 'chief' (*omuhona*) of the Ohorongo clan, was the wealthiest leader, at least until renewed raids by Nama-Oorlam and Herero commandos erupted in the 1880s. In the late 1860s his cattle wealth was estimated at around 10,000 head,[73] and in the 1870s around 40,000 (Duparquet 1953: 13). Irle counted seventy-six large homesteads (*ozonganda*), including stock posts, under his authority, each one comprising of about 300 people. In the 1880s, Kambazembi (died 1903) was regarded as the wealthiest chief. In 1876 Irle counted sixty large homesteads under his authority, each one comprising of about 200 people. His following increased when he inherited a major share of *omuhona* Kavingava's stock and followers after 1879. In the early 1890s his cattle wealth was regarded as exceeding some 70,000 head of stock, with more than 4,000 cows.[74] Even if these figures are exaggerated, they give an indication of the large wealth accumulated by some, though not all, 'chiefs' in the preceding twenty to thirty years.

Irle lists the Zeraua clan as the third wealthiest group. This has to be regarded as a major achievement given the fact that this group, as outlined above, was completely impoverished in the 1850s, and in the 1870s was living mainly in the drier western areas of central Namibia. Irle counted some 4,200 followers of Zeraua (died in 1876). Other influential clans were those of *omuhona* Kandjii (died in the 1890s) to the north-east of Okahandja, with some 2,400 people, *omuhona* Kukuri (died in 1903) in the same region with some 2,000 people and 'much stock'. Irle regarded the Mbanderu group, of which the most influential leader was Kahimemua (killed in 1896), as comprising some sixty-four larger homesteads of 200 people each. As we have seen, this group raided a major share of the Nama-Oorlam stock in south-eastern central Namibia in the 1860s. Irle could not gather much information for the region north of Omaruru and the Waterberg. Here, *omuhona* Kamureti ua Kaoko (died in 1887), seems to have been the wealthiest leader, seemingly with some 6,000 followers.

These figures indicate that marked differences existed amongst Herero chiefs (*ovahona*) as well as between individual stock owners. Irle remarked: 'Somebody who is rich, like Maharero [Kamaharero], has 200 herds; less rich people than Kukuri … 100 herds; somebody with only some wealth owns 50 herds … and somebody who is termed poor, has 10; a very poor one, who owns nothing, still has 5 herds or 30 to 50 cattle' (Irle 1906: 32). These figures are problematic as well. Whilst Irle regarded five 'herds' as comprising thirty to fifty head of cattle, other sources suggest that one herd comprised usually some 100 to 150 head.[75] Also, there were different interpretations of who was rich or poor. Thus *omuhona* Katjiriange, who owned some 600 head of cattle, was regarded as being poor 'in comparison to other Herero'.[76] Whilst such data is hence only of limited value and also does not tell us what actually is meant by cattle ownership,[77] it does indicate one important aspect, namely the very heterogeneous accumulation of stock amongst stock owners, despite the general process of (re-)pastoralisation amongst Herero.

The Cattle Economy and Cape Mercantile Capitalism

Whilst it is safe to state that raiding in the 1860s and herding, breeding and the (violent) appropriation of wells and additional labour in the 1870s facilitated the rapid increase of cattle herds, and thus the consolidation of Herero society, one aspect has so far been virtually overlooked in historiography, namely, the simultaneous and intensive trade by Herero in guns, ammunition and horses in the 1870s. Brigitte Lau has stated rather categorically that 'the Herero could not be drawn into the Cape trade network' (Lau 1987: 145). She bases her statement on the observation by the Cape Commissioner Coates Palgrave, who travelled extensively in central Namibia in 1876/77, and who noted with respect to Herero: 'Their long intercourse with white people has developed singularly few wants amongst them, so that the import trade of the country is but little affected by the majority. A universal desire to possess firearms and ammunition, which once existed, has been so fully gratified, that this trade may be said to have ceased' (Palgrave 1877: 84–85). As I shall now demonstrate, the 'universal desire to possess firearms' in fact has to be regarded as an indication of intense participation by Herero in the Cape trade network. Consequently, it is not a question of whether or not Herero 'could be drawn into the Cape trade network' (Lau 1987: 145). Rather, there is the need to explain why Herero pastoralists participated intensely in the mercantile trading system and as such supported its rapid extension throughout south-western Africa.

When Commissioner Palgrave toured south-central Namibia in 1876 he noted:

> The whole of this part of the country was studded over with cattle posts … All the men, and even boys of twelve, carried guns, and these mostly good rifles. The only flint guns I saw had been relegated by the original owners to the youngest of their sons able to carry them, and I may record here in passing that everywhere through the country the same sight is met with. (Palgrave 1877: 22)

Other observers noted the same, namely, that it was only 'poor stock owners' and children that did not posses guns.[78] A dramatic arming of the populace had taken place amongst the new pastoralists, which is mirrored in the trading figures of the large commercial ventures in central Namibia at the time. In the 1860s the trader Charles Andersson in Otjimbingue imported thousands of flints and muskets, as well as several hundred rifles, in addition to several tons of gunpowder. In the 1870s the inter-related trading companies of Axel W. Eriksson and Co. in Omaruru and A. Ohlsson and Co. in Cape Town dominated the importation of, among other things, guns and ammunition into south-western Africa, and the export of primarily ivory and ostrich feathers to Cape Town. In 1876 alone, some 6,000 guns, 40,000 pounds of gunpowder and 50,000 pounds of lead (to manufacture bullets) were imported through Walvis Bay, most likely by Eriksson and Co.[79] Other trading ventures such as the Mission-Handels-Gesellschaft of the Rhenish Mission Society participated in this trade as well, in addition to independent European and a few Griqua traders. The latter were primarily commercial hunters who either sold ivory and ostrich feathers to the large companies such as Eriksson and Co or were in some form or other engaged by the companies themselves on a basis of credit. From the middle of the 1870s, Ovamboland, southern Angola and Ngamiland became the prime hunting fields for these trader-hunters, with Omaruru being the central trading station for south-western Africa until the early 1880s.[80]

Herero men not only possessed the latest guns and purchased large amounts of ammunition: 'every Herero of some wealth' also bought a horse, and in the 1870s several hundred horses were imported annually to central Namibia.[81] This trade was mainly in the hands of Griqua, Tswana and Fingo from various South African colonies.[82]

The reasons for this extensive trade in guns, ammunition and horses are manifold and closely interwoven with the establishment of the cattle economy by the Herero. Guns and horses were regarded as new means of production, and Herero quickly integrated these, economically and symbolically, into the cattle economy. Already in 1860, the hunter and trader Frederick Green noted:

> All tribes soon became aware of the vast superiority of guns. They learn from experience of their neighbours that firearms not only protect them against the hostile feelings of others, but contribute greatly to the means of subsistence, as well as to enable them to retain their herds of cattle instead of being too often compelled to fall back upon them in the absence of other food. (Andersson 1989: 287–88)

Guns and horses were not only modern means of warfare, and hence allowed for a more secure accumulation of stock, they also enabled Africans to participate in the commercial hunting system. Whilst Green suggests that the participation of Africans in the new hunting system provided additional subsistence, game hunting in the case of Herero was mainly used to accumulate guns and additional stock. Already during the *omavita* raids of the 1860s, the *ovahona* Zeraua and Kamaharero sent out hunting parties and sold the ivory and ostrich feather they acquired for guns and ammunition.[83] In the 1870s Herero expanded their hunting interest into those regions neglected by other hunters, notably into the Omaheke. Such was their success that Palgrave in 1877 came to the conclusion that the sale of hunting products 'added considerably to the general wealth of their respective tribes ... So that the Herero when compared with other African tribes must be considered exceedingly wealthy' (Palgrave 1877: 19).

With regard to the organisation of hunting trips, 'chiefs' played a role and as such tried to extend their control over men and broaden their basis of income.[84] Such control, however, was only partially successful. Men living at the mission stations seem to have hunted independently from the 'chiefs' (*ovahona*). Hunters such as Katjimbonde, Katjimoho and January, a Zulu who had married into a Herero family, gained some status from their successes (Henrichsen 1997: 256). For the mission population, hunting trips were a vital means of income which allowed them to cover some of the costs of their increasing consumption of products such as clothing. A few like Kaende invested their hunting profits in oxen wagons, which in turn meant they could be engaged as a petty trader by Eriksson and Co.[85] Irle noted that 'every Herero of some wealth owns one or three wagons' (Irle 1906: 165), and in 1890 some 150 to 200 wagons were estimated to have been in the possession of Herero. These wagons, together with oxen and labourers, were in turn rented out to European traders, and 'for many natives especially on the mission stations [this meant] a new source of income, and the European traders started to send out individual traders as hawkers to their blood relatives' (Büttner 1884: 287). It thus seems that a few Herero could establish themselves as temporary petty traders due to their successes as hunters. No doubt many wealthy Herero in the 1870s, whilst being pastoralists, quickly took advantage of the Cape trade network in order to diversify their sources of income.

Most important was the fact that Herero initially paid for the bulk of their enormous purchases of guns, ammunition, horses and wagons with their hunting returns, and only partly with cattle. Guns, ammunition, clothing and wagons, according to missionary Büttner, 'initially were paid for by the natives with ostrich feathers and ivory' (Büttner in Loth 1963: 138). In fact, missionaries regarded the participation of Herero in the commercial hunting system as a 'major means of production' through which 'substantial amounts of ostrich feathers and ivory enter the trade'.[86] Palgrave added that, in 1877, 'the Damara [Herero] ... not infrequently used his wealth derived from the chase/hunting to increase the number of his already immense flocks and herds'.[87] Participation in the commercial hunting system thus provided Herero not only with the means of rapidly arming themselves without the initial need to sell cattle, but it also allowed for a direct additional accumulation of cattle. From the mid 1870s onwards Herero started to sell more cattle. Yet, as Palgrave pointed out, whilst some Herero sold cattle to acquire wagons and horses, others used their hunting returns in order to acquire cattle: 'Bullock wagons, and horses, are freely bartered for cattle, which change hands again for ostrich feathers and ivory' (Palgrave 1877: 84). The new trading system allowed for various forms of participation, all with the aim of stimulating the process of (re)pastoralisation even further.

Of course the regional economy was much more complex. Herero at the mission stations, for example, bought wire, ammunition and so on from the traders in order to sell them to 'the heathens with good profits – all this in order to increase their cattle [herds]'.[88] Apart from this, some men accepted working contracts with European traders and hunters 'until they accumulated a few cows or some small stock'.[89]

It is obvious from this analysis that Herero did not consolidate themselves as a cattle-based pastoral society simply because of an existing cattle culture and skilful stock management but because of a variety of possibilities that opened up during and after a period of stock impoverishment, of which the Cape trade network has to be regarded as the most important new possibility. Herero pastoralists participated actively in the Cape trade network, as hunters, as labourers, as wagon owners, and a few as temporary traders.

In addition to active participation in the Cape trade network, Herero were successful in controlling the terms in which the mercantile system operated. As Lau and Werner have noted, the effects of the trade network did not, as far as Herero were concerned, result in 'underdevelopment' and the vicious cycle of consumption, debt, raiding and increasing impoverishment, as was the case in southern Namibia in the 1850s (Lau 1987: 145; Werner 1998: 39–42). Herero largely determined the prices of the products they sold and bought. Missionary Brincker wrote in 1881:

> The Herero have become exceedingly rich in the past 8 years on cattle, so that sheer stock numbers should have kept prices low, but a Herero loves his cattle fervently ... The few cattle that are sold and bought after much discussion, mostly however after quarreling and wrangling, do not have ... the necessary weight (because Herero do not sell their large oxen).[90]

The control of the terms of trade was also based on the integration of the new means of production according to the Herero system of value and status. Thus, for certain products Herero sold only specific age classes of stock, for horses three-year-old oxen, for guns and ammunition four-year-old oxen.[91] The degree and terms of the participation in the Cape trade network was structured by the pastoralists, not by European traders.

Establishing a Gun Society

Simultaneously with the consolidation of a cattle-based society, Herero society developed into a gun society. By this I mean not only the massive amount of arms analysed above, but the fact that guns and gun owners were integrated into central social rituals in a way similar to cattle and cattle owners. As with cattle, owning a gun became a symbol of manhood and as such became part of the construction of masculinity. An important ritual in this respect was the *ombimbi*. *Ozombimbi* (pl.) had been described by missionaries since the 1840s as 'warrior dances', which were led by a specialist dressed in a peculiar way and who danced in front of a group of men, performing in a pantomime-like manner heroic deeds

(Hahn 1984/85: 339, 425).[92] In the 1860s *ozombimbi* played an important role in the mobilisation of hundreds of Herero men by the 'chiefs' (*ovahona*) for the *omavita* raids described above. Whilst Brigitte Lau (1987: 130–31) and Hildi Hendrickson (1992: 240–44) have exclusively concentrated on the mobilisation of Herero by a few European traders and missionaries,[93] one of the traders, Frederick Green, noted that, 'many of the [Herero] leaders might be seen on the march in front of their men going through the evolutions of the war dance and exhorting their men to show themselves in the coming strife worthy of being called men' (Green in Andersson 1989: 241).

Guns (*ondjembo*, sing.; *ozondjembo*, pl.) and gun owners formed part of these rituals from the very beginning.[94] The autobiography of Elia Kandirikirira, the well-known evangelist, provides a glimpse of the experiences of a young man with reference to the *omavita* raids of the 1860s and the importance of guns as a medium for constructing manhood. He recalls, for example, a frightening encounter in the 1860s which he experienced during a battle with a Nama-Oorlam who possessed a gun whilst he himself had none. Shortly afterwards he became the gun carrier for his elder brother. In the late 1870s he became the leader of a group of Christians fighting alongside their non-Christian brothers against Nama-Oorlam commandos, all of them possessing guns (Kandirikirira 1906: 6–8). Missionary Brincker observed that 'ever since Ovaherero have good fire weapons, they feel safer and increasingly abandon the wearing of their talisman',[95] thus attesting to the emotional and ideological power which guns exercised. Becoming a man increasingly meant not only to acquire stock but to become a gun owner as well.[96] This is not only testified to by the large scale acquisition of guns throughout Hereroland noted above, but also by oral traditions. Praise songs that mention Herero men who lived during the second half of the nineteenth century specifically make reference to a gun and/or a horse.[97]

Guns were not only integrated into specific male rituals such as the *ozombimbi* but also in the central rituals at the *okuruuo* (the so-called Holy Fire), which constructed the general social cohesion of a clan and the legitimacy and authority of the elders and *ovahona* (chiefs). Irle describes rituals at the *okuruuo* of Kamaharero in August 1870 in Okahandja on the occasion of the so-called peace talks between a large body of *ovahona* and various Nama-Oorlam *kapteins*. The ceremony Irle observed was the

> washing of all warriors and weapons according to Herero tradition. One morning all warriors and their women assembled at the *okuruuo* … All men and women approached the Holy Fire where Maharero sat. He consecrated the men and women with water … Not only the people, but also the guns, spears, kirris, bows, milk vessels, bowls and calabashes were consecrated and marked with white soil and ash before they were returned to their owners. (Irle 1906: 181–82)

These observations clearly indicate that guns, next to the central utensils of the cattle economy such as milk vessels and calabashes, had become part of the general system of meaning and belief amongst Herero, men and women alike. Bryan Pfaffenberger has stated, with reference to the anthropology of technology, that technology is not 'an independent, non-social variable that has an "impact" on society or culture. On the contrary, any technology *is* a set of social behaviour and a system of meanings' (Pfaffenberger 1988: 141, original emphasis). The extent of this is testified to by the fact that guns also became incorporated into rain rituals amongst Herero. The various meanings of guns allow me to speak of Herero developing a gun society.

The status of gun owners became institutionalised through the establishment of Herero commandos, albeit in a more limited way than was the case for the Nama-Oorlam. Despite the fact that guns were accumulated in a very decentralised manner and largely independent of control by *ovahona*, the latter started to organise gun owners in groups which Europeans called 'commandos'.[98] Kamaharero in particular did so, which is not surprising given his close association for decades with the commando of the Afrikaner Oorlam. He sent out groups of young men to the homesteads and districts of his region in order to enforce his authority and extract tribute. There are indications that age sets amongst young men[99] provided a base for the recruitment and organisation of such groups.[100] In the 1870s and 1880s they also controlled the movements of Europeans. Thus the Boer trader van Zyl mentioned a *kommando* of 150 armed men under the leadership of Wilhelm Maharero (alias Kauaita), the oldest son of Kamaharero, which followed the movements of the Trekboers in the Kalahari border region.[101] This commando in particular, which missionary Diehl called a 'section',[102] became an important institution in the power politics of Kamaharero and his military leader Riarua. But other *ovahona* and their military leaders also organised commandos apart from the *ad hoc* and highly effective efforts of mobilising hundreds of men with guns throughout the districts in the case of raids by Nama-Oorlam commandos from 1880 onwards. Increasingly Herero commandos were deployed by some *ovahona* in order to defend or reject the hegemonic politics of Kamaharero in western and northern Hereroland. As such, commandos became an important institution in the power system in central Namibia to support and legitimise the power basis of individual *ovahona*, as such adding to the cohesion of new chiefdoms.[103] For military leaders such as Riarua, men participating in these commandos were *ovasoldate*, a term he used when the first German troops entered central Namibia in 1889.[104] At that stage, the identification of men with guns and horses was part and parcel of a modern Herero-ness, as was the identification of Herero with a cattle-based pastoral society.

Conclusion

In this chapter I have argued that Herero society as a wealthy cattle-based society established itself in the 1860s and 1870s. Otjiherero-speaking groups engaged in a process of (re-)pastoralisation and employed a mixture of economic strategies in order to facilitate this process. It involved the skilful interaction of raiding, breeding and trading, which enabled them to restock their herds on a large scale and at the same time to expand their cattle economy. This process was supplemented by the active involvement of Herero in the Cape trade network, the terms of which they largely controlled themselves.

The process of (re-)pastoralisation had powerful consequences for the whole region. It implied a more rigid control over resources such as wells, pasture and labour, which was facilitated increasingly at a violent level and affected many other communities in the region, notably Damara communities. The process of (re-)pastoralisation thus deeply affected the lives not only of Herero but of many other people in the region as well. As such, the process of (re-)pastoralisation has to be regarded as the principal historical development in central Namibia in the second half of the nineteenth century. It simultaneously led to Herero becoming 'people with cattle' and 'people with guns' whilst affecting the social formation of all groups of central Namibia.

By this means, Herero established themselves as a modern pastoral society. Modernity for the majority of Herero became encapsulated in the system of meaning and behaviour attached to cattle, guns and, to a lesser degree, horses and ox wagons. Cloth-based clothing, which Hildi Hendrickson regards as an important new symbolic language for the legitimisation of authority and the 'ethnicization' of Herero in general (Hendrickson 1992: 97–101), despite its increasing importance to small groups on a few mission stations in southern Hereroland, did not affect the majority of Herero in the nineteenth century. Ethnic identity, insofar as this can be analysed for this period,[105] was not constructed around clothing or the so-called flags similar to the modern *otruppa* movement, but was based on powerful claims to land – that is, wells and pasture – and modern means of production such as guns and horses.

The powerful achievement of establishing a cattle-and-gun society led to what, amongst Herero of today, is regarded as 'the decade of the golden age'. After the genocide during the colonial wars between 1904 and 1907 and, more specifically, with the new possibilities that opened up with the establishment of reserves by the South African colonial government during the 1920s, this 'golden age' became the ideological backdrop on which Herero in general and Herero politicians in particular based their claims for land and cattle (Krüger and Henrichsen 1998). Thus an understanding of the consolidation of the cattle-based pastoral society during the 1860s and 1870s is central to the concerns and politics of Herero society in the twentieth century.

Notes

This chapter is a slightly revised and updated version of Henrichsen (2000), and draws together several key issues which I have analysed in greater depth in Henrichsen (1997). All translations from German into English are my own. Little new historical research has emerged for the region and period under consideration since then, an exception (though only for the modern Omahake region of Namibia) being Lindholm (2006). However, Lindholm has not made use of my thesis and seems not to have been aware of the extensive historical research on pastoralism, the topology of wells, and conceptualisations of territory and power in central Namibia as presented in Henrichsen (1997). For the latest update on historical research on nineteenth century central Namibia see now also Henrichsen (2011: 311–12, 316, 319–20).

1. When using the term 'Herero' in this chapter, I refer to a group of people who shared/share the Otjiherero language and socio-cultural similarities. These comprise Ovaherero, Ovambanderu and Ovatjimba, terms which possibly developed into ethnonyms in the late nineteenth century. See the discussion in Henrichsen (1997: 10–16) and below.
2. For a recent interpretation of Herero history, see Heywood et al. (1992).
3. See Hendrickson (1992: 97–99, 234–38, 284, 316). See also the conclusion to this essay.
4. The history of the Mungunda family, whose chief Kahitjene (died 1851) was a prominent member, is one example (interview with Katjiritja Mungendje, Windhoek, 8 May 1990). Copies of the interviews mentioned in this essay are accessible in the National Archives of Namibia (NAN), Windhoek, Henrichsen collection.
5. Oral traditions with reference to the Tjamuaha family in the first half of the nineteenth century provide an example (see Kaputu 1992b). Variations of this particular family history and their experiences were recorded by Heinrich Vedder (1981: 143).
6. The oral texts referred to are called *omitango*, 'praise songs' for people (*omutango*, sing.) and *omitandu*, 'praise songs' for places (*omutandu*, sing.). An *omutandu* with reference to Ovijombo ('big well') is associated with the sheep of Nambunga ua Tjikekenda. An *omutandu* with reference to Otjovisume mentions goats only. The Herero Kandase in one *omutango* is also linked to small stock only. The Herero Kahivesa is portrayed as a successful hunter. See my interviews with Uakaera Edda Hangeru, Otjitetue (5 December 1989); Ester Kandimbaukua, Gustafine Upendura and Rosalinde Kauari, Otjinene (7 April 1990); Katjiritja Mungendje, Windhoek (5 March 1990); Tjari Nicodemus Uazuu, Okonja, (10 April 1990) (NAN, Henrichsen collection). On Kahivesa, see also Muuondjo (1992: 163) and Njanekua (1992)
7. Tjamuaha (died 1861) and his father Tjirue lived at the beginning of the century at a place called Otjikuna ('garden', place where to plant).
8. See the remarks by a Damara as recorded by Willem von Reenen in 1792 in the area of Rehoboth (Moritz 1915: 193).
9. Already in 1835 missionary Cook wrote with reference to the mission station Warmbad in southern Namibia, which played an important intermediate role for the northern Nama-Oorlam groups: 'Our people have long been in the habit of giving in exchange for the … stolen cattle horses, guns, and ammunition, which have been used as the instruments of robbery and murder against the Damaras [Herero]' (School of Oriental and African Studies (SOAS), Wesleyan Methodist Missionary Society, South Africa (WMMS-SA), Box 304, Cook to Secretaries, 4 November 1835).
10. SOAS, WMMS-SA, Box 306, Cook to Secretaries, 8 December 1840. See also his 'Journal of a Visit from Nisbeth Bath to Ameral's Residence … 1 September 1840'. A similar account was already given by the traveller James Alexander (1838: 17).
11. SOAS, WMMS-SA, Box 308, Tindall to Secretaries, 4 February 1845; Tindall (1959: 59–60).
12. As reported by missionary Kleinschmidt: Archives of the Vereinigte Evangelische Mission (AVEM), Wuppertal (Germany), file 1.573, Kleinschmidt to Väter, 3 November 1844. See also Hahn (1984/85: 397).

13. NAN, A 83, Andersson Papers (AP), vol. 17, Diary of Frederick Green, 17, 19, 21 May 1857. See also NAN, A 83, Andersson Diary, 11 April, 6 July 1858, and Hahn (1984/85: 988, 991–92).
14. For an analysis of the geographical extension of the raids in the 1850s, see Lau (1987: 41) and Henrichsen (1997: 163).
15. See Hahn (1984/85: 251, 1010, 1089–91). San also were called Ovakuruha by Otjiherero-speaking people.
16. On the history and movements of the groups of Kahitjene and Tjamuaha in the 1850s, see Henrichsen (1997: 162, 307–16).
17. NAN, A 83, AP, Diary, 9 February 1853.
18. SOAS, WMMS-SA, Box 308, Tindall (to Secretaries), 1 January 1849.
19. Whilst some like the Herero Kaunakui lost their remaining stock in Ovamboland for reasons not given, others seemingly were engaged as herders in agro-pastoral Ovambo communities. On Kaunakui, see Hahn (1984/85: 991). Other references are found in Hahn (1984/85: 1048, 1226) and NAN, A 83, AP, 15 December 1859.
20. Hahn (1984/85: 139, 209, 313, 322–25, 362–63., 437, 447, 486–87, 513, 594, 1080); AVEM, 1.583, Kolbe to Pastor, 3 March 1849; AVEM, 1.581, Rath, Diary September 1850 to July 1851, 18 July 1851. More references are provided in Henrichsen (1997: 179 n.209).
21. South African Library, Cape Town (SAL-CT), MSB 290a, Kolbe, Diary, 23 June [1850]; AVEM 1.583, Kolbe, Diary No 6, January to April 1851, 14 March 1851.
22. AVEM 1.585, Schönbergs Tagebuch from September 1853 to 27 March 1854; 5 and 6 September 1853; also 18 September 1853.
23. Interview with Musengwa Lucius Mungendje, Okamatutjindo, 12 April 1990 (NAN, Henrichsen collection).
24. NAN, Quellen, Vol. 2, Mitteilungen von dem Herero Eduard Karokua, 12.
25. University College London (UCL), Galton Papers, 97, African Journal, pp.94–95. Another group of women which went to Ovamboland is mentioned in NAN, A 83, AP, 5 September 1858. Nothing is known about a possible stock transfer between Ovambo and Herero families in cases of marriage. On Herero women married to Ovambo, see also Finnish Missionary Society, Ph XX Viii.1., Rautanen diary, 25 January 1887; Cape Archives Depot (CA, Cape Town, today Western Cape Provincial Archives), A 2048, vol.7, Rautanen to Hahn, 29 January 1889.
26. SOAS, WMMS-SA, Box 306, 'Journal of a Visit from Nisbeth Bath to Ameral's Residence …', 9 August 1840.
27. Another prominent group that scattered was that of chief Mungunda in the early 1850s. See note 16, above.
28. Citations from AVEM, 1.585, Schönbergs Tagebuch No. 8, 25 October to 16 January 1853; Tindall (1959: 109); *Berichte der Rheinischen Missionsgesellschaft* (BRMG) (1852: 100); Hahn (1984/85: 217, 365, 990).
29. BRMG (1852: 100).
30. BRMG (1852: 100); see also Hahn (1984/85: 1120); SAL-CT, MSB 290a, Kolbe, Diary, 19 April [1850].
31. SAL-CT, MSB 290a, Kolbe, Diary, 19 April [1850]; Hahn (1984/85: 336).
32. On Kamuzandu, see Hahn (1984/85: 990. On Zeraua, see his statement recorded by the Cape Commissioner Coates Palgrave, in Palgrave (1991: 25). For more references, see Henrichsen (1997: 185 n.236).

33. See the statement by chief Zeraua recorded by Cape Commissioner Palgrave on the role of Captain Petrus Swartbooi and Missionary Rath (Palgrave 1991: 25).
34. AVEM, 1.585, enclosed in Schönberg to Väter, 7 August 1852.
35. See also Hahn (1984/85: 451).
36. On the history of the WBMC, see Andersson (1987), Lau (1987:, 95-98), and Henrichsen (1997: 165–68).
37. AVEM, 1.581, Rath, Tagebuch, 1 January 1855 to 7 May 1855, 22 March 1855; NAN, A 83, AP 2 March 1854, 28 August 1858, 1 February 1860; Andersson (1987: 88); Hahn (1984/85: 815, 1120).
38. See the references and discussion in Henrichsen (1997: 183).
39. The narrative is analysed in depth in Henrichsen (1997: 187–89).
40. AVEM, 1.594, Brincker to Inspektor, 7 February 1866; Lau (1987: 134).
41. Cited in AVEM, 1.597, Böhm to Inspektor, 20 October 1868. Vedder (1981: 442–43) claims that it was Herero who died of hunger whilst raiding in southern Namibia. This is not substantiated by the sources.
42. AVEM, 1.603, Viehe to Inspektor von Rohden, 1 August 1869.
43. I have analysed the political reshufflings and *omavita* in depth in Henrichsen (1997: 187–96). I draw on a wide-ranging corpus of sources not available to Brigitte Lau. Her arguments are to be found in Lau (1987: 128–34). They have since been uncritically adopted by Werner (1989: 36–37) and Hendrickson (1992: 240-44).
44. On the construction of alliances as a major power system of Kamaharero, see Henrichsen (1997: 252–59).
45. NAN, A 83, AP 28.10.1866. See also McKiernan (1954: 46).
46. AVEM, 1.597, Böhm to Inspektor, 20 October 1868.
47. AVEM, 1.594, Brincker to Inspektor, 29 August 1867.
48. See Henrichsen (1997: 195) for an analysis of the stock distribution.
49. AVEM, 1.590, Weber, Bericht von Gobabis vom April bis Juni 1864; AVEM, 1.594 Brincker to Inspektor, 7 February 1866, 18 October 1866.
50. Evangelical Church in Namibia, Windhoek (ELCIN), RMS I.1.8., Sheperd to Barkly, 28 November 1871; AVEM, 1.594, Brincker to RMG, 15 June 1870.
51. On the concept of 'Hereroland', see Henrichsen 1998.
52. AVEM, 1.594, Brincker to Inspektor, 25 October 1875. See also the remarks by the trader Grendon in Hahn (1992: 341).
53. AVEM, 1.603, Viehe to Inspektor, 25 September 1879.
54. Palgrave (1991: 208) cites the Mbanderu chief Japona with respect to disputes about 'land'. See also AVEM, 1.605, Irle to Fabri, 20 February 1882.
55. AVEM, 1.605, Irle to Väter, 18 September 1880. In 1906, Irle (1906: 32) stated that there was an eight-fold increase of stock during the 1870s.
56. For details, see Henrichsen (1997: 204, 266–67).
57. Otjovisume was a settlement of Herero pastoralists before the nineteenth century. During the 1850s there were only a few Herero/Ovatjimba living in the area. See Henrichsen (1997: 60–62).
58. The cattle of Kukuri 'mazeri, ko, kukure ko karui ketu kozongombe ozonduuombe omundu mariangera no mutjira uo horongo numbua ja kozonde za murongo, mazeri okujanda mozongombe omakondo zo nganda ja Ketjo. Zari zarimbara zeri ko kozondara za Tjorongo, zai rarimbara zeri ko ngombe ja Munina ja Tjapepa, zai … mozongombe za hengombe … Zeri kozongombe za Hanguvi, zeri kondjombo joo Hijavangere no Hijavazuvare'. NAN, Michael Scott Oral Records Project Manuscript No 24: Kukuri,

narrated by A. Kaputu, trans. E. Kapuuo. A slightly different version of the text has been published in Heywood et al. (1992: 117).

59. For an analysis of this concept, which played an increasingly important role with regards to claims of power and identity and was directed against claims by Damara people and early colonial agents, see Henrichsen (1997: 372–85).
60. Small stock herds can reproduce 20 to 40 per cent annually whilst cattle might only reproduce 3 to 5 per cent. An analysis of stock reproduction in Namibia is missing.
61. AVEM, 1.594, Brincker to Inspektor, 14 April 1873, f.2; Ludwig Koch, 'Report', Bundesarchive, Abteilung Potsdam (BAP), Potsdam-Berlin, RKA 10.01, 2128, Nels to von Bismarck, 4 June 1886, 20–22. On the role of small stock in the cattle economy, see also Büttner (1883: 491).
62. See the proverbs cited in Vedder (1930: 85).
63. A fascinating insight into the management of stock posts is provided by the son of Kukuri in his reminiscences, which refer to the late nineteenth century. See Dammann (1983: 6–7).
64. AVEM, 1.605, Irle to Fabri, 18 October 1876.
65. AVEM, 1.594, Brincker to Väter, 20 July 1872, p.6.
66. Büttner: 'Schulvisitationsreise 1879', enclosed in AVEM, 1.611, Büttner to Inspektor, 7 January 1879.
67. AVEM, 1.594, Brincker, 'Einige unmassgebliche Worte über Handel und Cultur im Hereroland (1881)'.
68. AVEM, 1.594, Brincker to Inspektor, 14 April 1873. See also McKiernan (1954: 127).
69. An analysis of these processes, and the incorporation of additional labour is provided in Henrichsen (1997: 72–73, 211–12).
70. This became shockingly visible during the droughts between 1879 and 1883, notably in the western parts of central Namibia. See Henrichsen (1997: 292–93).
71. It should be emphasised that territories were mapped by means of a complex enactment of historical claims over wells and pastures, and by virtue of the power of local and regional *ovahona*. However, individual families had space to negotiate their alliance with *ovahona* which in turn resulted in sometimes severe conflicts between various *ovahona* over people, wells and pasture. See Henrichsen (1997: 365–69).
72. The following analysis is based on Irle's list if not stated otherwise: ELCIN, RMS I.1.9., 1876 Beilagen: J. Irle, Referat 1876: 'Was können wir tun, um den zerstreuten Werften der Ovaherero, Ovambanderu und Ovazorotua das Evangelium nahe zu bringen'.
73. According to the Swedish trader Een, cited in Siiskonen (1990: 55).
74. NAN, A 71, Conradt, 300; von Francois (1899: 78).
75. See, e.g., AVEM, 1.594, Brincker to Inspektor, 14 April 1873, f.2; AVEM, 1.606, Diehl to Inspektor, 1 November 1880; AVEM, 1.593, Schröder to Deputation, 12 August 1880.
76. AVEM, 1.615, Dannert to Inspektor, 20 August 1880, f.3.
77. Each herd consisted of animals from various owners and represented multiple (historical) claims which were contestable.
78. Steinäcker to Deutsche Westafrikanische Gesellschaft, 20 June 1887, enclosed in NAN, ZBU G.VI2. Vol. 2381, Colonial Correspondence, 6 August 1887. See also Schinz (1891: 158).
79. CA, NA 1139, Palgrave to Barkley, 2 May 1876, p.8.
80. For a detailed analysis of the structure of the Cape trade network in central Namibia and imports and exports, see Henrichsen (1997: 215–51).

81. See AVEM, 1.594, Brincker to Inspektor, 25 October 1875; Irle (1906: 165). The fact that many horses perished of horse sickness (*Afrikanische Pferdesterbe*) explains in part the high annual figures of the horse trade.
82. See *inter alia* AVEM, 1.598, Böhm diary, 5 September 1864, 26; NAN, A 3, No. 3, J.J. Afrikaner to Kamaharero, 18 August 1871; NAN, A 3, No 5, Hermanus van Wyk to Kamaharero, 13 September 1872; AVEM, 1.616, Judt to Inspektor, 14 September 1880.
83. NAN, Quellen, Vol. 2, 'Mitteilungen des Herero Eduard Karokua', 35; NAN, A 83, AP 14, 25 July 1866; BRMG (1867: 207); BRMG (1868: 342–43); Hahn (1992: 316).
84. See, e.g., the statement by the trader Ayliff, in CA, Blue Book 14-81, 'Report of the Select Committee Appointed to Consider and Report on Damaraland Affairs, 1881', p.53, and the order by Kamaharero for the trader Redecker not to accept hunting products from 'Tjmba' as they should bring them to him 'because I am living from it' (NAN, Quellen Vol. 2, Maharero Tjamuaha to Redecker, 29 November 1886, 89–91, trans. H. Vedder and D. Henrichsen).
85. BRMG (1872: 358); Hahn (1837: 289); Palgrave (1991: 345).
86. AVEM, 1.603, Viehe to Inspektor, 27 July 1881. Irle (1906: 165).
87. CA, NA 287/1878, Palgrave to Secretary for Native Affairs, 1 July 1878.
88. AVEM, 1.603, Irle to Fabri, 18 October 1876.
89. AVEM, 1.594, Brincker: 'Einige unmassgebliche/unmaßgebliche? Worte über Handel und Kultur im Hererolande [1881]'.
90. AVEM, 1.594, Brincker: 'Einige unmassgebliche/unmaßgebliche? Worte über Handel und Kultur im Hererolande [1881]'.
91. NAN, A 71, Conradt, 234.
92. See also Hahn (1857: 147) and Brincker (1886: 130).
93. For a closer analysis of the arguments of Lau and Hendrickson, and their neglect of Herero agency, see Henrichsen (1997: 330–33).
94. See the description of an *ombimbi* by the missionary Viehe (1868: 190). The original meaning of the word *ondjembo* is related to a horn which is blown (Brincker 1886: 166).
95. AVEM, 2.635, Brincker: 'Die Ovaherero', p.9.
96. See also the observation of Palgrave as quoted above.
97. Men remembered as such are, for example, Kamaharero, Kukuri, Tjihundu ua Katjimune, Hirarapi (see Henrichsen 1997: 363–64).
98. Either spelled 'kommando' or 'com(m)ando' in contemporary sources.
99. Men who underwent circumcision together were called *ovakura*, those 'who grew up together'.
100. See, e.g., W. Njuanekua: 'Herero Clans and Customs', in Heywood et al. (1992: 142). See also the discussion by Henrichsen (1997: 349).
101. NAN, A 3, van Zyl to Kamaharero, 8 February 1876.
102. AVEM, 1.606, Diehl to Inspektor, 3 March 1873.
103. For a detailed discussion of the power system in central Namibia, the increasing conflicts between various Herero chiefs and the role of commandos, see Henrichsen (1997: 361–72, 392–93).
104. From Dutch/German: Solda(a)t, 'soldier'. For Riarua, German soldiers were 'fellow soldiers [*ovasoldate*]': see his letter in NAN, ZBU U.XIII.B.6, Vol. 1, Riarua to Nels, 22 January 1890, trans. E. Kapuuo.
105. It is more appropriate to speak of identifications which, given the problematic source basis, were the visible means of expressions by people which Europeans met and engaged with.

References

Alexander, J.E. 1838. 'Report of an Expedition of Discovery, Through the Countries of the Great Namáquas, Boschmans, and the Hill Dámaras, in South Africa', *Journal of the Royal Geographical Society of London* 8: 1–28.

Almagor, U. 1980. 'Pastoral Identity and Reluctance to Change: The Mbandern of Ngamiland', *Journal of African Law* 24: 35–61.

Andersson, C.J. 1856. *Lake Ngami, or, Explorations and Discoveries during four Years' Wanderings in the Wilds of Southwestern Africa.* London: Hurst and Blackett.

Andersson, J. 1987. *The Matchless Copper Mine in 1857: Diaries and Correspondence of Charles John Andersson*, Vol. 1, ed. B. Lau. Windhoek: Archives Service Division.

——— 1989. *Trade and Politics in Central Namibia 1860–1864: Diaries and Correspondence of Charles John Andersson*, Vol. 2, ed. B. Lau. Windhoek: Archives Service Division.

Anon. 1859. ''Damaraland. From Walwich Bay to Otjimbingue', *Cape Monthly Magazine*: 217–24.

Brincker, P.H. 1886. *Wörterbuch und kurzgefasste Grammatik des Otji-Hérero: mit Beifügung verwandter Ausdrücke und Formen des Oshi-Ndonga-Otji-Ambo.* Leipzig: T.O. Weigel.

Büttner, C.G. 1882. 'Aus Natur- und Völkerleben Südwestafrikas', *Das Ausland: Wochenschrift für Länder- und Völkerkunde* 55: 852–58.

——— 1883. 'Die Vieh-Wirtschaft der Herero', *Das Ausland: Wochenschrift für Länder- und Völkerkunde* 56: 489–94, 529–33, 550–56.

——— 1884. *Das Hinterland von Walfischbai und Angra Pequena.* Heidelberg: Carl Winter's Universitätsbuchhandlung.

Chang, C., and H.A. Koster (eds). 1994. *Pastoralists at the Periphery: Herders in a Capitalist World.* Tuscon: University of Arizona Press.

Chapman, J. 1971. *Travels in the Interior of South Africa: 1849–1863: Hunting and Trading Journeys from Natal to Walvis Bay and Visits to Lake Ngami and Victoria Falls.* 2 vols, ed. E.C. Tabler. Cape Town: Balkema.

Cohen, D.W., and E.S.A. Odhiambo. 1989. *Siaya: The Historical Anthropology of an African Landscape.* London: James Currey.

Comaroff, J., and J. Comaroff. 1992. *Ethnography and the Historical Imagination.* Boulder, CO: Westview Press.

Dammann, E. (ed.). 1983. *Herero-Texte: Erzählt von Pastor Andreas Kukuri.* Berlin: Reimer.

Dedering, T. 1997. *Hate the Old and Follow the New: Khoekhoe and Missionaries in Early Nineteenth Century Namibia.* Stuttgart: Steiner.

Duparquet, C. 1953. *Viagens na Cimbebásia.* Luanda: Museu de Angola.

Galaty, J. 1977. 'In the Pastoral Image: The Dialectic of Maasai Identity', PhD dissertation. Chicago: University of Chicago.

Gordon, R.J. 1992. *The Bushman Myth: The Making of a Namibian Underclass.* Boulder, CO: Westview Press.

Hagolani, E. 1967. 'Das Kulturmodell der Bantusprechenden Rindernomaden Südwestafrikas', PhD dissertation. Cologne: University of Cologne.

Hahn, C.H. 1857. *Grundzüge einer Grammatik des Herero (im westlichen Afrika) nebest einem Wörterbuche.* Berlin: Wilhelm Hertz.

Hahn, C.H. 1984/85. *Tagebücher/Diaries 1837–1860*, ed. B. Lau. Windhoek: Archive Services Division of the Department of National Education, SWA/Namibia.

Hahn, E.S. 1992. *The Letters of Emma Sarah Hahn: Pioneer Missionary among the Herero*, ed. D. Guedes. Windhoek: Namibia Scientific Society.

Hendrickson, H. 1992. *Embodied, Imagined and Perceived Power: The Language of Clothing and Ceremony in Colonial Namibia.* PhD dissertation. Evanston, IL: Northwestern University.

Henrichsen, D. 1997. 'Herrschaft und Identifikation im vorkolonialen Zentralnamibia: Das Herero- und Damaraland im 19. Jahrhundert', PhD dissertation. Hamburg: University of Hamburg.

——— 1998. 'Claiming Space and Power: The Relevance of Herero Praise Songs', unpublished paper presented at the African Studies Association conference, Chicago, 29 October.

——— 2000. 'Ozongombe, Omavita and Ozondjembo: The Process of (Re-)Pastoralization amongst Herero in Nineteenth Century Central Namibia', in M. Bollig and J.-B. Gewald (eds), *People, Cattle and Land: Transformations of a Pastoral Society in Southwestern Africa.* Cologne: Rüdiger Köppe Verlag, pp.149–85.

Henrichsen, D. 2011. *Herrschaft und Alltag im vorkolonialen Zentralnamibia. Das Herero- und Damaraland im 19. Jahrhundert.* Basel: Basler Afrika Bibliographien.

Heywood, A., et al. (eds). 1992. *Warriors, Leaders, Sages and Outcasts in the Namibian Past: Narratives Collected from Herero Sources for the Michael Scott Oral Records Project 1985–6.* Windhoek: Michael Scott Oral Records Project.

Irle, J. 1906. *Die Herero: Ein Beitrag zur Landes-, Volks- und Missionskunde.* Gütersloh: Bertelsmann.

Kandirikirira, E. 1906. *Lebensbild eines Herero-Evangelisten: Nach seinen eigenhändigen Aufzeichnungen und denen der Frau Miss. Eich.* Barmen: Verlag des Missionshauses.

Kaputu, A. 1992a. 'The War between the Nama and Herero', in A. Heywood et al. (eds), *Warriors, Leaders, Sages and Outcasts in the Namibian Past.* Windhoek: Michael Scott Oral Records Project, pp.8–10.

——— 1992b. 'Tjamuaha', in A. Heywood et al. (eds), *Warriors, Leaders, Sages and Outcasts in the Namibian Past.* Windhoek: Michael Scott Oral Records Project, pp.72–84.

Krüger, G., and D. Henrichsen. 1998. '"We Have Been Captives Long Enough, We Want to Be Free": Land, Uniforms and Politics in the History of Herero in the Interwar Period', in P. Hayes et al. (eds), *Namibia under South African Rule: Mobility and Containment 1915–46.* London, Oxford: James Currey, pp. 149–74.

Lau, B. 1987. *Southern and Central Namibia in Jonker Afrikaner's Time.* Windhoek: National Archives of Namibia (NAN).

Lehmann, R. 1955. 'Das Häuptlingstum der Herero in Südwestafrika. Eine kurze Übersicht über die Entwicklung der politischen Organisation und dem heutigen Naturbewußtseins der Herero', *Sociologus, new series vol.* 5: 28–43.

———. 1961. 'Jonkers, Afrikaner und die Herero-Missionare seiner Zeit als „Häuptlinge wider Willen" (1814-1861), Südwestafrika', *Beiträge zur Völkerforschung* 1961: 410–31.

Lindholm, K.-J. 2006. Wells of Experience. A Pastoral Land-Use History of Omaheke, Namibia (Studies in Global Archaeology 9), PhD dissertation. Uppsala: University of Uppsala.

Loth, H. 1963. *Die christliche Mission in Südwestafrika. Zur destruktiven Rolle der Rheinischen Missionsgesellschaft beim Prozess der Staatsbildung in Südwestafrika.* Berlin: Akademie-Verlag.

McKiernan, G. and P. Serton (ed.). 1954. *The narrative and journal of Gerald McKiernan in South West Africa 1874-1879.* Cape Town : The Van Riebeeck Society.

Moritz, E. 1915. 'Die ältesten Reiseberichte über Deutsch-Südwestafrika', *Mitteilungen aus den deutschen Schutzgebieten* 29: 136–253.

Muuondjo, K. 1992. 'Tjiponda, Kahivesa, and the Wars of the Herero', in A. Heywood, et al. (eds). *Warriors, Leaders, Sages and Outcasts in the Namibian Past. Narratives Collected from Herero Sources for the Michael Scott Oral Records Project.* Windhoek: Michael Scott Oral Records Project, pp.154–73.

Njanekua, W. 1992. 'Herero Clans and Customs', in A. Heywood, et al. (eds), *Warriors, Leaders, Sages and Outcasts in the Namibian Past.* Windhoek: Michael Scott Oral Records Project, pp.134–49.

Palgrave, W.C. 1877. *Report of W. Coates Palgrave, Esz., Special Commissioner to the tribes north of the Orange River, of his mission to Damaraland and Great Namaqualand in 1876.* Cape Town: Ministerial Department of Native Affairs.

Palgrave, W.C. 1991. *The Commissions of W.C. Palgrave, Special Emissary to South West Africa, 1876–1885*, ed. E.L.P. Stals. Cape Town: Van Riebeeck Society.

Pfaffenberger, B. 1988. 'Festished Objects and Humanised Nature: Towards an Anthropology of Technology', *Man* 23(2): 236–52.

Schinz, H. 1891. *Deutsch-Südwest-Afrika: Forschungsreisen durch die deutschen Schutzgebiete Gross-Nama- und Hereroland, nach dem Kunene, dem Ngami-See und der Kalahari, 1884–1887.* Oldenburg: Schulzesche Hof-Buchhandlung und Hof-Buchdruckerei.

Siiskonen, H. 1990. *Trade and Socioeconomic Change in Ovamboland, 1850–1906.* Helsinki: Societas Historica Fennica.

Spear, T., and D. Waller (eds). 1993. *Being Maasai: Ethnicity and Identity in East Africa.* London: James Currey.

Tindall, B.A. 1959. *The Journal of Joseph Tindall: Missionary in South West Africa 1839–1855.* Cape Town: Van Riebeeck Society.

Vedder, H. 1930. 'Herero-Sprichwörter', *Afrikanischer Heimatkalender* 1930: 85–86.

——— 1934. *Das alte Südwestafrika: Südwestafrikas Geschichte bis zum Tode Mahareros 1890. Nach den besten schriftlichen und mündlichen Quellen erzählt von Dr. H. Vedder.* Berlin: Martin Warneck.

——— 1938. *South West Africa in Early Times, Being the Story of South West Africa up to the Date of Maharero's Death in 1890.* London: Oxford University Press.

——— 1981. *Das alte Südwestafrika. Südwestafrikas Geschichte bis zum Tode Mahareros 1890. Nach den besten schriftlichen und mündlichen Quellen erzählt.* Windhoek: SWA Wissenschaftliche Gesellschaft.

Viehe, G. 1868. 'Aus dem Hereroland', *Der kleine Missionsfreund* 14: 189–92.

Vivelo, F. 1977. *The Herero of Western Botswana: Aspects of Change in a Group of Bantu-speaking Cattle Herders.* St. Paul: West Publishing.

von François, C. 1899. *Deutsch-Südwest-Afrika.* Berlin: D. Reimer.

Werner, W. 1980. 'An Exploratory Investigation into the Mode of Production of the Herero in the Pre-colonial Namibia to ca. 1870', BA thesis. Cape Town: University of Cape Town.

——— 1998. *'No One will become Rich': Economy in the Herero Reserves in Namibia, 1915–1946.* Basel: P. Schlettwein Publishing.

Wilmsen, E. 1989. *Land Filled with Flies: A Political Economy of the Kalahari.* Chicago: University of Chicago Press.

CHAPTER 7

The Emergence of Commercial Ranching under State Control and the Encapsulation of Pastoralism in African Reserves

Christo Botha

Introduction

An environmentalist recently argued that colonialism in Namibia did not only result in Apartheid and skewed development objectives, it left behind 'a vast environmental debt'. This debt is revealed in a stark history of environmental neglect: the collapse of Namibia's marine fisheries, mainly through overfishing, and a steady decline in the productivity of the country's agricultural rangeland as a result of environmental degradation, 'evidenced by a loss of biotic diversity, bush-encroachment, deforestation, overgrazing, soil erosion and loss of fertility of croplands. Many of Namibia's limited wetlands and water resources have been over-exploited' (Brown 1999: 15, 17). In stark contrast to this view, European settlers and colonial officials tended to stress notions of toil and battle to tame a hostile landscape in their retrospective assessments of European colonisation. In the minds of settlers, the early history of South West Africa (SWA) essentially revolved around the occupation by relatively small numbers of people and their livestock of a few areas where permanent water was available, and endemic conflict between these groups over such resources. Little awareness was displayed of 'indigenous impacts' and signs of 'non-European occupation'. The 'apparent absence of indigenous improvements', which helped to 'justify the removal of indigenes' from tribal lands in the New World (Lowenthal 1997: 234), arguably assisted in constructions of a white settler identity and helped to justify occupation of what eventually constituted the commercial farming area of SWA.

After the brief German period, which featured promising environmentally sensitive agricultural, water conservation and forestry experiments and initiatives, the first phase of the South African era appeared to be more concerned with maximising resource exploitation. This is particularly evident in the most ambitious project during the early phase of colonial rule in SWA, namely white land settlement, where warnings about signs of environmental stress were always counter-balanced by the expressed determination to ensure the success of this enterprise. In SWA, German and South African colonisers were both aware of the political and economic marginality of the territory and its heavy dependence on respective metropolitan countries. They had to contend with environmental constraints imposed by a territory of which 55 per cent is semi-arid or arid (du Plessis 1992: 132) and in which prospects for development appeared rather limited, at least before the 1950s. However, instead of inducing caution as far as the utilisation of natural resources were concerned, the operative paradigm continued to be improved management for more effective exploitation. Mackenzie's observation concerning ecological approaches in the imperial environment, namely that 'expert opinion was (always) highly attuned to the political contexts it served' (Mackenzie 1997: 224), referring to white settlers, is therefore particularly relevant for SWA. Expressions of concern about the awkward trade-off that constituted extractive policies were occasionally aired by colonial officials, a dilemma well-captured by Lowenthal's observation that 'men in charge of colonial lands faced agonising choices between market-driven extractive enterprise and precautionary stewardship' (Lowenthal 1997: 232). Unfortunately, a balance between the two imperatives was hardly ever struck in SWA. In addition, just as British imperialists viewed colonies primarily as 'locales for extracting commodities to be exported' to the home market (ibid.: 230), SWA, a dependent periphery of the periphery that South Africa constituted before industrialisation, was largely reliant on the export of unprocessed raw materials to South Africa, the mandatory power. Within the context of the colonial economy of SWA, livestock farming was considered crucial; consequently, significant investments were made in the white commercial farming sector. In contrast, African pastoralists were encapsulated in reserves, effectively constrained to a life of subsistence production. This chapter will attempt to explore the trajectory of these developments.

Nineteenth-century Namibia

Relatively little is known about pastoralism in SWA before the mid nineteenth century when the first reports about the territory were published by European missionaries and traders. Archaeological evidence is scarce, though excavations at Geduld in north-western Namibia suggest that sheep had been integrated into a hunter-gatherer economy, but according to Smith this did not signal the advent of pastoralism (Smith 2000: 62–64). In this respect Smith differs from Kinahan, who has reconstructed the settlement history of small stock herders (sheep and goats) in the Hungorob Valley in the arid Namib fringe, covering a period of 700 years (Kinahan 1991: 40–68). Apart from Kinahan, researchers generally agree that there is little evidence of the existence of a cattle-based economy before about 1700. According to Bollig and Gewald, 'northern and central Namibia remained a stronghold for pastro-foragers far into the second millennium' (Bollig and Gewald 2000: 12–13). After about 1700, Herero cattle herders utilised a complementary system of artificial and natural reservoirs in central Namibia to provide water for the dry periods of the year. This is clear evidence that they possessed the technological and organisational skills to build and maintain such a system (Henrichsen 1997: 53).

From the mid nineteenth century onwards, the indigenous pastoral nomadic systems of the Herero and Nama peoples were increasingly subjected to the transformative impact of the expanding merchant capitalist system radiating outwards from the Cape Colony. Brigitte Lau has argued that the socio-economic bases of the Oorlam and Nama peoples of southern Namibia were effectively destroyed in the process. The Oorlam peoples who entered SWA in separate migrations after 1800 were the principal initial agents responsible for carrying the trading frontier northward. They dominated and exploited Nama and some Herero groups for the purpose of obtaining cattle, which were exchanged for horses locally and guns, ammunition, wagons and consumer goods in the Cape Colony. The Cape-based trade also featured game products, especially ivory and ostrich feathers, which were highly prized. This led to an intensification of hunting to the extent that huge raiding parties were organised, either in cooperation with European traders or as independent ventures. An enormous toll was exacted on game – by general consent game in southern and central Namibia had been driven to the edge of extinction by the time Germany declared a protectorate over Namibia in 1884.[1] These hunting and raiding exploits impacted disastrously on the productive systems of the Nama, and to a lesser extent the Herero. The Nama responded by largely abandoning their productive system of pastoral nomadism in favour of the Oorlam 'commando' system of raiding to supply the Cape market. According to Lau this extractive economic approach undermined the earlier sustainable utilisation of resources through a pastoral-nomadic system finely attuned to the semi-arid environment (Lau 1987: 52–56).

Most Herero groups were less willing to sell their cattle and this helped to shield them to some extent from the depredations of Oorlam and later, Nama raiding groups. The end of Oorlam hegemony in southern and central Namibia by the 1860s, effected by a combined military campaign driven by Herero, missionaries and traders, coincided with the gradual collapse of the Nama/Oorlam cattle raiding and trading economy as well as the remnants of their former pastoral economy. The Herero by contrast managed to engage in a successful process of repastoralisation after 1860, rendered especially significant since it was, unlike Oorlam domination, not constructed on the back of stock raiding and hunting only but comprised stock raising and the maintenance of extensive socio-political networks (Henrichsen 1997: 197–265). By the mid 1880s the rise of the powerful Witbooi Nama under their charismatic leader Hendrik Witbooi checked this process and the fortunes of Herero stock owners were dealt a decisive blow by the rinderpest epidemic in 1897, which wiped out whole herds of cattle. The settlement of whites on land that was alienated from Nama and Herero after the war of 1904 to 1908 proved to be the final development that led to the emasculation of these previously self-sufficient African polities.

Although he does not discount the general effects of the Oorlam presence and merchant capitalism in south-western Africa, Dedering has cautioned against deploying the concept of a fundamental transformation of Nama, and to a lesser extent Herero, socio-economic formations, especially in the first half of the nineteenth century. Images of the Nama as a 'cattle-rich, self-sufficient African society that perpetuated itself in a state of ecological balance' is not supported by sources (Dedering 1997: 27). The picture that emerges from contemporary sources, mostly missionary and to a lesser extent traveller's accounts, depicts conditions in which an essentially 'transhumant structure of a fragile herding economy' obtained. Small-scale stock farming communities practising a nomadic lifestyle were subject to the vagaries of the highly irregular rainfall of the territory. Most people were poor and cattle, though important, were possessed mainly by a few prominent men. Dedering's observations serve as a cautionary tale against employing simplistic concepts of degradation and engaging in tendencies to read back into the past images of pristine harmony. However, available evidence does suggest that proto-colonialist forces decisively contributed towards proscribing socio-economic options and limiting resource use for various African polities in SWA after 1850. During the colonial period this process was accelerated, not only for the southern and central parts, but also the northern region, particularly the former Ovambo.

The German Period, 1884 to 1915

With the onset of German colonialism, the process of impoverishment described above for the peoples of southern and central SWA was accelerated by the purchase of African land, but the decisive thrust occurred with the wars of 1904 to 1908 waged by Germany against the Nama and Herero peoples. In northern Namibia, home of the Oshivambo speakers, increasing pressure was exerted by Portuguese and German colonialism, resulting in restrictions on human and livestock mobility. Increasingly heavy tributary exactions imposed by Ovambo rulers proscribed economic options and induced a growing reliance on migrant labour, which was reinforced by legislation during the colonial era. Ovambo consequently became a labour-exporting peasantry by the turn of the nineteenth century (Moorsom 1995: 13).

Germans arrived in SWA with a confident belief in their ability to adapt to and transform the environment, and they subsequently immersed themselves in various experiments and investigations into aspects of the natural environment. A can-do spirit prevailed among German-speaking Namibians ever since. This is reminiscent of the way in which John Flynn, a 'reformer of outback life' in Australia, envisioned 'a nation dependent on the pioneer fringe for its vitality and future development' (Hains 1997: 158). The manner in which he and a fellow adventurer, Douglas Mawson, embraced a 'blend of nostalgia for the organic, land-holding society allied to an intense enthusiasm for technology and scientific expertise' (ibid.: 164) captures very well the spirit of early German colonisers. Cartographers produced maps of SWA that continued to be used well into the 1960s of the South African era. Experiments were conducted concerning water conservation, crop cultivation, forestry and dairy farming. Lau has argued that these initiatives were environmentally friendly and ecologically sound, and had the potential of making SWA self-sufficient in food production and ensuring an equitable supply of water to all its inhabitants (Lau and Reiner 1993; Stern and Lau 1990). Most if not all of these projects were either terminated or suffered neglect under the South African administration. Dairy production is a good example of a local enterprise started during the German period that continued to produce substantial amounts of milk, butter and cheese until the late 1960s. According to Lau, for the twenty-four year period between 1935 and 1958, butter production averaged nine million pounds per annum (not even counting the not inconsiderable amount of farm butter and farm cheese, about 20,000 pounds per year for both products).

By then local products could not compete with cheaper imports from South Africa, the way being prepared by a host of restrictive regulations and controls over the local industry. Even in African reserves, dairy production was effectively destroyed.[2] Forrest endorsed Lau's position that German water conservation experiments were indeed 'long-term in nature and … took into account high-efficiency use of natural resources' and also made use of African water preservation techniques. He questioned, however, the assertion that German plans made provision for Africans, arguing that their schemes essentially aimed to provide water for white settlers, mines and urban areas (Forrest 2000: 5–7). The premise informing Lau's arguments is that the South African government deviated from the German-era search for self-sufficiency because it did not fit its plans for making SWA a captive market for its own products. It also transformed the territory's economy from mixed farming to the narrow pursuit of 'karakul or meat monocultures – heavily capitalised industrial ranches' (Lau and Reiner 1993: 58).[3] This did indeed occur, but only after 1950. South African officials were certainly sceptical about the potential viability of small settlements, yet there is also evidence from the German period suggesting that a transition towards large-scale ranching was, if not already underway, at least seriously contemplated. The *Deutsch Züdwestafrikanische Zeitung* argued in its editorial on 14 July 1906 that small settlements should be managed as subsidiary enterprises, next to stock farming. The *Windhuker Nachrictung* carried an article on 9 May 1908 in which support for stock farming was expressed on account of the expenses involved in establishing a small settlement system. After his visit to Namibia in 1908, German Colonial Director Dernburg declared that insufficient capital was available for small settlements, and he considered the transport infrastructure to be inadequate and the white population too small to serve as a market for local produce (de Kock 1965: 174). Engineer Von Zwergern was quoted in the *Frankfurter Zeitung* of 7 September 1911 as being of the opinion that the small settlement system did not fulfil expectations, mostly due to serious water problems, the expenses involved and the fact that settlers did not possess the required skills to manage a small settlement (ibid.: 174). A prominent colonial official, Dr P. Rohrbach, stated that 'whoever will live in German SWA must live essentially by stock farming' (Schrank 1974: 94). Governor Leutwein himself reportedly envisaged SWA eventually becoming a cattle-farming country able to compete on the world market (ibid.: 99).

German officials and settlers alike drew obvious satisfaction from their perceived ability to apply their expertise and enterprise to Namibia, whether in the field of industry, farming or environmental stewardship. In fact, it might be considered a leitmotiv for German settlement that they considered themselves uniquely qualified for assuming the role of guardians of an undeveloped, arid country. Pride in a methodical approach to farming, environmental control and a scientific or experimental approach designed to produce solutions to problems and guidelines for development is evident in memoirs of settlers and officials, numerous publications of a technical or scientific nature, and the contribution by Germans to the general cultural and economic development of the colony. There is a wide-spread belief in the German community that this contribution had been actively ignored.[4]

The South African Era, 1920 to 1950

South West Africa's conservation history did not feature the environmental activism displayed by British scientists and environmentalists so characteristic of British colonies, for example the Cape Province in South Africa towards the end of the nineteenth and beginning of the twentieth century (Grove 1997: 141).[5] Until the 1960s, research undertaken in SWA featured either South African or German academics who took their findings along with them: 'Research carried out in South Africa's colony both used the resources of the mandatory power and fed the creativity of its academic institutions' (Silvester, Wallace and Hayes 1998: 44). That said, however, it is clear that the development of an environmental and ecological awareness in Namibia was largely shaped by discourses in South Africa that benefited substantially from the expertise of British emigrant scientists and environmentalists. In the British Empire a 'common language and core culture has allowed an easy circulation of expertise and experts', according to Dunlap. He argued that ecology travelled from the centre to the periphery, but it did not come 'as a developed theory or even a discipline, but as a perspective that shaped research' (Dunlap 1997: 84).

Until 1950 the minds of colonial administrators, Lands Department officials and the small number of technical experts were largely occupied with South Africa's extensive land settlement programme for white settlers which had officially commenced in 1920, as well as with identifying appropriate farming practices for the territory. How maximum value could be extracted from the natural resources of SWA without destroying the resource base is the guiding principle that structured thinking in four significant official reports in the first thirty years of South African administration: the Drought Investigation Commission of 1924, the Farm Industry Commission of 1927, the Land Settlement Commission of the same year, and the influential long-term Agricultural Policy Commission of 1948. Seen together, they not only influenced economic policy by stressing the primacy of stock farming but also structured thinking about the environment along lines that continued to exert an influence on farming practices and policy well into the 1980s. Though recommendations emanating from these reports constituted the template on which official policy was to be based for the colonial period, it was never fully able to effect a bridging of the gap between the ideal and the realities of a harsh environment, difficult economic conditions and short-sighted, profit-orientated approaches to farming.

Prospects for the expansion of African pastoralism were effectively nipped in the bud after a short interval between the end of German rule and the beginning of formal South African administration in 1920, when Africans harboured hopes of restoring their herds and obtaining renewed access to land. The Report of the Native Reserves Commission of 1928 made it clear that the objective with African reserves was to ensure that it would serve as an abode for 'the native where he can spend his old age in peace and rest'. Rather than engage in stock farming, able-bodied males, considered to be persons below the age of fifty-five, were expected to render, to the SWA administration and white settler community, the more important service of wage labour. The Commission did not support a suggested further limitation on livestock numbers, given the fact that they envisaged an increase in pasturage fees and stricter control over livestock numbers and the movement of stock into reserves.[6] Given the nature of the mentioned artificial constraints imposed on African pastoralists, the frequent exhortations by officials that Africans can easily improve their position by selling surplus stock are more a reflection of official cynicism, ignorance and dogmatism than expressions of genuine concern with farming and ecological conditions in reserves. It therefore comes as no surprise that official policy and support concerning the stock-farming industry during the next thirty-five years were almost exclusively aimed at and supportive of the white stock-farming industry. Virtually all the commissions appointed prior to the Odendaal Commission in the early 1960s had as their major brief investigations into the well-being of the white settler community. Even investigations into the state of the meat industry in the 1950s failed to consider the situation of stock farmers in African reserves.

The Drought Investigation Commission

The 1924 commission gave the lie to the popular belief that drought was caused by progressive desiccation in Namibia and recommended sound farming management practices, such as fenced camps, distribution of water points across farming units, a system of rotational grazing, as well as improved market conditions and higher prices for cattle. Expert technical advice was strongly recommended. The commission also supported granting subsidies to farmers to promote the cultivation of supplementary feeding, such as fodder crops.[7] Significantly, it advocated exclusive focus on cattle breeding and rejected the mixed-farming approach pioneered during the period of German rule. The establishment of tree plantations was rejected as inappropriate, since it would reduce the amount of land available to stock farmers. Crop cultivation for human consumption was considered a waste, the commission argued, as food crops could be obtained from South Africa at costs lower than the cost of production in Namibia. If crop cultivation was to be contemplated at all, it had to be directed at fodder for animals.

The Farm Industry Commission and Land Settlement Commission

Both these 1927 reports endorsed the administration's land settlement policy, but the Farm Industry Commission noted with concern that the 'possession of certain assets', rather than knowledge of farming, appeared to have been the deciding criterion when allocating farms.[8] The Land Settlement Commission conceded the difficulty of farming in SWA, but was generally upbeat about its prospects and argued that the land settlement scheme would prove beneficial to the country. It did acknowledge criticism, though, that it was a costly undertaking without the required controls to avoid waste and inefficiency. Mention was made of the fact that a comparison of settlement schemes in New Zealand, Canada and Australia with that of SWA revealed that in none of them is the measure of assistance given to the settler for the purchase of stock and implements so generous, nor the rates of interest charged so moderate, nor the control exercised in respect of the settler so trusting and so lenient, as is the case in South West Africa.[9]

The SWA Commission of 1935, appointed in the wake of the drought, recession and political instability of the early 1930s, had as its brief the task of investigating the appropriateness of South African political and economic arrangements towards SWA. Significantly, it warned against continuing with the land settlement programme on two accounts: it was argued that whatever money was available for land settlement 'will be required to consolidate the position of the settlers already on the land, to save them from ruin and the administration from losing much of the money which has already been spent in placing them there'. It also pointed to the insecurity of farming in the territory as an argument to induce caution and recommended research to enable the authorities to be better informed about agricultural policies.[10]

The Long-term Agricultural Policy Commission

The 1948 report of the Long-term Agricultural Policy Commission (LAPC) can be considered the culmination of the experimental phase of South African colonial rule in SWA, during which attempts were made to configure the parameters for economic development and appropriate ecological approach to the limits imposed by unpredictable and irregular rainfall. Its recommendations repeated, with greater insistence, the warnings and observations of earlier reports. The

LAPC was particularly concerned about soil degradation. In this it was partly influenced by a steady accumulation of cautionary reports about soil and vegetation degradation over a period of a quarter of a century, but probably more decisive was the fact that its chairperson was a noted South African environmentalist (a forestry expert) who chaired the commission at a time that doomsday reports about the negative effects of soil degradation in South Africa dominated environmental discourses.[11] The LAPC was especially concerned about whether the threshold separating a relatively healthy environment from one irreparably damaged was in danger of being breached.

The LAPC concluded that a combination of poor farming practices, such as overgrazing and the absence of an appropriate camping system equipped with dispersed water points, as well as irregular and often low rainfall, combined to exhaust farmland. It recommended that the land settlement programme be terminated and remedial measures implemented to address problems of overgrazing and soil degradation. White youth, the report concluded, had to be assisted to turn to technical training instead of setting out to become farmers. Considered as a whole, the various reports referred to above contained expressions of concern about inappropriate farming practices, but all tended to endorse land settlement and stock farming. By the early 1950s a transition was under way from the more or less mixed-farming approach of the pre-1950 era to a more exclusive focus on either cattle or karakul farming.

The South African Era, 1950 to 1990: Soil Conservation and Scientific Farm Management

The LAPC's report contained two crucial observations that informed its other recommendations and provided the conceptual framework within which policy to promote sustainable commercial farming would be shaped for the rest of the colonial period. The first was to suggest that land settlement had reached saturation point, and the second that 'the limit in carrying capacity of land under occupation has been reached on present methods of farming'. The remedies the commission proposed were to terminate the official land settlement programme and conduct research to guide the selection of better quality stock, appropriate stocking rates and soil conservation and reclamation programmes.[12] As far as the first recommendation was concerned, an inauspicious start was made in the 1950s. When the local National Party took control of the Legislative Assembly in 1950 it set out to correct what was perceived to have been discrimination against its supporters in the allocation of settlement farms in the immediate postwar era.[13] Farms continued to be allocated into the early 1960s along the western escarpment and even into the Namib Desert, as well as in the region north-east and east of Grootfontein. Chairperson of the LAPC, J.D. Keet from South Africa, stated in 1948 that the biggest mistake the local administration had made was to allocate farms along the desert margins, while the director of agriculture in SWA noted his opposition to this practice in a meeting of the Executive Committee in 1957, all to no avail.[14] Warnings about soil degradation, overstocking and inappropriate farming methods were, however, taken more seriously by the local administration.

Dr J.C. Ross, a South African soil conservation expert on a visit to SWA in 1951, was of the opinion that the implementation of the LAPC's recommendations concerning soil conservation would place the country on a stable footing and that it would in fact be catastrophic if its recommendations were not heeded. He proposed that sufficient water points and camps be provided and overstocking avoided.[15] Ross also crucially cautioned that sound farm management was a prerequisite for success. In a letter on behalf of the Southern African Regional Committee for the Conservation and Utilisation of the Soil, Ross argued that throughout the region pasture deterioration was taking place. This was due to the fact that improved facilities have been provided without the application of improved management techniques. These facilitated greater exploitation rather than serving as an aid to better management.[16]

The local administration's response was to initiate what appeared to be a fairly comprehensive programme aimed at soil conservation and sound farm management. This was to be achieved through the Fund for the Promotion of Farming Interests, established by Ordinance 29 of 1952.[17] Its aim was to promote conservation farming through the creation of soil conservation committees in every farming district, the provision of funds to facilitate the establishment of a camping system suitable to local conditions, water conservation through dam building, the promotion of more appropriate stocking rates on farms, and efficient management. Locals and foreigners alike appeared to have been impressed with the scheme. By 1956, two thousand farms had already been planned and the Soil Conservation Board expressed the conviction that soil conservation policy had thus far been a success. R. Logan, a professor of geography in California, was of the opinion that the administration's soil conservation activities had progressed much further than was the case in the United States.[18] The measures to facilitate conservation and sustainable farming reflected the contemporary state of expertise and scientific knowledge in the Southern African region, South Africa in particular. However, in SWA several indicators soon suggested that the initial optimism may have been premature. The optimistic belief that increased investment in capital works (dams, fences, camps, windmills) and an effective extension service inducing sound farm management practices would remedy the situation proved to be misplaced.

The only concrete outcome of the stipulations of the Fund for the Promotion of Farming Interests was to enable farmers to secure cheap credit, either through subsidies, rebates or loans, to invest in farming operations benefiting the farmer, irrespective of whether soil conservation was promoted. The situation here almost exactly paralleled conditions in South Africa, particularly the Transvaal during the 1940s and 1950s, where investment in capital works enabled farmers to increase their herds, and when drought struck they appealed to the government for assistance. Soil conservation programmes in South Africa were undermined by political considerations which aimed to maintain a white farming population on the land, especially after 1948. Here too, many farmers utilised state support to enhance their farming operations, and if soil conservation works were introduced these were often neglected or fences and dams even removed when the maintenance costs proved to be too burdensome (Delius and Schirmer 2000: 736–42). Archer's study of results obtained by the introduction of camping systems in the Karoo region of South Africa confirms these points. The aim of promoting biodiversity was hamstrung by profit considerations, and the social costs of farming practices were generally not adequately accounted for (Archer 2002). In SWA too, it appeared that wide-ranging official assistance in times of drought effectively destroyed the discipline required to maintain ecologically sensitive farming practices. Evidence provided by successful farmers suggests that a great deal of self-discipline was required to reduce stock numbers in order to facilitate soil conservation and the restoration of vegetation.

A severe drought in SWA, lasting from 1959 to 1962, not only demonstrated the fragility of the country's resource basis but also exposed the superficial commitment of white farmers to conservation. The SWA administration responded to the drought with a comprehensive set of emergency measures. Fed with emergency rations, too much stock was retained on farms, inflicting a heavy toll on the soil and vegetation. A ban on sales for more than a year as a result of a severe epidemic of foot-and-mouth disease in 1962 further contributed to the overstocking of farms. Conservation farming was neglected, and though the revised Soil Conservation Act (1969) was applied to SWA in 1971, its provisions remained a dead letter. In the early 1970s about 800 farmers were identified as having contravened the provisions of the Act, but no action was taken against them. The Soil Conservation Committees were largely inactive, and isolated attempts – such as the 'master farmer competition' and the stock reduction scheme (where subsidies were paid to enable farmers to reduce stock to levels considered to be in line with carrying capacity and to allow grazing to recover) – failed to attain their stated objectives (the master farmer competition envisaged the promotion of excellence in farming by annually selecting a master farmer). Many farmers did participate in the scheme, but at the termination of the stipulated period the scheme was abandoned. The prevailing drought, subsequent irregular rainfall, farms too small to facilitate effective rotational grazing and the constant pressure exerted on farmers to maximise returns on farms, were all factors that militated against soil and grazing management strategies (Visser 1988). The extent to which soil conservation has failed is evident from bush encroachment, considered to have reduced the carrying capacity of commercial farms in the central-northern parts of Namibia by at least 30 per cent, and in some cases up to 50 per cent.[19] According to a local agriculturalist, 'rangeland degradation, in the form of bush thickening, remains one of the major structural problems handicapping optimal animal production in the central and northern regions of Namibia' (Bester 1995: 97). Inappropriate farming methods are also credited with contributing towards the growing phenomenon of desertification, especially along the western and eastern regions of Namibia.

A retrospective assessment of the failure of soil conservation schemes makes it clear that the original recommendations of the 1950s, though laudable in intent, failed to take into account the extent to which farming is affected by a variety of ecological conditions. Though self-serving attitudes surely played an important role, ignorance about realistic adaptation strategies in environments with highly variable rainfall also significantly contributed to the failure of conservation farming. Farmers tried to bring about improvements in line with the general soil conservation philosophy, but vital information was lacking. A report into farming in the Outjo district in 1981 mentioned that about 35 per cent of farmers had implemented a 'good' system of grazing management, while 30 per cent had a 'reasonable' and 35 per cent a 'poor' system in place. Tellingly, though, it is conceded that farmers underestimated the frequency of periodic droughts or depended too much on average rainfall figures, and consequently overestimated carrying capacities with 'unconscious' overstocking being the result.[20] Ironically, these were precisely the matters that the pre-1950 commissions, especially the Drought Investigation Commission and the Long-term Agricultural Policy Commission have warned against.

What earlier prescriptions lacked was the evidence required to assess the condition of open grassland, a vital prerequisite for decisions on rotating stock between camps. Information about vegetation, its variability, vegetation succession, the carrying capacity and quality of open grassland, the reaction of vegetation to grazing and the growth processes and life cycles of plants, was largely absent. In the former Rhodesia studies from the 1960s proposed the short-term grazing method, to be achieved by the Savory, or holistic resource-management approach. It was found, contrary to received wisdom, that the absence of animals negatively affected the relationship between plants, animals and soil because they are required to promote plant succession, achieved by breaking the soil crust and consequently promoting the establishment of seedlings and water infiltration (Versfeld, Bester and Stehn 1988: 53–58). These were matters about which information was not available in SWA before the late 1960s. It was also only in the 1980s that the concept of biomass (net weight of stock on land), rather than the received wisdom of adhering to set stocking rates, started to gain currency. Interestingly though, there were individuals who grasped the essentials underlying these ideas and applied it to their farms.[21] Furthermore, such farmers tended to have sound financial, stock and labour management systems in place, obviously facilitated by sufficient financial resources. Concrete steps to make loans and subsidies to farmers conditional upon the implementation of sound grazing management strategies, though already contemplated as early as 1957, were never implemented.[22]

Forty years after soil conservation measures were officially introduced in SWA, a former agricultural extension officer argued that the 'first nail in the coffin was drilled when the first iron pole was put in the ground, the first fences and camps constructed and boreholes drilled under the pretext of conservation farming. This so-called conservation farming was for our comfort only. We tapped and harvested even though we did not sow' (Steenkamp 1993: 25).[23] According to Archer, referring to conditions in South Africa, there was another side to the problem, namely the fact that commercial farming was rendered uncompetitive through tax breaks, subsidies, price controls and distorted property rights, with property prices being artificially inflated by state support for the farming sector.[24] Ever since the late 1920s the idea was floated in official and selective farming circles about tying state assistance to sound management practices on farms. At best, very general guidelines had been provided and attempts were made to gradually instil conservationist attitudes, but by the 1980s it was readily admitted that various measures introduced to promote conservation farming had failed, and it was suggested that future state aid had to be tied to adherence to such guidelines.[25] Another Afrikaner agricultural extension officer was of the opinion that a failure to adhere to these technical guidelines in itself does not account for the lack of progress, and that one should look for more deeply-rooted cultural traits to account for the failure of conservation farming.

Soil conservation was never part of the Afrikaner's culture. Opening up the wild, the taming philosophy and trekking towards better grazing were the cornerstones of traditional stock-farming practices. Droughts and periods of below normal rainfall were not accepted as normal climatic conditions; land users in most cases overestimated the potential of resources and consequently over utilised it; it was generally accepted that farmers were reluctant to adapt correct stock-farming practices optimally suited to the potential of the resource (soil); implementation of conservation measures were fragmentary and not coordinated; the word 'conservation' may have been interpreted, due to efforts by government and extension services, as a prohibition on use, whereas it actually meant sustainable utilisation (Visser 1988: 9–10).

The practice of continuous trekking with stock, so characteristic of Afrikaner farmers, did not simply cease with the introduction of an officially sanctioned land settlement system. Some farmers continued to sporadically engage in trekking, and until the 1960s white farmers were allowed to utilise emergency grazing on government land or private farms. This practice in effect largely resembles the old pastoral nomadic system employed by Africans.

The discussion so far would seem to confirm the general thrust of criticism directed against settler farming in South Africa and SWA in particular, in that it was less efficient than previously assumed and often environmentally damaging. However, more recent scientific research on commercial rangelands provides a possibly partial corrective to the uniform condemnation of colonial-era farming practices. A report published in 1998 indicates that since 1960 the number of cattle on white commercial farms has decreased from a peak of 2.5 million head to between 1 and 1.3 million by 1994. Prior to that date, herd productivity on commercial farms was maintained by an increase in new land, and when this was not possible any more by opening up more water points on farms through the construction of a camping system. Such a drastic drop in numbers would have had potentially disastrous consequences if it was not offset by improvements in management and herd productivity. Though farmers reduced their herds, they managed to maintain annual off-take by mass per hectare. This, it is suggested, is due to an assessment by farmers of the adverse consequences induced by too high stocking levels on farms, especially in times of drought. Improvements, such as higher calving rates, compensated for the lower overall cattle numbers. However, the report indicated that beef production per unit of land has remained static and herd productivity has not attained the potential standard for the beef industry. SWA has not attained the potentially high levels of 75 to 80 per cent in herd productivity, measured by calving rates. Though it is conceded that the sub-optimum performance of the beef industry may be due to a decline in effective rainfall and range degradation due to bush encroachment, there is currently no reliable information to accurately measure ecological changes due to these and other variables induced by tourism, wildlife farming and mixed livestock/wildlife ranching (Lange, Barnes and Motinga 1998: 569). McCann rightly points out that 'direct, one-to-one cause and effect relationships between climate and human action had provided a false lead' and that more sophisticated technologies are needed to account for those relationships (McCann 1999: 261). Referring to Africa's Sahel region, he mentions that though there is evidence that human action and population growth has indeed changed the physical landscape by altering vegetative cover, 'the meaning and direction of that change contradicts the human degradation narrative' (ibid.: 271). Desertification here, according to new evidence, rather points to the global warming that occurred during the second half of the twentieth century rather than localised human action.

During the last two decades of colonial rule, and at an accelerated pace since independence, academic and officially sanctioned research about the state of commercial farmland had been generated. More sophisticated assessments of the impact of animals on farmland and appropriate stocking and de-stocking practices resulted from these investigations. Received wisdom concerning stocking rates, the optimum number of camps and specified grazing patterns underwent adjustment. Ironically, some of the new evidence echoes aspects of formerly maligned African farming practices, such as moving stock to benefit from the availability of seasonal grazing and water resources. Rothauge's research into grazing practices on commercial farmland points to an approach aiming at 'opportunistic management' of grazing resources on commercial farms, realised in the state-and-transition model, which essentially entails a situation where farmers can quickly respond to changing conditions affecting open grassland quality, such as drought, by either rapidly stocking or de-stocking (Rothauge 2000: 49–55).

Such advances in ecological understanding do not necessarily translate into more successful – that is, profitable – farming practices. Research has revealed that many farmers simply abstained from implementing prescribed conservationist farming practices because it required an almost impossible degree of supervision of the impact of animals on grazing, the exact measurement of rates of increase in animal weights and large numbers of experiments to ascertain which grazing strategies are the most effective (Versfeld, Bester and Stehn 1988: 53–58). Düvel argues that despite decades of information, legislation and support, 'the adoption of recommended veld [i.e., open grassland] management practices is not worthwhile for the majority of farmers', ascribing it to insufficient economic advantages (Düvel 1990: 421). These difficulties are compounded by the semi-arid environment and fluctuating rainfall, making accurate assessments even more problematic. Recent environmental debates for Africa indeed indicate the extent to which the validity of concepts, such as a stable ecological balance and consistent patterns of degradation, have been contested. The accepted wisdom concerning soil degradation, the harm inflicted by interventionist colonial environmental policies, and the existence of once pristine environments, has been challenged. Beinart and McGregor rightly call for 'an examination of the moral and cultural filters though which concepts about the natural world have been passed'. Questions are now being raised about the extent to which the environment should be considered as an independent variable in assessing ecological change and the extent to which social relations determine environmental change (Beinart and McGregor 2003: 2).

Rangeland Transformations in the Commercial Farming Area and African Reserves after 1960

A superficial scrutiny of official policy clearly illustrates the extent to which white commercial farming was favoured over and above the interests of Africans in reserves. Cattle numbers on white farms were only limited by the farmer's financial position and a guideline stocking rate of from 10 to 15 hectares per head of cattle. The size of white ranches ranged from 4,000 to 12,000 hectares in the cattle farming area, which means that a farmer could have from 400 to 1,200 head of cattle, a decidedly superior position to that of Africans for whom the limit was set at 100 head or 300 head of small stock. It has already been noted that cattle numbers on white farms significantly declined after 1960, a phenomenon that is associated with the termination of the officially sponsored land settlement system and drought. Since then farmers had to contend with recurring droughts or years of irregular and below normal rainfall, conditions that continued unabated into the 1990s. Depressed prices for cattle and political pressure on the white farming sector after 1990 further contributed to uncertainty concerning the sector's long-term prospects. It is in this context that the emergence of game farming proved to be of particular significance.

The transformation of game into a potential commercial resource occurred in 1967 when an official proclamation conferred ownership of wildlife on farms to owners or occupiers, who gained the right to dispose of game and game products in the manner they saw fit. At the time, though, the prospects for such a turn of events offering a realistic alternative to or supplementing stock farming seemed rather remote. This was especially true if expectations were measured against prevailing attitudes towards the utilisation and preservation of game in the proto-colonial and colonial era proper. Game was slaughtered in the nineteenth century and though numbers recovered in the period between 1890 and 1940, it was then again threatened by the introduction of the Jeep motor vehicle in SWA and the rapid expansion of land settlement into the eastern fringes of the Namib Desert. Afrikaners were widely perceived to have been destroyers of wildlife by their German compatriots, a perception that may indeed have been solidly grounded in historical reality. Game meat constituted an important source of protein, a matter of some significance considering the fact that before 1950 most white farmers were considered poor – according to the census of 1946 only 8 per cent of white farmers had an annual income higher than £1,000 and only 12 per cent higher than £500, the former amount being considered the minimum to meet obligations and 'be a credit to the community'.[26]

It is against this background that the Game Preservation Commission in 1949 rejected with contempt the idea of conferring ownership of game on white farmers, a decision that undoubtedly reflected an awareness of white hunting proclivities.[27] It was only in 1955 that the regulations of the Game Preservation Ordinance of 1951 were applied to African reserves, to be administered by officials of the Department of Bantu Affairs, thereby cementing in law a de facto ban on hunting in these areas. In contrast to the growing public identification with game conservation in game reserves in South Africa during the course of the twentieth century, the general public and white farmers in Namibia in particular were not too enthusiastic about the prospect of expanding game reserves and conservation. Many perceived game, or at least certain species, to be a negative element on farms. Game competed with cattle and small stock for grazing, particularly where large herds of zebra, springbuck, oryx and wildebeest were found.[28] Certain game species, not necessarily vermin, considered a 'real or potential danger to the farming community' were identified and farmers were permitted to destroy them.[29] In the 1960s, game such as kudu was suspected as being carriers of foot-and-mouth disease and there were requests that farmers be allowed to destroy them. In 1962 the Secretary for SWA responded to a item on the agenda of a National Party Congress concerning the possibility of game being foot-and-mouth carriers by pointing out that no game had yet been identified as a carrier of the disease and that large numbers of kudu had been decimated by drought, more so than any organised extermination campaign could have accomplished.[30] The extermination of certain vermin species continued throughout the colonial period. A report in 1961 listed six vermin species that farmers were allowed to kill: jackal, hyena, wild cats, leopards, wild dogs and lynx. In the preceding year more than 6,000 jackals were killed. Even elephants and lions, considered fence breakers and stock killers respectively, were often shot in large numbers by farmers.[31]

The Etosha game reserve, established by the German administration in 1907 as one of three such reserves, did not attain the status of a national symbol comparable to the Kruger Park in South Africa, which was projected as a unifying symbol to facilitate closer relations between English speakers and Afrikaners (Carruthers 1995: 48). Etosha was consequently vulnerable to political considerations that favoured using game-reserve land for purposes such as white land settlement or additional land for African reserves. Pressure from white farmers in the 1950s resulted in the southern part of Etosha being set aside for farmland, while an Executive Committee decision of 1955 resulted in the western portion of Game Reserve No. 1 (east of Grootfontein) being excised, to be converted into thirty farms. In 1961, amidst the foot-and-mouth epidemic among cattle and a severe drought, it was decided to de-proclaim the game reserve in its entirety.[32] Game Reserve No. 3, later to be significantly enlarged as the Namib-Naukluft Park, was used for emergency grazing for white farmers in the period 1950 to 1970.

Two events proved to be of great import, not only for the conservation of game but the commercial farming industry too. The first occurred when the Odendaal Commission recommended a drastic reduction of the surface area of the Etosha National Park, to establish consolidated 'homelands' for the Kaokoland and Damaraland African reserves. In 1947 Etosha comprised 93,240 square kilometres, but several subsequent reductions resulted in the reserve being confined to a surface area of 22,270 square kilometres by the late 1960s.

After battling to save a small but important area in the south-western portion of Etosha considered crucial for the movement of game, the Director of Nature Conservation, B. de la Bat, set out to convince the administration to purchase several white farms in the marginal area along the plateau, next to the Namib Desert, to be merged with the Namib-Naukluft Park. Elsewhere, commercial farms with few prospects of profitable farming operations were bought up by the administration and combined to create other parks. Within a decade more than thirteen such nature parks were created (Barnard 1998: 43–45).[33] Game capture operations were introduced to relocate game from Etosha to the newly created parks or to private game reserves that had gradually begun to emerge in the country. According to de la Bat, moving game to numerous game farms and parks led to an awareness of the need to 'farm with game' and to conduct research into the various ecosystems in which they were situated (Schoeman 2007: 52).

The second event occurred a few years later, in 1967, when game was proclaimed as the property of a farm owner or occupier of commercial farmland. It rapidly led to the commercialisation of game hunting and farming in SWA and served as a spur to the embryonic tourist industry in the country. Trophy hunting became an increasingly lucrative enterprise and the number of game farms featuring game animals and the spectacular landscapes of the country multiplied. Many farmers, even those that did not contemplate converting their farms into private game reserves, bought game animals made available by the Department of Nature Conservation from stocks considered superfluous to the reserves. Trophy hunting also served to stimulate the conservationist initiative.[34] The Namibian trophy-hunting market was estimated to have captured about 12 per cent of the African market in 1985. Already in the mid 1970s a total of 92 hunting farms had been established. It is a highly profitable enterprise compared to other

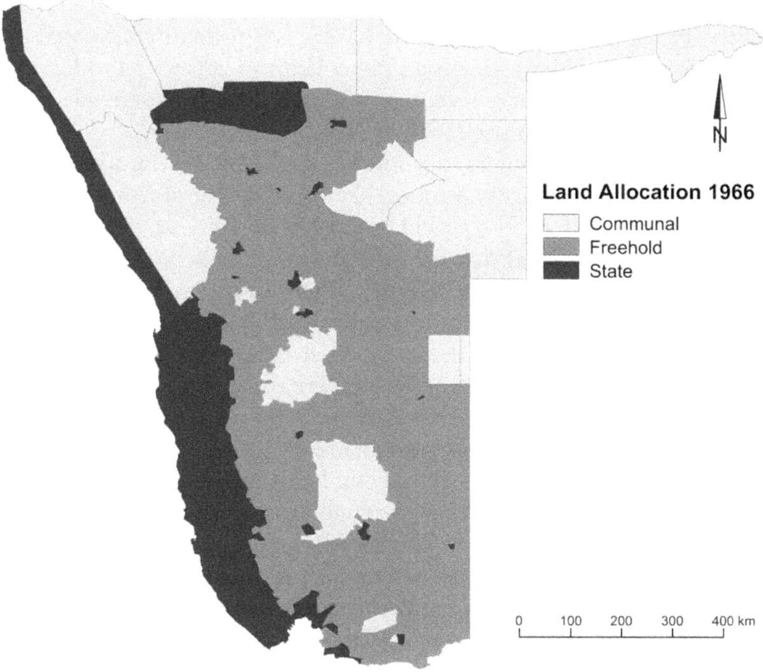

Figure 7.1: Land allocation in Namibia according to the planning of the Odendaal Commission.

wildlife utilisation activities, but capital investment can be very high and trophy hunting is therefore often combined with other land-use enterprises (Barnes 1996: 102).[35] Game farming's significance from an ecological perspective lies in the option being offered to commercial and communal farmers alike of broadening their income base of and promoting ecologically sounder farming activities (Schneider 1992: 82).

African Reserves

Although there are many parallels to approaches elsewhere in Southern Africa as far as official policies and practices towards African reserves were concerned, the absence of interventionist strategies aimed at promoting conservation sets SWA somewhat apart from other areas in the region. Before 1950 there was hardly any attempt to intervene in the farming and land-use systems of reserve dwellers, as happened for example in Rhodesia (Native Land Husbandry) and South Africa ('betterment').[36] In Rhodesia and South Africa , interventionist strategies before 1960 usually occurred to rectify the perceived degradation of soil and grazing reserves. Pankhurst distinguishes between northern SWA, where similar colonial conservationist interventions were largely absent, and the southern reserves, which did experience a degree of intervention, reflected in limits on stock numbers. However, interventions were hardly ever primarily informed by conservationist concerns, with the intention rather being to balance reproduction of reserves from their own meagre resources with the need to secure adequate supplies of labour power (Pankhurst 1996: 417).

More obvious than conservationist concerns was the South African administration's singular lack of any urgency to promote the development of reserves before the 1960s. The droughts of 1959 to 1962 and 1979 to 1982 severely affected African reserves and also revealed that the official policies of benign neglect and ignorance had a negative effect on these reserves. Whereas white farmers could rely on government support and more land to restrict stock losses in drought, these options were not open to Africans, and their often meagre herds were usually wiped out. In Kaokoland, cattle losses of 50 per cent (from 120,000 to 60,000) occurred between 1958 and 1960, and losses of about 90 per cent between 1980 and 1982 (Bollig 2000: 277). Herero farmers on proximate reserves in central and eastern SWA managed to limit losses by moving herds between reserves (Bollig and Gewald 2000: 28). The first, limited drought aid rendered to communal farmers occurred in 1961, prompted in part by foreign criticism of the lopsided assistance granted to white farmers and its absence for black ones. Indicative of the precarious position of most reserve dwellers is the fact that most herds featured multiple owners and stock were often entrusted to relatives living elsewhere, often reflecting responses to marginal environments, such as the former Damaraland with its highly variable rainfall patterns (Rohde 1999: 43).

It is therefore not surprising that the first comprehensive attempt to address the problem of underdevelopment in African reserves, the Odendaal Commission of 1962, was conceived as an essentially political solution to problems of underdevelopment.[37] Though 'development' in reserves was supposed to evolve within the confines of a modernisation paradigm, official policy was characterised more by extreme paternalism than by interventionist strategies. Unable to effect the desired improvements aimed at establishing commercial farming enterprises, officials effectively continued policies of neglect. Pankhurst argued that the Odendaal Commission's report, which remained a 'source of ideas for policy instruments' until the end of the colonial period, was much more an intervention into politics than land and agricultural production per se: about how to manage problems rather than how to solve them (Pankhurst 1996: 418).

The principles of commercial farming were to be selectively introduced, to operate in often fragmented political and socio-economic environments. In the Herero 'homeland', for example, by 1970 the idea was to make 4,000 hectare land units available to persons possessing 400 cattle. Extreme caution was displayed by officials in achieving these objectives, a clear sign of how political considerations impacted on issues at ground level.[38] The construction of fences in Hereroland was undertaken in the 1970s, and a growing number of farmers requested that the administration fence off land in order to facilitate farm planning. In Okamatapati region, land units of 7 kilometres by 7 kilometres were allocated, constricting farmers who were used to unlimited access to grazing on surrounding pastures. Soon a farmer had to share his land unit with five other households. The result was that enclosure of traditional pastures took place at the expense of small and marginal farmers (Werner 2000: 265–68). Consequently, the 'commercialization of livestock production has accelerated differentiation in pastoral communities in all parts of Namibia' (Bollig and Gewald 2000: 33). In the more cattle-rich former Herero reserves, the trend revealed an increase in absolute numbers of poor households, as well as a decrease in the number of rich households (see Werner 1998: 187–89). In other reserves in central and southern Namibia the transition from communal to commercial farming was much more limited and far fewer farmers were able to effect the transition (Rohde 1999; Botha 2002). How little the ambitious claims of the Odendaal plan affected the economic well-being of most reserve dwellers is evident from statistics derived from the former Damaraland. In 1991 a total of 1,841 farmers were registered, with average herds of 14 cattle, 99 small stock units (goats and sheep) and 3 donkeys. In areas with more favourable grazing reserves, such as Grootberg, the average increased to 22 cattle, 80 goats and 22 sheep, and in more arid areas such as Okombahe only 5 cattle, 70 goats and 25 sheep (Rohde 1999: 42–43). It should be kept in mind that figures reflect multiple owners of stock, further reducing individual ownership figures. A study undertaken shortly after independence confirms that the many projects proposed for Damaraland in the wake of the Odendaal Commission's report largely failed to address the structural features of poverty and underdevelopment. People utilised kinship and exchange relations to employ multiple economic and subsistence strategies to cope with drought and the lack of economic prospects (ibid.: 54).

Most communal farmers continue to be hamstrung by limited stock numbers, lack of quality grazing land, inadequate support services and recurrent drought. For most communal-area herd owners, life was and still is about survival, constantly engaged in cycles of stock rebuilding after periodic droughts. Rohde employs the concept of 'common sense' to describe Damara pastoralist's traditionally efficient and pragmatic adjustment to arid environments and ecological stress. This involved the aforementioned kinship and exchange relations, as well as opportunistic movements of stock to areas where better grazing and water are available. McCann refers to Iliffe's observation that 'Africa's agricultural systems were historically mobile, a strategy to adapt to the environment and the climate rather than transforming them'. Everywhere, Africans displayed finely tuned judgements about seasonal movements of stock and agricultural production cycles (McCann 1999: 274). In the various semi-arid lands of Namibia, the prospects for engaging in such seasonal and opportunistic movements were progressively limited by colonial encapsulation and the fencing of communal land since the 1970s. That soil and vegetation in communal areas often appear to be degraded is no surprise: these often marginal areas are subject to environmental stress induced by people and stock with few options other than to squeeze whatever they can from these lands.

Beinart points to a crucial distinction between African and European thinking about land:

> Both colonial and African practices saw land as to some extent divisible by its function. But colonial ideas, drawn from an industrialised and capitalist Europe, laid far more stress on rigid spatial division between lands set aside for different purposes. This applied at the national level so that certain areas could become reserves for exclusive functions like forestry, or game reserves, or farming, while multi-purpose common land was greatly diminished. Colonial ideas could then simultaneously accommodate the transformation or destruction of former ecosystems on private land, while attempting to preserve or reconstruct them in pockets of state land. (Beinart 1989: 158)

Such compartmentalised conceptions of land use are characteristic of colonial-era environmental approaches. The colonialist mindset that ascribed to reserve farmers an inherent conservatism and a reluctance to engage with the market is belied by the huge increase in the commercialisation of livestock farming. Many Herero rural households presently rely to a large extent on the sale of livestock (Bollig and Gewald 2000: 33).

A significant change, to an extent similar to the introduction of commercialised game farming on white commercial farms, occurred in the 1980s in African communal areas. Steps were taken to give Africans a stake in the management of natural resources, including wildlife, in their reserves. NGOs proved to be a key player in the transformation that has taken place. It was through the establishment of conservancies and introducing the concept of a sustainable use of resources in their regions, as well as allowing them the opportunity to benefit from tourism, that the long alienation that Africans had suffered as far as the natural environment was concerned began to be reversed. An investigation into 'customary law and nature conservation' in 1998/99, revealed accounts of African hunting rituals and practices that have virtually ceased to exist due to the prohibitions and restrictions of the colonial period (Hinz 2003: 35–49). According to the study, chiefs performed a key role in such rituals and hunting conventions, and respondents held strong opinions about the vital role that game used to play in African societies in Namibia. Legislation provided for communities to benefit from joint ventures and agreements with tourist operators. Trophy hunting, previously exclusively driven by private initiative and the search for profit, as well as ecotourism, began to offer the potential for job creation and income-generating activities for local communities. As these benefits become more tangible, there is evidence of a change in people's attitude towards conservation. 'The advantages to communities are considerable, in terms of economic diversification, financial buffering against drought, job creation, use of local skills and knowledge, and training in new skills such as management, hospitality, and languages' (ibid.). It is especially in the former Damaraland, the scene of pre-independence doomsday forecasts of wildlife extinction, that a remarkable transformation had been wrought. Amendments to the Nature Conservation Ordinance enabled rural communities to re-establish 'rights to manage, use and benefit financially from wildlife' (Barnard 1998: 43).

Conclusion

This chapter has tried to illuminate the extent to which a harsh environment imposed strict limits on humans and animals alike and constrained economic development. It is entirely appropriate that the Odendaal Commission chose in 1963 to depict the achievements of colonial rule in Namibia in terms of 'Western' civilisation, modernisation and development in casting a retrospective glance over Namibia's history. For white settlers and officials, the territory's history was synonymous with 'having tamed the land'. This they did without, however, always being aware of the long-term consequences of their actions. Though the European impact on the environment, measured in terms of overgrazing, soil degradation and bush encroachment is real, accurate assessments of the extent of the damage is difficult to arrive at. Factors to consider, as Beinart has pointed out, include the potential resilience of the veld and the fact that 'ecologists are now increasingly uneasy about cyclical, equilibrium models of vegetation change' (Beinart 1996: 71). He further argues that 'once the view from history and political economy is included, it becomes even less helpful to measure environmental change in terms of movement away from a pristine environment and call the result degradation' (ibid.: 71). Radical range ecologists, for example, question the old paradigm of degradation caused by intensive grazing, arguing that there is no stable ecological balance and considers disequilibrium as central to ecological history. There is, however, no denying the fact that Europeans derived from pastoralism, the major branch of their agricultural activity, far more benefits than was the case with their African counterparts. The latter were encapsulated in reserves of ever-diminishing economic viability, deprived of the extensive official support network available to white farmers, constrained by official policy and unable or unwilling to make the difficult and costly transition to partial, or full-scale commercial farming. An observation by the former Director Nature Conservation de la Bat, that 'nature has served us well', meaning that Namibia has too fragile a resource base to ever allow too much development and too much human and developmental stress to be induced, might well strike Namibia's communal-area dwellers as a rather ironic comment on their plight during the colonial period.

Notes

This is a substantially revised version of 'People and the Environment in Colonial Namibia', *South African Historical Journal* (2005). I am indebted to the editorial board of the journal for permission to use it here.

1. For overviews of the way game was destroyed in the nineteenth century, see Lau (1987: 56–58), Graf zu Castell Ruedenhausen, memo: 'Wild en Natuurbewaring in Suidwes-

Afrika', 12 May 1966 (translation from German); von Wietersheim (1985: 403–12). According to these accounts southern Namibia was stripped bare by 1860 and central Namibia in 1880. These regions once boasted a wealth of wildlife including rhinoceros, giraffe, hippopotamus, lion, buffalo, elephant and various antelopes.

2. For a discussion of the progressive deterioration in the fortunes of the local dairy industry, to the extent that even milk was imported from South Africa, see Lau and Reiner (1993: 15–20).
3. Karakul are an Asian breed of sheep imported into SWA by the German administration.
4. The studies by Lau and Stern (1990) and Lau and Reiner (1993) were conceived to highlight the achievements of Germans in the fields of agriculture, water and environmental conservation, and research tied to developmental issues. This focus serves to provide a contrast with the essentially extractive and exploitative policies and practices perceived to have characterised the South African colonial period.
5. According to Grove, evolving landscape and environmental sensibilities in Scotland had been shaped by evangelical ideas and opposition to 'Englishness and English rule' (Grove 1997).
6. NAN, AP 5/2/2, 'Natives Reserves Commission' (1928), p.2.
7. NAN, 'Report of the Drought Investigation Commission' (1924), pp.413–45.
8. NAN, 'Report of the Farm Industry Commission' (1927), p.6.
9. NAN, AP 5/7/2, 'Report on Land Settlement' (1927), p.17.
10. NAN, 'Report of the SWA Commission' (1935), pp.167–69.
11. NAN, AP 5/7/2: Soil Conservation Numerous articles from South African newspapers and journals are kept in files in the National Archives of Namibia dealing with soil conservation (AP 5/7/2). A typical example is headed: 'Reeds sedert 1774 smeek ons bodem: red my!' ('Already since 1774 our soil pleaded: Save me!', Landbouweekblad, 28 Maart 1945). An article titled 'The Future of Farming in South Africa', comprising a 'Summary of Report No. 4 of the Social and Economic Planning Council of South Africa' (1944), stated that bad farming practices, the result of 'ignorance, neglect and economic pressure', drastically reduced soil fertility. It quoted two 'renowned experts' who wrote that South Africa was facing a catastrophe due to soil erosion.
12. NAN, 'Long-term Agricultural Policy Commission' (1948), pp.68–71.
13. For a discussion of the extent to which political considerations influenced land settlement, see Botha (2000). The SWA Legislative Assembly introduced a scheme aimed at assisting returning soldiers after 1945, which the National Party (NP) stridently criticised as benefiting United Party (UP) supporters only. The continuation of the officially sanctioned land settlement scheme until at least 1960 should be seen as an attempt to 'correct the balance' by ensuring that NP supporters also benefited. The administration did, belatedly, implement the LAPC's recommendation that young men be advised to increasingly turn to technical occupations instead of farming, but that only started to occur during the course of the 1960s.
14. NAN, Minutes, 'Memorandum to Executive Committee', acting Secretary for SWA, 2 July 1957, p.5.
15. NAN, SWAA 1068, A138/22, C. Ross and J.C Fick, 'Report on visit to South West Africa', Section Soil Conservation and Extension, 31 July 1951.
16. NAN, SWAA 1078, A138/30, Ross to Secretary for SWA, 13 January 1954.
17. NAN, SWAA 1216, A158/190, 'Extraordinary Official Gazette', 25 June 1952.
18. NAN, SWAA 1069, A138/22/4, 'Minutes of Meeting of Soil Conservation Board', 6

February 1957; NAN, SWAA 1069, A 138/22/4, 'Report on Discussions with Chief Extension Officer, D. Louw', February 1957.

19. Mr. J. de Wet, former president, Namibia Agricultural Union (personal communication, 16 June 2002).
20. NAN, BB/0044, 'Verslag van die Departementele Komitee van Ondersoek na knelpunte in die boerdery en aanverwante aangeleenthede in Outjo distrik', Administrasie vir Blankes, 1981, p.13.
21. The master farmer for Namibia in 1993, Claus Düvel, stressed knowledge of vegetation, growth cycles, and plant succession as key elements in his approach to veld management: 'Meester se hoe hy dit doen', *AgricForum*, July 1993, Vol. 4(7), p.26. For virtually the whole of the colonial period, stocking rates of 1 large stock unit per 10 hectares, later adjusted to 15 hectares, were accepted as the norm in the northern commercial farming area, and 1 to 20 in the southern farming districts.
22. NAN, BB/0044, 'Verslag van die Departementele Komitee van Ondersoek na knelpunte in die boerdery en aanverwante aangeleenthede in Outjo distrik'; NAN, 'Grondbewaringsmaatreels', (soil conservation measures); NAN, 'Minutes of the Executive Committee of SWA', 1957, p.ix.
23. The statement was made by Leon Van Wyk, a former agriculture extension officer in southern SWA, who in effect, and rather unusually, questioned the whole edifice on which commercial farming in Namibia rested, such as generous subsidies, additional feeding, overstocking to compensate for lean years, inflated and unrealistic living standards and wide-ranging government support in times of drought.
24. Walker, quoted in Archer (2002: 137).
25. NAN, BB/0044, *Verslag van die Departementele Komitee van Ondersoek na knelpunte in die boerdery en aanverwante aangeleenthede in Outjo distrik*, Administrasie vir Blankes, 1981.
26. NAN, 'Long-term Agricultural Policy Commission (1948), § 235, 236.
27. NAN, Ordinance 11, 1951.
28. African reserve dwellers, for example in Otjohorongo Reserve, also complained bitterly that thousands of springbuck were destroying the grazing and requested the welfare officer to drive the game away (NAN, SWAA 475, A 50/188/10, 'Annual Report by Welfare Officer for Otjohorongo Reserve', 1944).
29. NAN, SWAA, 1349, A 205/1, 'Report of the Game Preservation Commission' (1948). The report recommended that farmers should decide whether to preserve or shoot (in whole or in part), zebra, wildebeest, ostriches and warthogs'.
30. NAN, NTB 1/41, A 50/2, Secretary for SWA to Mr. W. S Tromp, 24 September 1962.
31. NAN, SWAA. *Game Preservation Ordinance Amendment Ordinance*, 1928. Though 'royal' game such as lions and elephants could not be shot, except under a license issued by the administrator, before 1950 this regulation was by and large honoured in the breach thereof. Ordinance 11 of 1951 (articles 1 and 11) made provision for protected game too, to be shot if they proved to be threat to 'grazing or cultivated land', provided a police officer or game warden accompanied the hunt.
32. NAN, 'Minutes of the Executive Committee' (1961), p.1. Memo submitted to Administrator-in-Executive Council.
33. By the mid 1990s there were twenty-one parks and reserves, making up about 13.8 per cent of Namibia's land area. There were 148 private nature reserves totalling 7,642 square kilometres, or 0.9 per cent of the land area, registered with the Ministry of Environment and Tourism (this figure excludes private game farms).

34. For a discussion of guidelines structuring the trophy hunting industry in Namibia, see Barnes (1996).
35. A recent estimate for the income from trophy hunting is N$ 500 million (*Republikein*, 11 March 2008).
36. Considerable importance was attached to interventionist measures in African reserves in South Africa. In 1945 a New Era of Reclamation was proclaimed and the systematic implementation of betterment and rehabilitation was begun (Beinart 2002).
37. NAN, 'Commission of Enquiry into South West Africa Affairs' (1964), p.429.
38. NAN, NTB 2/111, 'Report of the Planning Commission for Hereroland', commissioned by the Department of Bantu Administration and Development, 30 November 1971.

References

Archer, S. 2002. 'Technology and Ecology in the Karoo: A Century of Windmills, Wire and Changing Farming Practice', in S. Dovers, R. Edgecombe and B. Guest (eds), *South Africa's Environmental History: Cases and Comparisons.* Cape Town: David Philip, pp.112–38.

Barnard, P. (ed.). 1998. *Biological Diversity in Namibia: A Country Study.* Windhoek: Namibia National Biodiversity Task Force, Ministry of Environment and Tourism, UN Environment Programme, Global Environment Facility and Namibia Nature Foundation.

Barnes, J.I. 1996. 'Trophy Hunting in Namibia', in P. Tarr (ed.), *Namibia Environment*, Vol. 1. Windhoek: Ministry of Environment and Tourism, pp.100–3.

Beinart, W. 1989. 'Introduction: The Politics of Colonial Conservation', *Journal of Southern African Studies* 15(2): 143–62.

——— 1996. 'Soil Erosion, Animals and Pasture over the Long Term: Environmental Destruction in Southern Africa', in M. Leach and R. Mearns (eds), *The Lie of the Land: Challenging Received Wisdom on the African Environment.* Oxford: James Currey, pp.4–72.

——— 2002. 'Environmental Origins of the Pondoland Revolt', in S. Dovers, R. Edgecombe and B. Guest (eds), *South Africa's Environmental History.* Athens: Ohio University Press, pp.76–89.

Beinart, W., and J. McGregor. 2003. 'Introduction', in W. Beinart and J. McGregor (eds), *Social History and African Environments.* Cape Town: David Philip.

Bester, F.V. 1995. 'Drought and Rangeland Management', in R. Moorsom (ed.), *Coping with Drought: Drought Impacts and Preparedness in Namibia.* Frankfurt am Main: Brandes and Apsel, pp.91–104.

Bollig, M. 2000. 'Production and Exchange among the Himba of Northwestern Namibia', in M. Bollig and J.-B. Gewald (eds), *People, Cattle and Land: Transformations of a Pastoral Society in Southwestern Africa.* Cologne: Rüdiger Köppe Publishers, pp.271–98.

Bollig, M., and J.-B. Gewald. 2000. 'People, Cattle and Land: Transformations of Pastoral Society', in M. Bollig and J.-B. Gewald (eds), *People, Cattle and Land: Transformations of a Pastoral Society in Southwestern Africa.* Köln: Rüdiger Köppe Publishers.

Botha, C. 2000. 'The Politics of Land Settlement in Namibia, 1890–1960', *South African Historical Journal* 42: 232–76.

——— 2002. 'Development, Survival and Adaptation in the Former Damara "Homeland" of Namibia', Geography and Environmental Studies Series 1. Windhoek: Department of Geography and Environmental Studies, University of Namibia.

——— 2005. 'People and the Environment in Colonial Namibia', *South African Historical Journal* 52(1): 170–90.

Brown, C. 1999. 'The Outlook for the Future', in P. Tarr (ed.), *Namibia Environment, Vol. 1.* Windhoek: Ministry of Environment and Tourism, pp.15–20.
Carruthers, J.T. 1995. *The Kruger National Park: A Social and Political History.* Pietermaritzburg: University of Natal Press.
Dedering, T. 1997. *Hate the Old and Follow the New: Khoekhoe and Missionaries in Early Nineteenth-century Namibia.* Stuttgart: Franz Steiner Publishers.
de Kock, A.J. 1965. 'Die Kleinhoewe Stelsel in Suidwes-Afrika Gedurende die Duitse Bewind, 1884–1914', MA thesis. Stellenbosch: University of Stellenbosch.
Delius, P., and S. Schirmer. 2000. 'Soil Conservation in a Racially Ordered Society: South Africa 1930–1970', *Journal of Southern African Studies* 26(4): 719–42.
Dunlap, T.R. 1997. 'Ecology and Environmentalism in the Anglo Settler Colonies', in T. Griffiths and L. Robbin (eds), *Ecology and Empire.* Edinburgh: Keele University Press, pp.76–86.
du Plessis, W. 1992. 'In Situ Conservation in Namibia: The Role of National Parks and Nature Reserves', *Dinteria* 23: 132–41.
Düvel, G.H. 1990. 'The Role and Limitations of Economic Incentives in Conservation Farming', in C. Csáki et al. (eds), *Agricultural Restructuring in Southern Africa.* Windhoek: IAAE, pp.421–41.
Forrest, J. 2000. 'Comments on the History of Water Policy in Namibia, 1895 to 1995', unpublished paper presented at New England Workshop on Southern Africa, 27–29 October.
Grove, R. 1997. 'Scotland in South Africa: John Croumbie Brown and the Roots of Settler Environmentalism', in T. Griffiths and L. Robin (eds), *Ecology and Empire: Environmental History of Settler Societies.* University of Natal Press, Pietermaritzburg, pp.139–53.
Hains, B. 1997. 'Mawson of the Antarctic, Flynn of the Inland: Progressive Heroes on Australia's Ecological Frontiers', in T. Griffiths and L. Robbin (eds), *Ecology and Empire.* Edinburgh: Keele University Press, pp.154–68.
Henrichsen, D. 1997. 'Herrschaft und Identifikation im Vorkolonialen Zentral Namibia, das Herero- und Damaraland im 19. Jahrhundert', PhD dissertation. Hamburg: University of Hamburg.
Hinz, M. 2003. *Without Chiefs There Would Be No Game: Customary Law and Nature Conservation.* Windhoek: Out of Africa Publishers.
Kinahan, J. 1991. *Pastoral Nomads of the Central Namib Desert: The People History Forgot.* Windhoek: Namibia Archaeological Trust and New Namibia Books.
Lange, G.-M., J.I. Barnes and D.J. Motinga. 1998. 'Cattle Numbers, Biomass, Productivity and Land Degradation in the Commercial Farming Sector of Namibia, 1915–1995', *Development Southern Africa* 15(4): 555–72.
Lau, B. 1987. *Namibia in Jonker Afrikaner's Time.* Windhoek: National Archives of Namibia.
Lau, B. and P. Reiner. 1993. '100 Years of Agricultural Development in Colonial Namibia', *Archeia* 17 (National Archives of Namibia, Windhoek).
Lowenthal, D. 1997. 'Empires and Ecologies: Reflections on Environmental History', in T. Griffiths and L. Robbin (eds), *Ecology and Empire.* Edinburgh: Keele University Press, pp.229–36.
McCann, J.C. 1999. 'Climate and Causation in African History', *International Journal of African Historical Studies* 32(2/3): 261–79.
MacKenzie, J.M. 1997. 'Empire and the Ecological Apocalypse: The Historiography of the Imperial Environment', in T. Griffiths and L. Robbin (eds), *Ecology and Empire.* Edinburgh: Keele University Press, pp.215–28.

Moorsom, R. 1995. 'Underdevelopment and Labour Migration: The Contract Labour System in Namibia', History Research Paper 1. Windhoek: Namibia History Trust.

Pankhurst, D. 1996. 'Similar But Different: Assessing the Reserve Economy Legacy of Namibia', *Journal of Southern African Studies* 22(3): 405–20.

Rohde, R. 1999. 'Afternoons in Damaraland: Common Land and Common Sense in One of Namibia's Former "Homelands"', Occasional Paper 41. Edinburgh: Centre for African Studies, University of Edinburgh.

Rothauge, A. 2000. 'New Ecological Perceptions of Arid Rangeland', *Agricola* 11*:* 49–55.

Schneider, H. 1992. 'Agriculture in Namibia: Opportunities and Constraints', in M.B. Schneider (ed.), *Namibia: Views and Perspectives since Independence*. Windhoek: Konrad Adenauer Foundation and Namibia Institute for Democracy.

Schoeman, A. 2007 'Laying the Foundation in Etosha', in R. van Schalkwyk (ed.), *Celebrating One Hundred Years of Conservation*. Windhoek: Venture Publications.

Schrank, G. 1974. 'South West Africa, Social and Economic Aspects of its History, 1884–1915', PhD dissertation. New York: New York University.

Silvester, J., M. Wallace and P. Hayes. 1998. 'Trees Never Meet: Mobility and Containment, An Overview, 1915–1946', in J. Silvester et al. (eds), *Namibia under South African Rule: Mobility and Containment, 1915–1946*. Oxford: James Currey, pp.3–50.

Smith, A.B. 2000. 'The Origins of Pastoralism in Namibia', in M. Bollig and J.-B. Gewald (eds), *People, Cattle and Land: Transformations of a Pastoral Society in Southwestern Africa*. Cologne: Rüdiger Köppe Publishers, pp.55–76.

Steenkamp, P. 1993. 'Liefde vir Grond, Spoor Bewaringsboerdery Aan', *Agriforum* 4(2).

Stern, C., and B. Lau. 1990. 'Namibian Water Resources and their Management: A Preliminary History', *Archeia* 15 (National Archives of Namibia, Windhoek).

Versfeld, W., F.V. Bester and H.C. Stehn. 1988. 'Benaderings tot Veldbestuur', *Agricola* 5: 53–58. Published by Administration of Whites.

Visser, C.J. 1988. 'Die Mens en sy Gesindheid Teenoor die Landbouhulpbronne', *Agricola* 6: 9–11. Published by Administration of Whites.

von Wietersheim, A. 1985. 'Im Jagdrevier zwischen Kalahari und Namib', in K. Becker and J. Hecker (eds), *Vom Schutzgebiet bis Namibia 1884–1984*. Windhoek: Interessengemeinschaft Deutschsprachiger Südwester, pp.403–12.

Werner, W. 1998. *'No One Will Become Rich': Economy and Society in the Herero Reserves in Namibia, 1915–1946*. Basel: P. Schlettwein Publications.

——— 2000. 'From Communal Pastures to Enclosures: The Development of Land Tenure in Herero Reserves', in M. Bollig and J.-B. Gewald (eds), *People, Cattle and Land: Transformations of a Pastoral Society in Southwestern Africa*. Cologne: Rüdiger Köppe Publishers, pp.265–68.

Chapter 8

Land, Boreholes and Fences
The Development of Commercial Livestock Farming in the Outjo District, Namibia

Ute Dieckmann

Introduction

This chapter explores the history of commercial agriculture in the Outjo district of northern-central Namibia from the onset of colonisation until the independence of the country in 1990. This is foremost a history of European – or white – livestock farming.[1]

After a short description of the methodology and the field site, I will sketch the progressive European occupation of the land. The economic strategies employed by farmers during the colonial period will be analysed, as well as major challenges, risks and crises.[2] Special attention will be paid to two aspects vital to the history of colonial farming in Namibia: state support to farmers both in the development of farms as well as in times of crises; and labour conditions and labour relations.

This chapter is based on recorded semi-structured interviews with farmers and ex-farmers in the Outjo district and various other key informants, such as extension officers, stock inspectors, and representatives of the Namibian Agricultural Union. Furthermore, informal conversations (especially on more sensitive topics) were also part of the research. I interviewed around forty farmers, ex-farmers and experts between 2005 and 2007.[3] Adult men constituted the majority of my interviewees. Women, as long as their husbands were alive, referred to them and their expertise with regard to the research topic.[4] Earlier research carried out in the region focusing on colonial history with regard to the Hai||om also fed into the present chapter. Additional research was carried out in the National Archives of Namibia on historical developments and in the Windhoek Deeds Office for every farm in the district. The farm registers of all the farms in the Outjo district were consulted to trace the history of settlement, ownership patterns and transaction sequences. In order to document all the transactions of every farm in the district, a dataset with a list of all the farms in the district was created. This dataset includes the name of the farm, the farm size, the dates of transactions, the full names and the birth dates of the farm owners (buyers and sellers) from the beginning of official transfer.

Due to the quality of the material collected– above all interviews with white commercial farmers – this article reflects their perspective on the settlement process and has to be understood within the socio-political context of independent Namibia. Land reform, land redistribution and land expropriation are major political issues, and white commercial farmers tend to legitimise their one-time role of establishing commercial pastoralism in Namibia.

The Site: A Landscape of Farming Units

The Outjo district covers an area of 38,722 square kilometres and is situated in the commercial farming area in the south-east of the Kunene region, which comprises the entire north-west of Namibia.[5] The mean annual rainfall in Outjo between 1913 and 1998 amounts to 405 millimetres.[6] The town of Outjo has been a municipality since 1944 and constitutes the commercial centre for the surrounding farms. In 2001, the town itself had 6,013 inhabitants[7] while the whole district had 8,947 inhabitants. The main economic activities are nowadays livestock farming, charcoal production and tourism (Anon. 2001). Characteristic of the whole commercial farming landscape in Namibia, the district is divided into farming units, nowadays separated from each other by fences; low fences for livestock farms and high fences for lodges and farms with wildlife.

According to data from the Deeds Office, around 430 farm units were originally given out in the present-day district over a period of around sixty years. The first farms were allotted in German colonial times around the turn of the twentieth century. Around 1960, the entire area had been divided into farming units, with an average size of 5,829 hectares.[8] In 1896, when the Outjo district was established as an administrative unit, it covered a much larger area, including the Kaokoveld (Zimmerer 2000: 27). Many farms in the western part were bought in the 1960s and early 1970s by the government from private owners and allocated to the Damara 'homeland'.[9] These farms have not been included in the dataset since they were situated outside of the present-day district.

From the 430 original farms, 291 have remained basically unchanged in size, 115 have been divided and 54 have been consolidated. Note that farms could be divided and consolidated over the years as well. Indeed, 30 of the original farms were divided and parts were either consolidated with another farm or consolidated with portions of other farms into new farms.[10] Of the original farms, 14 were bigger than 10,000 hectares, and the biggest was 26,107 hectares in size. According to the Deeds Office, no plot given out in the beginning was less than 100 hectares. The town lands had already partly been portioned up and sold in the 1940s and 1950s, and some plots consisted of less than 100 hectares.

According to ownership data provided by the Namibian Agricultural Union (NAU) and the Ministry of Agriculture, Water and Forestry (MAWF), the Outjo district consisted of 567 farming units in 2006, a figure which includes the smaller plots of the town lands of the two towns in the district, Outjo and Kamanjab.[11] An average farm was about 4,399 hectares in size. Five farms were bigger than 10,000 ha, the biggest being 34,704 hectares, and 20 farms covered less than 100 hectares. With regard to ownership, 78 per cent of the farms were in private hands, while 8 per cent belonged to the government and 14 per cent were owned by different kinds of companies and corporate entities.

Looking at the history of farm ownership in the district, more than 60 per cent of farm owners were male individuals, with women being owners in less than 10 per cent of cases. Companies were involved in around 12 per cent of farm transactions and the government in around 16 per cent. These figures also include the time after independence in 1990 up until 2006. The developments of land reform changed transaction patterns without a doubt – that is, the government increasingly acts as a buyer of farms in order to use them in their resettlement programme. In retracing the colonial history of white settlement in the district, I next look at how this particular farming landscape came into being.

The European Occupation of the Land

South West Africa (SWA) was officially proclaimed a German protectorate in 1884. A serious move towards expansive colonisation began only after 1894. In 1896, German military stations were erected at Otavifontein, Grootfontein and Outjo (Köhler 1959: 20–21; Dierks 1999: 48). In 1903, the protected area (*Schutzgebiet*) consisted of six districts: Windhoek, Keetmanshoop, Otjimbingwe, Omaruru, Swakopmund and Outjo. The Outjo district was mainly populated by Khoekhoegowab- and Otjiherero-speaking pastoralists and foragers.[12] A few white people from South Africa had already entered the area, also following a rather nomadic lifestyle.[13]

In 1901, only nine farms in the whole district were reported to be occupied by about thirty-nine settlers, originating from Germany, Transvaal, the Cape and England. It was not until the end of the wars of 1904 to 1907 that significant white settlement took place in the entire territory. Indeed, the war and the regulations introduced afterwards (including the expropriation of former 'tribal' land) laid the basis for the establishment of white-settler agriculture (Schmokel 1985: 96). In August 1907 new enactments concerning 'natives' were issued as a direct consequence of the war. Their aim was to order and regulate the 'natives' lives in an autocratic manner. These included the Pass – or Registration – Law (*Verordnung betreffend die Paßpflicht von Eingeborenen*), regulations relating to the 'control' of 'natives' (*Verordnung betreffend Maßregeln zur Kontrolle der Eingeborenen*) and laws relating to labour contracts (*Verordnung betreffend Dienst- und Arbeitsverträge mit Eingeborenen*).[14] The Pass Law required that all Africans over the age of seven (with the exception of the Rehoboth Basters) had to register with the authorities. They were given a metal badge which served as a pass (and had to be displayed prominently on the person) and a service book (*Dienstbuch*). The law empowered every white person to arrest Africans without passes and to hand them over to the police. The order for the 'control' of 'natives' stipulated that no African could obtain right or title to fixed property, or own cattle and horses, without the consent of the governor. Certain exceptions were made to the ban on the ownership of livestock after 1912. Also, it was forbidden for more than ten 'native families' to stay on any one farm or property outside the 'native reserves'. The law relating to labour contracts laid down that all 'natives' without visible 'means of support' must be employed (Emmett 1999: 74). These enactments formed the legal framework of German 'native policy' for the remaining time of the occupation. They were evidently aimed at a complete control of the mobility and settlement practices of local inhabitants and went hand-in-hand with the settlement policy with regard to European farmers.

By the end of the German colonial era in 1915, around sixty-seven farms were allotted in the Outjo district (see Figure 8.1, below). Up to then, European settlement in the Outjo district had taken place close to open springs along the main traffic lines, mainly ox-wagon tracks, such as Outjo-Okaukuejo, Outjo-Fransfontein and Outjo-Otjitambi-Sesfontein.[15] Despite the fact that at least approximately 2,700 'natives' were estimated to live in the area of the former district (Külz 1909: 112), the land was not considered to be densely inhabited in the minds of the incoming settlers. The colonisers imported ideals of private landownership and agriculture, which had evolved in the European context, namely the notion of demarcated farms with well-defined borders and incontestable individual proprietors. Land tenure and land-use practices of the former inhabitants did not conform to European concepts, and land was not 'settled' and possessed according to the European understanding. The Germans did not yet have the power to impose their ideals of landownership on the landscape on a permanent basis (see also Silvester 1998: 99), but the landscape's 'emptiness' already helped to justify the European appropriation of the land (see also Botha, this volume).

The system of control, which was intended by the rigid legislation, was never completed during the German period. Due to their high mobility, the population groups falling under the category Bushmen posed a special problem to the administration.[16] Attacks on migrant workers and conflicts with farmers were reported in the district. This had the effect that the police stations of Outjo district were staffed with the highest number of officers (thirty) in the whole territory of SWA.[17]

Under the military rule of the South African era from 1915 to 1918, the legislation with regard to 'natives' was liberalised to a certain degree, since South Africa had a strong interest in the territory and sought to persuade the world that its efforts as a colonising country were justified and that its rule would be far better than that of the Germans. With the establishment of the mandate, South Africa imported a body of new laws to regulate the flow of labour and to control the indigenous inhabitants (Silvester, Wallace and Hayes 1998: 23). Several proclamations were enacted which played a role with regard to the facilitation of European settlement. The Vagrancy Proclamation of 1920 made it an offence for black people to move around in the police zone unless they could show 'visible lawful means of support'.[18] As with the case of native reserves, this legal tool made provision for the supply of labour to the colony. The 'means of support' were set at either ten cattle or fifty head of small stock. During the South African military period (1915–1918), the Pass Law had apparently made provisions for this matter, and was only now formulated in a proclamation (see also Emmett 1999: 76).[19] The Vagrancy Proclamation was amended further in 1927, and among other things prison terms for vagrancy were increased from three to twelve months.[20] Another important regulation connected to the control of mobility was Proclamation 11 (1927), which sought to prevent squatting by further limiting the number of people allowed to reside on a farm to five 'native families'.

The settlement of land within the Outjo district continued – by and large – from south to north. In 1920, the Land Settlement Act of the Union of South Africa was applied to Namibia with some alterations to meet local conditions.[21] With this, a rapid programme of settlement by white South African citizens began.[22] In the beginning only very poor people ventured into the uncertainties of farming in Namibia, but by 1926 the Advisory Council decided to prohibit the entrance of poor whites as they were considered a drain on resources and tended to be constantly on the move. By the beginning of the 1930s it became official policy to give preference to South West African applicants when farms were allocated (see, e.g., Silvester 1998: 106; Botha 2000).

At least until the 1930s, farm occupancy was not restricted to whites in the Outjo district. Black Namibians were officially allowed to occupy farms, although the terminology differed: farms allotted to black Namibians within the police zone were called reserves. The farm Aimab in the Outjo district, for example, was used as a 'native reserve' until the 1920s.[23] The farm Otjeru was a 'native reserve' from German times until the late 1930s, and included until the 1920s two more farms.[24] The reserve Fransfontein, nowadays outside Outjo district, was also part of the district.[25] In the 1930s, the provisional reserves in the district were dissolved and the inhabitants had to move to Fransfontein, 'native reserves' in other districts, or outside the police zone unless they took up regular employment (Schnegg 2007). However, it is likely that former inhabitants of the district were still able to squat illegally on unoccupied farms or un-surveyed areas, at least for limited periods of time until the area was completely divided into farms, allotted to European settlers and finally fenced in the 1960s.[26]

The European settlement of the district did not progress continuously but rather occurred in surges, dependent on political developments, natural conditions and the development of the necessary infrastructure. In 1928, for example, about 1,900 Afrikaners, who had earlier trekked from South Africa to Angola, were offered the possibility of returning to Namibia. The majority of them were resettled in the so-called Osire Block, east of Otjiwarongo. By then, farming conditions there were far from what was required to build up successful farms, and the effects of drought and recession in the early 1930s put additional stress on the incomers. Although supported with extra subsidies by the government, their situation did not improve considerably and the administration later decided to resettle them. In the second half of the 1930s the administration made land available in the north-west of the Outjo district for a number of these so-called Angola Boers.[27] In the Annual Report Land Resettlement of 1937, it was mentioned that:

> The rate of progress of land settlement at present cannot be maintained much longer, as most of the land suitable for settlement purposes has been disposed of. There are un-surveyed areas in the Outjo, Swakopmund, Maltahöhe and Warmbad districts which it is proposed to cut up into farms during the course of this year, and these holdings will be made available for settlement purposes. When these have been disposed of there will remain very little land for further settlement.[28]

Despite the findings, that there was little suitable farm land available in the respective areas, a fact that had already been mentioned at the end of the 1930s, more land in marginal areas was made available for settlers until the 1950s.[29]

While the allotment or purchase of farms in the whole territory was put on hold during the Second World War, extensive provision was made after the war for the support of war veterans. Ex-soldiers were given land and could qualify – on recommendation of the Discharged Soldiers' Assistance Board – for additional loans for such things as building houses and to purchase breeding stock. A large amount of land in the western part of the Outjo district – formerly one huge farm of 247,346 hectares – was made accessible to settlers. Aruchab, as the farm was called, had been allotted to the Imperial Cold Storage and Supply Company in 1924.[30] The company had used it for cattle – an inspection report of 17 January 1927, for example, stated that 6,203 cattle were kept on the farm.[31] In the second half of the 1940s, this piece of land was surveyed and divided into about forty farms. Most of them were allotted immediately afterwards. Additionally, part of the Etosha game reserve, which was situated on the northern border of the district, was made available for settlement; boreholes were drilled and grazing licences could be obtained by interested settlers.[32] Apart from the ex-soldiers, settlers from the southern regions of Namibia moved to the district since the south had suffered from enduring drought. The ensuing settlement continued during the 1950s. Mainly formerly landless grazing-licence holders and farmers who were sharing a piece of land with male kin received their own farms during these years. This strategy reflected recommendations of the Long-term Agricultural Policy Commission at the end of the 1940s.[33] Sixty farms in the Outjo district were gazetted in 1951, and fifty-seven farms were gazetted in 1955, most of them in the western part, the so-called Kaross Block. In 1960, 740 surveyed farms were situated in the Outjo district.[34] All the land of the present district had been surveyed and allocated to settlers by that time. By the early 1970s, 233 farms in the west of the district were cut off and incorporated into Damaraland following the recommendations of the Odendaal Commission and the size of the district decreased a great deal (RSA 1964 [Odendaal Report]: 89). This explains why the district comprised considerably fewer farms in 2006 than in 1960. Later changes entailed mainly the division or consolidation of already existing farming units. Figure 8.1 is compiled from data for every farm in the district from the Deeds Office in Windhoek and reflects the progression of registered farm ownership. The map shows the progress of settlement according to title deed holdings. However, the mapping does not reflect the very dynamic dimension of land settlement. This becomes clear when taking a closer look at the whole process of land allocation under the land settlement programme: after a farm was officially made available for disposal in the government gazette for a period of two months, during which applications on prescribed forms were received, the process went as follows:

> Holdings are allotted for a probationary period of one year or more, and on confirmation of the allotment, a lease, for a period of 5 years, is issued with the option of purchase at any time during this period or a subsequent extension thereof … During the first year of the lease term no rental is payable, but a rental of 2 percent per annum, calculated on the purchase price, is payable during the second and third years, and 3 percent per annum during the fourth and fifth years. Subject to satisfactory development of holdings, remission of rent may be granted to the end of the fifth year of the lease period … [W]hen a lessee exercises his option to purchase, the purchase price becomes payable in half yearly instalments extending over a period of thirty years, interest being charged at the rate of 4 percent per annum.[35]

The official registration of a transaction often took place more than a decade after the first allotment. Moreover, the first person allotted the land was not in every case the first holder of the title deed. Furthermore, the title deeds were sometimes registered much later than the actual transaction took place. Additionally, in the early years of settlement, grazing was often exploited without official allotment, since the existence of unfenced boundaries made trekking rather unproblematic.[36] These circumstances once again provide evidence that the settlement of the area was a rather flexible and partly informal process during the early years. Many settlers did not succeed in making a living on the piece of land they had applied for and left the location voluntarily, sometimes exchanging the farm for a piece of land in another area. Others could not pay back their debts or did not fulfil the conditions under the scheme – for example, permanent occupancy – and had to return the farm to the government. Farms were then gazetted again years later.[37] The reasons for failures were manifold: soil quality, lack of water, lack of capital to invest, drought, personal incapability, or just 'bad luck' (an incalculable combination of different factors).

Two cases are presented here in order to illustrate the process of settlement. The first farmer, O.L., is an Afrikaans-speaking man, who farmed on the border of the

Figure 8.1: First transaction of farm ground (Deeds Office, Windhoek) and German owners in 1915 (G.P. Kruger, 'Outjo 1885–1960', n.d., unpublished manuscript, Swakopmund Library, Namibia).

Etosha National Park in 2006. The second, E.L., is a German-speaking woman who had farmed mainly in the south-western part of the district during her life.

Case Number 1

O.L.'s great-grandfather came to the Outjo area around the 1850s from the Cape in South Africa. In 1915, O.L.'s grandfather moved to Angola and joined the Angola Boers who had settled there in the twentieth century. O.L.'s father was born in Angola in 1917. In 1927, when the South African administration offered the Angola Boers favourable conditions to start farming in Namibia, the family came back minus their cattle along with many other Angola Boers. They first moved to the Grootfontein area and got some cattle from the administration to start farming. In the mid 1930s, the family moved from there to be with a son-in-law, who farmed in the Outjo district.

O.L. stressed that people moved a lot in former times, they didn't stay long on a farm. 'They stayed at a place for two to three months, then six months at another place. They did not own land. They leased this piece of land and then another piece of land'. The family moved later to another farm, which had been part of the huge Aruchab ranch mentioned above. Although the Deeds Office data document that Aruchab was officially divided into smaller farming units in 1948, the process must have started earlier. The first development – the drilling of boreholes – had been initiated and grazing licences given out for portions of the ranch by the beginning of the 1940s.

After O.L.'s father got married, O.L.'s father and his family continued to stay and move together with his father's family for a long time. They stayed temporarily on at least three different farms, all part of the Aruchab ranch, until 1948. O.L. himself was born in 1943.

In 1948, two rows of farms were cut off from Game Reserve No. 2 (nowadays the Etosha National Park) and added to the commercial farming area. A drilling machine was pulled through the area and boreholes were drilled about every 5 kilometres. According to O.L., it took one-and-a-half to two months for one borehole (depending on the level of the water table and the soil quality). His father and grandfather applied for one of these new farms. They waited on the neighbouring farm where the brother-in-law was staying until the drilling machine had completed its work. The oldest brother of O.L.'s father also farmed a neighbouring farm.

In 1957, O.L.'s grandfather died and O.L.'s father bought the farm from the grandfather's widow (this is the first entry in the Deeds Office about the farm), and she moved to Outjo. Thus, almost ten years after moving to the farm, the father could farm alone. O.L. went to primary school in Outjo in the 1950s, and later to a technical high school in South Africa. In the 1960s, he worked in South Africa for some years in the construction business, where he met and married his wife. After his return to Namibia, he changed between working for private companies and working with his father for the state, erecting fences in Namibia. At the beginning of the 1970s, O.L. bought the farm belonging to oldest brother of his father, neighbouring his father's farm, with a loan from the Namibian Land Bank. While O.L. was working during the week, his father looked after the farm.

In 1981, after he had paid off the loan for the farm he acquired in the early 1970s, he bought the farm belonging to his father, again with a full loan from the Land Bank. He gave up his non-farming work as a condition of the Land Bank's loan was that he work full-time on the farm. His father sold all his cattle and moved to Outjo. In 2006, O.L. had two farms and was exclusively farming cattle. Between 2006 and 2011, he sold his farms to the government who allocated them to the Hai||om as resettlement farms.

Case Number 2

In 1887, E.L.'s mother was born in Otjimbingue. Her parents were Germans from Westphalia and Silesia, and had a store and a hotel in Otjimbingue.

E.L.'s father P.H. was born in Germany in 1884. He came as a cabin boy to Namibia in 1902, driven by his family's poverty. He went to Otjimbingue where he learnt farming and bricklaying. During the war of 1904 to 1907, P.H. was a soldier and rode patrols in the Outjo area. After the war, he bought a farm in the south-western part of today's Outjo district from the German government; as a former soldier he obtained favourable conditions. Thanks to his patrols, he knew that there was a natural spring on the farm. It is unknown if the farm was surveyed after he had chosen that area to farm on it or beforehand. According to E.L. he developed the farm, opened the spring with dynamite; later he started to grow wheat and citrus fruits.

E.L.'s parents married in 1910, and in 1918 E.L. was born, the fourth daughter of five children. According to Deeds Office entries, her father bought a second farm in 1928. E.L. went to school in Swakopmund in 1935, after being taught by her mother for some years at home. Around 1936, E.L. went to Germany, where she went to a commercial school and also worked in a hospital and in a mine. In 1938, E.L. returned to SWA. A couple of months later the Second World War broke out. During the war, although most German farmers were in internment camps in South Africa, her father escaped this fate because he was growing cereal and fruit on his farms. As many farm managers were also imprisoned in the camps, E.L. managed her father's two farms during the early years of the war. Being without family on the farm, E.L. had a piano to fight the solitude. Her father was already working a third farm, which was, according to Deeds Office data, officially sold by the government to his wife only in 1944.

In 1941 to 1942, due to a serious drought, E.L.'s father bought farms in the Otavi area, which had formerly been Crown land, from the administration. Following an interview with her, E.L. answered some of my questions in a letter:

> 1941–42 were years of drought, it had just rained 32 mm during the year. All cattle, with the exception of one milk cow and a calf, and all the sheep had to trek via Outjo almost to Otavi, where my father and I had bought three raw farms. This was only possible via another official buyer (since we as Germans were enemies and we were not allowed to buy any land). The farms were only marked on the map, otherwise we had to look for the set *baken*.[38] There was no path, no water, no roof. The sheep trekked behind the cattle. A few days earlier, the drilling machine was pulled by donkeys to our new grazing area. My brother had already gained some experience in drilling. Fortunately, the sites, which had been indicated by the divining rod, proved successful at little depth. The livestock had only to stay for a couple of days on neighbouring farms. We stayed in tents, a matron was our cook. I had to count all the sheep of my father and us, his children. Only after two years, there was again enough rain and grazing that we could trek home again. It was always advisable to have another farm in another area.[39]

During the war, E.L. bought another farm, from a private owner, neighbouring the family farms in the Outjo area with a bank loan. When her father died in 1943, E.L. sold the farm in the Otavi area and paid off the loan on her farm in the Outjo area (according to the Deeds Office, the transaction took place in 1946), and she moved to the latter farm. Her brother started farming at his father's first farm and her mother moved to Schoenau. In 1945, her mother handed over the farm Persephone to her.

In 1947, E.L. married Dr L. who was, according to her, not a 'born farmer'. He had bought a farm a year before. The farm he had stayed on for years, according to E.L., was only officially transferred to him a year after his marriage, according to Deeds Office data. In 1948, E.L. sold her karakul sheep and thereafter farmed mostly cattle and a few goats and sheep. In the 1960s when foot-and-mouth disease and a serious drought coincided, E.L. bought a farm east of Outjo (according to Deeds Office entries, the transaction took place in 1974). According to her, it was impossible to lease land during those years. (Supposedly the demand was too high, although other farmers reported having trekked to state-owned land.) Some of her younger livestock trekked to the Waterberg plateau.

In 1979, she transferred the farm east of Outjo to one of her daughters. In 1981 (according to Deeds Office records) her brother sold his father's original farm. In 1982, E.L. sold two farms to her second daughter, who also inherited a farm from her father. In 1984, E.L. sold the farm she had inherited from her late husband to her first daughter.

In 2006, E.L. was partly staying in Swakopmund, where one daughter lived, and partly staying with her other daughter on the farm, which is nowadays used as a hunting farm.

In the context of the Deeds Office records, these cases illustrate the fact that Deeds Office data and actual settlement patterns are far from being consistent. The cases also indicate the almost nomadic lifestyle of some of the early farmers when the land was not completely surveyed and divided into farm units. During these times, farming involved a trial-and-error approach and required moving between various (recently surveyed) farms, a strategy frequently applied by potential settlers in the hope of finding a 'better' farm. Regardless of the high mobility of the early days, both cases furthermore illustrate that mobility – the possibility of trekking – was an important means of maintaining farming operations in Namibia, especially during droughts.

The examples also show the importance of family networks. The joint farming of father and son (with their families) as in the first case was rather the rule than the exception. Both cases demonstrate clearly that farming was an extended family business in which resources – land and livestock – were to a certain degree jointly managed. Selling the farms at a later stage in life to descendants was the way of obtaining a pension. The older farmers then often moved to town, where the infrastructure was better.

Another crucial aspect mentioned in the first case is that the interviewed farmers often started to farm full-time rather late in their lives and that they had temporal or regular employment elsewhere for quite some time. In the next section I will look in more detail at these changing and diversified economies.

Changing Economic Strategies

Although livestock farming remained an important component of the farming sector throughout the twentieth century, the economic strategies employed by farmers changed considerably. From the beginning, it was obvious that the land was mostly suited for animal husbandry. In 1907, Carl Schlettwein, a German farmer from the Outjo district wrote a guide book for incoming settlers, in which he stated: 'It is well known that effective crop production or horticulture can only take place with regular irrigation in South West Africa, the exception being the north of the territory. Therefore, it goes without saying that the settler who is coming as a farmer to South West Africa has to start with animal husbandry for his livelihood' (Schlettwein 1907: 5).

By 1902, about 3,000 cattle were reported to be in the Outjo district.[40] From the beginning, district officers and farmers undertook experiments with breeding to improve the adaptability of the imported foreign breeds, such as Simmenthaler, Pinzgauer, Vogelsberger, Hollaender and Short-horn (ibid.: 77). During the German period, a number of karakul sheep were imported into the country. These were significant in so far as droughts did not affect the karakul economy as much as the cattle economy, since karakul lambs were slaughtered for their pelts directly after birth. As elsewhere, state-run and private experiments with the production of cash crops were started – such as tobacco, cotton and fruit – but by and large they did not meet with satisfactory results. Small-scale horticulture provided vegetables and staple crops for household consumption.

During the South African period, stock farming continued to be the predominant mode of agriculture in the district, and for a long time dairy farming was at least as important as meat production. Still, milk and meat production was for a long time accompanied by horticulture with vegetables, as well as a smaller number of hectares under staple crops, mainly for subsistence. Thus, a certain degree of diversification was evident. Indeed, in the annual report for 1938 of the land branch, it was recommended that farmers should combine karakul pelt farming with the rearing of stock for meat and the production of cream.[41] Farming karakul sheep became increasingly significant from the 1930s on due to international markets and the adaptability of the breed (Schmokel 1985: 100). The dryer western part of the district was especially suitable for sheep, and the majority of farmers also kept karakul sheep. By 1946, karakul pelts made up half the value of the territory's total exports (Emmett 1999: 185). Besides the beef and pelt industry, the marketing of milk and milk products played a vital role in the economy of the territory as a whole and the Outjo district in particular. In the 1930s, butter was not only exported from SWA to South Africa, but also to England, Rhodesia, the Belgian Congo and French Equatorial Africa (Lau and Reiner 1993: 17). Apparently, almost every farmer in the district supplied milk for the market. The lorry transporting the milk served several purposes, as one farmer recollected:

> There was always a lot of excitement when, once a week, the so-called Milch-lorry [dairy lorry] passed the farm. It also brought food and drove further but I don't know where … I don't know how far the lorry went, but it came back later the same way. At that time there was not a lot of traffic, that's why it was always exciting. Is the car coming to the farm or is it just passing? And as a child, I took a lift with the Milch-

> lorry to the neighbour's farm to play with their kids, and the next day I went back home, with the same lorry, that passed the one day and returned the next.[42]

The dairy economy ensured the farmer with a regular, though small, income, while the beef industry in particular demanded more long-term planning and management of financial resources. Both cheddar and sweet-milk cheese were manufactured by the SWA Cold Storage and Stock Farmers Company in a factory in Outjo (RSA 1964 [Odendaal Report]: 281).

From the 1950s onwards, the combination of various lines of production declined and beef and karakul pelt farming became the dominant strategies. As one farmer told me:

> [I]f you want to produce milk, you've got to have enough green ... and energy food (*kraag*). In those days, we did not have it, we could not import it from anywhere. We had to rely on our own natural green grass. And the moment the grass turns yellow, the milk yield of the cows drops dramatically. And ... the other problem is a management problem, your cows run all over the farm; you had a few camps there, in which you kept the cows, and during the milking season you had to bring all the cows to your homestead or a central point, and within a couple of weeks there was no grass left around your house. Then the cows had to go long distances in order to get grass. It was difficult to manage the situation. It led to many farms ... being overexploited a few kilometres around the house that led to bush encroachment. The bush had no competition from the grass ... And you see it [bush encroachment] around many houses ... Then, extension services said we have to focus on something else, we don't have the natural resources to sustain milk production. [In] about ... 1954, '56 ... the government realised it had to encourage farmers to move away from milk production in our area.[43] And from that time on, the creameries and the cheese factories struggled to survive; and even today there are, almost 60 years later on, the dairy farmers in this country, there are about twelve of them left ... in the whole country ... Anyway, after, I think, 1956 ... the emphasis shifted to beef production.[44]

Lau and Reiner speak of 'karakul or meat monocultures – heavily capitalized industrial ranches' (Lau and Reiner 1993: 58), mainly due to South Africa's market interests in the country. Several ordinances were passed by the Administration, following the 1948 Long-Term Agricultural Policy Commission (e.g. the Soil Conservation Ordinance, the Promotion of Farming Interests Ordinance, the Preservation of Trees and Forests Ordinance, all promulgated in 1952) and subsidy policies introduced (e.g. for the establishment of camps, fencing, extra cattle feed), which mainly focused on cattle farming (ibid.: 20). The importance of the dairy industry decreased drastically in the 1960s because of lower dairy production costs and rising beef prices in South Africa (see ibid.). Thereupon, farmers concentrated solely on cattle and karakul farming. The cheese factory in Outjo hung on, heavily subsidised by the South African Ministry of Agriculture, and operated at a loss until the late 1970s (ibid.: 21).

Prosperous as karakul farming might have been for several decades, it came to an almost abrupt end.[45] With the complete collapse of the international karakul pelt market at the beginning of the 1980s, the majority of farmers switched to beef production only, complemented by a couple of meat sheep as well as goats, mainly for household consumption.[46]

Thus, while in the first half of the colonial period, a diversified farming economy was rather the rule, an increase in state-induced specialisation took place during the second half. This was due to South Africa's interest in the territory, international markets and ecological reasons. However, the shift towards a single economic strategy was never complete. Despite the generous conditions for loans and advances – such as for stock, implements and machinery – provided for in the land settlement programme (see Schmokel 1985; Silvester 1998; Botha 2000), it seemed difficult if not impossible for a lot of the farmers to develop and maintain their farms without additional income.[47] My interviews with farmers in the Outjo region revealed that many of them earned additional income either temporarily or continuously, especially in times of drought or if they had to pay off loans:

> O.L.: All those years, I wanted to farm. And then [in 1972], I got the chance to buy the farm. I had a little money, all the money I had earned I had invested in cattle already, which I had on leased ground on other farms. I had also leased a piece of land from my father. And in 1972, I made an application to the Land Bank to get a loan to buy the farm. I got it and that was how I got the farm.
>
> U.D.: How did it work with the loan?
>
> O.L.: I got the whole amount from the Land Bank and I had to pay it back annually. It was difficult. I went to work in between. The first ten years I worked.[48] Later I just farmed. But during the time I worked, my father kept an eye on the farm during the week … I had a loan to pay off in twenty-five years. And I needed exactly twenty-five years to pay it off.[49]

Farmers worked as drivers, road construction labourers, cattle inspectors; they built fences, drilled boreholes (when they owned drilling equipment) or helped out on other farms:

> What was a very bad time, during the time of my father, during my childhood, were those extreme years of drought, at the beginning of the 1960s, '63. The extreme drought with the foot-and-mouth disease … I still remember, my father … he had never learnt a real profession; he had been a farm trainee. Now, all the farms were closed! No meat, nothing was permitted to be transported anywhere. Then, the farmers, they had to earn income somewhere … Then the government started to construct a north–south and a east–west fence, through the whole country! A game-proof fence, east–west, north–south, to prevent further outbreaks. That time a lot of farmers did the fencing work. A lot of farmers got enrolled in that work. My father got along with the vet at Otjiwarongo and asked him for a job. And the vet said: 'Yes, you can come to work here!' They constituted the vaccination team. They drove through the whole country and worked out programmes. And every single farmer knew at

> what date the vaccination team was to come to his farm. That day, you had to have all the cattle in the kraal. And every single head of cattle was vaccinated against foot-and-mouth disease.[50]

Some had permanent employment in town and came to the farms only at weekends, leaving the farms to others during their absence:

> W.R.: [In] 1969, I went to Windhoek to start an apprenticeship.
>
> U.D.: To learn what?
>
> W.R.: Kind of diesel mechanic. Panel beater, something like that …
>
> U.D.: Why did you learn that?
>
> W.R..: In order to be able to repair things by myself. I *must* have a profession. My father stipulated it. And I agree. It has to be like that. You need a proper profession. If you are living on a farm and you have extreme droughts, you need to look for work somewhere else.[51]

Challenges, Risks and Crises: 'Starting from Scratch'

During the early period of agricultural development in the Outjo district, the potential settlers – coming from Germany or South Africa – were confronted with an alien environment, unpredictable and highly seasonal rainfall, wild animals, a low population density and not too much visible human impact on the land. In short, many of them thought of themselves as pioneers in a 'wilderness', or what J.M. Coetzee has called 'dream topographies' in which whites converted 'a howling wilderness into the productive garden of settler nationhood', thinking of 'South Africa as a vast, empty space', thereby ignoring former (black) inhabitants entirely (Coetzee, cited in McDermott Hughes 2006: 270).

The major challenge for the incoming European farmers was the development of a farm, which often only consisted of a surveyed area with corner stones or other signs indicating borders, lacking housing and sometimes even without any water:

> U.D.: Thus, your father built up the farm [which he obtained in 1907]. What did it look like in that area, were there farms around?
>
> E.L.: There was just bush. Maybe a path perhaps. Fransfontein was a police station. Otherwise, there were no farms. There was nothing! No fence, no path, no water, no house, nothing! And he lived in a tent in the beginning. And then he started to build. Then he burnt bricks. Clay was there. The oven was still there when I was a kid.[52]

The experience was not too different forty years later in the area, which had just been cut off from the game reserve and opened up for incoming settlers after the Second World War. One difference, however, was that those settlers applying for land during the later period could not choose spots with natural springs; water had to be drilled for:

> We came on the farm in 1947 … There was nothing. The land was not yet surveyed; we had to drill for water every five miles or so. There was no water, nothing. The first night, we slept under a *mopanie* tree; there were spiders, those things, but we slept on stretchers. The next morning we stretched a canvas. But that was the first place, where the drilling machine stopped, where we tried to get water. Unfortunately, the drilling man had drilled crooked. He had to try at another spot again. Later we got another man to drill. And it took us eight-and-a-half months to get the first water! And … you cannot build a house, you can do nothing without water. After eight-and-a-half months, we got the first water![53]

The techniques used to obtain water changed over time. Farmers have vivid memories how wells were dug or how dynamite was used, first by the settlers themselves and later by specialists. *Waterwysers* (experts using a divining rod, as above) as well as geologists were consulted to find the right spots. When the land of the game reserve was cut off and allotted to farmers in the second half of the 1940s, drilling machines were in use (see above).

The construction of houses, drilling of boreholes, instalment of wind pumps, building of dams and camps with the necessary fences for the animals and so on were the first steps to be taken:

> And if you would ask old Gert L. and old Danie de T. [farmers in the same area] … which one was the hardest piece of land on the 'red line',[54] they would without hesitation answer, 'Mooiplaas'; Mooiplaas is the hardest piece of land. I had worked at Graniet [farm], but really, the rock of Mooiplaas is the hardest I have ever encountered. I dug holes there to put in poles for the fence and so on; there are litres and litres of blood from Flip R. [the speaker] round those holes where the poles were put up, and those holes are so deep they look like they have been drilled, but I dug them with my bare hands. Those poles have stood the test of time, they don't move; where they stand, they stand.[55]

Incoming settlers frequently lived the first years in improvised shelters, the development of the other infrastructure – such as that for the livestock – being more urgent.

The defence of the herds – cattle, sheep and goats – against stock diseases, many of which had not yet been clearly defined and could not be successfully combated, was a major problem for the stock farmer. Poisonous plants constituted a further risk for the health of the animals. Archival records give evidence of the difficulties some farmers faced during these early years of settlement. For example, in 1934, a veterinarian visited several farms in the district because the settlers on these farms had formerly complained that their farms were 'unhealthy' and they had asked for land in exchange. The veterinarian compiled a report, documenting the occurrence of poisonous plants on the farms as well as the quality of the soil. He recommended that most farms should just be used for grazing

Figure 8.2: An improvised farm settlement at Gamkarab.

for a couple of months and that some supplements such as calcium phosphate or bone meal should be fed to the animals. In these records both the lack of scientific knowledge about animal diseases such as botulism (*Oesteomalacia*) and the absence of vaccination, as well as some individual farmer's ignorance of available scientific information, is obvious.[56]

> U.D.: And in those times, your father kept the livestock at Success [a farm in another area]?
>
> J: Until 1933, then he sold the farm.
>
> U.D.: Why did he sell it?
>
> J: There was botulism, and there was no medicine against it. He went bankrupt but not officially, he lost all his animals because of botulism since there was no medicine in those times. Lack of phosphate … [T]hat area of Success is called Sandveld, the soil is very sandy; there is little phosphate and the animals don't get enough phosphate and die.[57]

The protection of livestock against wild animals posed a further problem to the farmer. Farmers have vivid memories about the first days of settlement and the loss of livestock due to predators:

> We were the first settlers in that area, the very first settlers. It was still entirely wild. The first night – we had one hundred sheep and four cows – the first night, the lions came and killed two of the cows … It was full of lions; when you slept during the night, you could hear them roaming everywhere. Jackals and hyenas as well. That was a wild world … but we were young: I was 17, 18 years old, my brother a bit younger

> than me. And that was a nice adventure that time ... we were unmarried, just the two of us. We lived from meat and *papp* [porridge]; that was our food in those times.[58]

Under such unfamiliar circumstances and conditions of unpredictable and variable rainfall, the support of the state was an important precondition for starting and continuing the farming enterprise.

State Support as a Crucial Given

With the beginning of a systematic settlement policy after the war of 1904 to 1907, the German administration stressed the need for settlers to possess 10,000 marks as start-up capital, temporary housing and the required technical information. Schlettwein (1907: 62–67) calculated in detail the capital needs and profitability of a farm in the first ten years of operation. He stated that a farmer needed 92,000 marks to buy a farm and get it operating immediately, in a way comparable to practices in Germany (with housing, water, 250 head of cattle, 1,000 head of small stock, horses, pigs and chicken). Loans totalling 6,000 marks were provided to such farmers by the German government, to be repaid after six years. Failures were apparently common. In 1913, the Land Bank was established (Botha 2000: 235). The German government – well aware of the capital needs for starting a successful farming operation – was more limited in their financial assistance than the South African administration after 1920.[59]

Throughout the time of the South African administration, state support for commercial farming played a crucial role in the farm economy. With the establishment of the Mandate in 1921, the land settlement laws in force in South Africa were applied to SWA, and a land board was established to smooth the progress of settlement (Emmett 1999: 93). Very generous conditions – such as low minimal capital requirements, loans for building homesteads, advances for camps, kraals, reservoirs and dipping tanks – were provided for new settlers. Besides these general measures aimed at encouraging and facilitating the settlement of white farmers in the country, the state also helped with emergency programmes in cases of severe crisis. For instance, in 1935, after the disastrous droughts of the first half of the 1930s, which coincided with the Great Depression, relief measures were adopted to assist the farmers, such as the remission of interest and rent which became due and payable during the years 1932 to 1934. According to the Annual Report of the Land Branch, the relief measures involved a financial sacrifice by the administration of £144,733. All reductions and remissions took effect from 1 April 1935.[60] In 1952 the Farmers Interests Fund was introduced. It provided measures to combat soil degradation and additional support measures, such as subsidies and loans for dams, boreholes, water installations and fences (Botha 2000: 266). During the major crisis of 1959 to 1963, when the combined effects of prolonged drought and the outbreak of a foot-and-mouth epidemic affected farming operations in a devastating way, the administration assisted in various ways to buffer these effects, for example, with the provision of money for the repayment of obligations to private creditors and for livestock feed, and special rehabilitation loans for restocking.[61] Also during the serious drought of the early 1980s, which coincided with the collapse of the karakul market and a drop in cattle prices on the South African market, the administration granted comprehensive assistance to farmers (see Schmokel 1985: 106; Botha 2000). Schmokel, who looked into the economy of commercial agriculture in Namibia during colonial times, concluded that 'objectively seen, the farmers as a group, for all their hard work, were pensioners of the state, which placed and maintained them on the land for political reasons' (Schmokel 1985: 106). Not surprisingly, the farmers themselves did not feel like pensioners of the state. In my interviews state support was mentioned in passing without further discussion:

> The state provided the material for the people, to fence the border [with Etosha]. That was 1955 – 1954/55, or 1955/56. That time, the state provided the material, and each Boer put up his fence from corner to corner, the landmarks were already there ... The lions had killed the cattle, the Boers shot the lions until 1955/56 when the state provided the material.[62]

> Our farm was close to the red line, the border with Etosha. The biggest problem of course was the lions. The lions came in one night, just a week after – we didn't have anything yet, we didn't have money either – after we had bought twenty-five adult cows, one of them was calving in that particular night. In that night, nine lions came in, and they killed all the cows, twenty-four cows! The one calving was not together with the others. Then my husband went to borrow money from the Land Bank and bought another twenty-five cows and a bull. The third night, after my husband had come back with the cattle, the lions came again. They killed the bull and three of the cows![63]

State support was central to surviving the unpredictable incidences in this unfamiliar country and to achieve success as a farmer. Even so, the interviewed farmers stressed their own efforts to develop functioning farm enterprises as well as the difficulties they faced saving the money to pay back their loans. In short, they focused on the 'hard life' they had to cope with. The favourable conditions and the largely supportive attitude of the state were not really taken into account when these farmers assessed their former situation of farming in Namibia.

A Second Crucial Given: The Configuration of Labour Relations

The second essential precondition for farming in Namibia was the employment of a considerable number of farm workers, often with several families staying on a *werft*, an extra settlement or 'hamlet' on a farm. In the beginning, African people (above all Hai||om Bushmen, Damara and Herero) mainly provided the necessary labour force on the farms. For the early period of German rule, Schlettwein mentions that strict and fair treatment (combined with the vagrancy laws) helped to ensure that farmers always had enough workers (Schlettwein 1907: 178). Ethnic stereotypes often determined which workers were suitable for specific tasks. Schlettwein recommended so-called 'bastards' – born of Europeans and locals – as foremen (*Vormänner*), Herero as shepherds, while Damara and Ovambo should be used for horticulture, crop cultivation, brick making and so on (ibid.: 175–78).

As early as the late 1930s, with the expansion of farming, especially in the regions where karakul sheep were reared, migrant workers from the north were no longer restricted to the mining sector in the south but were also being channelled into the farming system (Schmokel 1985: 101; Silvester, Wallace and Hayes 1998: 25). Thus, contract workers from the northern regions complemented and partially replaced local farm labourers:

> U.D.: Did you have workers on the farm [where interviewee grew up]?
>
> M.A.: Bushmen, always Bushmen, Herero very few, but no Ovambo. Later we could get Ovambo from Ovamboland but you had to order him [the worker]. He came from Ovamboland to Grootfontein with SWANLA [South West African Labour Organisation, recruiting migrant workers from the north]. He got a card there: Outjo, the Glen [a farm], Otjikondo [area and police station]. And he was without clothes; you had to spend about two dollars for clothes and one dollar for a blanket. And then, he came with his new clothes, shirt and trousers and his new blanket and his card to the farm.[64]

Farms were 'microcosms of Apartheid' (Schmidt-Lauber 1998: 74), small systems in themselves. Some commentators have argued they might best be understood as 'total institutions' with little outside interference, whereby the farm owner was the ultimate authority (du Toit, cited in Suzman 2000: 56). Still, the single-farm system was, of course, never entirely independent from the political conditions under which it had emerged. The concept of *baasskap* is crucial to the understanding of social relations on farms in the entire region of Southern Africa (e.g., Guenther 1986; van Onselen 1992; Sylvain 1999; Suzman 2000). Guenther characterised *baasskap* as 'European economic domination combined with the racist values and practices of Apartheid' (Guenther 1979: 135). The institution of *baasskap*, with the concentration of patriarchal power in the hands of a single *baas* 'father figure', enabled the farmer to exercise a great deal of personal power, shaped by a paternalistic stance often combined with racial-evolutionary attitudes about the various degrees of civilisation in regard to the ethnically heterogeneous farm labourers (Sylvain 1999: 227). The farmers understood their role with respect to their workers as comprising one of training, education, assistance and, in the final instance, punishment and discipline.

With progressive settlement, farm work became one of the very few options for making a living for many local Africans. Labourers were expected to undertake a wide variety of skilled and unskilled tasks. These included working with livestock (for example, herding, milking, feeding), agricultural activities (such as planting and harvesting maize), garden work, construction and maintenance (especially fencing). Local women meanwhile were mostly engaged in domestic work. Remuneration for migrant workers was prescribed at fixed rates, consisting of a balance of rations and wages. Remuneration for local workers was more flexible, sometimes nothing more than small rations of food and tobacco.[65] Low payment was often justified in terms of workers' alleged laziness, especially that of Bushmen farm labourers. The most common ration package included weekly (or daily) supplies of mealie (maize) flour, sugar, coffee or tea, tobacco, sometimes milk and occasionally meat. Reliable local workers who had been on the farm for over a year were sometimes given livestock by the farmers as a bonus.

Due to the size of farms, farmers were highly dependent on their farm workers. Shepherds in particular were essential for successful farming operations. The strategies farmers used to keep their labour force operating differed considerably, and behaviour varied: corporal punishment or threats of it on the one hand were common; handing out benefits (better rations or salaries) and efforts to create a relationship of mutual trust were used as motivating methods on the other.[66] The 'reliability' of farm workers varied to a similar degree. They had some – albeit very limited – means for coping with the situation on farms. High mobility, theft and slaughtering livestock, shirking and refusals were part of the repertoire.[67]

In my interviews, farmers talked about their workers mainly in terms of complaints about them, often expressed along ethnic lines. Anecdotes about the funny, stupid or troublesome behaviour of individual workers were prevalent too. Thereby, ethnic stereotypes about farm workers were widespread but by no means consistent. It appears that they were rather drawn from personal experience with particular individuals and generalised according to ethnicity. The importance of farm workers and farmers' evident dependency on them were not mentioned. The topic was presumably avoided in order to prevent awkward discussions of unequal power relations.

Notwithstanding the favourable conditions white farmers enjoyed regarding applying for loans and subsidies, as well as the availability of a cheap labour force, farming in the Namibian environment remained a difficult and risky enterprise over the years.

Help Mekaar: Farmers' Associations and Support Networks

Social networks played a crucial role in providing assistance to farmers. The low population density combined with the effects of the Apartheid system, which restricted contact with non-white people, were factors which – along with the vulnerable condition of the farming system – augmented the significance of mutual assistance between farmers. Due to the long distances to the nearest town and the lack of infrastructure, one's direct neighbours were an important source of help on the farm with regard to transport and sharing information.

Networks between members of the same language or cultural group – Germans on the one hand, Afrikaners on the other, as well as a few English-speaking people – were the closest. Political affiliations also played a role. Some Afrikaners had left South Africa after the British victory at the beginning of the twentieth century and felt closer to the Germans:

> There was one common factor, and that was that most of them were Afrikaans speaking. And the division was not so strong along those lines then. But before the Second World War, you had the two separate strong opposing groups, the pro-German and the anti-German. The pro-German were not always so much pro-German than they were sometimes more anti-English … and therefore they had left South Africa because they did not like the English colonisation of South Africa.[68]

Most Germans could speak Afrikaans and most Afrikaners could at least understand German, and the dependency on neighbours was considerable. Language boundaries were permeable and networks (sometimes also marriages) cut across these boundaries.

The most institutionalised organisations were the local farmers' associations (*Boere Verenigings, Farmervereine*).[69] In the Outjo district there were at least five farmers' associations.[70] Most, if not all, of the farmers were members of a farmers' association. First of all, the regular meetings of farmers' associations served predominantly for information exchange. National and international experts were invited to inform them on specific topics, such as stock breeding, veterinary recommendations, rangeland management and so on. Alternative access to new information and the latest research results – either through regular visits to town and mouth-to-mouth diffusion or via newspapers, the internet and so on – was limited in former times. The farmers' associations were members of the Namibian Agricultural Union (Namibian Landbou Unie). They constituted lobby groups and also held political sway. Additionally, the associations were mutual support networks. The members were organised and prepared to help each other. For example, there was a clear allocation of responsibilities and obligations in the case of bush fires.

The Predictably Unpredictable: Droughts and Ways of Coping

Over the years, droughts have been one, if not the, major risk for livestock farmers. According to the interviews, the worst droughts before independence remembered by the farmers were those of the early 1930s, which coincided with the Depression, the drought at the end of 1950s and beginning of the 1960s, which was accompanied by an outbreak of foot-and-mouth disease in the second year, and the droughts in the early 1970s and the early 1980s.[71] As rainfall is very patchy in Southern Africa, it could happen that a farmer received almost no rainfall in one year while a neighbour had sufficient rain, a major risk typical for freehold farmlands in arid regions. Working temporarily to earn additional income and selling cattle during drought periods (since the grazing was not sufficient) were common strategies employed by farmers. Because prices were low, selling cattle during droughts meant a considerable loss for farmers. Those farmers, who produced oxen for the meat market, were frequently forced to sell their oxen and had to start the whole production cycle from scratch.

> W.B.: I farmed with oxen. The drought forced me to sell the oxen, I had to sell all of them, just the cows were left over. That was around 1980. Then I started again to raise oxen [it takes some years for the oxen to reach the required size], then again there was a drought. I had to sell and sell and sell again. And I'm not yet back to oxen.
>
> U.D.: Is that a cycle which repeats again and again?
>
> W.B.: Yes. Okay, you can survive a drought with oxen, but then you need to lease additional grazing. Otherwise it doesn't work, you *have* to sell.[72]

The movement of livestock to other areas was one strategy for coping with the lack of rainfall (see above). During drought periods, a considerable number of farmers trekked to un-surveyed Crown land and other areas, such as east of Kamanjab, east of Grootfontein and the so-called Mangetti Block, as well as to the Etosha game reserve (for example, in 1962), sometimes trekking over several hundred kilometres. The outbreak of foot-and-mouth disease at the beginning of the 1960s during the time of a severe drought had the disastrous consequence that farmers were neither allowed to sell nor to move their cattle anymore. Those who had their cattle at emergency grazing places in other areas had to stay there until the outbreak was under control.

Trekking was not without risk, and some farmers reported suffering from a considerable loss of animals, in particular due to poisonous plants in the respective area.[73] This is one farmer's recollection of the situation in 1959:

> M.A.: [B]ecause of the drought, all the people from the Outjo area trekked to the government farms at Grootfontein.
>
> U.D.: All the farmers?
>
> M.A.: Many farmers. My brother and I as well. We went with sixteen-hundred head of cattle to Grootfontein and came back with thirteen-hundred.
>
> U.D.: Why?
>
> M.A.: Poisonous plants. *Maakgou* [*Dichapetalum cymosum*, Gifblaar family]. The cattle born in that area, they did not feed on those plants, they knew. But incoming animals, it is poison for the heart. That *maakgou* and the lions, there were many lions catching the cattle in that area. There were a lot of lions and no fences. Our Bushmen workers had to move behind the cattle the whole day long, so that the animals could eat, and they had to bring them back to the kraal in the evening again.[74]

Trekking was done on foot, with lorries and, sometimes, partly by train. At times, emergency grazing was supplied by the government on un-surveyed Crown land.[75] On occasion, farmers hired grazing from private owners.

U.D.: What else did you do during droughts [except getting employment]? Did you also lease grazing?

W.R.: Yes, I also leased. Seven years in the 1980s. Seven years, I did not have a single head of cattle on this farm![76]

In retrospect, droughts were perceived as the main risk, threatening the whole business of a farmer, which could force the individual farmer to start from scratch again with regard to livestock. Other risks were particular to specific periods of time. Just to mention a few: The instability of markets for some products such as karakul pelts was one of them, changing over time. The Namibian liberation war also had some impact on farming, and the entrance of SWAPO (South West African People's Organisation) liberation fighters into the farming area was perceived as a risk during the 1970s and 1980s, though some areas in the Outjo region were less affected than others.[77] Bushfires constituted another risk. Fire breaks and cleaning bush around dwellings, as well as the tough organisation of farmer networks to help each other, were parts of the coping repertoire. The major crises were a combination of droughts and other political, ecological or economic factors.

Consolidation and Thereafter

During the years of settlement, adaptation to the unfamiliar environment progressed. Veterinary services had already been established in the early years to deal with animal diseases and related problems (RSA 1964 [Odendaal Report]: 269). With increasing scientific knowledge, medication and vaccination became available. For instance, according to the interviewees, vaccination against botulism (*Lamsiekte* in Afrikaans), one of the major threats to livestock, became available in the early 1940s. Stricter legislation and control measures with regard to foot-and-mouth disease also showed its effects after the outbreak of the disease at the beginning of the 1960s. The development of infrastructure on farms helped farmers in many ways. Fences and/or keeping guard helped to protect livestock against wild animals. Improved rangeland practices – the camp system – replaced the former free-run system (Bähr 1968: 73).

Archival material, including correspondence between farmers and the Land Branch of the administration for instance,[78] gives evidence of the struggles farmers went through to build up their farm from scratch and to pay back their debts. Also, the frequent changes in farm ownership evident in the title deeds data indicate that many farmers indeed failed in their attempts to build up a sustainable farming business. The return of one farm to the government and the leasing or buying of another one, also visible also in Deeds Office data, was another attempt to survive as a farmer during the early years of settlement when farmland became progressively opened up for settlement. Interviewees also stressed the thread and existence of failure:

> You have to make plans, how to survive, how to cope. Making plans, otherwise you get stuck. Not all farmers pulled through. Maybe around twenty started to farm at the same time as me. I don't believe that more than ten survived as farmers. The others left the farms, were seeking employment – the ones worked here, others worked there, road construction for example. They simply could not make it. If you don't have the ability to adapt, if you are not able to make a plan, then you'll sit there afterwards.[79]

Reasons for failure or differences in performance and varying degrees of success in farming were due to the dynamics of manifold aspects. A lack of knowledge of the environment and less favourable conditions regarding soil, water availability, poisonous plants and so on, and hazards such as droughts and animal diseases, were important factors. Access to financial and physical capital and the personal characteristics of individual farmers, such as long-term planning skills, agricultural knowledge or experience, constituted further vital factors. Especially at the beginning, for many farmers the most urgent problems concerned their daily needs. Thoughts concerning longer-term issues were barely on the agenda; long-term planning was often not possible. Therefore, a solid portion of 'luck' was important at the beginning.

Those people who continued to subsist as farmers had 'adapted' to the alien environment, they had appropriated and cultivated the wilderness and had developed strategies to cope with specific risks and crises. According to my understanding, the concept of adaptation is central to many farmers' ideas of how to survive. It implies a considerable degree of flexibility for reacting to unpredictable or unexpected events and crises. During the colonial period, the segregationist system, and later Apartheid, formed the essential background for the successful operation of farming in offering generous assistance to farmers on the one hand and providing cheap labour on the other.

With independence in 1990 and the termination of Apartheid legislation, major changes in the political economy of Namibia occurred that had profound effects on white – previously advantaged – farmers. Most importantly, key shifts took place in policies regarding land and commercial agriculture. With the implementation of the Agricultural (Commercial) Land Reform Act of 1995 (since amended several times, most recently in 2003), and the formulation of related policies, such as the National Land Policy of 1998 and the National Resettlement Policy of 2001, the new government has tried to address the inequitable ownership of commercial land. The most essential component of the Land Reform Act for already operating commercial farmers is the provision that any commercial farm offered for sale has to be offered to the government for the purposes of resettlement first (LAC 2005: 6). Furthermore, the Act, as amended in 2003, prescribes that any farm identified as suitable for the resettlement of the landless and poor may be expropriated following certain procedures (see ibid.: 18–19). Although expropriation up to 2007 was limited to three farms,[80] it is perceived as a new risk by many farmers. The passing of the Labour Act (1992) also played a crucial role in these changes.

A lot of established white farmers find it difficult to adapt to the new circumstances. Above all, land reform is still in progress and the objectives are overambitious and not well defined. Additionally, land reform lacks a systematic approach in various aspects, as well as transparency.[81] Thus, it is difficult for farmers to anticipate which specific circumstances they have to adapt to. During all periods, unpredictability causes the most stress. On the other hand, the problem of adapting is at times also due to the observable defiance of some farmers, with their specific political opinions and glorification of the past.

Conclusion

It should have become clear that the European appropriation and cultivation of land in the Outjo district – and to some extent this holds true for the whole country – was a major challenge for the new settlers and, from the current perspective, is presented as an important achievement. Dangers and risks in colonial times mainly concerned struggles with nature, droughts, animal diseases, bushfires and so forth. The whole settlement process took place in an unfamiliar environment for each and every farmer,[82] and involved new diseases, new grazing conditions (including dangerous plants) and the extreme unpredictability of rain. It was mainly approached with a trial-and-error attitude, an approach which was – due to colonial interests and policies – possible during the early years of European settlement when land for settlement was still abundant and when the state played a crucial role in assisting white farmers.

Later on, the unfamiliar became familiar. Those who survived as farmers found ways to adapt to the new situation and to cope with regularly occurring risks. Over the years, farmers learned to anticipate and to prepare. The crucial role which structural factors played in supporting people's coping strategies (in particular, state support and legislation) was rather neglected by farmers of European origin during the interviews. This is due to the current position of European farmers within independent Namibia, where affirmative action for formerly disadvantaged people is promoted and taken. The previously 'advantaged' farmers have an interest in stressing that they were not as previously 'advantaged' as they are perceived to have been by the rest of the Namibian public. They are keen to ensure that it is acknowledged that they built up their farms with their own hands and did not get everything for free during the periods of German and South African rule. They are in a position in which they need to justify their right to possess a farm.

Notes

The research for this chapter was undertaken between 2005 and 2007 within the framework of the special research programme 'Arid Climate Adaptation and Cultural Innovation in Africa (ACACIA)' (Sonderforschungsbereich 389) at the University of Cologne, funded by the German Research Foundation (Deutsche Forschungsgemeinschaft).

1. I will use the terms 'white' farmers or settlers and settlers of 'European origin' synonymously here, including the 'whites' who came to Namibia from South Africa. During colonial times, and especially during the Apartheid Era, skin colour was a distinguishing feature of politics and society. Nowadays, the terms 'previously advantaged' or 'established' farmers refer to the same group of people. Opposed to these are the 'previously disadvantaged', 'emerging' or 'black' farmers.
2. For an overview of definitions of risk, see Bollig (2006).
3. Transcripts from these interviews are cited below; all translations are my own.
4. Remarkably, the 1952 farm register contains 137 women (14 per cent of all cattle brand owners) who had their own brands at their disposal, compared to 849 men. I did not get to know any women who had not been married running and owning a farm of their own, although widows or divorced women sometimes continued to manage the farm after their husband had died or had left.
5. The data on the area of Outjo district comes from: http://www.statoids.com/una.html (retrieved 5 April 2007).
6. Data provided by the Meteorological Service of Namibia, Windhoek.
7. Population figures are drawn from the National Planning Commission's data set from 2001 (http://www.npc.gov.na/census/census_localities.htm, retrieved 5 August 2004).
8. It is difficult to calculate a precise number with respect to the first farming units. Farms were allotted gradually, and it might be that when the last farm in the district was allotted some of the first farms had already been divided or consolidated. Thus the number does not refer to an exact date but to the first farms and the size of these.
9. This was done following the plans of the Odendaal Commission, which aimed at a spatial segregation of the various ethnic groups in Namibia, similar to the Apartheid system in South Africa; see RSA (1964) [Odendaal Report].
10. It would go beyond the purpose of this chapter to analyse changes in colonial and postcolonial land and settlement policy to explain these consolidations and divisions.
11. Plots are small allotments, situated on town lands, the latter falling under the Local Authorities Act (1992); rules and regulations are determined by the municipality or the town council. Farming units fall under the Agricultural (Commercial) Land Reform Act (1995, amended several times since then).
12. For a settlement history of the Fransfontein area, which was originally part of the district, see Schnegg (2007).
13. Swakopmund Library, Namibia (SL), G.P. Kruger, 'Outjo 1885–1960' [unpublished manuscript] (n.d.), p.15.
14. National Archives of Namibia (NAN), ZBU W III A.1.
15. See the 'Kriegskarte von Südwestafrika', originally published in 1904 and republished in 1994 by the NAN.
16. Nowadays, the groups formerly called Bushmen are referred to as San in official discourse.
17. NAN, ZBU WII O2, Landespolizei Windhoek an den Kaiserlichen Gouverneur, 15 April 1912.
18. Under German rule, police protection was confined to those areas falling within the sphere of influence of the railway lines and main roads. The 'police zone' separated those parts of Namibia that were being progressively settled by white farmers from those areas where indigenous land-use systems were left largely intact; these included the Kaoko, Ovambo, Kavango and Caprivi regions (Werner 1993: 139). Under South African martial law, Proclamation 15 of 1919 was issued, proscribing the crossing of the line marking the border of the police zone without official permission. This became known as

the 'red line'. The boundary shifted numerous times during the period of South African rule as white land holdings in the north-central areas expanded (Silvester, Wallace and Hayes 1998: 3). For a history of the 'red line', see Miescher (2009).

19. Referring to 'the means of support' at the 1923 Permanent Mandates Commission meeting of the League of Nations in Geneva, Lord Lugard, the British colonial administrator and member of the Commission, remarked that this meant that all 'Bushmen' were de facto vagrants and thus subject to forced labour (Gordon 1998: 52).
20. NAN, SWAA, A50/27, Proclamation No. 32 (1927).
21. The Proclamation of 1920 was replaced by the Proclamation 310 of 1927; see Deutsche Uebersetzung des ausserordentlichen Amtsblattes fuer Südwestafrika, Windhoek, 22 December 1927.
22. This option was only open to poor people. One could also acquire a farm through the Land Bank, whose interest rates were slightly higher, or by private purchase (Botha 2000: 249).
23. NAN, LAN 579, 1379, Klein Omburo nr. 148, Outjo: General File, 18 August 1920, Magistrate, Outjo, to Secretary, Windhoek.
24. See Miescher (2006) for more details, including the prosperity of the farming business in the reserve.
25. Fransfontein (initially 10,000 hectares, later increased to 36,188) was one of the smaller 'native reserves' (RSA 1964 [Odendaal Report]: 69; Werner 1998: 79); many others covered several hundred thousand hectares (for example, Berseba, Epukiro, Aminuis, Otjohorongo) before the reserves were consolidated into homelands following the Odendaal Commission (RSA 1964 [Odendaal Report]).
26. For an example of illegal squatting in a neighbouring district, see Dieckmann (2007: 139–43).
27. NAN, LAN 1/1/89 31, 53 vol. III: Annual Reports Land Settlement.
28. See the Annual Report Land Resettlement of 1937.
29. For more details, see Miescher (2009: 265–336).
30. In 1922, Imperial Cold Storage concluded an agreement with the SWA administration for the erection of cold storage works at Walvis Bay. In terms of the agreement, the latter was granted the sole right concession for the export of livestock for three years in addition to a fifteen-year term to export meat from the cold storage works. The SWA Cold Storage and Stock Farmers Company was a subsidiary of Imperial Cold Storage's farming and manufacturing operation. See Rawlinson (1994: 182).
31. NAN, LAN 957, 3082.
32. Grazing licences could be obtained from the local magistrate to graze animals on named segments of Crown land within each district, and these were issued on a monthly basis. The number of livestock was stipulated in the licence and the costs of these varied accordingly (Silvester 1998: 102; Miescher 2009: 270).
33. NAN, LAN 1/1/89 1534, file 4649.
34. NAN, LAN 1436, 4293, vol. III, IV.
35. NAN, LAN 1/1/89 31, 53 vol. III: Annual Reports of the Land Branch of the Administration.
36. For southern Namibia, see Silvester (1998: 105).
37. Boskop farm, for example, was gazetted in 1927 and again in 1937. The later owner had a grazing licence from 1935. The government title deed was finally issued in 1953 (NAN, LAN 1047 3365/1, Boskop, No. 235).

38. *Baken* were signs, used in the early years, to mark the corners of the farms. The proportions and material of these were exactly determined in Proclamation 78 (1920), 'Deutsche Uebersetzung des Amtsblattes für das Schutzgebiet von Süd-West Afrika', Windhoek 28 February 1921, p.44.
39. Letter to the author, January 2007.
40. SL, Kruger, 'Outjo 1885–1960', p.16. It is not clear if the number refers to cattle belonging to European farmers only or to Africans as well.
41. NAN, LAN 1/1/89 31, 53 vol. III: Annual Reports Land Settlement.
42. Interview with I.B. (b.1935), 9 May 2006.
43. Apparently this was not limited to the Outjo district. See Lau and Reiner (1993: 22) for a discussion of the political motives behind this national move.
44. Interview with J.B. (b. 1943), 24 August 2006.
45. Karakul fleeces received some bad press in Europe and the US at this time.
46. In recent years the karakul economy has grown; see, e.g., coverage of this in the *Economist* (http://www.economist.com.na/content/view/702/53/, retrieved 13 November 2008).
47. Interest rates were set at 4 per cent, compared to rates of 11 to 12 per cent offered by private banks; the purchase price had to be paid back in thirty years (NAN, LAN 1/1/89 31, 53 vol. III: Annual Reports Land Settlement).
48. O.L. worked for a South African construction company which had a branch in Namibia as well.
49. Interview with O.L. (b. 1943), 11 November 2006.
50. Interview with H.R. (b. 1949), 22 August 2006.
51. Interview with W.R. (b. 1949), 22 August 2006.
52. Interview with E.L. (b. 1918), 12 July 2006.
53. Interview with B. de B. (b. 1922), 1 November 2005.
54. This particular farm borders the 'red line' which in this part of the country is also the border of Etosha National Park.
55. Interview with F.R. (b. 1929), 22 June 2006.
56. NAN, LAN 579, 1378/2 (vol. II), Aimab No. 124, Outjo Allotment File, ceded to V.G. Harding.
57. Interview with J.L. (b. 1929), 22 May 2006.
58. Interview with W.B. (b. 1917), 20 October 2005.
59. Among other policies, the South African administration sought to solve the 'poor white problem' in South Africa with its settlement programme in Namibia. At least at the initial stage, little emphasis was put on ensuring that these emigrants became successful farmers in the mandated area.
60. NAN, LAN 1/1/89 31, 53 vol. III: Annual Reports Land Settlement.
61. E.g., NAN, LAN 53/1, Budget Speech 1961, 1962; see also Schmokel (1985: 105). In 1961, a cordon fence of 2,720 kilometres was erected within seventeen months by farmers and farm workers. The administration supplied the materials (Rawlinson 1994: 98).
62. Interview with O.L. (b. 1943), 11 November 2005.
63. Interview with B.B. (b. 1924), 1 November 2005.
64. Interview with M.A. (b. 1938), 3 November 2005.
65. E.g., NAN, LGR 3/1/7 2/20/9, Magistrate, Grootfontein, to the Secretary for SWA, Windhoek: Annual Report on Native Affairs, 1946. There is no reason to believe that the situation in the Outjo district differed from that in the neighbouring Grootfontein district.

66. For the relevance of trust in the principal–agent relationship, under which the farmer–labourer relation can be subsumed, see Ensminger (2001).
67. For more details, see Dieckmann (2007: 217–30).
68. Interview with J.B. (b. 1943), 24 August 2006.
69. Farmers' associations have a long tradition. The first was established in 1898 with its headquarters in Windhoek and had branches throughout the territory (Rawlinson 1994: 20).
70. Geographical area was the central criterion for the various farmers' associations; thus they were also neighbour associations.
71. These must have been the most devastating droughts in the territory (Rawlinson 1994: 98–100, 138).
72. Interview with W.B. (b. 1945), 22 November 2005.
73. This was reported by a number of my interviewees. See also, e.g., NAN, LAN 1/1/89 53, Annual Reports Land Branch, vol. 5.
74. Interview with M.A. (b. 1938), 3 March 2005.
75. E.g., NAN, LAN 1/1/89, SU 32, 53/1, Budget Speech. The administration's general policy of keeping farm land for emergency grazing was already under pressure in the 1940s (Rawlinson 1994: 24). However, in 1959 emergency grazing was provided to farmers (NAN, LAN 1/1/89 31, 53 vol. III: Annual Reports Land Settlement, Afdeling Lande: Statistike I.V.M. Droogtetoestande, BL. 103, Tydperk 1 January 1959–30 November 1959).
76. Interview with W.R. (b. 1949), 22 August 2006.
77. SWAPO fighters had special 'corridors' when entering the police zone from the north. The area immediately south of the Etosha Pan was reportedly less affected than other areas. Again, social networks played a crucial role in this regard. So-called commandos were called upon to assist farmers, on whose farms liberation fighters had been reported.
78. NAN, LAN 579, 1378/2 vol. 1, Aimab 124, J.A. Lindeque, Allotment file.
79. Interview with W.B. (b. 1917), 20 October 2005.
80. See Harring and Odendaal (2007) for an analysis of the expropriation process.
81. For further detail on the progress and problems of land reform, see Dieckmann (2011).
82. This also holds true for those South African farmers who tried to start from scratch in Namibia – the circumstances were not familiar either to them or their animals.

References

Anon. 2001. 'The History of Outjo Town'. Outjo: Municipality of Outjo.

Bähr, J. 1968. *Kulturgeographische Wandlungen in der Farmzone Südwestafrikas*. Bonn: Ferd. Dümmlers Verlag.

Bollig, M. 2006. *Risk Management in a Hazardous Environment: A Comparative Study of Two Pastoral Societies*. New York: Springer.

Botha, C. 2000. 'The Politics of Land Settlement in Namibia, 1890–1960', *South African Historical Journal* 42(1): 232–76.

Dieckmann, U. 2007. *Hai||om in the Etosha Region: A History of Colonial Settlement, Ethnicity and Nature Conservation*. Basel: Basler Afrika Bibliographien.

——— 2011. 'The Hai||om and Etosha: A Case Study of Resettlement in Namibia', in K. Helliker et al. (eds.), *Land Struggles and Civil Society in Southern Africa*. New Jersey: Africa World Press.

Dierks, K. 1999. *Chronology of Namibian History: From Pre-historical Times to Independent Namibia.* Windhoek: Namibian Scientific Society.
Emmett, T. 1999. *Popular Resistance and the Roots of Nationalism in Namibia, 1915–1966.* Basel: Basler Afrika Bibliographien.
Ensminger, J. 2001. 'Reputations, Trust, and the Principal Agent Problem', in K. Cook (ed.), *Trust in Society.* New York: Russell Sage Foundation, pp.185–201.
Gordon, R. 1998. 'Vagrancy, Law and "Shadow Knowledge' Internal Pacification 1915–1939', in P. Hayes et al. (eds), *Namibia under South African Rule: Mobility & Containment 1915–1945.* Oxford: James Currey Ltd, pp.51–76.
Guenther, M. 1979. *The Farm Bushmen of the Ghanzi District.* Stuttgart: Hochschul Verlag.
_______ 1986. *The Nharo Bushmen of Botswana: Tradition and Change.* Hamburg: Helmut Buske Verlag.
Harring, S.L., and W. Odendaal. 2007. *"No Resettlement Available": An Assessment of the Expropriation Principle and its Impact on Land Reform in Namibia.* Windhoek: Legal Assistance Centre.
Köhler, O. 1959. *A Study of the Grootfontein District.* Pretoria: Government Printer, Ethnological Publications.
Külz, W. 1909. *Deutsch-Südafrika.* Berlin: Wilhelm Süsserott.
LAC. 2005. 'Our Land We Farm: An Analysis of the Namibian Commercial Agricultural Land Reform Process'. Windhoek: Legal Assistance Centre.
Lau, B., and P. Reiner. 1993. *One Hundred Years of Agricultural Development in Colonial Namibia: A Historical Overview of Visions and Experiments.* Windhoek: National Archives of Namibia.
McDermott Hughes, D. 2006. 'Hydrology of Hope: Farm Dams, Conservation, and Whiteness in Zimbabwe', *American Ethnologist* 33(2): 269–87.
Miescher, G. 2006. "The Ovambo Reserve Otjeru (1911–1938): The Story of an African Community in Central Namibia', Basler Afrika Bibliographien Working Paper 2006/1.
——— 2009. 'Die Rote Linie: Die Geschichte der Veterinaer- und Siedlungsrenze in Namibia (1890er bis 1960er Jahre)', PhD dissertation. Basel: Universität Basel.
Rawlinson, J. 1994. *The Meat Industry of Namibia 1835–1994.* Windhoek: Gamsberg Macmillan.
RSA. 1964. 'Report of the Commission of Inquiry into South West African Affairs' [Odendaal Report]. Pretoria: Government Printer, Republic of South Africa.
Schlettwein, C. 1907. *Der Farmer in Deutsch-Südwest-Afrika: Ein Leitfaden für Anfänger.* Wismar: Hinstroff'sche Verlagsbuchhandlung.
Schmidt-Lauber, B. 1998. *'Die verkehrte Hautfarbe': Ethnizität deutscher Namibier als Alltagspraxis.* Hamburg: Dietrich Reimer Verlag.
Schmokel, W.W. 1985. 'The Myth of the White Farmer: Commercial Agriculture in Namibia, 1900–1983', *International Journal of African Historical Studies* 18(1): 93–108.
Schnegg, M. 2007. 'Battling Borderlands: Causes and Consequences of an Early German Colonial War in Namibia', in M. Bollig, O. Bubenzer, R. Vogelsang and H.P. Wotzka (eds), *Aridity, Change and Conflict in Africa.* Cologne: Heinrich-Barth-Institut, pp.247–64.
Silvester, J. 1998. 'Beasts, Boundaries and Buildings: The Survival and Creation of Pastoral Economies in Southern Namibia 1915–35', in P. Hayes et al. (eds), *Namibia under South African Rule: Mobility and Containment 1915–1946.* Oxford: James Currey, pp.95–116.
Silvester, J., M. Wallace and P. Hayes. 1998. '"Trees Never Meet": Mobility and Containment, An Overview 1915–1946', in P. Hayes et al. (eds), *Namibia under South African Rule: Mobility and Containment 1915–1946.* Oxford: James Currey, pp.3–48.

Suzman, J. 2000. *'Things from the Bush': A Contemporary History of the Omaheke San*. Basel: Basler Afrika Bibliographien.

Sylvain, R. 1999. '"We Work to Have Life": Ju|Hoan Women, Work and Survival in the Omaheke Region', PhD dissertation. Toronto: University of Toronto.

van Onselen, C. 1992. 'The Social and Economic Underpinning of Paternalism and Violence on the Maize Farms of the South-western Transvaal, 1900–1950', *Journal of Historical Sociology* 5(2): 128–60.

Werner, W. 1993. 'A Brief History of Land Dispossession in Namibia', *Journal of Southern African Studies* 19(1): 135–46.

——— 1998. *'No One Will Become Rich': Economy and Society in the Herero Reserves in Namibia.* Basel: Basler Afrika Bibliographien.

Zimmerer, J. 2000. *Deutsche Herrschaft über Afrikaner*. Münster: LIT.

Chapter 9

The Political Ecology of Specialisation and Diversification

Long-term Dynamics of Pastoralism in East Pokot District, Kenya

Michael Bollig and Matthias Österle

Introduction

Highly specialised livestock herders shaped the pastoral economies and societies of East Africa in the nineteenth and twentieth centuries. Spatial mobility, subsistence orientation, egalitarian relations between household units, common property resource management, age sets and/or generation set organisation, extensive networks of stock loans and gerontocratic political organisation have been key elements of East African pastoral livelihoods (Dyson-Hudson and Dyson-Hudson 1980; Fratkin 1997; Spencer 1998). High degrees of spatial mobility, subsistence mainly dependent on livestock, the symbolically prominent role of livestock-based food ways, and the dominant position of livestock, namely cattle, in all social exchanges and rituals were deemed as essential defining principles of this form of specialised pastoralism. Specialised pastoralism can be differentiated from other forms of pastoralism practised in East Africa. The range includes, for example, a more sedentary mode of pastoralism as was practised by the so-called Sirikwa of the Western Highlands of Kenya between the thirteenth and the eighteenth century (see Lane, this volume), diverse forms of pastro-foraging as documented in ethnohistorical and archaeological accounts (also Lane, this volume) or combinations of livestock husbandry and agriculture as described for the Sebei (Goldtschmidt 1976) and the Kamba (Tiffen, Mortimore and Gichuki 1994). With the emergence of highly specialised pastoralism in the nineteenth century, ethnic boundaries became visualised through complex styles of clothing and beadwork.[1] Although ethnic differences have been emphatically played out in visual symbols, ties cross-cutting ethnic boundaries have also been essential and ethnic shifting has also taken place regularly. While archaeological findings

(Robertshaw 1990; Smith 1992) suggest that such adaptations were long-term adjustments to the environmental challenges of the East African savannah, and that there has been long-standing cultural continuity in specialised pastoralism in East Africa, recent anthropological accounts suggest that this very kind of pastoral specialisation is more recent than previously thought and only developed during the past 200 or 300 years. According to oral traditions of typical representatives of specialised East African pastoralism, such as the Maasai (Galaty 1993), the Turkana (Lamphear 1988), and the Pokot (Bollig 2006), contemporary pastoral communities became specialised nomadic herders only about 200 years ago.

Recent accounts of East African pastoralism underline the fact that specialised herding communities are presently undergoing a number of groundbreaking changes: sedentarisation and a trend towards agriculture is widespread, but livestock husbandry continues to be a key economic strategy, and communal pastures are often dissected into village territories or even freehold farms. There are several reports of pastoral communities opting for sedentary agriculture.[2] Labour migration to urban centres and/or to rapidly expanding agro-industrial complexes are further alternatives to herding (Fratkin 1997). Livestock husbandry is now in many cases organised from a sedentary farmstead, often implying a shift of focus from cattle to small stock. Economic changes are paralleled by rapid political changes: East African pastoralists are thoroughly integrated into the political structures of the state nowadays; since the late 1990s political parties have been campaigning in pastoral areas, and a growing and often aggressively propagated ethnicity is fostered by incipient political elites in the former pastoral peripheries. In many instances, global concerns over nature conservation have also had an immediate impact on the political ecology of East African rangelands.

By combining Bollig's fieldwork (1987 to 1989, and several short-term studies in the 1990s) with the restudy of Österle (2003 to 2005) and a larger interdisciplinary project starting in 2010 (Greiner 2012), we will make use of the advantages of longitudinal studies. By combining the results of Bollig's study, which looked at long-term economic and social change, and Österle's and Greiner's focus on recent socio-economic dynamics we will cover a span of about 200 years. While many ethnographic studies focus either on the emergence of specialised pastoralism (e.g., Lamphear 1976) or at its recent changes (e.g., Fratkin and Roth 2005), we would like to detail one case here, sketching the emergence, expansion, involution and demise of specialised pastoralism among the Pokot of northwest Kenya. The story we want to tell describes major economic and cultural ruptures and phases of swift reorganisation. Our theoretical interest lies in considerations pertaining to complex socio-ecological dynamics. What are the causes for change and what (and/or who) are its main drivers? At what points do systems change and in which direction do they change?

The Emergence and Development of Specialised Pastoralism

Oral traditions of the Pokot which depict the history of the group in general and/or the history of specific clans and lineages (*kililyet*) suggest that the emergence of specialised nomadic pastoralism and the ethnogenesis of the Pokot took place in the late eighteenth or early nineteenth century. The recording of these oral traditions took place in the late 1980s and early 1990s. These traditions imply that these processes may have started in the last quarter of the eighteenth century and came to an end during the second quarter of the nineteenth century.

The reconstruction of the eighteenth and nineteenth century history of the pastoral Pokot is based on oral traditions. These were recorded with a number of elders in many sessions. Whereas some interviews (about twenty) were mainly focused on the reconstruction of lineage histories, with a limited number of key informants a much wider range of topics was elaborated. Informants generally belonged to the oldest generation set, Chumwö, or to the oldest men of the following generation, Koronkoro.[3]

Becoming Pokot

Many ancestors of the elders interviewed in the late 1980s were neither pastoralists nor necessarily Pokot according to interviewees' recollections. While the dating of pastoralisation is still problematic in the absence of archaeological data, tentative dating via generation sets and genealogies suggests that a transition towards pastoralism took place some six to seven generations ago. Assuming that a generation lasts some twenty-five to thirty years, this process should have taken place about 150 to 210 years ago. At that time the community, later to be known as the pastoral Pokot (*pipö tich*, 'cattle people' in contrast to *pipö pagh*, 'grain people'), emerged from a number of others. According to oral traditions most of these communities had diversified economies with varying combinations of small stock keeping, cattle husbandry, small-scale rain-fed agriculture and hunting. Many narratives suggest a main region of origin between Mount Sekerr, the Cherangani Hills, the Suam River and Mount Elgon (Bollig 1990: 74). Other ancestors were identified as Turkana-speaking hunter-gatherers and horticulturalists from the Turkwell River, but also as pastoral Turkana, Sebei, Dodos, Karimojong and Rendille. Several lineage traditions report disastrous famines, earthquakes and/or lost animals, which triggered a move from highland settlements towards the lowlands or (in a pastoral context) away from well-established ranges. These ancestors joined an incipient pastoral community which had 'learnt' how to be pastoralist – including a livestock-based livelihood, high degrees of mobility, a specific way to express ethnicity and also the formation of age-sets – from the Karimojong. Apparently, the label 'Pokot' was attractive and inclusive at the time as many ancestors became Pokot within a generation: while the earliest ancestors were often still depicted as Sebei, Dodos or Kurut, their offspring were designated as Pokot. This, of course, entails the telescoping of complex processes of ethnogenesis and shifting between communities. The process of change is described as rapid and revolutionary. Similar processes of growing identification with pastoral labels such as Maasai (Galaty 1993), Turkana, Jie (Lamphear 1994), and Rendille (Schlee 1989) are typical for this historical phase of pastoral consolidation. The general picture which emerges is that pastoralisation does not only affect the economic system but is also instrumental in fostering ethnic homogeneity in style and outward appearance, with ethnically specific beadwork, hairstyles and ornamentation being adopted. Whereas pre-pastoral ancestors are said to have worn Maasai or Turkana beadwork and hairstyles (despite claiming Pokot, Kurut or Kasauria identity), the emergent pastoralist community developed an elaborated style, which was ethnically specific in some fields (such as women's beadwork and coiffure) and bore direct resemblances with other ethnic communities in other fields (such as male hair styles). This change is described as a process of rapid learning and intense borrowing from neighbouring pastoral communities, mainly the Karimojong: young men absconded from the homesteads of their fathers, temporarily joining Karimojong pastoral camps. After grave conflicts with Karimojong, they left their Karimojong hosts and rejoined their fathers cherishing a new cultural style. Whereas ethnohistorical accounts based on oral histories and ethnolinguistic reconstructions for the time before pastoralisation report diverse ethnic identities, these were subsumed by one pastoral identity. Communities such as the Oropom and Siger, for example, which still existed as distinct communities in the early nineteenth century (Lamphear 1988), became Turkana, while Mtia and Kasauria became Pokot.[4]

Historical paths of early ancestors moving to become pastoral Pokot were vividly remembered in oral traditions gathered in the 1980s and early 1990s. Accordingly, ancestors had moved from different directions to the recent areas the pastoral Pokot lived in and they raided other peoples' livestock in order to start a pastoral existence. Many clan histories start like this:

> The Kakirop clan[5] was started by a man called Sitir. Sitir belonged to the Merkutwö generation. He lived in Kipnakey, on the slopes of Mount Elgon. He decided to migrate to Chebleng country [the land of Hill Pokot] because he had had bad fortune in his original home country for many years. As most of his animals had died he went without anything. In Chebleng he settled in a place called Koipirir. At Koipirir, Sitir started his gardens where he cultivated millet and sorghum … During the time of the Maina generation set, another drought struck Sitir. His herds died again and he moved without anything towards Tiati [in the land of the pastoral Pokot]. He stayed there living on hunting and gathering fruits and roots. Sitir's sons went to raid Turkana at Kolossia near Katilu and brought cattle with them; they also stole a Turkana girl which one of Sitir's sons married later on.[6]

The *kililyet* songs give highly condensed and consensual (in the sense that there was little or no variation in these songs) lineage and/or clan-specific information about such constitutive migrations. Many Pokot clans/lineages claim ancestry from afar, from Turkana or from Rendille country, from the Cherangani Hills or from the Laikipia Plateau. The *kililyet* songs are sung by women to welcome guests at a ritual ground. Early in the morning women 'storm' onto the festival ground and loudly sing the *kililyet* of the descent group organising the ritual. To take one example, the *kililyet* of the Kamadewa clan:

> This is the day of the man from Nachi,
> this is the day of the man from Kwelkwel,
> this is the day of the man from Nayap,
> this is the man coming from Dodos,
> if not for the woman Chepöchepunyö,
> we would still live at Mount Ima.[7]

Nachi, Kwelkwel, Nayap and Mount Ima are places in Turkana District. The song can be read like a route map describing how Kamadewa ancestors migrated from Dodos country in north-eastern Uganda, through Turkana lands to the Baringo Plains, where they finally became Pokot. The reference to 'the woman Chepöchepunyö' refers to the Pokot lineage which adopted the Kamadewa clan. The paths of the past are recalled with some ambition. If not much is remembered of early ancestors, the paths of their migration are intensely recalled and the genealogy of ancestors is linked to these paths.[8]

Pokot historical paths establish a spatialised history, the group's history inscribed onto the landscape of north-western Kenya and the adjacent regions of Uganda.[9] While many more recent paths are described in terms of cattle paths, these early paths of human ancestors are frequently ventured along by people who are famished due to the loss of their livestock. In many instances traditions claim that these ancestors were not pastoralists at all. In other instances, however, it is a camel which leads people from Rendille country to the Baringo Plains or an ox that leads its owners from Mount Elgon to the Kerio. Prehistory is thus constituted by a myriad of paths, and pastoral Pokot history begins where they converge. These paths circumscribe the social and geographical universe of the pastoral life-world and signal that ethnic boundaries were well developed but porous, and identity shifts were frequent.[10]

Pastoralisation and Social Change

Pastoralisation was accompanied by a fundamental transformation of hitherto existing social structures. The clan-based territoriality of largely sedentary communities and the social organisation of foraging communities became obsolete, both replaced by an ethnically or 'tribally' defined access to the patchy and volatile key resources of highly mobile pastoralism. As a consequence, clans and lineages became less important as principles of social organisation, too. The generation-set system was combined with newly created age sets. Elders and their age sets were vested with power and authority, largely irrespective of individual wealth in animals. Gerontocracy also meant that property rights in livestock were concentrated in the households of elders, and that elders maintained extensive networks of bond partners. At the same time the adoption of age sets may have allowed for more effective mobilisation of young men coordinated in 'warrior age sets' (see also Galaty, this volume). The pastoral Pokot took possession of the areas between the Kerio Valley, the Leroghi Plateau and Lake Baringo, which were inhabited by Laikipiak Maasai, in Pokot traditions referred to as Lobbeyok. These rangelands were extensively used, and according to traditions mobility and resource tenure depended as much on military as on ecological considerations.

In summary, Pokot pastoralism entered a stage of rapid expansion and specialisation towards the end of the eighteenth century, and towards the end of the nineteenth century Pokot were highly specialised nomadic herders. Von Höhnel and Count Teleki are the first European travellers crossing Pokot territory, and they describe them as wealthy pastoralists (von Höhnel 1897).

Economic Change and Climatic Perturbations

What triggered this rapid move towards pastoralisation? Recent lake coring analysis gives a plausible explanation for this scenario. Sediment analyses of several lakes in the Rift Valley provide a fresh picture of past climatic conditions in the wider region, including climate fluctuations across decades (Verschuren, Laird and Cumming 2000; Verschuren 2004; Kiage 2007). Most notably, sediment analyses of Lake Baringo and Lake Naivasha, located about 150 kilometres south of Lake Baringo, allow for a detailed reconstruction of climatic conditions over the past 1,100 years (see Figure 9.1). According to Verschuren, Laird and Cumming, five multi-decade droughts occurred during this time, and Lake Naivasha dried up almost completely during each prolonged drought. Kiage, meanwhile, reports a 'general decrease in the percentages of major fungal spore taxa at 252–165 cm (circa AD 1750–1840s)' in a core from Lake Baringo (Kiage 2007: 146), which he sees as evidence for severe desiccation in the Lake Baringo area. During that period of time, Lake Baringo evidently dried up completely. Strikingly, the last Naivasha drought of 1760 to 1840, according to Verschuren, Laird and Cumming (2000), and Kiage's depiction of a period of severe desiccation, both coincide with the emergence of the pastoral Pokot towards the end of the eighteenth century as reconstructed from oral evidence. Kiage interprets the reappearance of certain spores in the early nineteenth century as evidence of increased anthropogenic activity within the Lake Baringo catchment, and concludes 'that after AD 1830 the area around Lake Baringo was probably occupied by pastoralists and some agriculturalists who must have been actively impacting on the ecosystem' (Kiage 2007: 147).

The ecological and economic consequences of this multi-decade drought were dramatic: The savannah vegetation changed drastically and cultivation in the Lake Naivasha catchment was disrupted (Verschuren, Laird and Cumming 2000). The severe desiccation of the region may have encouraged the use of rangelands with mobile herds. In the first half of the nineteenth century, fairly different

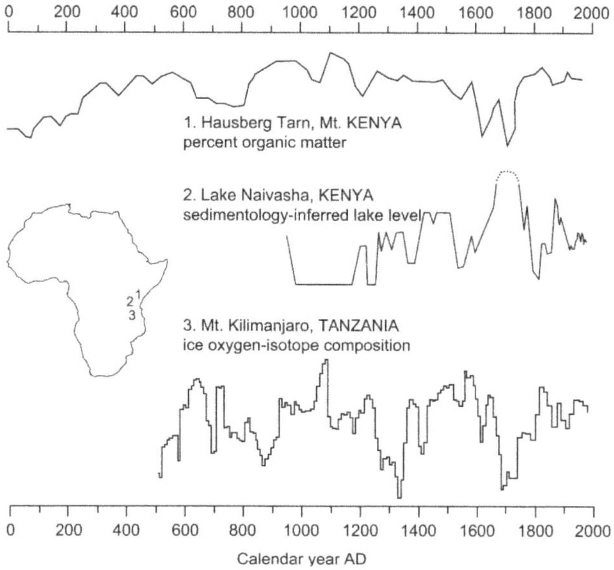

Figure 9.1: Environmental fluctuations in East Africa during the last two millennia (source: Verschuren, Laird and Cumming 2000; Verschuren 2004).

but nonetheless related forms of specialisation occurred in the Cherangani Hills and on the southern shore of Lake Baringo. In both cases, elaborated systems of irrigated agriculture developed in egalitarian local communities (Anderson 2002; Östberg 2004). Most likely, the dry spell between 1760 and 1840 not only motivated a move towards specialised pastoralism but also triggered the emergence of irrigated agriculture among Pokot and Marakwet communities settling on the slopes of the Cherangani Hills. At this stage – and in the absence of further *in situ* excavations – we may only speculate on the linkages between a decades-long drought and economic specialisation. Was the move towards pastoralisation and irrigated agriculture an adaptation to prolonged drought as small-scale rain-fed agriculture became too risky and hunting only yielded subsistence in some lesser occupied areas? Do the frequent hints at disasters such as earthquakes and droughts represent metaphorical translations for major social-ecological ruptures? These are pertinent questions which will only be answered once we have a better understanding of the region's history over the past 200 to 300 years.

The Elaboration of Specialised Pastoralism in the Nineteenth and Twentieth Centuries

The drought came to an end in the 1840s (Verschuren, Laird and Cumming 2000; Kiage 2007) and by that time the pastoral economy was well developed. Wetter conditions and improving rangeland conditions did not lead to a reversal of pastoralisation but rather motivated rapid pastoral growth. Oral accounts indicate that bridewealth payments increased during this time: whereas hoes and beehives were previously part and parcel of such presentations, now several heads of livestock had to be paid.

Much of the territory claimed by Pokot herders at that time was contested between pastoral Pokot, Turkana and Maasai. Pokot oral traditions report that borderlines shifted considerably due to changing success in raiding. During periods of peace, communities with differing ethnic backgrounds used pastures conjointly. For instance, until the first decade of the twentieth century, Pokot herders shared the better parts of the Baringo Plains with Turkana herders (Hobley 1906; Dundas 1910). This meant that resource tenure problems had to be resolved across ethnic boundaries if they were to be solved at all. According to traditions a basic differentiation between settlement areas (*kor*) and border zones (*serim*) was of importance. Specific rituals were created to ensure peace in border areas. The differentiation rather describes varying degrees of resource control and safety than a dichotomy between rainy-season and dry-season pasture. Herds were moved between safe core settlement and unsafe, outlying dry-season grazing areas. Whereas the use of pastures in the core settlement areas was arranged by neighbourhood councils (*kokwö*), the management of herds in outlying areas was in the hands of young men who first of all aimed at ensuring the security of their herds. The vulnerability of precolonial herding due to raids is depicted in oral traditions reporting major shifts of Pokot settlements within short periods of time: whereas oral traditions claim that precolonial Pokot herders advanced onto the Leroghi Plateau (some even maintain that their ancestors' herds reached the proximity of Mount Kenya), at other times Pokot were repelled westwards to the Cherangani Hills and even Sekerr Mountains. Pokot pastoralism in the second part of the nineteenth century was largely conditioned by short-time negotiations of resource use and/or violent conflict with competing pastoral groups.

The pastoral system came under severe stress during a period of catastrophes in the late 1880s and early 1890s. This period of disasters is summarised by the Pokot under the label *asis* ('sun'). Rinderpest, several droughts and also a chicken pox epidemic devastated herds and communities (Bollig 2006: 146–47). For the Pokot the period of disasters commenced with an epidemic of contagious bovine pleuropneumonia (von Höhnel 1897: 768). Then rinderpest (*kiplok*) struck Pokot herds in 1891 causing dreadful losses. The rinderpest epidemic was followed by a drought, by mange (*simpiriyon*), a contagious goats' disease, and by various human diseases. An elder reported on the disaster:

> During *asis* people suffered a lot. There were no cattle. The Chebleng sold their women for milk and did not even ask for cattle. If Pokot men who still had one or two cows gave some milk they obtained a woman. People like the Sapiny [Sebei] were giving women for a hide. The Pokot – among themselves – gave one cow for a woman during *asis*. In bridewealth negotiations they agreed to give four cattle, but many people had

> to give just one cow. And if you had no cattle at all you could dig edible roots and exchange them for a woman.[11]

The Pokot had to adapt their social system rapidly to a massively decreased number of livestock. Even rich people were said not to own more than five head of cattle. Bridewealth payments were reduced to one or two head and often not paid at all, compensation in case of murder was postponed and arrangements for livestock friendships changed. Oral traditions describe the breakdown of social standards rather than the disaster itself or its causes.

As a consequence, former herders were reduced to foraging; they had become *pich chole mey*, 'poor people', the Pokot equivalent of the Maasai term *il torrobo*. However, neighbouring agricultural communities fared even worse. Traditions indicate that the pastoral Pokot in those days were joined by destitute people from agricultural and agro-pastoral communities. Women from other communities were married for substandard bridewealth. However, the contrary seems to have happened as well: ancestors of the Kapchepkow lost their livestock during the disaster and temporarily joined the Cheptulel Pokot in the Cherangani Hills. Others fled to the mountains within Pokot land where they could survive on hunting and gathering. Starving people dug for tubers such as *kela* (*Vatovaea triphylla*) and *tapodin* (*Lannea triphylla*). At Korossi people dug for the *chowit* root (unidentified). Traditions report that to find *chowit* roots people had to dig holes that were so deep a man could stand in them.

By the turn of the century conditions had normalised, and by 1904 the Pokot had to pay taxes to the British colonial government. Within a short period of time the Pokot had succeeded in reconstituting their depleted herds. Beech, who administered the district in 1907, described the Pokot as wealthy herders 'exceptionally vain and exceptionally generous' (Beech 1911: 26). Recovery was apparently swift, and already during the first decade of the twentieth century, when the first British officers ventured through Pokot land, the Pokot were described as rich pastoralists. In 1902 the British established a station in the region, first in the Kerio Valley and then at Mukutani near Lake Baringo. In 1911 headquarters were relocated to the Tugen Highlands, and colonial impact on the Pokot pastoral community remained rather ephemeral until the early 1950s.

The pastoralisation of the Pokot was accompanied by a significant increase in productivity, and specialised herders soon possessed much more livestock than those Pokot who continued an agro-pastoral mode of production.[12] Accounts of cattle numbers and numbers of people given by colonial tax enumerators between 1920 and 1940 suggest an average number of fifteen head of cattle fell per person in East Pokot.[13] In the late 1940s Schneider noted a cattle:human ratio of 5:1 among nomadic pastoralists, while merely 1.2 cattle per person were noted for the agro-pastoral Pokot (Schneider 1957).

Barton reports on substantial differences in bridewealth payments between different subgroups of the Pokot, at that time still labelled Suk in colonial jargon. Bridewealth varied 'from a few goats with the Hill Suk [chiefly agricultural] to up to 80 head of cattle with the pastoral Suk' (Barton 1921: 96). According to Goldtschmidt (1974: 315), an increase in livestock ownership among the adjacent Sebei of Mount Elgon resulted in rising bridewealth payments. In total, Goldtschmidt presents a sample of twenty-seven pastoral communities, which clearly shows a positive correlation between wealth in livestock and size of bridewealth payments. In addition, the high polygyny rates among specialised pastoralists cause a relative scarcity of marriageable women, which contributes to rising bridewealth payments (see also Spencer 1998: 52–60). The pastoral Pokot fully complied with Goldtschmidt's prognosis: bridewealth payments increased with growing pastoral wealth.

Growing demands for bridewealth not only motivated specialised pastoralists to further enlarge their herds but also ensured the circulation and reallocation of livestock within the group, and thus prevented lasting stratification and cross-generational accumulations of wealth. So-called 'stock friendships' between persons, who were not related with each other, further facilitated the circulation of livestock within the group (Bollig 1997; 2006: 286–87). Although stock friendships also existed among the agro-pastoral Pokot of the Cherangani Mountains, they were of much less economic importance than among the nomadic pastoralists. In fact, stock friendships are apparently common with highly specialised herders organised along egalitarian lines and rarely occur among poor pastoralists and highly stratified agro-pastoral communities such as the Tswana (Johnson 1990; McCabe 1997; see also Galaty, this volume). Herren (1991), for example, notes that stock friendships were virtually unknown among the comparatively poor Mukogodo (Kenya), and White (1997) tells us that stock friendships were once very common among the Wodaabe Fulbe (Niger) but then declined in importance due to increasing poverty.

The social organisation of the pastoral Pokot around 1900 was based on patrilineal descent groups, age and generation sets, and individual networks. Every person was a member of a patrilineal descent group by birth, and both men and women remained members of one descent group for life. There were twenty-two Pokot clans, which in turn consisted of three to seven lineages. Clans were exogamous and were united under a specific symbol, such as rain (the Ngusur), frog (Kiptinkö), or fire sticks (Koimö). Cattle brands and a set of symbols and songs (*enwait*) were also clan-specific. Beyond that there was little formal organisation at clan or lineage level. Although numerous non-Pokot lineages had been integrated into the descent system, there was no ranking of lineages or clans. Due to the fragmentation of descent groups, lineages were numerically rather small and were spatially distributed. The Pokot did not put much emphasis on descent ideology. Communal rituals and celebrations were connected rather to the dynamics of age-set and generation-set organisation than to issues concerning the descent system.

Through separate initiation rituals each man became member of an age set (*asapantin*) and of a generation set (*pïn*). Membership of one set involved egalitarian, almost brotherly relationships with one's peers on the one hand, and submission under an age-based hierarchy on the other. Initiation into generation sets only took place every twenty-five to thirty-five years, when numerous young men and boys were circumcised together and from then on constituted one generation set (Bollig, 1990).[14] At any one time, three generations were alive: a junior or warrior generation, a generation of seniors and a generation of elders (Bollig 1990; 1992: 85).

By contrast, membership in an age set was attached to a fixed age. Boys of seventeen to twenty years underwent the *sapana* ritual (Bollig 1990). In this celebration they had to slaughter an ox in a ritually prescribed way, and were 'washed' with the contents of the stomach of the ox which they had slaughtered. Later they were formally endowed with the signs of manhood: the men's colourful headdress. Each initiation set – that is, those boys who were initiated in one particular year – was given a name which was reminiscent of a special event of that particular year or initiation period. After some ten to twenty years, between four and seven sets were moulded into one set and the set with the most members usually gave its name to the united set.

Age sets and generation sets constituted a complex gerontocratic system. Men achieved status not as members of descent groups but as members of age groups. During each and every festivity, even if it was minor, the gerontocratic hierarchy was depicted in the seating within the half circle (*kirket*) where all men sat according to seniority. Age sets competed for prestigious symbols: junior age sets had to fight for certain feathers and colours and senior age sets had to ask elders for the grant of ritual powers symbolised in adornment (such as feathers and specific beads) and colours accepted for the men's headdress (see Figure

Figure 9.2: Pokot elder displaying the typical adornment of his generation set.

9.2). Each set had to undergo a prescribed set of rituals in order to climb up the gerontocratic hierarchy.

Extended personal networks were a third essential feature of pastoral Pokot social organisation. Friendship between men (*kongot*) always involved the transfer of animals (Schneider 1953: 240). Some paternal and affinal relatives and some friends were invited for bridewealth distributions and were given some heads of livestock – always with the expectation that the donor would be invited to similar occasions in the future. While the gerontocratic age-based system was founded on the principle of respect (*tekotön*), solidarity (*tilyontön*) was the basic ideological principle of friendship.

Pastoral specialisation and increasing productivity were accompanied by plenty of new socio-economic opportunities, such as the intensification of trading activities, and the development of internal livestock-based exchange networks. At the same time, though, they involved novel hazards and uncertainties. These included the devastating loss of livestock due to drought, disease and raids; the degradation of pasture land because of overgrazing; and dependency on external traders and cereal food staples from elsewhere.

Pokot Pastoralism under Colonialism

When Baringo District was created in 1902, the British colonial authorities established administrative structures which directly affected the pastoralists in East Pokot. However, colonial officers soon discovered that there was not much to gain in East Pokot for the colonial state: 'there is little in the district to attract the European settler, and this area may be definitely looked upon as a native reserve', Hobley (1906: 472) reported. A few 'big men' were subsequently appointed as chiefs among the Pokot, who numbered no more than 7,000 people at the time (Bollig 1992: 48). The administration's major task was the collection of taxes during the first two decades of colonial rule. This, however, they mainly achieved by putting pressure on some of the big men acclaimed as chiefs by the local administration, requiring them to bring in the taxes of 'their people' to the administrative headquarters. Besides interfering with local socio-political structures in order to implement indirect rule, the colonial regime brought about four major changes in the savannahs of north-western Kenya: first, armed conflicts were reduced from about 1920 onwards due to the *Pax Britannica*; second, demarcations of ethnic territories curtailed pastoral mobility; third, ethnic segregation and fixed territories also constrained social mobility and fostered the 'hardening' of ethnic categories; fourth, in the northern Baringo Plains, Somali herders traded livestock and commodities from the early twentieth century. While inter-regional trading activities generally increased, the administration simultaneously constricted livestock trade in the pastoralist zones in order to protect British settlers and their products (Kerven 1992).

What is perceived as the traditional system of pastoral Pokot resource management – by outside observers as much as by local actors – only came into existence in the early twentieth century as a result of colonial administration and the creation of a 'tribal reserve' exclusively defined for people who identified themselves as Pokot. While the pastoral Pokot settled between the Kerio River and the Leroghi Plateau were administered within the framework of Baringo District, internal boundaries between ethnic communities were demarcated in the 1920s and 1930s, and each ethnic community was ensured the exclusive use of 'its territory'. By about 1910 colonial authorities intervened on behalf of the Pokot, and Turkana were summarily expelled from the newly created Baringo District and repatriated to Turkana District. Colonial forces guaranteed a stable peace for some decades, and the Pokot were allotted a tribal reserve with uncontested land rights and fixed boundaries. Colonial files of the late 1920s are filled with the tiresome reports of land surveyors who fixed ethnic boundaries on maps and made them congruent with administrative boundaries. Colonial officers strongly believed that a reduction of competing claims to natural resources was brought about by clearly defined boundaries, and that such boundaries would result in a reduction of interethnic violence and facilitate administration.

Pastoral tenure resting on controlled access to dry-season and rainy-season pastures developed in a highly institutionalised manner once violent competition for resources had been reduced, and the fixing of boundaries stipulated that grazing was a limited resource – that is, access to additional resources could not be gained through violent strategies. Whereas military preparedness and the maintenance of peaceful relations across ethnic boundaries had previously been essential, pastoral resource management could now take account of long-term perspectives. The institutional foundations of an access regime which was internally regarded as fair and just and aimed at a sustainable use of pasture emerged. The transaction costs of resource tenure institutions – such as monitoring mobility and sanctioning early grazing in dry-season grazing resorts – were, however, born internally by the council of men of one neighbourhood (*kokwö*) in particular.

The newly established system of resource protection rested on a general differentiation between rainy-season and dry-season pasture. The mountainous areas of Paka, Korossi and Silali as well as the plains between Paka and Silali were defined as dry-season grazing reserves. No grazing was allowed there in the rainy season and in the early dry season (from May to September/October), thus allowing the predominantly annual grasses to produce sufficient seed for sustainable reproduction. The protection of Mount Paka was of special significance since the Pokot believed that the extinct volcano was the 'home of the rain'. While other pastures were continuously burned at the end of the dry season to encourage grass growth, it was considered sacrilege to burn pastures on Mount Paka. Interviews in the 1980s showed that burning was used to combat bush encroachment, to lower tick infestation and to ensure good grass growth. The burning of pasture was a collaborative effort, and neighbourhood councils agreed beforehand which tracts of land would be burned and which ones would be spared for late-dry-season grazing. The burning of pasture was prohibited, however, by colonial authorities and later by the respective Kenyan authorities. However, the ban on burning was apparently not consistently followed up, and in the 1980s Bollig could still witness how fire was skilfully used to manage pastures. Besides the creation of dry-season pasture, some smaller areas were specifically defined as dry-season grazing areas for calves. Obviously it was of benefit for a community to have easy access to pasture for calves. This allowed herders to stay in the major settlement areas for long into the dry season and graze their calves on nearby grounds which were well protected. Usually a close-by hill, which was renowned for its good cover of highly nutritious grasses, was chosen for this purpose. A neighbourhood (*kor* or *korok*; see also Conant 1965) constituted a non-kinship-based group of resource users. Membership was flexible and community decision making generally operated through the council of adult men (*kokwö*).

Before 1900 Pokot mainly engaged in small-scale trade with neighbouring agro-pastoralists, but from the first decade of the twentieth century onwards they entered into more wide-ranging trade connections with mobile Somali traders who rapidly developed trade links in the pastoral areas of northern Kenya and eastern Uganda. Those traders initially took heifers to East Pokot in order to trade them for goats and/or sheep, which they later sold for slaughter in the provincial capital Nakuru. Remarkably, the Pokot at first were neither interested in foreign commodities nor in the exchange of livestock products for cereals, but rather used trade with itinerant Somalis predominantly to enlarge their cattle herds.[15] In the early 1920s itinerant traders then started to bring commodities such as maize, tea, sugar and textiles to trade for livestock (see Anderson 2002: 147). After the colonial administration inhibited the settlement of Somali traders in the region until the 1930s, from the late 1930s onwards a small number of shops was established by these traders. While other groups increasingly took up wage labour in order to pay taxes levied by the colonial regime, very few Pokot tried to find a job outside their territory. Most people concentrated their energy on livestock husbandry instead, and apart from infrequent vaccination campaigns and the occasional coercive reduction of the steadily growing livestock population, the administration did not interfere with herding in any serious way. The preferred Pokot response to administrative measures which they considered unfavourable was passive resistance (Schneider 1959: 149). To avoid taxation, for instance, temporary migration to remote areas proved an adequate means. The only significant open confrontation with the colonial power occurred in 1950 when many Pokot were inspired by an anti-colonial, millenarian movement known as *Dini ya Msambwa*. The conflict culminated in the so-called Kolloa Affray, in which several Pokot and some colonial officers lost their lives (see Wipper 1977; Bollig 1992).

The Crises of Pastoral Specialisation and the Transformation of Pastoralism

The phase of pastoral growth ended in the 1930s/1940s. While the population grew consistently, livestock herds did not increase accordingly. Prolonged curtailment of pastoral mobility and denial of access to pasture around Lake Baringo and the Laikipia Escarpment beyond the borders of Baringo District had negative consequences for the pastoral economy. Large herds of livestock packed in an administratively restricted area involved permanent overgrazing and finally resulted in massive bush encroachment and pasture degradation.

How did the Pokot attempt to adapt to these novel challenges, such as demographic growth, state impact on access to resources and the decline of grazing potential? In many respects Pokot intensified social exchange: where an intensification of their economic system was inhibited by colonial administration and ecological circumstances, internal exchanges of livestock in ritual expanded, and an apparently growing number of rules stipulated how livestock slaughtered during rituals should be divided, in such a way that afterwards often little remained for each recipient.

'Big men' lost their influence and power in the 1950s and 1960s, and the community apparently emphasised egalitarian relations. Whereas for the first half of the twentieth century there had been numerous big men who were remembered to have married many wives (usually between six and twelve) and had their herds herded by shepherds, such big men did not feature importantly in the second half of the twentieth century. Instead, reciprocal relations and intra-community solidarity were emphasised. A great number of rituals underpinned solidarity and ethnic identity. Geertz applied the concept of involution 'to describe culture patterns which, after having reached what would seem to be a definite form, nonetheless fail either to stabilise or transform themselves into a new pattern but rather continue to develop by becoming internally more complicated' (Geertz 1963: 80). Indeed, pastoral Pokot culture in the second half of the twentieth century may serve as an example of pastoral involution: the development of ethnic identity played out vociferously in a myriad of ornamental patterns of beadwork and coiffure, a plethora of rituals to capture internal conflicts at an early stage, and a great number of instances to engage in livestock-based exchanges.

Economically, however, the Pokot were progressively meeting problems of supplying an ever-growing population with sufficient food. The ratio of people to livestock constantly declined and ranges degraded massively. Interethnic raiding resumed in the 1970s (Bollig 1992). Large areas of grazing land on the boundary between Pokot and Turkana were given up for fear of livestock theft.[16] Violence towards outsiders was as much motivated by relative deprivation as by culturally prescribed male career paths: many of the young men engaging in raiding saw that they would probably never own the plentiful herds of their fathers and their grandfathers. Externally directed violence and severe problems with solving problems of collective choice exacerbated the situation. During the 1980s and 1990s Pokot were actually most successful in situations where internal solidarity and egalitarian social organisation proved operative in terms of risk and crisis management. Yet, they often ran into serious problems when they tried to regulate resource use in a sustainable manner, and/or facilitate innovations in common resource management (see Bollig and Österle 2008). The dense networks that Pokot created and maintained via intensive internal exchange often proved rather inappropriate or even counter-productive in the face of the new challenges which emerged in the second half of the twentieth century.

In the years after Kenya's independence, Pokot, like other pastoralists in the country, were confronted with the agendas of a postcolonial state, and after the severe droughts of the 1970s and 1980s they became a target for international development cooperation. After maize from the USA was first given to the needy in a drought in the mid 1960s, food aid became a regular occurrence in Pokot country. Both government and NGOs focused on the sedentarisation of mobile herders, and propagated the need for the fundamental 'development' of nomadic pastoralists; this involved the establishment of livestock markets, the introduction of improved breeds, and the construction of boreholes and small dams. At the same time, cattle raids and armed conflicts with neighbouring groups resumed, and these had turned into a state of chronic low-intensity warfare by the beginning of the 1990s (ibid.).

Since the late 1990s, pastoralism in East Pokot has entered a phase of rapid and radical transformation (Österle 2008a, 2008b). The interplay of a rapidly growing human population (Figure. 9.3), the progressive degradation of pasture (Bollig 2006: 77–83) and the aforementioned escalation of armed conflicts (Bollig and Österle 2007; Österle 2007) has reduced the availability of key resources and increased the vulnerability of nomadic households. As Pokot herders are no longer able to manage pastoral risks and crises adequately, their subsistence is increasingly threatened by hazards such as droughts and floods, cattle raids and livestock diseases (Galvin et al. 2001; Little, Mahmoud and Coppock 2001). Similar scenarios have been reported for other groups of East African pastoralists (Markakis 1993; Spencer 1998; Fratkin and McCabe 1999; Fratkin 2001; Gray, Leslie and Akol 2002; Desta and Coppock 2004; Schwartz 2005). As specialised nomadic pastoralism can no longer sustain an increasing number of Pokot households, present-day pastoralism is characterised by the progressive diversification and transformation of the livestock economy (Österle 2008b).

The Shift to Small Stock

There is no doubt that the cattle:human ratio changed dramatically during the second half of the twentieth century: While an average of fourteen cattle fell to every Pokot in the 1920s, less than two cattle per person remained at the turn of the millennium. The population density was at about fourteen people per square kilometre around 2000 – a density which is rather typical of agro-pastoral sedentary communities (Bollig 2006: 367). Obviously the cattle population could not keep up with the growing human population under the given socio-ecological circumstances, but kept oscillating around 100,000 animals instead. Recently, a massive shift towards small stock husbandry (goats and sheep) occurred. In fact, Pokot pastoralism is no longer dominated by cattle (see Schneider 1957) but has transformed itself into a market-oriented small stock economy (Österle 2008b). This transformation finally allowed them to counter the fatal decline of the

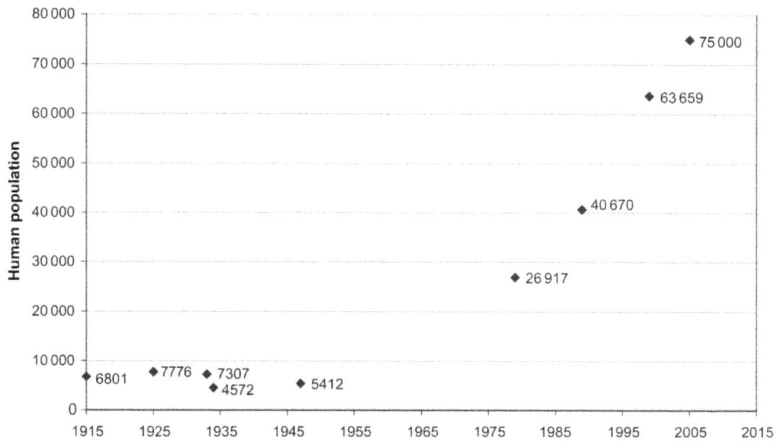

Figure 9.3: Population change in East Pokot, 1915 to 2005.[17]

tropical livestock unit (TLU) per person ratio.[18] After the cattle:human ratio had fallen below 1:1 at the beginning of the 1990s, and the TLU per person ratio had simultaneously dropped to merely 1.9 (Bollig 2006: 68–70), this critical value has meanwhile risen to 3.3 due to formidable increases in small stock.

Small stock numbers rose from an estimated total of 100,000 goats and 20,000 sheep in 1988 to around 600,000 goats plus 100,000 sheep in 2005 (Figure 9.4). Thus the small stock:cattle ratio is currently 6:1, compared to 3:1 in 1950. This corresponds with the 'Maasai model', which 'suggests that households will shift their emphasis to include more small ruminants as resource constraints change and market involvement increases' (Desta and Coppock 2004: 479). Bush encroachment on former grasslands, the trend towards sedentarisation (which is particularly enhanced by a growing number of artificial waterholes), the availability of many children as an ideal workforce for small-stock herding, and above all the growing demand for goat meat in Kenya's cities are the decisive factors which have pushed this development. In addition, the markets in East Pokot are characterised by the positive development of the caloric terms of trade (CToT) of currently 6:1 between maize and meat – that is, the sale of one goat generates six times more calories from the maize that can be purchased with it than would have been generated by direct consumption of the meat (see also Zaal and Dietz 1999: 169). While huge small stock herds solve the cash problems of many Pokot households, they create other problems as well. There is no doubt that the huge flocks of small stock now held currently impact vegetation dynamics profoundly.

It is worth noting that the camel population in East Pokot has been growing significantly in recent years too. Most likely camels were first integrated into Pokot pastoralism at the beginning of the twentieth century (McGovern 1987: 2), after Pokot had supported a punitive expedition against neighbouring

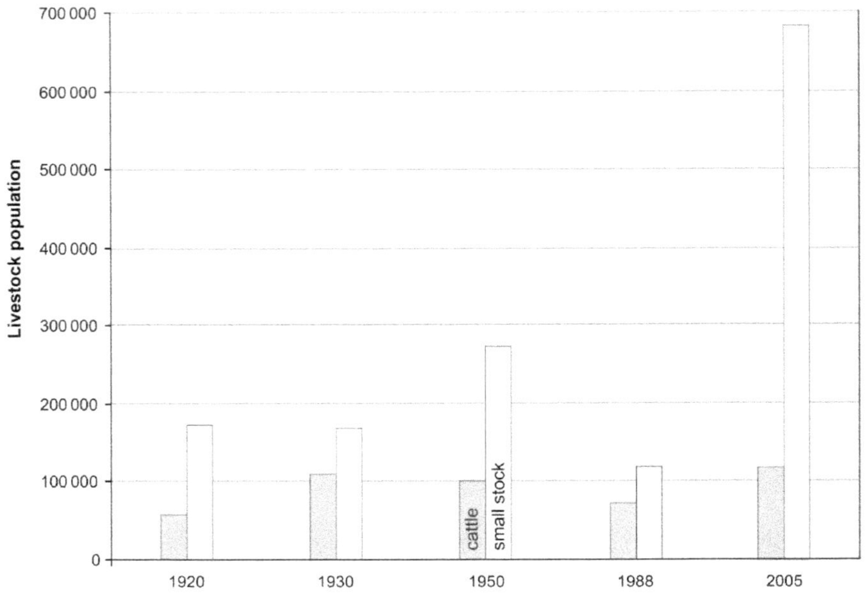

Figure 9.4: The development of cattle and small stock populations in East Pokot, 1920 to 2005.[19]

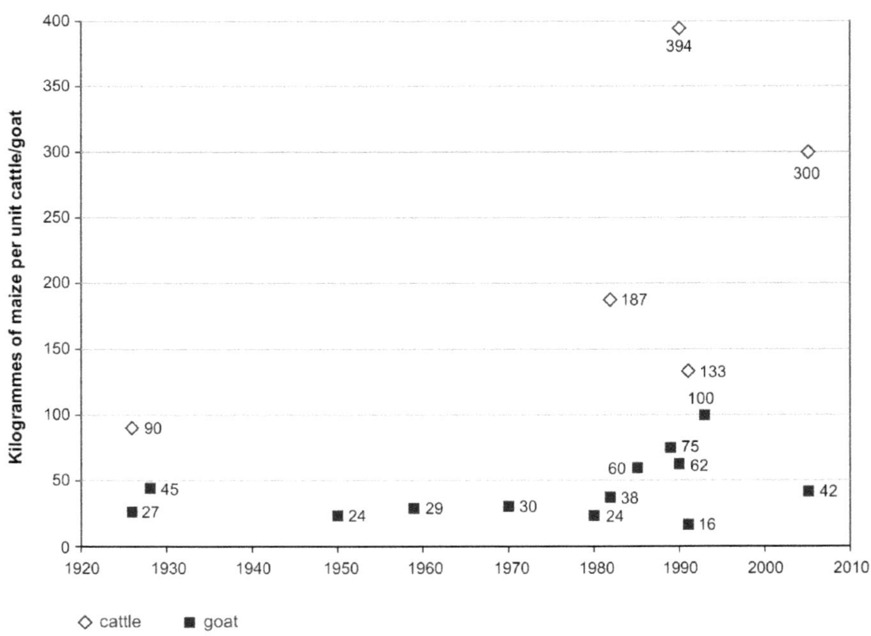

Figure 9.5: Terms of trade between livestock and maize in East Pokot, 1926 to 2005.

Turkana in 1917/18 (Collins 2006). The few available records suggest a total average number of around 1,200 animals between 1928 and 1932. According to Pokot informants, the number of camels started to rise continuously in the 1950s. The marked increase in recent years is also due to the promotion of camels by development projects, which began to restock the camel population already in the 1980s. At that time the total population was put at a maximum of 3,500 animals (Saltlick 1991: 24). In the past twenty years the population has increased more than twofold and amounts to more than 8,000 animals currently. Desta and Coppock (2004) observed even more striking rates of growth among the Boran of southern Ethiopia, where about one camel statistically fell to every tenth household in 1985. Only twelve years later every household had 1.7 camels at its disposal – a drastic increase, explainable only by a combination of biological reproduction and theft from the neighbouring Gabra. It is possible that the theft of camels from the neighbouring Turkana has also contributed to the substantial growth of the camel population in East Pokot.

Livelihood Diversification

Recent data from 120 households document the process of livelihood diversification in East Pokot in detail. Whereas a general trend of diversification is observable, households living in a 'pastoral zone' near Mount Paka diversified their economic activities quite differently from households living in Chemolingot, a small savannah town. A growing number of households in the pastoral zone cannot sustain a livelihood based on specialised pastoralism because their herds are far too small for it. These households have no choice but to implement other economic activities in order to survive. It is evident, however, that not only the poor but also many successful and wealthy households currently opt for diversification and are no longer 'pure' pastoralists' (see also Sperling and Galaty 1990: 73). In fact, while all sixty households in this sample have at least a few animals, only one-third (21) concentrates solely on livestock husbandry. The majority (39) relies on at least one additional economic activity. It is amazing that more than half of the households (31) in this particular area are farming nowadays while only a few years back any kind of cultivation was frowned upon and regarded as not befitting the status of a Pokot. In 2004/5 the size of fields near Mount Paka ranged from 200 to 6,000 square metres, and the average field measured between 400 and 500 square metres. Several households cultivated more than one field at the same time.

Beekeeping was implemented as an additional activity by twelve of the studied farming households. In total fifteen out of sixty households in the pastoral zone engaged in beekeeping, which is a rewarding but seasonal enterprise. External traders buy the honey in East Pokot and later resell or process it elsewhere in Kenya. Both Christian churches and NGOs have invested in the improvement of local beekeeping in recent decades. They provided several thousand industrially produced beehives in order to facilitate more sustainable and efficient ways of producing honey. As the production of alcoholic beverages is illegal, none of the households acknowledged brewing as an economic activity, although it is evident that some households are brewing in order to make a living.

Livestock trading is engaged in by 10 per cent (6) of the households near Mount Paka, while only 5 per cent (3) have a job or work for money occasionally. Livestock trading is particularly popular with young men who are trying to use the opportunities offered by growing markets for livestock everywhere in East Pokot. They model themselves on some extremely successful livestock traders. These famous men have convincingly demonstrated that it is possible to advance from humble beginnings as a small stock trader to the respected position of a rich cattle dealer, one who is able to marry many wives, keep huge flocks of livestock and own property such as houses and cars. Like most of the young men who try to tread in their footsteps, they have not gone to school but are living proof that one can succeed without formal education.

There are no significant correlations between wealth and diversification as recently hypothesised by Little et al. (2001) in their comparison of diversification patterns among East African pastoralists. Following Barth (1964), Little et al. claim that, 'diversification mainly affects the wealthy and poor herders and not middle-wealth groups of pastoralists who may lack sufficient motivation and/or pressures to diversify' (Little et al. 2001: 405). Our data clearly show that households from all wealth groups, including the middle strata, currently opt for diversification (Figure. 9.6).

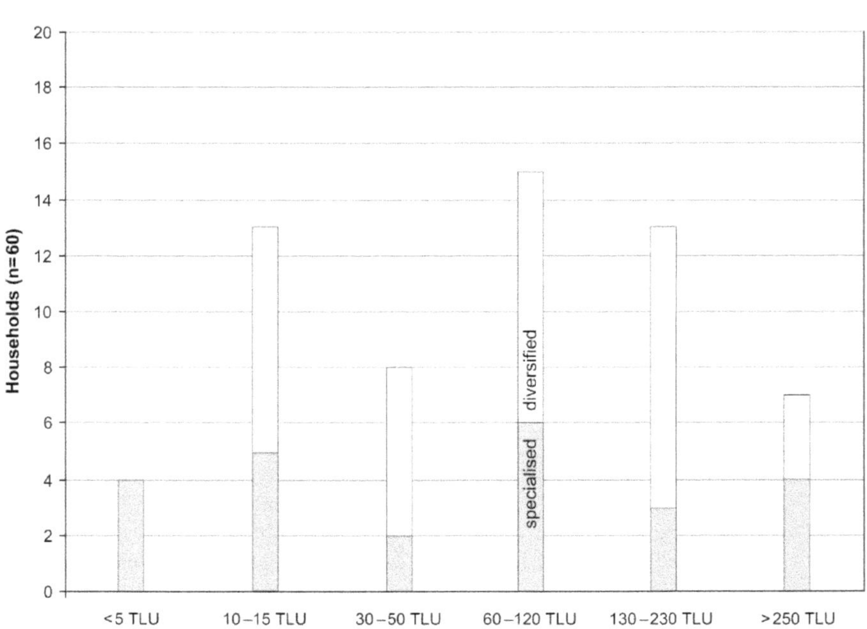

Figure 9.6: Correlations of wealth and diversification in East Pokot, 2005.

In contrast to the pastoral zone near Mount Paka, the small savannah town of Chemolingot, which has now become the district headquarters of the newly created East Pokot District, is characterised by depastoralisation and livelihood diversification (see Dietz 1993: 85). The majority of households no longer subsist on pastoralism, and sixteen out of sixty households in the Chemolingot sample did not have a single animal at their disposal. Therefore most families are actually forced to diversify, though farming is not a viable option around Chemolingot, which is one of the driest places in East Pokot. Only in years with plenty of rain, as in 2005 for example, do some households lay out small gardens next to their homesteads. Dealing in water and/or firewood are the most frequent activities, implemented by one-third (21) of the 60 households in the Chemolingot sample. Fourteen households draw money from wage labour and 12 openly admit that they produce alcoholic beverages, while the real number is probably twice as high. Most female-headed households (15 out of 20) rely upon brewing beer and/or distilling hard liquor, but due to the illegality of these activities only 5 out of 20 households openly admit that they are following this business. In some female-headed households the production of alcoholic beverages is combined with prostitution. At the time of the study around 7 households in Chemolingot were engaged in so-called 'food-for-work projects', run by churches and NGOs. Another 6 households produced charcoal, 4 households ran retail shops in Chemolingot, and another 4 collected stones and/or wood for building houses. Others specialised in making headdresses, fixing punctures, and loading and unloading lorries.

A rapidly growing number of impoverished households have settled down in or nearby villages such as Chemolingot, Nginyang, Tangulbei, Kolloa, Churo and have tried to make a living there (see Fratkin and Roth 2005). Unlike their Turkana neighbours (Dyson-Hudson and Meekers 1999; Lokuruka and Lokuruka 2006), stranded Pokot pastoralists normally do not opt for permanent out-migration to other regions of Kenya but try to make a living from the resources available in their home area. Thus the small centres in East Pokot District are characterised by growing inequalities. At one pole there is a small number of affluent households: members of this elite earn most of their living from businesses and/or salaried jobs. The opposite pole constitutes a huge number of more-or-less destitute households, trying all sorts of things in order to win their daily bread. Both rich and poor households have in common that they try to keep as much small stock as they can manage and/or afford within their permanent settlements, a practice which is leading to severe environmental degradation in the immediate vicinity of such centres.

Conclusion

The Pokot example shows both the dynamics and the resilience of East African pastoralism: Specialised nomadic pastoralism reached its limits in the 1950s and 1960s, some 100 to 150 years after its emergence, due to both political and ecological effects. The colonial regime created administrative districts with fixed borders, which curtailed nomadic movements and prohibited the further expansion of Pokot pastoralism. Does highly specialised pastoralism need this perspective of permanent growth and territorial expansion? Which approaches are identifiable as adapting pastoral strategies to growing population numbers not through expansion but through intensification? These are questions only a comparative project could answer.

Former grassland turned into thorny bush savannah and became less suitable for cattle keeping throughout the twentieth century. The escalation of armed conflicts since the late 1960s and the massive droughts of the 1980s made specialised and cattle-centred pastoralism even more difficult to sustain. Things worsened again noticeably in the 1990s, and Pokot finally had to look to alternative livelihood strategies to cope with the apparent crisis of nomadic pastoralism. Small stock keeping and camel husbandry have become the new foci of Pokot pastoralism over the past two decades, favoured by comprehensive market integration and growing external demand for meat. Patterns of mobility have changed accordingly: large flocks of small stock browse and graze in the vicinity of permanent homesteads while cattle-centred nomadism has transformed into transhumance – that is, the animals are taken to distant pastures in the dry season and graze at home during the rainy season. Only young men migrate to dry-season pastures while the rest of the household stays in one place all year round. All of a sudden, farming and livestock trading have become no longer slanderous activities but instead important options of livelihood diversification.

While Harold Schneider described Pokot reluctance to change in the 1950s (Schneider 1959), we have shown that Pokot economy and culture changed rapidly in the 1990s. Not only were pastoral strategies re-orientated but agriculture and trade were also swiftly integrated into household economies. It is striking how quick and efficient changes have been. It is as if an entire community had decided to change its economy and social system. Of course, such a decision was never taken, and it is a great number of individual decisions which persuaded the pastoral Pokot to change. Nevertheless, how individual decisions rapidly translate into community-level structural changes is still very little understood.

Notes

1. See, e.g., Klumpp and Kratz (1993) for Maasai and Okiek.
2. See, e.g., McCabe (1997) on the Maasai, and contributions on the Rendille, Ariaal and Samburu in Fratkin et al. (2005).
3. I tried to interview elderly women on ethnic origins and precolonial socio-economic change. This did not result in any usable information. Early on women claimed that this field of knowledge is entirely the domain of elderly men. Women, however, were champions of clan-specific songs, which portray family histories in an abbreviated way.
4. A move towards agricultural intensification may tentatively be dated to the same time. Agricultural communities such as the Njemps, Marakwet and agricultural Pokot exchanged goods intensively with the pastoral Pokot. Kurita (1983) gives a detailed

account of exchanges between pastoralists and agriculturalists on Chesegon market. We do not know anything though on the emergence of such markets at the boundary between specialised economies.

5. While there is a clear conceptual differentiation between clan (*lilö*) and lineage (*or*) when Pokot talk abstractly about their social system, in daily practice such boundaries are rather blurred. Whether, for example, a taboo only refers to a lineage or an entire clan is negotiable; the same holds true for marriage restrictions. In the same vein, tales of origin often do not clearly differentiate whether they are linked to a clan or a lineage.
6. Narrated by Cheretoi wero Naoro, 10 December 1988.
7. In the original: *kayech chito Nachi, / kayech chito Kwelkwel, / kayech chito Nayap, / kayech chito Dodos, / anta mökichepöchepunyönye, / tökimitecha Ima.*
8. It is noteworthy that while prose versions of ancestral migrations were the domain of men, the *kililyet* were mainly (though not exclusively) the domain of women.
9. For a similar relation between ancestral paths and ethnogenesis among the Turkana, see Broch-Due (1999).
10. There is no account of the precolonial history of the agricultural Pokot living in the Cherangani highlands as yet. Davis (2008) started archaeological and ethnoarchaeological work with agricultural Pokot. He is particularly interested in the emergence of irrigation agriculture among the agricultural Pokot.
11. Narrated by Todokin, September 1993.
12. See Bonte (1991) and Henrichsen (this volume) for similar cases in other regions.
13. Figures on the livestock population and the corresponding human population are available for East Pokot.
14. With regard to the interval between generation sets, other Kalenjin groups differ. Among the Hill Pokot and Marakwet, circumcision-based initiations into the generation-set system take place approximately every seven years (see Peristiany 1951; Kipkorir and Welbourn 1973).
15. This, however, does not indicate that they were not interested in exchanging millet or sorghum with agriculturalists. Such exchanges were frequent along the boundary between Tugen and Pokot as well as along the boundary between Pokot on the one hand and Marakwet and agricultural Pokot on the other.
16. Ssee McCabe (2004) for an account of the effect of raiding on the southern Turkana.
17. All data are census figures, both from colonial and postcolonial records, except for the data for 1915 and 1935, which are based on counts for tax collections carried out by the colonial regime. The figure of 75,000 persons in 2005 was generated by using our own data on the fertility of Pokot women, which suggest an average annual population growth of 2.4 per cent. The strikingly low figures for 1934 and 1947 are due to temporary migration to remote areas, which a number of Pokot households undertook in order to avoid taxation by the colonial regime.
18. TLU are calculated thus: 1 cow equals 1 TLU; 1 camel equals 1.25 TLU; 1 goat/sheep equals 0.125 TLU (see Galvin and Little 1999: 128).
19. The figures from the 1920s and 1930s are results of livestock counts in the context of tax collections by the colonial administration. The numbers from the 1950s are estimates by the government agency in charge of livestock. The data from 1988 were collected during a vaccination campaign and are to be found in Saltlick (1991). The actual livestock figures from 2005 are based on personal data collections among 120 households. The total livestock population was subsequently extrapolated, using the total number of 12,500 Pokot households mentioned in the latest government census (RoK 2001).

References

Anderson, D.M. 2002. *Eroding the Commons: The Politics of Ecology in Baringo, Kenya, 1890s–1963*. Oxford: James Currey.

Barth, F. 1964. 'Capital Investment and the Social Structure of a Pastoral Nomadic Group in South Persia', in R. Firth and B.S. Yamey (eds), *Capital, Savings and Credit in Peasant Societies*. London: Oxford University Press, pp.415–25.

Barton, J. 1921. 'Notes on the Suk Tribe of Kenya Colony', *Journal of the Royal Anthropological Society* 51: 82–99.

Beech, M.W.H. 1911. *The Suk: Their Language and Folklore*. Oxford: Clarendon Press.

Bollig, M. 1990. 'An Outline of Precolonial Pokot History', *Afrikanistische Arbeitspapiere* 23: 73–92.

——— 1992. *Die Krieger der Gelben Gewehre: Intra– und Interethnische Konfliktaustragung bei den Pokot Nordwestkenias*. Münster: Lit Verlag.

——— 1997. 'Bridewealth and Stockfriendship: The Accumulation of Security through Reciprocal Exchange', in H. Kreutz (ed.), *Leben und leben lassen: Die Fundierung der Marktwirtschaft durch symbolischen Tausch und Reziprozität*. Opladen: Leske und Budrich, pp.57–72.

——— 2006. *Risk Management in a Hazardous Environment: A Comparative Study of Two Pastoral Societies*. New York: Springer.

Bollig M., and M. Österle. 2007. '"We Turned our Enemies Into Baboons": Warfare, Ritual and Pastoral Identity among the Pokot of Northern Kenya', in A. Rao, M. Bollig and M. Böck (eds), *The Practice of War: Production, Reproduction and Communication of Armed Violence*. New York: Berghahn, pp.23–51.

——— 2008. 'Changing Communal Land Tenure in an East African Pastoral System: Institutions and Socio–economic Transformations among the Pokot of NW Kenya', *Zeitschrift für Ethnologie* 133(2): 301–22.

Bonte, P. 1991. 'To Increase Cows, God Created the King: The Function of Cattle in Intralacustrine Societies', in J.G. Galaty (ed.), *Herders, Warriors, and Traders: Pastoralism in Africa*. Boulder, CO: Westview, pp.62–86.

Broch-Due, V. 1999. 'Remembered Cattle, Forgotten People: The Morality of Exchange and the Exclusion of the Turkana Poor', in V. Broch-Due and D.M. Anderson (eds), *The Poor Are Not Us. Poverty and Pastoralism*. Oxford: Currey, pp.50–88.

Collins, R.O. 2006. 'The Turkana Patrol of 1918 Reconsidered', *Ethnohistory* 53(1): 95–119.

Conant, F.P. 1965. 'Korok: A Variable Unit of Physical and Social Space among the Pokot of East Africa', *American Anthropologist* 67(2): 429–34.

Davis, M.I.J. 2008. 'An Applied Archaeological Approach and Anthropological Study of Intensive Agriculture in Northern Cherangani Hills, Kenya', PhD dissertation. Oxford: University of Oxford.

Desta, S., and D.L. Coppock. 2004. 'Pastoralism Under Pressure: Tracking System Change in Southern Ethiopia', *Human Ecology* 32(4): 465–86.

Dietz, T.1993. 'The State, the Market, and the Decline of Pastoralism: Challenging some Myths, with Evidence from Western Pokot in Kenya/Uganda', in J. Markakis (ed.), *Conflict and the Decline of Pastoralism in the Horn of Africa*. London: Macmillan, pp.83–99.

Dundas, K.R. 1910. 'Notes on the Tribes Inhabiting the Baringo District', *Journal of the Royal Anthropological Institute* 40: 51–52, 56, 59.

Dyson-Hudson, R., and N. Dyson-Hudson. 1980. 'Nomadic Pastoralism', *Annual Review of Anthropology* 9: 15–61.

Dyson-Hudson, R., and D. Meekers. 1999. 'Migration across Ecosystem Boundaries', in M.A. Little and P.W. Leslie (eds), *Turkana Herders of the Dry Savannah: Ecology and Biobehavioural Response of Nomads to an Uncertain Environment*. Oxford: Oxford University Press, pp.303–14.

Fratkin, E. 1997. 'Pastoralism: Governance and Development Issues', *Annual Review of Anthropology* 26: 235–61.

——— 2001. 'East African Pastoralism in Transition: Maasai, Boran, and Rendille Cases', *African Studies Review* 44(3): 1–25.

Fratkin, E., and T. McCabe. 1999. 'East African Pastoralism at the Crossroads: An Introduction', *Nomadic Peoples* 3(2): 5–15.

Fratkin, E., and E.A. Roth (eds). 2005. *As Pastoralists Settle: Social, Health, and Economic Consequences of Pastoral Sedentarization in Marsabit District, Kenya*. New York: Kluwer Academic/Plenum Publishers.

Galaty, J.G. 1993. 'Maasai Expansion and the New East African Pastoralism', in T. Spear and R. Waller (eds), *Being Maasai: Ethnicity and Identity in East Africa*. London: James Currey, pp.61–86.

Galvin, K.A., and M.A. Little. 1999. 'Dietary Intake and Nutritional Status', in M.A. Little and P.W. Leslie (eds), *Turkana Herders of the Dry Savannah: Ecology and Biobehavioural Response of Nomads to an Uncertain Environment*. Oxford: Oxford University Press, pp.125–46.

Galvin, K.A. et al. 2001. 'Impacts of Climate Variability on East African Pastoralists: Linking Social Science and Remote Sensing', *Climate Research* 19: 161–72.

Geertz, C. 1963. *Agricultural Involution: The Processes of Ecological Change in Indonesia*. Berkeley: University of California Press.

Goldtschmidt, W. 1974. 'The Economics of Brideprice among the Sebei (Uganda) and in East Africa', *Ethnology* 13: 311–33.

——— 1976. *Culture and Behaviour of the Sebei: A Study in Continuity and Adaptation*. Berkeley: University of California Press.

Gray, S., P. Leslie and H.A. Akol. 2002. 'Uncertain Disaster: Environmental Instability, Colonial Policy, and the Resilience of East African Pastoral Systems', in W.R. Leonard and M.H. Crawford (eds), *Human Biology of Pastoral Populations*. Cambridge: Cambridge University Press, pp.99–130.

Greiner, C. 2012. 'Unexpected Consequences: Wildlife Conservation and Territorial Conflict in Northern Kenya', *Human Ecology* 40(3): 415–25.

Herren, U. 1991. 'Socioeconomic Strategies of Pastoral Maasai Households in Mukogodo, Kenya', PhD dissertation. Bern: University of Bern.

Hobley, C.W. 1906. 'Notes on the Geography and People of the Baringo District of the East Africa Protectorate', *Geographical Journal* 26: 471–81.

Johnson, B.R. 1990. *Nomadic Networks and Pastoral Strategies: Surviving and Exploiting Local Instability in South Turkana*. Ann Arbor: UMI Research Press.

Kerven, C. 1992. *Customary Commerce: A Historical Reassessment of Pastoral Livestock Marketing in Africa*. London: Overseas Development Institute.

Kiage, L.M. 2007. 'Vegetational Change and Land Degradation in the Lake Baringo Area, Kenya, During the Late Holocene: Evidence from Paleorecord and Remote Sensing', PhD dissertation. Baton Rouge: Louisiana State University.

Kipkorir, B. and A. Welbourn. 1973. *The Marakwet. A Preliminary Study*. Nairobi: East African Literature Bureau.

Klumpp, D., and C.A. Kratz. 1993. 'Aesthetics, Expertise, and Ethnicity: Okiek and Maasai Perspectives on Personal Ornament', in T. Spear and R. Waller (eds), *Being Maasai: Ethnicity and Identity in East Africa.* London: James Currey, pp.195–222.

Kurita, K. 1983. 'Marakwet at Chesegon Village: A Preliminary Report on Economic Activities of the Pokot and the Marakwet', in B.E. Kipkorir, R.C. Soper and J.W. Ssennyonga (eds), *Kerio Valley, Past, Present, and Future: Proceedings of a Seminar Held in Nairobi at the Institute of African Studies, University of Nairobi, May 21-22, 1981.* Nairobi: Eleza Services, pp.61–74.

Lamphear, J. 1976. *The Traditional History of the Jie in Uganda.* Oxford: Clarendon Press.

——— 1988. 'The People of the Grey Bull: The Origin and Expansion of the Turkana', *Journal of African History* 29: 27–39.

——— 1994. 'The Evolution of Ateker "New Model" Armies: Jie and Turkana', in K. Fukui and J. Markakis (eds), *Ethnicity and Conflict in the Horn of Africa.* London: James Currey, pp.63–92.

Little, P.D., H. Mahmoud and D.L. Coppock. 2001. 'When Deserts Flood: Risk Management and Climatic Processes among East African Pastoralists', *Climate Research* 19: 149–59.

Little, P.D., et al. 2001. 'Avoiding Disaster: Diversification and Risk Management among East African Herders', *Development and Change* 32: 401–33.

Lokuruka, M.N., and P.A. Lokuruka. 2006. 'Ramifications of the 1918 Turkana Patrol: Narratives by Ngturkana', *Ethnohistory* 53(1): 121–41.

McCabe, T. 1997. 'Risk and Uncertainty among the Maasai of the Ngorongoro Conservation Area in Tanzania: A Case Study in Economic Change', *Nomadic Peoples* 1: 54–65.

——— 2004. *Cattle Bring Us to Our Enemies: Turkana Ecology, Politics, and Raiding in a Disequilibrium System.* Ann Arbor: University of Michigan Press.

McGovern, S. 1987. 'The Last Straw: Camels in East Pokot', unpublished manuscript.

Markakis, J. (ed.). 1993. *Conflict and the Decline of Pastoralism in the Horn of Africa.* London: James Currey.

Östberg, W. 2004. 'The Expansion of Marakwet Hill-furrow Irrigation in the Kerio Valley of Kenya', in M. Widgren and J.E.G. Sutton (eds), *Islands of Intensive Agriculture in Eastern Africa.* Oxford: James Currey, pp.19–48.

Österle, M. 2007. 'Armed Economies, Militarized Identities, Excessive Violence: Automatic Rifles and the Transformation of Nomadic Pastoralism in Northwest Kenya', in M. Bollig et al. (eds), *Aridity, Change and Conflict in Africa.* Cologne: Heinrich–Barth–Institut, pp.193–222.

——— 2008a. 'Innovation und Transformation bei den Pastoralnomadischen Pokot (East Pokot, Kenia)', PhD dissertation. Cologne: University of Cologne.

——— 2008b. 'From Cattle to Goats: The Transformation of East Pokot Pastoralism in Kenya', *Nomadic Peoples* 12(1): 81–91.

Österle, M., and M. Bollig. 2003. 'Continuities and Discontinuities of Warfare in Pastoral Societies: Militarisation and the Escalation of Violence in East and North East Africa', *Entwicklungsethnologie* 12: 109–43.

Peristiany, J.C. 1951. 'The Age-Set System of the Pastoral Pokot', *Africa* 21: 188–206.

Robertshaw, P. 1990. 'Early Pastoralists of South-western Kenya', *Memoirs of the British Institute in Eastern Africa* 11: 16–35.

RoK. 2001. 'The 1999 Population and Housing Census: Counting our People for Development'. Nairobi: Government of the Republic of Kenya.

Saltlick, A. 1991. 'Baseline Data Survey in the Nginyang and Tangulbei Divisions of Baringo District, Kenya', unpublished manuscript.

Schlee, G. 1989. *Identities on the Move: Clanship and Pastoralism in Northern Kenya.* Manchester: Manchester University Press.

Schneider, H.K. 1953. 'The Pakot (Suk) of Kenya with Special Reference to the Role of Livestock in their Subsistence Economy', PhD dissertation. Ann Arbor: University of Michigan.

——— 1957. 'The Subsistence Role of Cattle Among the Pakot and in East Africa', *American Anthropologist* 59: 279–300.

——— 1959. 'Pokot Resistance to Change', in W.R. Bascom and M.J. Herskovits (eds), *Continuity and Change in African Cultures.* Chicago: University of Chicago Press, pp.144–67.

Schwartz, J.H. 2005. 'Ecological and Economic Consequences of Reduced Mobility in Pastoral Livestock Production Systems', in E. Fratkin and E.A. Roth (eds), *As Pastoralists Settle: Social, Health, and Economic Consequences of Pastoral Sedentarization in Marsabit District, Kenya.* New York: Springer, pp.69–86.

Smith, A.B. 1992. *Pastoralism in Africa: Origins and Development Ecology.* London: Hurst.

Spencer, P. 1998. *The Pastoral Continuum: The Marginalisation of Tradition in East Africa.* Oxford: Clarendon Press.

Sperling, L., and J.G. Galaty. 1990. 'Cattle, Culture, and Economy: Dynamics in East African Pastoralism', in J.G. Galaty and D.L. Johnson (eds), *The World of Pastoralism: Herding Systems in Comparative Perspective.* New York: Guilford Press, pp.69–98.

Tiffen, M., M. Mortimore and F. Gichuki. 1994. *More People, Less Erosion: Environmental Recovery in Kenya.* Chichester: Wiley.

Verschuren, D. 2004. 'Decadal and Century-scale Climate Variability in Tropical Africa During the Past 2000 years', in R.W. Battarbee, F. Gasse and C.E. Stickley (eds), *Past Climate Variability through Europe and Africa.* Dordrecht: Springer, pp.139–58.

Verschuren, D., K.R. Laird and B.F. Cumming. 2000. 'Rainfall and Drought in Equatorial East Africa During the Past 1,100 Years', *Nature* 403: 410–14.

Von Höhnel, L. 1897. *Zum Rudolph-See und Stephanie-See: Die Forschungsreise des Grafen Samuel Teleki in Ost-Aequatorial-Afrika, 1887–1888.* Berlin: Alfred Hölder.

White, C. 1997. 'The Effect of Poverty on Risk Reduction Strategies of Fulani Nomads in Niger', *Nomadic Peoples* 1: 90–107.

Wipper, A. 1977. *Rural Rebels: A Study of Two Protest Movements in Kenya, 1942–1962.* Nairobi: Oxford University Press.

Zaal, F., and T. Dietz. 1999. 'Of Markets, Meat, Maize and Milk: Pastoral Commoditisation in Kenya', in D.M. Anderson and V. Broch-Due (eds), *The Poor Are Not Us: Poverty and Pastoralism in Eastern Africa.* Oxford: James Currey, pp.163–98.

Chapter 10

Social-ecological Change and Institutional Development in a Pastoral Community in North-western Namibia

Michael Bollig

Changing Land Tenure in North-western Namibia

Pastoral nomadism in north-western Namibia has been depicted as well-adapted to the hazards of an arid environment (Malan 1972, 1995; Steyn 1977). Society and culture have been often interpreted as being determined by the vicissitudes of seasonal and inter-annual resource scarcity.[1] Drought is generally conceptualised as the major risk within this semi-arid environment (Malan 1995). However, accounts which argue this glance over the fact that human–environment relations in north-west Namibia have changed rapidly throughout the past one hundred years due to governmental measures. Herding in north-western Namibia was formed by and impacted on by the political formation constituting the segregationist South African state until the 1990s. The colonial encapsulation of the pastoral community in the 1920s and 1930s starkly diminished the role of trade and exchange with neighbouring communities but apparently led to a thorough pastoralisation of north-western Namibia's population (Bollig 1998a, 1998b). The heavy-handed control of pastoral mobility conditioned early colonial land-use patterns significantly (Bollig 1998b; van Wolputte 2004). The pastoral land-use pattern which is nowadays seen as the traditional way of Himba pastoralism came into being only in the 1960s when a great number of newly drilled boreholes enabled the expansion of pastoralism and gardening.

A Short Note on Fieldwork

Between 1994 and 1996 I gathered information on pasture management among the Himba between Okangwati and the Kunene River. The management of common pool resources seemed to be fairly settled and successful at this time. This was not only my impression but also that of other specialists working in the wider area (Behnke 1998a; 1998b: 23; Schulte 2002: 101). I described Himba coping mechanisms as traditional and long-lasting (Bollig 1997). In 2004/5 and 2009 I restudied the communal management of natural resources, and to my astonishment not much of the stability I had observed a decade earlier had remained. Uncontrolled access to pasture, pasture degradation and the assumed propensity of the government to intervene in grazing matters were listed as major risks by local herders. In-depth research on the history of resource management showed that the 'resilient' system of resource tenure and risk management I had observed in the mid 1990s and which informants had dubbed as 'traditional' had only existed for a few decades. What I had perceived as a traditional and successful state of common pool resource management now seemed to be a transient phase of stability within a larger sequence of social-ecological change characterised by ruptures as much as by phases of continuity.[2] The data analysed in this contribution come from archival research and fieldwork carried out between 1994 and 2009 (Bollig 2006).

Land Tenure and Mobility before the 1950s

The pre-1950 system of grazing rotation and land tenure was strikingly dissimilar to the system observed in the 1990s, as interviews conducted on the change of settlement patterns all over the northern Kunene Region showed: livestock camps moved out to graze distant pastures during the rainy season (and not during the dry season as today) when pans and other seasonal water sources offered enough water. Cattle grazed in orbits around these seasonal pools and once these rain-dependent water resources fell dry, the mobile cattle camps had to retreat to settlements near permanent water along rivers and permanent wells. In the Omuramba area (Figure 10.1), households from Ombuku River migrated to Omuramba with their large cattle herds during the rainy season and on to Ondova with their oxen herds.

Once the water sources in these areas had dried up they returned to the Ombuku River, where water could easily be found in the river's sandy deposits. The group of users of specific resources was clearly defined. Households held tenure rights in specific places that had permanent water. Large tracts of land were not used at all.

Places with plentiful permanent water sources were often dominated by powerful leaders. Their power rested on their wealth in cattle, their possession of guns, the number of their clients and their claim to land. These pastoral big men acted within an economy in which many still subsisted partially on small stock only or on combinations of small-stock husbandry and foraging. Scientists and administrators travelling the region in the late 1910s and early 1920s depicted the population as small stock breeders and emphasised foraging strategies (Kuntz 1912; Hartmann 1941: 32).[3] Pastoral *ovahona* also maintained gardens to sustain their households. Vedder met the local leader Tjongoha at Kaoko Otavi's permanent well and reports that Tjongoha was an acknowledged and wealthy leader in

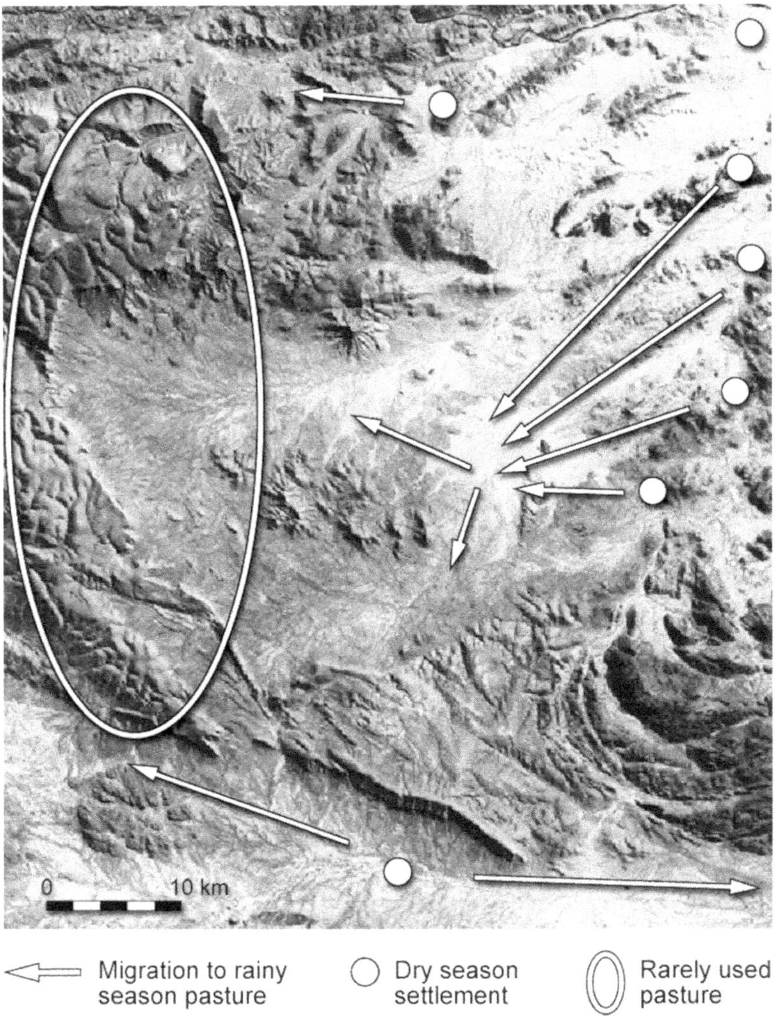

Figure 10.1: Map of pastoral tenure and mobility in wider Omuramba-Epupa before 1960.

the environs of the place. He addresses him as the 'owner' of the affluent Kaoko Otavi well and mentions that he borrowed cattle from the 'king' of the neighbouring Ovambo polity, Uukwaludhi.[4] Big men migrating into the Kaokoveld either from Angola or from the settler frontier in the Kamanjab/Outjo region targeted permanent water points in the Kaokoveld to settle there. Habitually, an acknowledged leader was surrounded by relatives and clients. Leaders mentioned in oral traditions and archival sources for the 1920s and 1930s typically had either been mercenary leaders in Portuguese Angola, had been engaged in the Portuguese colonial administration in other ways (such as Adrian Karipose), had close contact with the leading clique in neighbouring Sesfontein (Kakurukouye, for example), or originated from prominent Herero families fleeing the genocidal war of 1904 to1907. Gaerdes gives an account of a visit to Vita Tom's settlement in the early 1920s (see K'Ozombaze 1970): he describes lush gardens in the surroundings of the settlement and hints at numerous horses. Leaders competed for the better wells in the area, sometimes violently and sometimes in battles of words and through intrigues in which they tried to involve the colonial administration from the late 1910s onwards (Rizzo 2007).[5]

I will shortly address salient features of the pre-1950s common pasture management system, which I reconstructed mainly on the basis of oral evidence: Households 'owned' specific places and a clearly defined and numerically small number of households managed pastures together. The heads of these place-owning households were addressed as 'owners of the earth/land' (*oveni vehi*, pl.; *omuni wehi*, sing). The concept *omuni wehi* is of old Bantu stock and stresses the ritual affiliation of a senior man to a place. An *omuni wehi* is not necessarily rich in livestock and his ritual affiliation does not provide him with exclusive access rights. While he should be asked for the permission to settle by new-comers, he can rarely deny such a quest in a straightforward manner. It is suggested however, that his ritual power provides those living in the place with ancestral blessings and diverts hazards from their livelihood. The more affluent water points were usually also associated with a big man (*omuhona*, sing.; *ovahona*, pl.). The term *omuhona/ovahona* denotes a person who is wealthy and who wields power through his wealth in livestock, clients and possibly weapons. The typical *omuhona* is also a vociferous and gifted speaker at public meetings.[6] While the access rights of *oveni vehi* to land were mainly based on ancestral beliefs, ritual linkages of living household members to ancestral graves and tied to specific places, the access rights of *ovahona* were based on wealth and power. While both concepts were clearly not identical, there was the quest of place-owning *ovahona* to gain acceptance as *oveni vehi* or to liaise closely with an *omuni wehi* in an attempt to bring ritual power and material dominance together. In this way, claims to specific wells were legitimised through the prior dwelling of ancestors and the presence of ancestral graves in the area.

Big men established settlements which were constituted by close patrilineal and matrilineal relatives and clients, who often also claimed some kind of kinship relation with the local big man. Typically clients were poorer herders and/or foragers who often acted as herdsmen to more endowed households. It is especially the larger water points which allowed for some agriculture and sedentary living. Major wells such as Otjiyandjasemo and Kaoko Otavi sustained small villages with a population of a few hundred people (see, e.g., K'Ozombaze 1970; Stals and Reiner-Otto 1999; see also Figure 10.2). For the larger wells, the political domination of leaders (through the possession of 'many guns' and 'good relations with the colonial administration') is emphasised in oral accounts; for other water sources (especially in the Himba context), ritual domination (including the 'presence of ancestral graves') is emphasised in explanations of why some people have access rights to a well and others do not. The absence of water in large tracts of land significantly reduced choices for mobility during the major part of the year.

During the rainy season, specific pastures around pools were used commonly by herders from that village.

We know little about rules of resource protection for the pre-1950 period. Several informants claimed that such rules were not necessary at all as there were very few people and much less livestock than today. Additionally we may add that the water sources in the rainy-season grazing areas were usually exhausted a few months after the rains. This brought a rapid end to the exploitation of these pastures and there was no need to develop rules of protection or equality of access to them.

What role did the colonial government play in this context? Botha (2005: 183) reports that before 1950 there was hardly any effort to intervene in the farming of reserve dwellers. During the first thirty years of South African rule the impact of the colonial government on resource management in north-western Namibia was moderate or, to be more precise, solely targeted pastoral mobility and the maintenance of boundaries. Three big men were nominated as chiefs and were given reserves with more or less fixed boundaries in the northern parts of the Kaokoveld. Migrations across chiefdom boundaries were controlled for some time. For about twenty years (between 1925 and 1945), settling at the Kunene River was prohibited. In the southern parts of the Kaokoveld a no-go zone of about 100 kilometres in width was established between a line crossing in an east–west direction through Ombombo and the commercial farming zone in 1929. In order to free this zone from population, around 1,200 people with about 10,000 head of livestock had to be relocated (Bollig 1998b: 511) to the central parts of the Kaokoveld. Mobility between the westernmost areas of the Ovambo polities of Uukwaludhi, Uukuolonkadhi and Ongandjera was also strictly controlled. Here, too, a buffer zone of 10 kilometres in width was established. Herders

Figure 10.2: Vita Tom and Muhona Katiti in the 1920s.

found settling in these prohibited zones were punished summarily: habitually one or two animals in their herds were shot and they had to evacuate the place immediately. While the delineation of chiefdom boundaries did not have a great impact on mobility and was circumvented frequently, the prohibition of settlement along the Kunene severely upset transhumant cycles of those living in the vicinity of the river. Generally, the South African administration acknowledged and sustained what they perceived as local authority structures. Where no such structures existed they were ambitious in pushing for them. However, the administration did not impact greatly on resource management as such. Such attempts at modernisation would have contradicted the conservative ideology of the administration. The long-serving Native Commissioner for the region, Cocky Hahn, was adamant that what he thought to be the traditional set up should be retained to maintain 'a healthy' tribal situation (Hayes 1998, 2000). Such a situation was thought to be the basis for a disciplined African labour force which could then be used in the mines and the commercial ranching area to the south.

What were the risks of this system and how were pastoral strategies adapted to such risks? A major risk of crucial importance today – the non-availability of fodder in dry-season pastures – was simply structured in another way. Homesteads returned to permanent wells during the dry season and left outlying pastures once the last pools there had dried up. They did not have to care whether these pastures were still productive and offered enough grazing for the dry season: the system was grafted upon the premise that such pastures were only used during the rainy season. The risk of lack of pasture was more specifically connected with dry-season grazing around the few permanent wells. There was a number of risks directly linked to them: first, insufficient grazing due to high grazing pressure as livestock had to be fed on these restricted pastures for many months (throughout the 1950s cattle numbers rose considerably from around 50,000 at the end of the 1940s to 120,000 at the end of the 1950s; see Figure 10.4); second, the denial of access to key resources due to a lack of stable social links to people 'owning' the resource; and third, the eviction of one's own group by another, militarily stronger and/or politically better connected group of resource users.

There was a number of risk management strategies directly answering these challenges. A major percentage of the Kaokoveld's population continued to survive well into the twentieth century as foragers or as pastoralists with few livestock. The fact that only a small number of homesteads in each community owned substantial numbers of cattle made control easier. Access to a permanent waterhole and its surrounding resources were guaranteed by social strategies effectively tying individuals into patron–client networks. Oral traditions often mention such strategies. Kinship ties to patrons were eagerly forged. While local authority structures since the 1950s were based on administrative chiefs who controlled a specific territory, this earlier political system was structured by patrons who controlled numbers of clients. The threat of being outbalanced by other groups could be countered by a number of strategies: first, enlarging the number of clients, as the larger the number of clients of a big man, the less the chance that the community could be driven off the land; second, improving the military strength of the community, as cattle and especially ivory were invested in weapons and horses; third, creating reliable linkages to the administration, or 'capturing' the resident Native Commissioner in a patron–client network by making him the super-patron; four, creating symbolic ties to the land through ancestral graves which in conflicts could serve as reference points for claims of legitimacy.

Only in the 1920s was a pastoral mode of production firmly re-established in the Kaokoveld after the region had been used by foragers for some decades after earlier pastoral communities had fled or had been robbed of their livestock. Throughout the 1920s a majority of the population lived on small-stock herding and foraging strategies, and only a minority of politically dominant big men had major cattle herds. Throughout the 1930s and 1940s the number of cattle grew: a first survey in the late 1920s, of dubious accuracy, estimated some 6,000 to 9,000 cattle; in 1942, 36,200 cattle and 56,600 small stock were counted (Bollig 2006: 71–72), while in the late 1950s the number of cattle rose above 100,000 for the first time. The pastoral mode of production which emerged in the 1920s hinged upon the control of permanent water points. Livestock herds were kept in the vicinity of these permanent water sources during the long dry season and then were taken to access outlying pastures in the rainy season. Local big men and a small number of chiefs dominated the economy and local politics: they were prosperous in cattle and established patron–client networks through loans of cattle. They also brokered contacts with the colonial administration. Through such exchanges, channelled within patron–client networks, the society of the Kaokoveld became more pastoral as more and more people gained access to cattle. Whereas in the 1920s only a minority of households kept large cattle herds, by the end of the 1950s the Himba and Herero of the Kaokoveld were truly cattle keepers.

Was the social-ecological system of the 1920s to 1950s resilient? In many ways, yes: its basic structure could be maintained despite severe droughts in 1915, 1929 to 1932 and 1941, and throughout the century we witness an intensification of livestock husbandry in the region. The system also withstood the imposition of colonial rule. A researcher working with the resilience concept in the 1930s would probably have regarded local pastoralism as highly resilient. Major subsistence crises in the wake of droughts were countered by a number of buffering mechanisms, such as reliance on drought foods (roots, tubers, berries) or far-ranging migrations (ibid.: 192–196). But pastoral production and the tenure regime changed massively when the availability of grazing resources was enlarged when a comprehensive borehole drilling programme began in the late 1950s.

Post-1950 Attempts at Modernisation: The Borehole Drilling Programme

After the official introduction of Apartheid in South Africa in 1948, policy towards the African reserves also changed profoundly in Namibia. The Long-term Agricultural Policy Commission (LAPC) of 1948 found that 'the limit of carrying capacity ... has been reached' (cited in Botha 2005: 177) and recommended soil conservation, appropriate stocking rates and improved livestock breeds. Reports of catastrophic degradation of Southern African savannahs due to over-exploitation haunted the public media (ibid.: 176). Reserve economies were targeted for modernisation, so that they produced enough food for a growing population. Botha reports that 'modernisation theory ruled and went with disdain for any form of local, especially African, knowledge' (ibid.: 170–71). Traditional authorities were regarded as important partners in the policy of modernisation.[7] The decision on how and where to start modernisation programmes was given to experts, who emerged as a new group of administrative actors: where a Native Commissioner had ruled until then, he now had to share his powers with experts sent from the centre. Whereas the Kaokoveld had been cordoned off beforehand, after 1950 a number of scientific studies were undertaken in the area: geological and hydrological (e.g., Abel 1954), anthropological (van Warmelo 1951; Malan 1974) and zoological (Viljoen 1988). In 1957 the Kaokoveld became the 'independent' reserve of Kaokoland under the direct authority of the chief Native Affairs Commissioner in Windhoek (van Wolputte 2004: 163).

Scientists, especially geologists, hydrologists and agriculturalists, became influential and grounded much of the political decision-making process with data. While in the 1930s a few waterholes had been blasted, a regular borehole programme was launched in the mid 1950s. In a letter to Native Commissioners, the chief Bantu Commissioner in Windhoek argued in 1955 that development (*ontwikkeling*) in reserve areas was necessary and that borehole drilling was an important step in that direction. He asked Commissioners to identify areas which could be 'opened up' for human exploitation by boreholes, and in the late 1950s and early 1960s boreholes were drilled in considerable numbers.[8] The borehole programme was to fulfil two purposes: it was to modernise local livestock husbandry, and to win the 'hearts and minds' of local leaders at the same time.

In 1963 the Odendaal Commission went into the Kaokoveld with the task of preparing plans for giving the area semi-independence as a 'homeland'. The Odendaal Commission's report (RSA 1964: 183 [Odendaal Report]) then recommended boreholes as the key to economic development in the vast semi-arid savannah of the Kaokoveld and argued that drilling boreholes would mean that pastures rarely used until then could be fully exploited. Water was regarded as a key resource which had to be exploited with new technological means (ibid.: 411). Throughout the latter part of the 1960s and through the 1970s and 1980s, boreholes were drilled in large numbers in the Kaokoveld. In the 1950s some 43 boreholes were drilled; these were all around the homeland capital Opuwo and to the south-east. During the 1960s another 136 boreholes were drilled, with numerous others in the ensuing decades: 128 in the 1970s, 57 in the 1980s, and some 40 in the 1990s some (Figure 10.3). Generally, local chiefs applied to the Department of Water Affairs for the placement of a borehole in their area, and they were then consulted by local officials to identify its optimal placement.[9] At the very beginning of the borehole programme, chiefs and their councillors openly resisted the initiative: they alleged that the government was keen to drill boreholes in order to extend its control over the pastoral population, a claim that was not far from the truth.[10] Some leaders advocated that boreholes should only be drilled with money collected by the herders themselves, while others vociferously demanded that pastoral lands that had been allocated to Etosha National Park needed to be returned before any other meaningful development

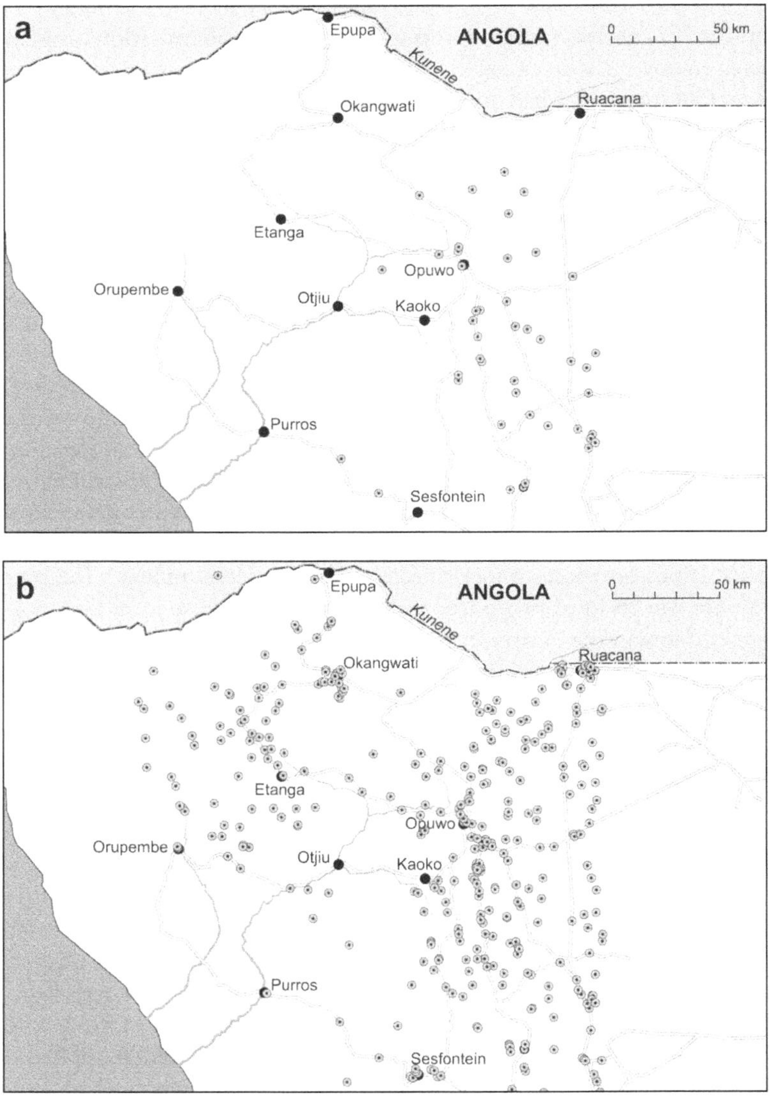

Figure 10.3: Borehole drilling in Kaokoland. a: boreholes drilled in the 1950s; b: boreholes drilled until the 1990s.

could take place. From the mid 1960s, however, the programme ran smoothly and traditional authorities apparently realised that the government had given them a very powerful means of control and dominance: they could determine where boreholes were placed, thereby expanding opportunities for themselves and their followers. During the mid 1960s a lot of meetings consisted of formal applications by chiefs for boreholes. Boreholes were carefully registered and fully maintained by the Department of Water Affairs. The drilling of boreholes, however, also entailed the rhizomatic extension of the state, which penetrated the local economy and pastoral landscape effectively and irrevocably.[11] Boreholes and a network of roads connecting them constituted the pastoral landscape and largely impacted upon settlement and mobility.

The development of a network of boreholes in former dry-season grazing areas between the 1960s and the 1970s led to a reversal in mobility patterns. Herds could now stay in former rainy-season areas during the dry season as they were no longer dependent on rainwater. Much wider areas than before were thus made accessible for grazing. A look at the increase in cattle numbers suggests some relation to the borehole drilling programme (see Figure 10.4). During the mid 1950s, cattle numbers increased to above 60,000 head and then quickly rose to about 120,000 in 1958. While it is true that the borehole drilling programme began at that time, the presence of additional water sources cannot explain the rapid increase alone: most of the 1950s were very good rain years, with annual precipitation well above the long-term average (Bollig 2006: 112–13). A major drought at the end of the 1950s, which ran into the early 1960s, brought cattle numbers down again to 65,000. Unfortunately we lack livestock counts for most of the 1960s, but by 1973 cattle numbers had reached an all-time high of some 140,000 head. I surmise that the rapid increase in livestock numbers was only possible because of the expansion of boreholes.

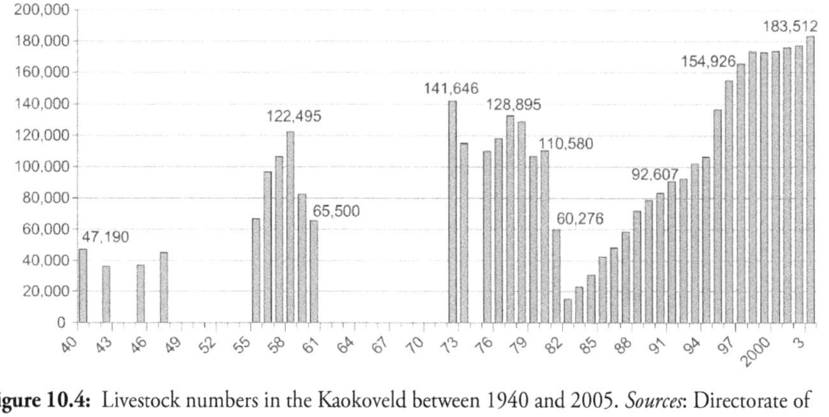

Figure 10.4: Livestock numbers in the Kaokoveld between 1940 and 2005. *Sources*: Directorate of Veterinary Services internal files; Page (1976); van Warmelo (1951).

A new set of rules had to be developed to address questions pertaining to the regulation of access to pastures and to the sustainability of their use. The herds stayed at permanent water points along rivers during the rainy season and then moved out to outlying pastures during the dry season where water was now provided by boreholes. This also made more labour investment in agriculture possible. Whereas, prior to the drilling of boreholes, during the rainy season most people congregated close to the rivers, and thus far from arable land, when planting could take place, now they could converge on these lands during the rainy season. Sander, Bollig and Schulte (1998) as well as Welle (2007) document the rapid increase in gardens in the Omuhonga basin since the mid 1970s (Figure 10.5).

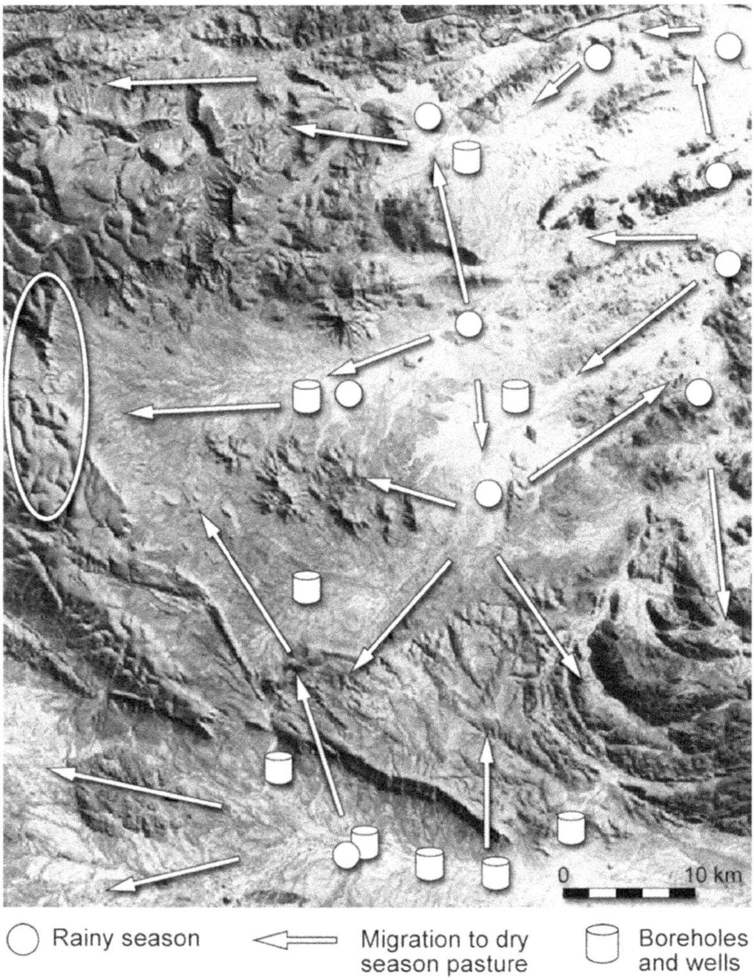

Figure 10.5: Pastoral tenure and mobility in the 1990s.

The situation also changed in the wider Epupa region with the development of water resources in the area (see Figure 10.5). Boreholes were drilled at Omuramba, Otjikango and Omuhandja in the 1960s and 1970s, and a dam was built at Okombanga near Omuramba. At the same time further boreholes were installed at other places in the region (the Omuhonga Valley, Ominyandi). What had once been rainy-season pastures now became dry-season pastures. However, this also meant that whereas they had previously been grazed for three or four months they were now used for a much longer time. The major focus of settlement now became Omuramba, replacing Ombuku, a place with permanent water at a perennial river. Only during the height of the dry season would people fall back to the Ombuku Valley where they had dwelt before over the entire dry season. The change in the grazing regime probably took place as a consequence of borehole drilling in the area.[12]

After the drought and subsequent heavy livestock losses of the early 1960s, cattle herds increased rapidly once again: within roughly a decade cattle numbers climbed from some 60,000 to about 140,000. The rapid increase of livestock was made possible by continued borehole drilling and the opening up of pastures for dry-season grazing since the mid 1950s. Whereas before the mid 1950s, with cattle numbers in the Kaokoveld not exceeding 50,000 head, there had been vast areas of unused pasture every year, now a regional cattle herd exceeding 100,000 head needed all the available grazing. Whereas cattle numbers had only exceeded 100,000 for two years before the major drought of 1960/61, livestock numbers stayed well above 100,000 over a period of ten to fifteen years before the drought of 1980/82. The intensification of mobile cattle husbandry and the accumulation of livestock led to a general increase and distribution of wealth. In the 1960s and 1970s all inhabitants of the Kaokoveld subsisted on more or less extensive livestock herds. All households owned at least some cattle and wealthy herders, such as the Himba chief Munimuhoro, may have owned well above 10,000 head distributed over a number of livestock camps.

Apparently the perennial layer of grass rapidly gave way to annual grasses, which in turn were much more susceptible to drought. The droughts of the mid 1970s and especially the major drought of 1980 to 1982 showed how vulnerable the system had become through the intense increase in herds made possible by the government-endorsed borehole programme. We may conclude that in the long run the hydrological revolution of the 1960s and 1970s made local pastoralists richer but also more vulnerable to drought.

Other risks were brought about by the borehole programme and the opening up of vast stretches of land for dry-season grazing. Competition for access to dry-season pastures became a potential source of conflict. While in the past, when these outlying pastures had only been grazed when water was available there, now everybody who could make primary use of such dry-season pastures had distinct advantages. Whereas grazing such stretches during the rainy season did not bring about the need for cooperation because pastures grazed during the rains still had considerable regrowth, now stretches grazed during the dry season could only be reused the next year. The coordination of grazing efforts and a detailed set of rules during the dry season was the only way to circumvent such conflicts. But how did herders and their communities react to these changes in the mode of production?

Changing Patterns of Mobility and Tenure as a Way to Lower New Risks

The community that managed pastures was fairly clear-cut. All households that were under one chief were allowed to use the pastures that fell within that chiefdom. Behnke (1998b: 4) observed that most Himba chieftaincies in the early 1990s were organised along drainage systems and combined river frontage and upland grazing areas. A chieftaincy in this sense formed a large and loosely integrated herding-and-resource-management unit (ibid.: 12). The administration divided the Kaokoland homeland into different subunits (*wyke*) and appointed headmen to administer them. In contrast to the prior state, a clear link between individual households and specific grazing areas was replaced by a clearer specification of the political constituency dealing with a resource – a water point was no longer under the control of a local big man but rather part and parcel of a chiefdom with rather clear-cut administrative boundaries. Increasingly chiefs and their councillors as paid semi-officials and intermediates between local communities and the state became prominent. Membership in a resource-managing community was still established via kinship links. Graves and rituals conducted at them gave proof of the legitimacy of tenure rights in a specific place.

The rules of good grazing (*ozondunino yomaryo*) apparently evolved in reaction to the implementation of the borehole programme and the altered structure of subsistence risks: first, grazing in the dry season grazing areas is prohibited during the rainy season; second, cattle camps must move a considerable distance away from the main settlement areas; third, livestock camps must move together in a group; four, herders should look for dry season grazing near to their main settlement; five, once settling in a specific place, cattle-camp owners decide in which 'direction' each herd is herded – that is, in the morning each herder takes his herd out of the camp in a specific direction; six, too much movement between camp sites (*okukandakanda*) is not appreciated; and seven, special areas should only be used during droughts and not during normal dry seasons.

These rules first of all tried to keep conflicts over grazing at bay. For example, the prohibition on settling ahead of others ensured that all herders had the same chance to exploit an area of pasture (see Behnke 1998a: 15). Rules narrowed down the choices for cattle camps and made moves more predictable. A second goal was to maintain grazing near main settlements to ensure the supply of milk to elderly folks and children during the dry season when the majority of cattle were taken to outlying pastures; a third intention may have been to ensure a sustainable management of pastures.

Grazing guards (*ovatjevere vomaryo*) were appointed. These were chosen by a meeting of the community. The fact that these men were also addressed as 'grazing police' (*ovapolise vomaryo*) shows that they were regarded as an extension of the homeland bureaucracy. Formally these men were entitled to inspect the area for homesteads and camps which did not adhere to the rules. However, everybody could bring up complaints against a neighbour or herder they thought was breaking the rules. At first a meeting took place at the neighbourhood level. The accused was summoned and given time to explain their case in an ad hoc meeting. Usually no punishment was meted out at these meetings and, if possible, those involved tried to reach a consensus on how to change the situation for the better. If these minor meetings at a neighbourhood or cattle-camp level did not lead to a decision, the case was then referred to a meeting at the chief's home.

The South African government encouraged chiefs to punish wrongdoers 'according to tradition'. Chiefs occasionally resorted to flogging, which was rather in the tradition of the colonial administration than legitimised within the local authority system. Furthermore, trespassers had to pay a fine, usually an ox. When several oxen had come together, these were driven to the Native Commissioner's compound in Opuwo. In the 1990s physical punishment was no longer practised and fines were invariably paid in livestock. These were either paid to the neighbourhood council or to the men gathered in the meeting at the chief's home and no longer transferred to the administration, although such fines were still often addressed as 'government animals'. Usually animals paid as fines to the men gathered in the meeting were immediately slaughtered to provide those attending the meeting with meat.

Pastoral intensification, tightly linked to the expansion of boreholes, led to some changes in the pastoral mode of production. The availability of water in many places made permanent settlement possible in localities which until then had only served as seasonal grazing areas. This change in mobility pattern brought about an increased independence of cattle-camp owners. Whereas cattle camps had previously stayed away from permanent settlements for a few months only during the rainy season, now they were away from settlements for most of the year. In the 1990s I came across many cattle camps which had not joined their main households at all over the course of a few years. At the same time, labour allocation within households changed: the increase in gardening activities brought about more agricultural labour, especially for women. Most heavy manual labour connected to gardening was sourced out to labourers from southern Angola. In several places, prior main settlements were given up or declined in importance as central places and settlements shifted to the proximity of boreholes. This led to a wider distribution of a growing population across the landscape. Accounts of the deeds of and conflicts involving *ovahona* so prominent in the oral lore of the pre-1950s are conspicuously rare for the 1960s and 1970s. Whereas before the 1950s big men dominated the pastoral setting, since the 1960s pastoral wealth became more widespread. For the later period, the term *omuhona* either denoted the administrative 'traditional' authorities (chiefs and councillors) salaried by the government or local wealthy and politically influential herders. The continuous large livestock numbers herded in the region allowed for the expansion of intricate and expansive livestock exchange networks. Bollig describes how large numbers of cattle are invested in social exchange and leave the owner's herd as 'loaned cattle' (*ozondisa*) or 'presented animals' (*ozohakera*). Matrilineal inheritance also provides for the distribution of livestock herds in the region (Bollig 2006: 294).

Chiefs were co-opted by the colonial government and became salaried administrators, allotted wards as administrative units. Hence they relied less on power to sustain their dominance but rather fostered their links to the colonial administration. In 1995 the socio-ecological system appeared to be fairly resilient. It had gone through some major disturbances in the 1980s: civil war and military rule had negatively affected pastoral mobility, and a major drought from 1980 to 1982 had killed 90 per cent of the region's cattle – and still the system continued to exist without collapsing into a qualitatively different state, controlled by a different set of processes.

However, changes in the original mobility pattern brought about profound changes in rangeland vegetation. The pre-1950s system characterised by intense grazing on outlying pastures during the rainy season – that is, for only three to four months during the growing period of rangeland vegetation – favoured perennial grasses. In contrast, the new mobility pattern, which meant intense stress on vegetation during the dry season, disadvantaged perennial grasses and advantaged annual species. The new rules stipulated that outlying pastures should not be grazed during the rainy season, thereby ensuring that annual grasses were not disturbed during their main growing period. Through the rules of 'good grazing', annual grasses on dry-season pastures were protected until they had seeded, thus ensuring their reproduction. In contrast, perennial grasses were massively disturbed during a period of time when, due to a lack of moisture in the soil, they could not sufficiently recuperate. This led to a rather rapid change from pasture dominated by perennials to pasture dominated by annuals (see Figure 10.6). The major drought-induced livestock mortalities of the early 1980s – the number of cattle dropped from about 120,000 to about 15,000 animals in a period of only two years (ibid.: 166–67) – may have brought about some reversals in this trend, but from the early 1990s grazing pressure increased, and by the end of the 1990s cattle numbers had reached about 180,000 animals.

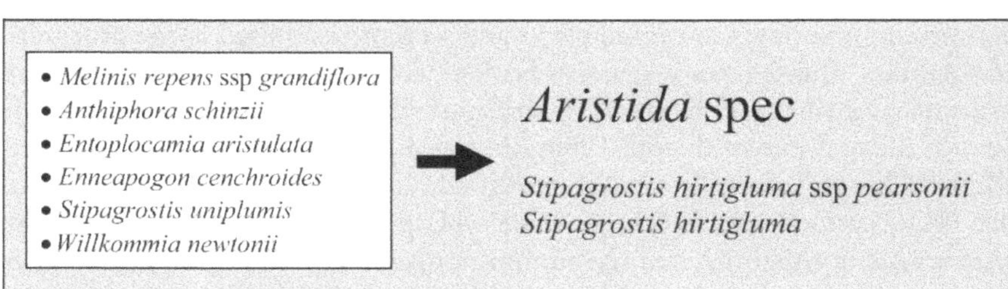

Figure 10.6: Degradation of pastures: from diversified pastures to pastures dominated by *Aristida* spec.

Consequently, a decline in the diversity of the grass and herb cover, as well as in the tree and bush cover, was observable in many areas (Bollig and Schulte 1999). In the late 1990s, remaining perennials such as *Stipagrostis uniplumis* and highly valued annuals such as *Willkommia newtonii*, *Entoplocamia aristulata* and *Anthephora schinzii* were in decline, whereas lower-valued annuals such as *Enneapogon desvauxii* and *Aristida adscenionis* were on the increase. The almost complete change from perennial to annual grasses made the local grazing system more vulnerable to drought: whereas perennial pasture may still yield some fodder in years of low rainfall, annual pastures rarely produce any significant fodder during drought years. Very high livestock numbers have altered the state of the vegetation further. Since the late 1990s a distinct trend towards the replacement of high-yielding annuals with more robust but low-yielding annuals has been observable.

With the independence of Namibia in 1990 a number of fundamental socio-political processes were set in motion which can only be touched upon briefly here (see also Miescher and Henrichsen 2000; Friedman 2005). Whereas the Kaokoveld had been a separate administrative unit with its own homeland administration, and had been isolated for many decades from the rest of the country, now linkages to actors and organisations beyond the Kaokoveld were intensified. Tourism in the region boomed and the number of tourist entrepreneurs from outside the region increased. NGO activities became prominent in various development projects. Probably due to the affiliation of some parts of the population with the South African administration and the South African army, many inhabitants of the Kaokoveld were opposed to the new independent government. When, in the mid 1990s, the government tried to push for a huge hydro-electric dam on the Kunene River against the will of the local population, this anti-government stand became even more pronounced (Friedman 2005). Nevertheless, especially since the latter part of the 1990s, a number of development initiatives have been inaugurated in the region.

Deinstitutionalisation and the Search for the Reorganisation of Common Pool Resource Management

In 2004/5 I again focused research on grazing management and, much to my surprise, a lot had changed. During interviews in 2004, respondents emphasised the high degree of detrimental vegetation change they had experienced during the past decade. It is interesting to note that ten years earlier, in 1994/95, herders had related the degradation of the vegetation to the imponderability of rainfall and rejected the idea that the phenomenon was related to high livestock numbers. In 2004, however, respondents were sure that high grazing pressure combined with institutional failure was causing severe forms of degradation in the grass and herb cover. The perceived trend becomes clear from herders' observations on specific localities: a diverse layer of perennial and annual grasses is replaced by low-yielding annual species such as *Aristida* species.

In several interviews informants claimed that the local system of pasture management had virtually crumbled. Several indicators for the demise of grazing control were given.

First, in several grazing areas there were disputes over who was allowed to make use of specific pastures. While on the one hand politically dominant figures succeeded in reserving pastures for their own herds (without fencing them), in other areas young herders had moved into areas which were far away from their regular settlement areas.

Second, Himba informants in the Epupa area stated that the sanctioning of misbehaviour had virtually been abandoned. Several reasons were given: one informant said that it was simply hard to put enough pressure on a culprit to pay their fine; others claimed that the death of some elders, who had dominated the local community as wealthy cattle patrons and who had been central to the handling of cases, had weakened grazing control; yet another informant stated that many people reasoned that it was not cattle-camp herders who should be fined but the owners of the herds in the camps who were frequently far away and had given orders to their herders to move into forbidden pasture land, or, at least, were the ones who profited from such moves. Such ambiguous considerations prevented the community from effectively defining and sanctioning misbehaviour.

Third, some significant shifts of pasture use had taken place. Some previous dry-season pastures were now permanently settled by households. The plains of Omuhandja and Ovizorombuku near the Kunene River, for example, had previously been only accessed during the dry season. Since the mid 1990s a number of households had settled there permanently exploiting the demand for 'Himba villages' for tourist entrepreneurs based in Epupa.

Several factors were apparently contributing to these institutional dilemmas. The number of cattle had nearly reached the 200,000 level, an all-time high, and to manage such large numbers of livestock within a restricted area was simply much more difficult. This increased the transaction costs of institutions: communal meetings took longer because of more community members being present and because a higher diversity of interests had to be taken into account. Informants mentioned the strain lengthy community meetings put on their time. Frustratingly, many such meetings ended without any clear decision. Large livestock numbers and the decline of grazing control had apparently contributed to a demise of pasture quality. Nomadic livestock husbandry in the region seemed to be in disarray. The demise of a grazing system which was perceived as successful in guaranteeing equal access to resources (and diminishing competition) and in safe-guarding pastures was deemed to be the major risk to communal grazing. The issue was much debated and many meetings were held on how to best curb the situation. Several solutions were proposed, some highly individualistic, some rather geared towards communal responses.

How did local herders react to the crisis? It is noteworthy that interviewed herders were unanimous in their judgement that the system was actually going through a crisis. They framed the degradation of pasture as 'the country has become weak' and then listed more detailed observations. They also reported on a great number of meetings which had taken place to find new institutional solutions to the grazing problem. Some, however, did not want to wait for new institutions to emerge. Over the few years prior to my return, a remarkable out-migration of households from the research area had taken place: some households had migrated south of Opuwo (some 150 kilometres away), some reaching the Sesfontein area (some 300 kilometres away), others going to the wider Ekoto area (about 300 kilometres); some households had left for Angola and a few had more or less permanently settled in a neighbouring mountainous area. I interviewed some of the heads of out-migrating households for the reasons for their decision. The answers were unanimous: the pastures had been bad in the research area, losses of livestock due to emaciation had been frequent, attempts at improving the situation had failed, and other areas seemed to offer better grazing. The out-migration of households from the wider Epupa area is paralleled by moves of Himba households from other areas to the southern parts of the Kaokoveld, areas which are mainly occupied by Herero herders. In a two-week-long survey we listed twenty-six Himba households in the southern parts of the Kaokoveld. It is hard to establish with what number of cattle they migrated there. In interviews, migrant herders said that they were attracted to these areas by better grazing conditions and by more stable institutions of resource tenure. Many Herero herders, however, saw these moves with some scepticism and alleged that migrant Himba herders often contributed to the demise of grazing by adding substantial numbers of cattle to the regional herd and by not adhering strictly to grazing rules.

In this situation local actors actively sought help from the government and NGOs in an attempt to reorganise resource tenure. Between 1996 and 2003 the Namibian government overhauled its legislation on natural resource management in communal areas. The direction of these legal changes were unidirectional: the state devolved authority over natural resources – game, forests, wells, pasture – to communities once such communities formalized membership, established democratic decision-making structures, clearly defined the boundaries of the resources they intended to exploit and protect, and formulated a management plan for the sustainable exploitation of key resources. Resource management communities – as conservancies, community forests, grazing associations or water-point associations – were also entitled to derive income from these resources. Conservancies could sell hunting licences to tourist entrepreneurs or lease land for the building of a lodge, and water-point associations could stipulate user fees for water points. Their major official aim was to conserve wildlife; the intangible benefits from establishing a conservancy, however, were also rated highly: tighter control over immigration, clearly demarcated boundaries and institutions for internal resource-related conflicts. The co-management of resources became an answer to the crisis of the pastoral sector; NGOs and governmental bodies were seen as allies in the search for solutions to problems which seemed unsolvable at the local level. The region's herders resorted to the conservancy movement in astonishingly high numbers and in an amazingly rapid manner: the first conservancy in the wider region was inaugurated in 1998;

only ten years later almost the entire region was parcelled up between already established or planned conservancies (see Figure 10.7).[13] At the time of writing, a number of planned conservancies in the centre of the Kaokoveld are seeking formal acknowledgement from the government, and in a few years' time probably the entire Kaokoveld will be covered by conservancies.

Many conservancies in the Kunene region are bounded by former wards which under South African rule were identical with chieftaincies: On the one hand it was much easier to resort to former chieftaincy boundaries than to define completely new ones when negotiating boundaries with neighbouring communities. On the other hand, traditional elites were keen to see their territorial structures transformed into entities of natural resource management. In order to prepare the documents for registration, boundaries of conservancies had to be clearly established using modern mapping equipment. The successful boundary demarcation of chiefdoms was frequently mentioned by traditional authorities as something beneficial. While traditional authorities had comparatively little

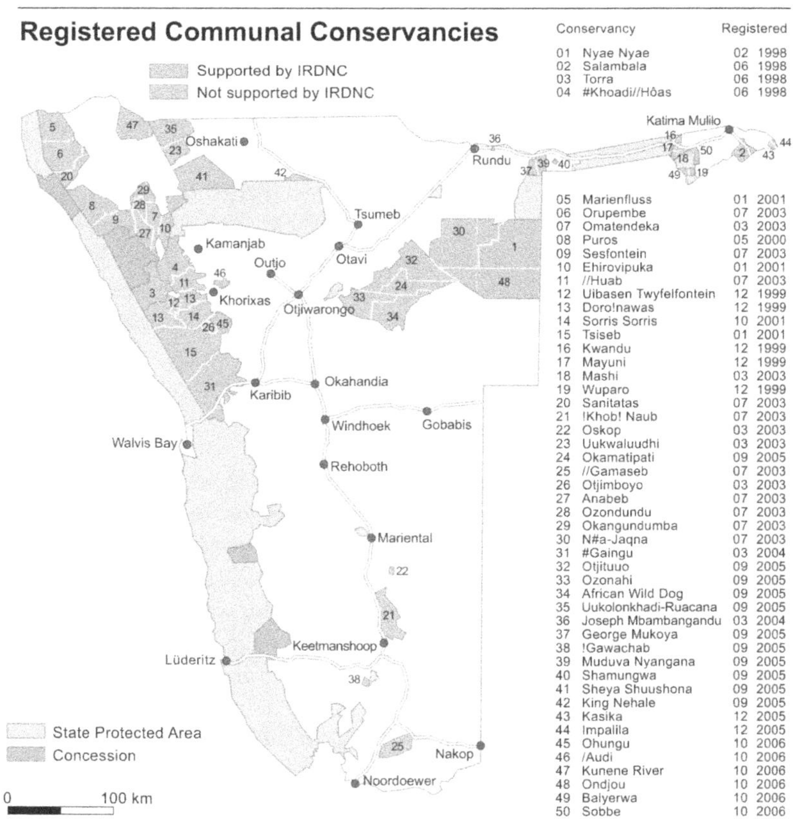

Figure 10.7: Namibia's communal conservancies (*source*: www.nacso.org).

say in governmental administrative procedures, they now became partners of NGOs. This engagement with the NGO sector also offered material rewards. While beforehand the South African government had paid chiefs' salaries, this procedure was disbanded as only two out of the Kaokoveld's thirty-six chiefs were acknowledged as traditional leaders. Through conservancies and agreements with the conservancy committees, some chiefs could divert some income from tourism into their coffers. Institutionally, they were also included in the reorganisation of resource tenure. The first issue a conservancy had to solve was defining the relation of its elected committee to the local chief and his councillors or sub-chiefs. Conservancies in north-west Namibia have solved this issue with a great enthusiasm for constitutional experiments. Conservancy constitutions have found very different ways to define the linkage with traditional authorities: in some committees the traditional authorities are *ex officio* members, in some they are only allowed to send representatives, and in others they establish a sort of second chamber together with their councillors. It is noteworthy, however, that there is not one constitution where traditional authorities are not of decisive relevance. Chiefs, however, are often ambivalent about the emergence of new resource-management units and clearly feel that new nodal points of decision making are emerging within the local community.

A land management plan in which all members of a conservancy agree upon a land-use plan which specifies pastures for seasonal grazing, settlement areas and core conservation areas is part and parcel of an application to the Ministry of Environment and Tourism for conservancy status. The land management plan then becomes an officially endorsed document which is deemed to be the blueprint of future land use. Many people interviewed voiced their hope that the reshaping of chieftaincies as conservancies, the combination of traditional and modern elites in committees and a governmentally endorsed land-use plan would make common pool resource management effective again.

Water management became a second instance of the co-management of natural resources. Until the end of the 1990s, rural water supplies had been fully managed by the Directorate of Water Affairs within the Ministry of Agriculture. Following international agreements, Namibia then changed its policy of rural water supply, and the government introduced legislation which gave more responsibilities to local resource users. In contrast to conservancies, non-governmental organisations were of lesser importance in the establishment of this policy. The new legislation stipulates that control over boreholes in rural areas should be handed over to rural communities.[14] Local water users are encouraged to establish water-point user associations whose members elect water-point committees to oversee the daily management of water resources and financial issues.[15] Water-point users associations are meant to make rules for the use of rural water supplies, prevent any person who does not comply with the rules or constitutions of a user association from using its water point, adopt measures to prevent the wastage of water, and plan and control the use of communal land in the immediate vicinity of the water point in cooperation with the respective communal land board and traditional authority. Not mentioned here, but of crucial importance for the overall application of the legislation, is that water-point user associations are to collect user fees and use them to manage their water point.

In some areas of the Kaokoveld, grazing schemes based on the principles of holistic range management (Savory 1998) have been inaugurated. Holistic range management rests on the assumption that high livestock densities can be maintained once intense grazing is concentrated in short time periods. In order to guarantee such conditions herds are amalgamated, preferably numbering no more than a few hundred animals, and put under a management committee. Herds then graze on clearly demarcated stretches of pasture for short periods of time. The introduction of such schemes involves a profound reorganisation of herd management for Himba and Herero households. Where such schemes are implemented several households come together and build one central cattle enclosure from which grazing is organised. Each household proposes one or two herders to manage the joint herd. In several instances milk is transported from this main enclosure to the cattle camps. Through most of the year the herd is herded together at varying distances from the main settlement.

What impact have these changes had on land management in the wider Epupa region? Several profound changes have taken place there over the past ten years. Firstly, conservancy status has been applied for. After lengthy boundary negotiations with a neighbouring proposed conservancy and even more difficult negotiations between the local chief and the committee of the proposed conservancy on the sharing of benefits derived from tourist businesses in the conservancy area, all necessary documents have been handed to the Ministry of Environment and Tourism. The land management plan attached to this application shows core conservation areas, areas designated for intense use through livestock herding and settlement areas. Currently the community is awaiting the Ministry's final decision. Within the area of the proposed Epupa conservancy a grazing scheme was launched some years ago which is supported by an NGO and which also renders support to the proposed conservancy. A number of herders (varying between three and six over the course of the years) came together to herd their cattle according to the principles of holistic range management in the Otjikango area. In order to allow them to water a larger herd of cattle permanently at the Otjikango well, the NGO drilled a new borehole there. The households established one big homestead in the area and from there they herd all their animals jointly through most of the dry season. During the rainy season they move their animals to Ondova where they join other cattle camps. There they do not herd according to the principles of holistic range management. They agreed with other community members and the local chief that other cattle camps should abstain from moving into the Otjikango area unless they join the grazing scheme.

Other changes of range management in the area can only be touched upon briefly here. In the Omuhandja area near Epupa, a village-like settlement developed. Households there have become fairly sedentary and only move their livestock camps to dry-season pastures. Households are apparently tied to Omuhandja as tourists from Epupa frequently visit households and bring presents in return for permission to take photographs. This villagisation process may be enhanced as a primary school was recently built there. In the centre of the grazing area, the local chief, wealthy in cattle, succeeded in reserving prime grazing lands for himself and his patriline. These chiefly grazing grounds are unfenced but virtually nobody else makes use of the grazing in Omuramba and Ombuku unless they have been explicitly permitted to do so by the chief.

In many respects patterns of resource exploitation have changed. It is as yet difficult to say in what direction the pastoral mode of production will change as new institutions of resource tenure are only emerging. At the moment the fragmentation of common resource management dominates: the group of resource users has become internally more differentiated and so too have the rules of resource use themselves. In our case study area the people of Omuhandja and the herders of Otjikango – for different reasons – try to establish their own commons and to stipulate their own rules for their management. In both cases the community sharing a resource has become restricted and certain management principles have been more narrowly defined. Outside agencies – NGOs and the government alike – are involved in the local search for new institutions: they either inform people of blueprints for conservation management or rules conducive for a smooth working of grazing institutions. Rules derived from such blueprints as well as older, locally established institutions are reconsidered locally and their potential combination is much discussed. Such developments seem rather typical all over the Kaokoveld, and whether or not new stable institutions will emerge for the organisation of resource management across the wider landscape remains an open question.

Closing Remarks

The preceding sections show that pastoral resource management in north-western Namibia has gone through phases of stasis and rapid development over the past century. While the pre-1950s set-up relied on home ranges, the second half of the twentieth century was characterised by the joint regulation of pastures within the framework of chieftaincies and neighbourhoods, and the recent period seemingly rests on management through smaller user units. Stability dominated these two earlier phases for some thirty to forty years and transitions have been swift. During stable periods of each mode of regulation a certain set of institutions was developed and supported by a specific configuration of social networks and power structures. Each mode of regulation can also be linked to specific vegetation structures in pasture land. Presently a number of alternative regulative approaches compete for influence and the local community is searching for a predictable and fair approach to the management of a set of common pool resources, which they regard as rapidly degrading due to uncoordinated use. Whether local communities will again succeed in adopting efficient and internally undisputed institutions of common pool resource management which appeal to a wider population remains to be seen.

For anthropological theory-building, it is highly interesting to consider that order is created rapidly and not incrementally, and that periods of rapid transformation are followed by periods of stability and cultural elaboration. It is these periods of stability which are characterised by specific modes of risk management. Due to the fact that most anthropological studies concentrate on a rather brief time period of one to two years, or – in some better cases – rely on two periods some twenty, thirty or more years apart, anthropologists usually have little to say about the speed of change and about the relation between phases of stability and those of transformation. Only long-term studies will be helpful in developing theories of change within complex social-ecological systems.

Notes

1. For a critical account of this discourse emphasising static human–environment relations in north-western Namibia, see Bollig and Heinemann (2002).
2. For a general introduction to the ecology, history and contemporary economy of north-western Namibia, see Bollig, Brunotte and Becker (2002).
3. NAN, J XIIIb5, H. Vedder. 'Reisebericht des Missionars Vedder an den Bezirksamtmann von Zastrow. Geographische und Ethnographische Forschungen im Kaokoveld 1900–1914', (1914). Shortridge reports 'The Kaokoveld is very sparsely inhabited by nomadic Ovashimba Herero tribes that live chiefly by raising a sufficient number of goats and sheep for their own requirements', and 'The Ovashimbas are a primitive race of nomadic Hereros that chiefly inhabit the Kaokoveld. Their chief occupation appears to be raising goats and the local breed of sheep' (National Archives of Namibia [NAN], SWAA 1331/198/6/2 [1923]).
4. NAN, J XIIIb5, H. Vedder, (1914). For similar remarks on the close connection between local big men and waterholes, see, e.g., Reitz (1943: 103).
5. On the political economy and organisation of these returnees from Angola, see Bollig (1998a).
6. Vansina describes the position of Herero *ovahona* as follows: 'those rich in cattle had many dependents. When nineteenth century rumours speak of owners of over ten thousand head of cattle, such persons probably had between one and two hundred client families, at least as many servants, and occupied a very large grazing territory to match' (Vansina 2004: 130).
7. See also Alexander (2001: 216) for similar developments in Zimbabwe.
8. NAN, SWAA BOP 5, Kantoor van die Hoofnaturellekomissaris, Windhoek SWA Naturellesake Omsendbrief Nr 17/1955, 26 August 1955.
9. Garth-Owen Smith, former agricultural officer (personal communication).
10. NAN, SWA/KC/7E/52. (RSA 1964: 623).
11. For a general account of the history of the 1960s and 1970s, see van Wolputte (2004).
12. There were other reasons for the move from Ombuku to Omuramba: a newly built road came close to Omuramba, which also lay in the midst of splendid grazing whereas Ombuku was placed at the edge of it.
13. See also Bollig (2004) for the process until 2003.
14. For the legislation, see RoN (2004, 2008).
15. See RoN (2005), especially part 5, section 16.

References

Abel, H. 1954. 'Beiträge zur Landeskunde des Kaokoveldes (Südwestafrika)', *Deutsche Geographische Blätter* 47: 7–123.

Alexander, J. 2001. 'Technical Development and the Human Factor: Sciences of Development in Rhodesia's Native Affairs Department', in S. Dubow (ed.), *Science and Society in Southern Africa*. Manchester: Manchester University Press, pp.212–37.

Behnke R. 1998a. 'Grazing Systems in the Northern Communal Areas of Namibia: A Summary of NOLIDEP Socio-economic Research on Range Management', unpublished manuscript. Windhoek.

——— 1998b. 'Range and Livestock Management in the Etanga Development Area, Kunene Region: Northern Regions Livestock Development Programme', unpublished manuscript. Windhoek: Ministry of Agriculture, Water and Rural Development.

Bollig, M. 1997. *'When War Came the Cattle Slept': Himba Oral Traditions*. Cologne: Köppe.

——— 1998a. 'Power and Trade in Precolonial and Early Colonial Northern Kaokoland', in P. Hayes et al. (eds), *Namibia under South African Rule: Mobility and Containment, 1915–1946*. London: James Currey, pp.175–93.

——— 1998b. 'The Colonial Encapsulation of the North-western Namibian Pastoral Economy', *Africa* 68(4): 506–36.

——— 2004. 'Landreform in Namibia: Landverteilung und Transformationen Kommunalen Landbesitzes', in L. Foerster, D. Henrichsen and M. Bollig (eds), *Namibia – Deutschland: Eine geteilte Geschichte*. Cologne: Minerva, pp.304–23.

——— 2006. *Risk Management in a Hazardous Environment: A Comparative Study of Two Pastoral Societies*. New York: Springer.

Bollig, M., E. Brunotte and T. Becker. 2002. 'Interdisziplinäre Perspektiven zu Kultur– und Landschaftswandel im Ariden und Semiariden Nordwest Namibia', Cologne. *Kölner Geographische Arbeiten* 77.

Bollig, M., and H. Heinemann. 2002. 'Nomadic Savages, Ochre People and Heroic Herders: Visual Presentations of the Himba of Namibia's Kaokoland', *Visual Anthropology* 15: 267–312.

Bollig, M., and A. Schulte. 1999. 'Environmental Change and Pastoral Perceptions: Degradation and Indigenous Knowledge in Two African Pastoral Communities', *Human Ecology* 27: 493–514.

Botha, C. 2005. 'People and the Environment in Colonial Namibia', *South African Historical Journal* 52: 170–90.

Friedmann, J. 2005. 'Making Politics, Making History: Chiefship and the Post-Apartheid State in Namibia', *Journal of Southern African Studies* 31(1): 23–51.

Gaerdes, F. [K'Ozombaze]. 1970. 'Ovita: Erlebnisse am Kunene', *SWA Scientific Society Newsletter* XI/6: 1–12.

Hartmann, G. 1941. 'Die Bevölkerung des Kaokoveldes', *Deutsche Kolonialzeitung* 53: 29–33.

Hayes, P. 1998. 'Northern Exposures: The Photography of C.H.L. Hahn, Native Commissioner of Ovamboland 1915–1946', in W. Hartmann, J. Silvester and P. Hayes (eds), *The Colonising Camera: Photographs in the Making of Namibian History*. Cape Town: Out of Africa, pp.171–87.

——— 2000. 'Camera Africa: Indirect Rule and Landscape Photographs of Kaoko, 1943', in G. Miescher and D. Henrichsen (eds), *New Notes on Kaoko: The Northern Kunene Region in Text and Photographs*. Basel: Basler Afrika Bibliographien, pp.48–76.

Kuntz, J. 1912. 'Ovatschimba im Nördlichen Kaokofeld (Deutsch-Südwestafrika)', *Petermanns Geographische Mitteilungen* 25: 206.
Malan, S. 1972. 'Dubbele Afkomsberekenings van die Himba. 'n Hererosprekende Volk in die Suidwest Afrika', PhD dissertation. Johannesburg.
——— 1974. 'The Herero-speaking Peoples of Kaokokland', *Cimbebasia B* 2: 113–29.
——— 1995. *Peoples of Namibia.* Wingate Park: Rhino Publishers.
Miescher, G., and D. Henrichsen (eds). 2000. *New Notes on Kaoko: The Northern Kunene Region in Text and Photographs.* Basel: Basler Afrika Bibliographien.
Reitz, D. 1943. *No Outspan.* London: Faber and Faber.
Rizzo, L. 2007. 'The Elephant Shooting: Colonial Law and Indirect Rule in Kaoko, North-western Namibia, in the 1920s and 1930s', *Journal of African History* 48: 245–66.
RoN. 2004. 'Water Resources Management Act', *Government Gazette of the Republic of Namibia* 3357 (23 December): 1–67.
——— 2005. 'Background Research Work and Findings of the Permanent Technical Team on Land Reform Studies'. Windhoek: Ministry of Lands and Resettlement, Republic of Namibia.
——— 2008. 'Water Supply and Sanitation Policy', file no. 7/3/78/P. Windhoek: Ministry of Agriculture, Water and Forestry, Republic of Namibia.
RSA. 1964. 'Report of the Commission of Enquiry into South West Africa Affairs' [Odendaal Commission]. Pretoria: Government Printers, Republic of South Africa.
Sander, H., M. Bollig and A. Schulte. 1998. 'Himba Paradise Lost: Stability, Degradation and Pastoralist Management of the Omuhonga Basin (Namibia)', *Die Erde* 129: 301–15.
Savory, A. 1998. *Holistic Management: A New Framework for Decision Making.* Washington, DC: Island Press.
Schulte, A. 2002. 'Stabilität oder Zerstörung? Veränderungen der Vegetation des Kaokolandes unter Pastoralnomadischer Nutzung', in M. Bollig, E. Brunotte and T. Becker (eds), *Interdisziplinäre Perspektiven zu Kultur– und Landschaftswandel im Ariden und Semiariden Nordwest-Namibia.* Cologne: Kölner Geographische Arbeiten 77, pp.101–18.
Stals, E.L.P., and A. Reiner-Otto. 1999. *Oorlog en Vrede aan die Kunene.* Windhoek: Capital Press.
Steyn, H.P. 1977. 'Pastoralisme by die Himba n'Kulturelle-Ekologiese Studie', PhD dissertation. Stellenbosch: University of Stellenbosch.
Vansina, J. 2004. *How Societies Are Born: Governance in West Central Africa Before 1600.* Charlottesville: University of Virigina Press.
Van Warmelo, N.J. 1951. *Notes on the Kaokoland (South West Africa) and its People.* Pretoria: Government Printers.
Van Wolputte, S. 2004. 'Subject Disobedience: The Colonial Narrative and Native Counterworks in North-western Namibia, c. 1920–1975', *History and Anthropology* 15: 151–73.
Viljoen, P.J. 1988. 'The Ecology of the Desert-dwelling Elephants *Loxodonta Africana* (Blumenbach 1797) of Western Damaraland and Kaokoland', PhD dissertation. Pretoria: University of Pretoria.
Welle, T. 2007. 'Adaptation Strategies on a Local Scale in the Context of Global Change', in O. Bubenzer, A. Bolten and F. Darius (eds), *Atlas of Cultural and Environmental Change in Arid Africa.* Cologne: Heinrich Barth Institut, pp.110–11.

CHAPTER 11

Pastoral Belonging

Causes and Consequences of Part-time Pastoralism in North-western Namibia

Michael Schnegg, Julia Pauli and Clemens Greiner

Introduction

We are sitting in front of Elias's hut in the hinterlands of Fransfontein, Namibia, awaiting the return of his cattle and goats to his kraal.[1] In the morning, Elias had made arrangements via mobile phone for a livestock trader to pick up two of his oxen. It is the beginning of the university year and Elias needs to pay N$2,500 in fees for two courses he wants to take with UNISA, South Africa's largest distance-learning institution.[2] While we are sitting and waiting for the animals to return he reflects on the investment he is about to make, lamenting the fact that this will significantly reduce his fifteen head of cattle at a time when prices are relatively low. Yet upon thinking about the future his mood swings. Now forty years old he wants to finish his studies within the next two years and start working as a teacher in one of the primary schools in the district. He imagines that once he has a steady income his herd will grow quickly, since he does not have to withdraw animals every time cash is needed in the household. Thus, in less then ten years from now he imagines his herd will be substantial. At that point Elias reflects that he could quit his job as a teacher and become a full-time pastoralist again. This time he would be almost as well-off as the commercial farmers of mainly European descent that live nearby. While for Elias these are future plans, others of his generation live this dream, combining wage labour in Namibian towns and cities with weekend or part-time pastoralism.

The convergence of pastoralism, migration and employment described in this vignette has received relatively little attention in Southern Africa (Rohde et al. 2006; Berzborn 2007; Berzborn and Schnegg 2007; Greiner 2008). Yet similar transformations have been observed in Eastern and Western Africa (Swift 1982; Little 1985; Hogg 1986) and other parts of the world (Barth 1964). Hogg (1986) coined this the 'new pastoralism' and argued that job opportunities which emerged in Kenya around independence had opened up new economic opportunities for those better educated and connected. These people moved to the cities where they found well-paid jobs in the growing public sector. However, employment could often not provide long-term emotional, financial and social security, and migrants were often reluctant to view their new city houses as their future homes and their new place of belonging.[3] As a result, intense linkages between the city and the rural hinterlands emerged (ibid.). Little made similar observations among the Il Chamus in Kenya, which led him to re-evaluate the question of whom to blame for the overgrazing and degradation of pastoral landscapes in Baringo District (Little 1985). Scientists and politicians had long identified local herders, trapped in Hardin's 'tragedy of the commons' (Hardin 1968), as the prime movers. In contrast to this popular discourse, Little argued that three distinct groups were to blame for the demise of pastures: first, ranchers coming from commercial farmland that entered territories of neighbouring groups who in turn entered the grazing area of the Il Chamus; two, livestock traders who kept animals in the area before they were brought to market; and three, townsmen who owned large herds on communal territories. Since all three groups were not entirely dependent on pastoralism they seemed to be less concerned with long-term conservation (Little 1985: 146; see also Anderson 2002).

Hogg and Little recognised at about the same time the emergence of a 'new' form of pastoralism in postcolonial societies, one that is linked to migration and people's integration into the labour market. Their works laid the foundation for a new way of looking at the pastoral mode of life, a vantage point that was largely informed by political economy and political ecology (Bryant 1992; Watts, Sheppard and Barnes 2000; McCabe 2004; Biersack 2006; Bollig 2006). While Hogg emphasised the critical role employment plays in the diversification of pastoralism, Little focussed on the ecological consequences of part-time farming for local ecosystems and their possible demise.

Their arguments rested on the then dominant scientific model to explain ecosystem dynamics. Until the 1980s it was commonly acknowledged that rangelands tend to approach a fixed equilibrium that is maintained through density-dependent feedback mechanisms. It was assumed that certain biological checks, including slower animal reproduction as vegetation gets scarce, ensured that livestock and vegetation balance around the 'carrying capacity' of an ecosystem. Within this line of thought, pastoralists were often identified as the main threat to ecosystem stability. Overstocking by indigenous groups was made responsible for irreversible degradation, decline in soil fertility, soil erosion and the like, and political interventions were directed to reduce the number of livestock on a given territory.[4]

These long-held assumptions were challenged in the 1980s (Wiens 1984), when it became increasingly evident that ecosystems are often far removed from an equilibrium and largely driven by stochastic and abiotic events. For rangelands the driving force is rainfall. While classical equilibrium thinking puts the emphasis on the link between herbivores and vegetation, non-equilibrium ecologists do not question that livestock require forage. However, they argue that population densities are often (if not always) below the level where density-dependent regulation occurs. They are kept at these very low levels through extreme droughts that reduce herd sizes independent of prior stocking rates. This shift in the ecological model has important consequences for understanding human activities. If animals have relatively little impact on vegetation, humans can hardly be blamed for its rescue nor its demise (Behnke, Scoones and Kerven 1993; Scoones 1999; Sullivan and Rohde 2002; McCabe 2004). Recent publications indicate that density-dependent (equilibrium) and density-independent (disequilibrium) regulation of herd sizes co-occur in many of Africa's rangelands (Vetter 2005; Gillson and Hoffman 2007). Density dependent regulation seems to be more likely during consequent years of 'normal' rainfall when high stocking rates reduce the vegetation and eventually the quality of the livestock. At the same time, severe droughts still decrease the animal population independent of its prior numbers during extreme and inter-annual droughts (Vetter 2005).

While these coupling mechanisms between different parts of the ecosystem are not entirely understood, it is well known that actors have quite different strategies at hand to respond to drought and other crises. Human vulnerability is not only a function of the extent of external (ecological) stress but is equally shaped by a household's means of coping with perturbations without damage and loss (Chambers 1989; Watts and Bohle 1993; Turner et al. 2003).[5] Pastoralists across Africa are known for their ample coping strategies. This includes both diversification within the pastoral mode of production as well as outside it by combining livestock keeping with other economic activities (Spencer 1998; Little et al. 2001; Bollig 2006; Berzborn 2007). Diversification outside livestock husbandry is common in Fransfontein, where many households are linked to the labour market through migration (Greiner 2008; Schnegg 2009).

This chapter goes beyond an economic and ecological analysis of the causes and consequences of part-time pastoralism offered by Hogg (1986), Little (1985) and others. Elias's dreams and desires about moving from being a poor, full-time pastoralist to a part-time pastoralist (and wage labourer) to a wealthy full-time pastoralist provide important insights. Not for love or money would Elias imagine himself without livestock. Thus, cattle, goats and sheep mean much more to him than mere economic assets. In contrast to the so-called commercial areas of Namibia where land is privately owned, in the Fransfontein communal area, private land ownership is largely impossible. In this situation the ownership of livestock is the most important way to create and express belonging to the territory and the community.[6] In this situation, part-time pastoralism is far more than a financial investment; it is also an expression of belonging, identity and group membership.

With great complexity, Appadurai has shown the ongoing centrality of locality production despite a 'dramatically delocalized world' (Appadurai 1996: 178). He thereby differentiates between locality defined as 'phenomenological quality' or 'dimension of social life' (ibid.: 178) from neighbourhood, which Appadurai conceptualises as 'the actually existing social forms in which locality, as a dimension or value, is variably realized' (ibid.: 179). Appadurai emphasises that locality – that is, a specific 'structure of feeling' (ibid.: 181) or sense of place and belonging – is always historically situated. Actors use a number of cultural techniques, such as rituals, and the building of houses, to produce locality and 'local subjects, actors who properly belong to a situated community of kin, neighbours, friends and enemies' (ibid.: 179).

In Fransfontein, practices and discourses associated with livestock husbandry are central techniques in the production of local subjects, people who properly belong to the area. For those who combine labour migration and part-time pastoralism, these techniques of locality production are especially relevant. For those who migrate, the sense of locality and belonging is often particularly fragile and contested (Lovell 1998; Turton 2005; Pauli 2008). This pastoral belonging must be understood within the context of German and South African colonialism and the violent land dispossession the people of Fransfontein and many other Namibian communities have had to endure. For Fransfontein's full- and part-time pastoralists, the meaning of locality and belonging is thus linked with livestock – even if (or maybe because) they have migrated to other Namibian regions.

The aim of this chapter is to explore the political and economic causes and historically situated dynamics of belonging and identity that have shaped part-time pastoralism(s) in Fransfontein and its social and ecological consequences. After a brief description of the ethnographic context and the methodology we explore in depth the political factors that have driven people into wage labour and part-time pastoralism in Fransfontein, factors that have also centrally shaped the migrants' sense of place and belonging. Next, we show the types of pastoral households that have emerged in this process. Remote-sensing data provide a means of understanding how these strategies impact upon the environment. The analysis leads us to ask what kind of stress the current farming system produces and how this shapes the vulnerabilities and coping capacities of all households involved. In closing we ask how far the desire for pastoral belonging by largely absent migrants has (probably unintentionally) severely reduced the well-being and flexibility of those left behind.

Ethnographic Context

The ethnographic focus of this study is Fransfontein, a community of 137 households in north-western Namibia, and its rural surroundings, which are home to another 161 households.[7] The semi-arid environment of communal pastureland is dotted with small settlements of 3 to 19 households. These settlements are clustered around drilled boreholes that provide water in a landscape with annual rainfall of approximately 200 to 250 millimetres (Schnegg and Welle 2007). Most inhabitants of the larger Fransfontein area consider themselves to be Damara, Nama, Ovaherero or Ovambo people (Dawids et al. 2007). Before contact with their German colonisers, the Damara were most likely hunter-gatherers who also engaged in significant amounts of small-scale trading and even iron working. At the same time, some of them worked as herders for other groups (Lau 1987; Barnard 1992; Sullivan 2004). The //Khau/-gôan (Swartboois) Nama arrived around 1880 in the Fransfontein area. At that time most Nama combined pastoralism with trading and raiding (Lau 1987). Under the supervision of German missionaries, the //Khau/-gôan started gardening in a protected area close to the community's fountain. Water must have been relatively abundant at the time and not only enabled growing tobacco as a cash crop but also wine for the missionaries (Schnegg 2007).

More than 100 years later, Fransfontein has a comparably good public infrastructure, part of which comprises remnants of its vital past as an outpost of the Rhenish (*Rheinische*) Mission. At present, the police station, a local government office, a church, a boarding school and a local health clinic are the most significant public institutions. The larger private enterprises in Fransfontein are bars, small grocery shops and the establishments of two car mechanics. While quite a number of households in Fransfontein own livestock, the proportion of livestock owners on the communal lands surrounding the community is much higher.

Data and Methods

Building on research conducted over several research stays, in 2003 and 2004 the two senior researchers, Julia Pauli and Michael Schnegg, carried out eighteen months of fieldwork in Fransfontein. The overarching aim of the larger project is to understand the kinds of new social, economic and cultural practices that people adopt to cope with a harsh and rapidly changing ecological, political and economic environment. This fieldwork included an ethnographic census of 328 households in the study area.[8] The survey collected information on household composition, property and social and economic strategies. In 2005/6 Clemens Greiner conducted thirteen months of fieldwork in the area. As part of Pauli and Schnegg's research project, Greiner focused on the causes and consequences of migration through a multi-sited study involving the Fransfontein area, the capital Windhoek and the booming harbour city of Walvis Bay.

Reconfiguring Pastoral Belonging: Reserves, Homelands and Migration

Both locality, as a sense of place and belonging, and neighbourhood, as the actual existing social forms of locality (Appadurai 1996), have been highly contested throughout the history of Fransfontein. Massive land dispossession, a trauma that all indigenous communities in southern and central Namibia went through (including Fransfontein), centrally configured and reconfigured both locality and neighbourhood.[9]

Land Dispossession, Colonial Encapsulation and 'Native' Reserves

The expansion of the white settler state had far-reaching consequences for the pastoral way of life in these areas. Focusing on the 1920s and 1930s, Fuller summarises the forced change from an economic system based on (agro)pastoralism to one centred on wage labour as follows: 'During the inter-war years indigenous Namibians in the central parts of the country traded one form of mobility for another. They swapped an agropastoral system of transhumance for tightly controlled entry into limited portions of the labour market' (Fuller 1998: 194). Farm owners of European descent – backed by the colonial administration – rapidly appropriated the land, created boundaries and marked them with fences. Thus, the beginning of privately owned land (and the expression of such in the marking of borders through fences) strongly contrasted with pre-colonial land use by the Damara, Nama, Herero and Baster communities of southern and central Namibia.

The white settlers' demand for land went hand in hand with a strong demand for cheap labour. In the 1920s, over 60 per cent of the 'non-white' population within central and southern Namibia worked on white farms in rural areas (Rohde 1997: 247; Botha 2003). During this time, in Western Namibia reserves such as Fransfontein functioned as labour reserves for the surrounding settler farming community (Rhode 1997: 261). This is clearly expressed in the high percentage of elder Fransfonteiners who have worked on white commercial farms (Pauli 2009a).[10]

In his analysis of the development of land-tenure systems in Namibia, Werner (2004) emphasises the connection between the massive land dispossession of indigenous Namibians, the white settler economy's labour needs and the establishment of 'native' reserves. With an area of only 36,188 hectares, Fransfontein was among the first six reserves established by the German colonial administration (ibid.: 295). After Germany surrendered to the Union of South Africa, South Africa continued this policy. South Africa's Native Land Act of 1913 served as a 'prototype' for the politics of 'separate development' in Namibia during the first years of South African colonial rule (Fuller 1993: 24; Werner 2004: 296). The establishment of reserves alongside the expansion of white commercial farms heralded the beginning of the dichotomy between 'communal' and 'commercial' land tenure still present today. Access to land directly determined the availability and cost of black workers for the colonial economy (Werner 2004: 292). Because of the limited amount of land left to the indigenous population, the people were not able to survive independently of the white economy and were forced to work for the settlers under intolerable conditions (Fuller 1993: 43–65; Werner 2004: 292). The costs for their employers were extremely low. Not only was there an ample supply of labour but the small amount of land still available to indigenous workers also buffered social security costs (such as health insurance) which neither the state nor the white commercial farmer had to pay (Fuller 1993: 55). Figure 11.1 shows the dynamics of white settler expansion and indigenous colonial encapsulation for Fransfontein through the first decades of the twentieth century.

Figure 11.1: The colonial encapsulation of Fransfontein.

The data in Figure 11.1 have been compiled from regularly updated farm maps that indicate what land was surveyed by the colonial administration. Even though this does not give precise information on when exactly the land was settled by farmers of European descent, it allows a first approximation of the process of land occupation and shows that significant parts of the farmland around Fransfontein were occupied by settlers in the first half of the twentieth century (Schnegg 2007).[11] The indigenous populations that inhabited these lands were left with few alternatives: one, to retreat to freehold land in the north and west; two, to start a living as workers for the new landowners; or three, to move to Fransfontein or other reserves and cities. As church records of the places where people where baptised during the first decades of the twentieth century indicate, many people continued to settle in the same place and started to work for the new landowners. Life histories of inhabitants of Fransfontein support this interpretation of the archival material.

Not only around Fransfontein was land increasingly occupied by European settlers. The same process forced a group of relatively wealthy pastoralists to leave Otjoruu, a settlement between Outjo and Omaruru that was granted to them under German colonial rule (Miescher 2006). After failed attempts to stay in their place of residence they decided to leave in 1938 in order to prevent eviction. According to oral traditions, their leader, Lazarus Amporo, sent a message to Petrus Swartboois, the head of the Fransfontein Reserve Board and leader of the Swartboois, asking him to grant his people refuge in Fransfontein.[12] Petrus Swartboois was in favour of the idea but pointed out that the reserve would be too small for all of them, so he made a petition to the South African colonial administration to significantly extend the boundaries of the reserve.[13] Yet despite a slight extension of the reserve's boundaries, the available land was far from sufficient to provide a sustainable living for all of its old and new inhabitants. With the arrival of the Otjoruu group, the population in the reserve roughly doubled and livestock numbers quintupled (ibid.: 19). Werner's observation that the 'native' reserves established by the South Africans from 1915 onwards were not created to foster a sustainable livelihood for indigenous Namibians but to fulfil the white settlers' need for cheap labour (Werner 2004: 292) describes these developments in Fransfontein.

The results clearly show how the comparably high stocking rates in Fransfontein have affected the landscape, such that much less vegetation can be found than on the freehold cattle and game farms in the surrounding area. In comparison to the Swartboois territory, the vegetation cover is also thin and contains little buffering capacity for a severe crisis. This difference may be due to a much lower settlement density, stricter rules of access and better institutional solutions. These observations support Klintenberg's findings that degradation occurs around densely populated key resources (Klintenberg and Verlinden 2008).

Vulnerability as a Social Consequence

Vulnerability has emerged as a dynamic concept for capturing the interplay between the stress households or individuals are exposed to and their coping strategies. The foregoing analysis has shown that absentee livestock owners possess a large share of the goats, sheep and cattle that graze in Fransfontein. These animals cause costs and stress that is external to individual households. As we have seen, they shape the availability of fodder for all livestock. While we worked in Fransfontein in 2003/4, the situation did not lead to a loss of animals due to insufficient fodder or water. However, even during these normal years the animals were notably underweight and provided insufficient nutrition and calories for consumers. In addition, informants reported that more lambs died than was perceived to be normal and slowed down the reproduction of the herd. While mechanisms that regulate the number and quality of livestock apply for all households, the part-time farmers have very different coping strategies to hand. To analyse these we must distinguish between 'normal' years with the stocking rates described above and times of extreme drought. Let us turn to the former first.

During normal years animals are sold to meet a household's consumption needs or they are directly consumed. Given their relatively low weight, the price per animal is comparably low. The commercial livestock traders that come to Fransfontein to buy large quantities of livestock are well aware of this and put the livestock on vegetation rich commercial pastures where they quickly gain weight. After three to six months these traders, mostly of European descent, sell the animals on the commercial meat market. The combination of this weight gain and the difference paid per kilogram in Fransfontein and in urban centres is so significant that the margins for the traders are enormous. Traders do not only profit from the low quality of livestock but additionally from price differences between Fransfontein and the town of Outjo, some 150 kilometres away, where the nearest national meat market is located.

Migration and Homelands

With the establishment of Apartheid in the 1960s and 1970s in Namibia, this colonial policy continued. The Odendaal Commission made recommendations on how to implement Apartheid politics and declared the area around Fransfontein to be part of the Bantustan 'Damaraland' (RSA 1964 [Odendaal Report]). For the creation of Damaraland, 223 farms were bought from white owners (Rohde 1997: 258). This land, some state land and three existing reserves – Okombahe, Fransfontein and Sesfontein – became Damaraland (ibid.: 259). Throughout the 1970s and 1980s many people classified as 'Damara' by the South African Apartheid regime were relocated into what had become 'their' homeland. But efforts to resettle Damara people were only partly successful. Already in the 1950s Damara were the most urbanised ethnic group, living in large numbers outside the territory of the later Damaraland (Peyroux 2004: 41), and in 1981 only 24,000 of the 76,000 persons classified as Damara by the South African authorities were living within the boundaries of 'their' homeland (Botha 2003). The high rate of urbanisation was, however, not an expression of choice. As during the earlier German and South African colonial periods, the manifold economic and ecological limitations of the 'homeland' forced people to leave and migrate to Namibia's towns and cities.

These causal relationships are explicitly named in the Odendaal Commission's report, which states that subsistence agriculture could feed only a minority of the population within Damaraland (RSA 1964: 93 [Odendaal Report]). Unlike the 8,500 hectares which were considered the minimum amount of land a 'white' farmer needed for farming, families in the newly created Damaraland had to survive on average on a meagre 250 hectares per family. Many Fransfonteiners were not able to live in the 'homeland' but had to live on white commercial farms or in the Windhoek township of Katutura (Barnard 1992: 213; Rohde 1997: 259). This situation was comparable to the situation in other homelands: 'As none of these artificially created homelands was economically viable in terms of subsistence agriculture, the inhabitants were forced to look for wage employment in order to provide for their families' (Jauch 1998: 28). This lived experience of violent land possession and the need to migrate in order to survive clearly shaped the kind of pastoral belonging that developed among part-time pastoral migrants. In the next section we will discuss the current pastoral configurations that have emerged from these complex historical processes.

Pastoralism in Fransfontein Today

The availability of land for the people of Fransfontein did not change significantly after independence in 1990, although the Namibian government started to buy commercial farms under a 'willing seller, willing buyer' scheme (Bollig 2004). When we began our field research in 2003 we thus found a situation of land ownership that was still largely circumscribed by the colonial encapsulation described above. Yet although the options for full-time pastoralism have not really improved, in many conversations the wish to become an autonomous full-time pastoralist was conspicuously expressed. During an interview with Julia Pauli in August 2005, 34-year-old Petrus, for instance, said that he wanted to 'become a farmer with a vision; then I know that I am on my own now', while in August 2004, Theresa, then fifty-four years old and having retired early from a well-paid government job, explained to Julia Pauli during a life history interview: 'All I want to do is to get a farm – not a farm on my own, but a place on a farm where I can go and start farming with goats and cows'. Remarkably, these pastoral desires and sense of belonging do not depend on gender, generation, location or class. Both men and women, elder and younger Fransfonteiners, wealthy migrants from Windhoek and local school teachers, name full-time pastoralism as an ultimate goal in their lives. These attitudes must be understood in relation to the violent Fransfontein history and forced dispossession briefly outlaid above. In this context livestock husbandry is not only a means to survive but also a means of creating belonging regarding a place and a social group.

Many Fransfontein households diversify their productive strategies. This form of combined pastoralism is well described by Hogg's term 'new pastoralism' (Hogg 1986; cf. Little 1985). Livestock herding is, and for a long time was, only one part of a complex bundle of strategies used to sustain the livelihood of the household, a bundle that also includes wage labour, remittances from migrants, welfare payments from the state and small-scale economic activity. Elsewhere we have described in more detail how these strategies are combined, how they vary with economic status and how they relate to the emergence of localised elites since the 1970s (Greiner 2008; Pauli 2009a, 2009b; Schnegg 2009). The following section builds on these insights and introduces three types of pastoral household that have developed in the light of these historical developments in the Fransfontein area.

Impoverished Rural Full-time Pastoralists and Their Children

Justus is an elderly man whose eight children live partly on the one farm and partly in the towns Khorixas and Outjo. Fifty years ago, Justus's parents moved into the area and settled on one of the communal Fransfontein settlements. He is well connected through marriage and descent with most other members of the community. When Justus was younger, he and his late wife Maria both lived on different white commercial farms in the area where Justus and sometimes his wife found employment. Their children, including the ones who have migrated, do not have well-paid jobs and there has been no investment in the parental house, which remains in rather poor condition. Justus owns a meagre ten goats and two cows, which he bought with the money he made as a wage labourer. These animals are kept in the kraal next to the house. The animals are not guarded by a herder when they go out to the field during the day. This means that the household is rather vulnerable to stock theft. In addition to their farming income, Justus receives a pension which provides them with cash.

Return Migrants with Economically Better-off Children

Emma's parents were the first Damara-speaking people to settle on a small farm settlement in the Fransfontein communal area. By the end of the 1960s, when she was in her thirties, and together with her husband, a local Herero, Emma left the Fransfontein area for the nearby town of Khorixas. Some years later they left for Okombahe, where Emma's sister's husband was employed as a janitor and cook in a local primary school. In the mid 1990s, and after her husband's retirement, Emma and her husband returned to the Fransfontein area and resettled on another communal settlement. Today, Emma lives in a traditionally built house on a sandy plot of land located not far from her sister's house. The main house is traditionally built from poles of mopane, clay and cow dung. Next to this house a smaller one has been built with cement and bricks. It belongs to Emma's daughter who is living in Windhoek. Emma's husband passed away years ago, as have five of their eight children. Currently, she is living together with five of her grandchildren and great-grandchildren. The household's income situation is dominated by Emma's monthly old-age pension. Apart from this regular revenue, income also stems from remittances sent by the children living and working in urban areas of the country, especially by the grandchildren's parents. Officially, the household is not in possession of any livestock because Emma has handed over the livestock she accumulated during her husband's employment to her surviving children as a *pre mortem* inheritance. However, the milk and sometimes also the meat of the animals are mostly consumed by the household. Emma's children send money to pay for a hired herder, animal medicines and water fees. The animals – a moderately sized herd of nine cattle and fifty-nine goats – are kept together in a kraal with the livestock of her sister's nearby household.

Successful Migrants and Part-time Pastoralists

The part-time pastoralist households clearly differ from the two types of household described above. John, today in his forties, was born in the Fransfontein area but left the region at an early age, going to school first in Khorixas and then in Windhoek. After independence he became a manager. Despite owning a big and luxurious townhouse and taking holidays on the Cape, John nevertheless continues to invest in a herd in the Fransfontein communal area. Furthermore, he spends many weekends in Fransfontein taking care of his cattle and observing the construction of his rural house. During the week, John's herd is watched by a brother and hired herders. In many conversations John stressed how central it is to him to belong to the Fransfontein area. He wants to have his cattle on his 'home soil', not at a commercial farm far away.

Like John, Junius has a well-paid and prestigious occupation. Junius has worked all his way up to become a senior operator at the Rossing uranium mine in Arandis. He started working in the open pit in the early 1980s. At that time, a disastrous drought was affecting the country, killing much of the livestock. Many young men like Junius were forced to leave their parental farms in search of employment. He spent a great deal of his early earnings buying new livestock, which he kept in his father's kraal. This helped the family to survive. Eventually, the family herd recovered and reached a size that could no longer be managed by family members alone. Wages in the mining sector rose quickly over the years and Junius and his brother started looking for a hired herder. When Junius's parents died in the 1990s, Junius and his brother inherited a considerable number of cattle from their parents. At the same time, Junius kept investing in the herd. His brother, meanwhile retired from the mine, took over the parental home, raising his own animals and supervising the worker who keeps Junius's livestock. During the drought that befell the country again in the early 1990s, Junius decided to take the better part of his goats to another communal farm in the Kamanjab area, where he hired another worker to look after animals. Junius feels, however, that his home and the land to which he belongs is the parental farm and the Fransfontein communal area. There, he plans to resettle with his wife as soon as he retires from work. As with John, Junius spends most of his weekends and holidays at his Fransfontein home, looking after his sizable herd, which amounts to around seventy mature animals.

Distribution of Livelihoods and Livestock

All these examples show that pastoralism is embedded in a wider range of economic strategies that often involve the mobility of people, resources and ideas between rural and urban livelihoods (Schnegg 2009; Greiner 2010). For those who have become successful away from Fransfontein, investment in livestock is both an attractive economic option and a central means of maintaining a sense of belonging. Livestock possessions are visible and highly prestigious.

Table 11.1: Distribution of different household types in Fransfontein (n = 130).

	Percentage of all households	Average number of LSU	Total number of LSU
Local full-time pastoralists	18.5	17.8	427
Households combining full- and part-time pastoralism	64.6	29.1	2442
Absent owners and part-time-only pastoralists	16.9	40.4	890
Overall average		28.9	
Total	100		3759

Table 11.1 shows the average and total number of livestock units (LSU) owned by three different groups among a sample of 130 rural households of the Fransfontein communal area.[14] To allow for comparison between different types of pastoralism, livestock numbers are commonly converted into LSU: one head of cattle and seven goats or sheep equal one LSU. As Table 11.1 shows, the 18.5 per cent of households that act only locally own fewer livestock than the overall average (17.8 LSU/household against 28.9 LSU/household). They also own only a small proportion of the total amount of livestock found in Fransfontein. Although the total number of entirely absent owners is equally small (16.9 per cent), these own many more animals (40.4 LSU/household). However, most households fall into the second category. They take care of their own and partly their relatives' property. Unfortunately, it is often very difficult to assess how many of the animals belong to locally residing people and how many belong to migrants. This would require a livestock census listing individual ownership. However, official livestock records usually aggregate figures at the household level, making it difficult to pin down livestock ownership to individuals. Even if such data were collected ethnographically, they might easily be biased as property rights are not entirely fixed and are open to negotiation. For Fransfontein we were able to achieve first approximations from case studies indicating that the largest part of the livestock population in those herds belongs to people who live outside the community. The second household type (full- and part-time pastoralism) combines multiple strategies at the household and family-network levels to profit from price differences between the cities and the rural areas: while wages are much higher in urban areas, meat is much cheaper in Namibia's hinterlands. Combining rural and urban livelihoods, households utilise this price difference by transferring meat from rural to urban places and money made through wage labour in the opposite direction. This price difference is also used commercially by some pastoralists. A donkey carcass, for example, has a value four to five times greater in the city than in Fransfontein (Greiner 2008; Greiner and Schnegg 2009).

Assessing the Consequences of Part-time Pastoralism

Ecological Consequences

Stocking rates are a common means of measuring and comparing the extent of pasture usage. They can be expressed either as hectares per LSU or as kilograms per hectare. Unfortunately relatively little data exists on stocking rates in Namibia.

Table 11.2 gives an overview of stocking rates that have been reported for comparable pastoral economies in Namibia. The figures indicate that the stocking rate in Fransfontein is very high in comparison with other communal and commercial farming systems in the country.[15] The following analysis reveals that this has effects on the vegetation. In a second step we will ask how this produces stresses on the livestock and consequently shapes the vulnerability of livestock owners.

Table 11.2: Stocking rates that have been reported among Namibian pastoralists.

Region	Rainfall (mm/year)	Hectares/LSU	Kilograms/ hectare[16]	Source
Otjimbingwe, communal		18.99	23.69	Ward et al. (1998: 359)
Otjimbingwe, commercial		40.38	11.14	Ward et al. (1998: 359)
Omaruru area, communal	100–200		4.07	Burke (2004: 394)
Omaruru area, commercial 1	100–200		3.70	Burke (2004: 394)
Omaruru area, commercial 2	200–300		8.20	Burke (2004: 394)
Fransfontein	200–250		31.73	Own data

Remote sensing techniques allow the monitoring of the vegetation cover of large geographical areas. Grasses and savannah trees are the typical vegetation of the Fransfontein area and are the fodder for the livestock held there. To interpret the results from Fransfontein we will compare them with commercial and communal management systems around Fransfontein that share many geo-biophysical characteristics. Firstly, there are the commercial farms on which white owners of European descent raise cattle (classified as 'cattle farms' in Figure 11.2 and Table 11.3) and game ('game farms'). Second are the communal conservancies that have been established in Namibia to protect wildlife and allow the conservation and marketing of natural resources throughout the community ('//Huab conservancy') and the communal territory of the Swartboois, which is largely occupied by relatively few Nama families ('Swartboois').

The data were taken from a Landsat ETM scene on 18 May 2001. To estimate and compare vegetation cover, the normalised difference vegetation index (NDVI) is an acknowledged parameter, with NDVI values commonly used to distinguish between vegetation and non-vegetation areas.[17] NDVI values ranges from –1.0 to 1.0, higher index values indicating higher levels of vegetation cover and lower or negative values representing clouds, bare soil, low vegetation cover or dead vegetation. In the present case, NDVI values were calculated and the interval scale measure was divided into ten steps (Schnegg and Welle 2007).

Figure 11.2 shows the NDVI values for the larger research area. The colour scale goes from green to brown, green indicating the highest NDVI values, hence most vegetation (see caption). Figure 11.2 reveals some contrasts between Fransfontein and the surrounding management systems. The vegetation cover is highest on the game farms and the Swartboois farms, whereas Fransfontein and the conservancy area are covered by the least vegetation. The low vegetation cover in the conservancy area may come as a surprise. There are two explanations

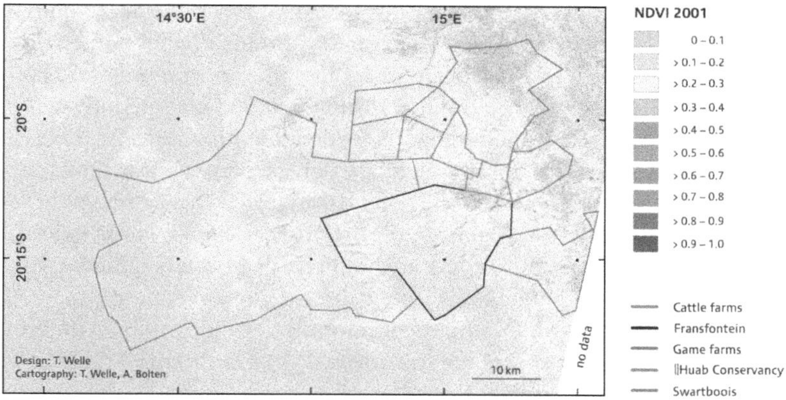

Figure 11.2: NDVI values for the larger Fransfontein area, 2001.

for this: Firstly, at the time of the data collection, the conservancy was not fully established and had not yet imposed any restrictions on the number of livestock a household could keep. The land was thus used as communal farmland just like the area around Fransfontein. Secondly, the geomorphologic structure of the landscape is slightly different. The rocky hills of the conservancy area are relatively infertile, and so one finds little vegetation. Commercial cattle ranches fall in between the two extremes. Table 11.3 summarises these differences numerically. It gives the average NDVI values for all five areas.

Table 11.3: Average NDVI values, 2001.

	2001
Fransfontein	0.227983
Swartboois	0.267786
//Huab conservancy	0.194228
Game farms	0.287077
Cattle farms	0.248414

Even though this affects households of all types, those that rely mostly on wage labour and have steady cash incomes at the individual or at the household level are less often forced to sell under these disadvantageous conditions. While large herd owners rent a truck to transport their animals to market, the small-scale farmer is left only with the opportunity of selling their animals to a trader who comes to Fransfontein. In the absence of real competition and with arrangements favouring the buyer, traders are in a favourable position. An example of live-weight prices for goats underlines this. At a livestock auction in the Fransfontein area in 2005, prices between N$8 and N$12 per kilo were offered. At that time the average national auction price was above N$20.[18]

Disequilibrium theories argue that during extreme years, livestock mortality is independent of its prior population density. This means that all households face equal ecological stresses. But the means of coping differ even more during critical times. However, as there was no drought during our fieldwork periods we do not have first-hand data that would allow us to describe how different households respond to inter-annual droughts such as the 1992/93 crisis. This limits the following analysis to an assessment of the potential strategies that actors could adopt. We have extracted these strategies from the literature and our own interviews, and will discuss how the three different household types can access different kinds of capital to react to crisis. The analysis reveals three dominant strategies.

Moving animals to another area: A common strategy for coping with drought is to move animals to areas that are less affected. In Fransfontein this can mean either simply moving livestock or moving the entire household. The three types of pastoral household introduced above have very different access to this strategy. The absent livestock owner who hires a herder to supervise his animals can relatively easily move his herd to a new location. Most of these part-time pastoralists have financial resources and also own cars which make remote areas relatively easily accessible to them. Their political and social capital is also likely to pave their way to being accepted as new members in the settlement they move to. The two other pastoral household types are much less flexible when it comes to moving animals and other assets. Not only do they often lack the financial capital to organise moving livestock, but a move will also disconnect them from the multiple neighbourhood networks that buffer daily risks.

Selling animals as long as prices are good: A second common strategy is the sale of animals. The sooner one anticipates the situation the better the price will be, as in drought years the market is flooded both by communal and by commercial livestock owners. As mentioned before, market prices in Fransfontein are significantly lower than in Outjo and Otjiwarongo. To rent a lorry to transport the animals to the market place requires social and economic capital that will only pay off if large numbers of animals are sold at once. Thus it is often not an option for more economically marginalised pastoral households (the first two types in Table 11.1). Once animals are sold the money is invested or used. However, in the extremely sparsely populated and poor area around Fransfontein in particular, alternative investment options are scarce. Also, the demands of others who see that a household has just sold most of its property will put it under pressure to give a fair share of the money it has obtained to kith and kin. In combination, both factors often lead to a situation in which most of the money is used up after a short period of time without any long-term effect on the household economy and its livelihood. This makes selling a very difficult strategy for locally embedded and rooted pastoral households.

Buying fodder: Buying hay or other fodder is a coping strategy often utilised by commercial farmers. Among communal households this is often not an easy option as it requires a lot of capital and also information, transportation and storage, usually unavailable to all three pastoralist household types. However, those who are more closely connected to urban centres buy fodder for lambs if the milk from their mothers and the available grazing is insufficient to feed them.

The intersection between two components – stress and coping strategies – shapes the vulnerability of households. In sum, the foregoing analysis indicates that while the effect of overstocking is the same for all animals, the declining livestock quality and the threat of loss concerns them quite differently. This has to do with the different coping mechanisms at hand. Household vulnerability thus increases as we move from local through mixed-migrant to absentee-pastoral farmers.

Conclusion

The sense of belonging among pastoralists in Fransfontein has been shaped by the complex historical trajectories of land dispossessions, reserves, homelands and migration patterns. These ruptures have led to specific forms of locality and neighbourhood (Appadurai 1996). The passionate desire, expressed by so many Fransfontein people who differ according to generation, gender, class and location, to live an independent life as full-time pastoralists is, on the one hand, an important economic strategy and, on the other hand, a central way to create and maintain belonging to the area. However, as we have shown, the social and ecological costs of the combination of these economic and coping strategies and the desire to belong strongly vary depending on the type of pastoralist household.

As in other parts of Africa, migrants from Fransfontein are reluctant to see their city dwellings as their home and place of belonging. Even very wealthy urban migrants maintain ties to their birth place, and they often invest heavily in livestock that is kept on the communal lands around Fransfontein. This investment serves multiple purposes. It is a risk-minimisation strategy against economic crisis. In the long run it can also serve as the basis for a rural livelihood at a later stage in life. And finally it serves as a means of creating identity and belonging. Through livestock keeping, migrants remain 'local actors' (Appadurai 1996). Yet these economic and symbolic strategies of absent wealthy migrants also have costs, albeit not for the part-time pastoralists but for those who stay behind and are forced to survive as full-time pastoralists on the Fransfontein communal lands. Compared to other areas in Namibia, the stocking rates are high and the landscape in Fransfontein is much less vegetated than its surroundings. This produces physical stress on all animals, leading to animals being relatively low in weight and having a slow rate of reproduction. The environmental costs produced by rich townsfolk are external and shape the availability of vegetation and the quality of livestock for all inhabitants. Economically marginalised, full-time pastoralists have fewer means of coping with this situation. This strongly shapes the vulnerability of pastoral households – especially during times of crisis. The message of these ecological, social and economic analyses is very clear: the better a household is embedded in the labour market, the better it is equipped to cope with a crisis. Thus, the wealthy part-time and weekend pastoralists living in

Namibia's cities are best prepared to survive the next crisis, most likely another drought. They will maintain both their sense of belonging to a pastoralist life-world and their livestock.

But how will the locally based and much more vulnerable full-time pastoralist households survive the next critical ecological event? In her analysis of Molepolole, a rural area of Botswana, Anne Griffiths classifies the local population into two basic social and economic forms: the 'peasantariat' and the 'salariat' (Griffiths 1997: 24–25). The peasantariat has to survive through a mixture of economic strategies, involving crops, cattle and unskilled wage labour (ibid.: 24), and occupy what Spencer had called an increasingly marginalised niche (Spencer 1998). The salariat, in contrast, constitute the small, educated elite with stable employment. In Fransfontein, the salariat consists of the wealthy absent part-time pastoralists portrayed here and the small local elite of politicians, civil servants and professionals such as teachers (see Pauli 2009a, 2009b). The two other types of pastoral household – locally based full-time pastoralists with varying successful migrating children and grandchildren – comprise a peasantariat.

There is a tragically high likelihood that with the next drought many poor, locally based full-time pastoralists will turn into something like a 'pastoraltariat', losing all their animals and being left with nothing more to offer than their physical labour as herders. Some impoverished households in Fransfontein have already emerged as such a group. They are mostly marginalised, consisting for instance of many needy single mothers (Pauli 2009a: 72–136). Households such as these may provide a glimpse of the future for many vulnerable and poor pastoral households that still possess some livestock but are unable to cope with an ecological crisis, such as a drought. Pastoralism, which has been the prime economic activity and source of belonging for a number of generations of people in Fransfontein, will then become more and more a fashion, a lifestyle of the urban and the rural elites.

Notes

The data presented here were collected during several research stays as part of the collaborative research project 'Arid Climate, Adaptation and Cultural Innovation in Africa' (SFB 389) and funded by the German National Science Foundation (DFG).

1. All names of informants are pseudonyms.
2. In 2004 when this incident happened the exchange rate between one Namibian dollar (N$) and one euro was approximately 8.5:1.
3. A very similar pattern was observed by Watson some decades ago among migrants on the Copperbelt (Watson 1958).
4. For a good discussion of the issues, see Anderson (2002) and Rohde et al. (2006).
5. The perspective of individual actors and classes of similar actors has often been neglected in ecological anthropology.
6. Currently (2008 onward) this is starting to change and pastoralists have to register for certain farming areas under the new land legislation.
7. These figures are based on census data collected by Pauli and Schnegg in June 2004.

8. The questionnaire was administered through ten local assistants who had been trained in ethnographic interviewing skills during a two-week workshop by the lead researchers.
9. Massive land dispossession did not occur in Namibia's northern territories, though. The 'settler state model' (Silvester, Wallace and Hayes 1998: 18–19) is thus not the appropriate model for understanding historic processes in all parts of the country. There is no one grand narrative of land dispossession in Namibia, as Silvester, Wallace and Hayes remind us (ibid.). However, for Fransfontein the 'settler state model' is the dominant model.
10. Until the uprising of 1904 to 1907, indigenous Namibians from southern and central Namibia had mainly been exploited as cheap sources of labour (Jauch 1998). After the genocide, the system of migrant contract labour with north Namibia was established (Moorsom 1995).
11. For a much more detailed analysis, see Dieckmann (this volume).
12. Reserve Boards were the political body established by the South African administration to represent the local inhabitants against the colonial power.
13. The colonial archives tell a slightly different story, though. According to the written records, the extension of the Fransfontein reserve was granted at about the same time, not to accommodate the 'Ovambo' settlers though but to make space for Nama migrants from Grootfontein and Walvis Bay.
14. The number of cases is lower than the number of households on the communal lands because households of herders who do not possess animals were excluded from the analysis.
15. According to our observations the processes causing these stocking rates are similar in many parts of central Namibia. The difference may indicate that they are more advanced here than elsewhere.
16. Calculations are based on 450 kilograms per cow (Ward et al. 1998: 359).
17. The NDVI value gives the difference of the reflectance between the near infrared (Landsat Band 4) and the red (Landsat Band 3) spectral ranges, which is divided by the sum of the reflectance between the near infrared and the red spectral ranges. Thus: NDVI value = near infrared – red / (near infrared + red).
18. The auction data here comes from our fieldwork, while national data was obtained from the Namibian Meat Board (http://www.nammic.com.na/stats.php).

References

Anderson, D. 2002. *Eroding the Commons: The Politics of Ecology in Baringo, Kenya, 1890s–1963*. Oxford: James Currey.

Appadurai, A. 1996. *Modernity at Large: Cultural Dimensions of Globalization*. Minneapolis: University of Minnesota Press.

Barnard, A. 1992. *Hunters and Herders of Southern Africa: A Comparative Ethnography of the Khoisan People*. Cambridge: Cambridge University Press.

Barth, F. 1964. 'Capital, Investment and the Social Structure of a Pastoral Nomad Group in South Persia', in R. Firth and B.S. Yamey (eds), *Capital, Saving and Credit in Peasant Societies*. London: Allen and Unwin, pp.69–81.

Behnke, R.H., I. Scoones and C. Kerven. 1993. *Range Ecology at Disequilibrium: New Models of Natural Variability and Pastoral Adaptation in African Savannas*. London: Overseas Development Institute.

Berzborn, S. 2007. 'The Household Economy of Pastoralists and Wage-labourers in the Richtersveld, South Africa', *Journal of Arid Environments* 70: 672–85.

Berzborn, S., and M. Schnegg. 2007. 'Vulnerability, Social Networks, and Resilience: Rural Livelihoods in the Richtersveld, South Africa', in B. Lohnert (ed.), *Social Networks in South Africa*. Saarbrücken: Verlag für Entwicklungspolitik, pp.115–47.

Biersack, A. 2006. 'Reimagining Political Ecology: Culture/Power/History/Nature', in A. Biersack and J. Greenberg (eds), *Reimagining Political Ecology*. Durham: Duke University Press, pp.3–42.

Bollig, M. 2004. ,'Die Landreform in Namibia', in L. Förster, D. Henrichsen and M. Bollig (eds), *Namibia – Deutschland: Eine Geteilte Geschichte*. Munich: Minerva, pp.304–23.

——— 2006. *Risk Management in a Hazardous Environment: A Comparative Study of Two Pastoral Societies*. New York: Springer.

Botha, C. 2003. '"Development", Survival and Adaptation in the Former Damara "Homeland" of Namibia', in *Geography and Environmental Studies 1*. Windhoek: Department of Geography and Environment Studies, University of Namibia.

Bryant, R.L. 1992. 'Political Ecology: An Emerging Research Agenda in Third World Studies', *Political Geography* 11: 12–36.

Burke, A. 2004. 'Range Management Systems in Arid Namibia: What can Livestock Numbers Tell Us?' *Journal of Arid Environments* 59: 387–408.

Chambers, R. 1989. 'Vulnerability, Coping and Policy', *Institute of Development Studies Bulletin* 20: 1–7.

Dawids, F., et al. 2007. *Living Together: Culture and Shared Traditions in Fransfontein, Namibia*. Cologne: Heinrich Barth Institut.

Fuller, B.B. 1993. 'Institutional Appropriation and Social Change among Agropastoralists in Central Namibia, 1916–1988', PhD dissertation. Boston: Boston University.

——— 1998. '"We Live in a Manga": Constraint, Resistance and Transformation on a Native Reserve', in P. Hayes et al. (eds), *Namibia under South African Rule: Mobility and Containment 1915–1946*. Oxford: James Currey, pp.194–216.

Gillson, L., and M.T. Hoffman. 2007. 'Rangeland Ecology in a Changing World', *Science* 315: 53–54.

Greiner, C. 2008. *Zwischen Ziegenkraal und Township: Migrationsprozesse in Nordwestnamibia*. Berlin: Reimer.

——— 2010. 'Patterns of Translocality: Migration, Livelihoods and Identities in North-west Namibia', *Sociologus* 60(2): 131–61.

Greiner, C. and M. Schnegg. 2009. 'Urban-Rural Lifestyles'. *Development and Cooperation* 36(6): 254–55.

Griffiths, A. 1997. *In the Shadow of Marriage: Gender and Justice in an African Community*. Chicago: University of Chicago Press.

Hardin, G. 1968. 'The Tragedy of the Commons', *Science* 162: 1243–48.

Hogg, R. 1986. 'The New Pastoralism: Poverty and Dependency in Northern Kenya', *Africa* 56: 319–33.

Jauch, H.M. 1998. *Affirmative Action in Namibia: Redressing the Imbalances of the Past?* Windhoek: New Namibian Books.

Klintenberg, P., and A. Verlinden. 2008. 'Water Points and their Influence on Grazing Resources in Central Northern Namibia', *Land Degradation and Development* 19: 1–20.

Lau, B. 1987. *Southern and Central Namibia in Jonker Afrikaner's Time*. Windhoek: National Archives of Namibia.

Little, P. 1985. 'Absentee Herd Owners and Part-time Pastoralists: The Political Economy of Resource Use in Northern Kenya', *Human Ecology* 13: 131–51.

Little, P.D., et al. 2001. 'Avoiding Disaster: Diversification and Risk Management among East African Herders', *Development and Change* 32: 401–33.

Lovell, N. (ed.). 1998. *Locality and Belonging.* London: Routledge.

McCabe, J.T. 2004. *Cattle Bring Us to Our Enemies: Turkana Ecology, Politics, and Raiding in a Disequilibrium System.* Ann Arbor: University of Michigan Press.

Miescher, G. 2006. *The Ovambo Reserve, Otjeru (1911–1938): The Story of an African Community in Central Namibia.* Basel: Basler Afrika Bibliographien.

Moorsom, R. 1995. *Underdevelopment and Labour Migration: The Contract Labour System in Namibia.* Windhoek: University of Namibia, Department of History.

Pauli, J. 2008: 'A House of One's Own: Gender, Migration and Residence in Rural Mexico', *American Ethnologist* 35(1): 171–87.

——— 2009a. 'Celebrating Distinctions: Marriage, Elites and Reproduction in Rural Namibia', Habilitation dissertation. Cologne: Universität zu Köln.

——— 2009b. '(Re)producing an Elite: Fertility, Marriage and Economic Change in North-west Namibia', in C. Greiner and W. Kokot (eds), *Networks, Resources and Economic Action: Ethnographic Case Studies in Honour of Hartmut Lang.* Berlin: Reimer, pp.303–25.

Peyroux, E. 2004. *Windhoek, Capitale de la Namibie: Changements Politiques et Recompositions des Périphéries.* Paris: Karthala.

Rohde, R. 1997. 'Nature, Cattle Thieves and Various Other Midnight Robbers: Images of People, Place and Landscape in Damaraland, Namibia', PhD dissertation. Edinburgh: Edinburgh University.

Rohde, R., et al. 2006. 'Dynamics of Grazing Policy and Practice: Environmental and Social Impacts in Three Communal Areas of Southern Africa', *Environmental Science and Policy* 9: 302–16.

RSA. 1964. 'Report of the Commission of Enquiry into South West African Affairs' [Odendaal Commission]. Pretoria: Government Printer, Republic of South Africa.

Schnegg, M. 2007. 'Battling Borderlands: Causes and Consequences of an Early German Colonial War in Namibia', in M. Bollig et al. (eds), *Aridity, Change and Conflict in Africa.* Cologne: Heinrich-Barth-Institut, pp.247–64.

——— 2009. 'It's the Combination that Counts: Diversification of Pastoral Livelihoods in North-western Namibia', in C. Greiner and W. Kokot (eds), *Networks, Resources and Economic Action: Ethnographic Case Studies in Honour of Hartmut Lang.* Berlin: Reimer, pp.229–48.

Schnegg, M., and T. Welle. 2007. 'Inequalities Inscribed in Landscape', in O. Bubenzer, A. Bolten and F. Darius (eds), *Atlas of Cultural and Environmental Change in Arid Africa.* Cologne: Heinrich Barth Institute, pp.100–3.

Scoones, I. 1999. 'New Ecology and the Social Sciences: What Prospects for a Fruitful Engagement?' *Annual Review of Anthropology* 28: 479–507.

Silvester, J., M. Wallace and P. Hayes. 1998. '"Trees Never Meet": Mobility and Containment: An Overview 1915–1946', in P. Hayes et al. (eds), *Namibia under South African Rule: Mobility and Containment 1915–1946.* Oxford: James Currey, pp.3–48.

Spencer, P. 1998. *The Pastoral Continuum: The Marginalization of Tradition in East Africa.* Oxford: Clarendon Press.

Sullivan, S. 2004. 'Natural Resources: Use, Access, Tenure and Management', in T. Bowyer-Bower and D. Potts (eds), *Eastern and Southern Africa.* London: Addison Wesley Longman, pp.118–66.

Sullivan, S., and R.F. Rohde. 2002. 'On Nonequilibrium in Arid and Semi-arid Grazing Systems', *Journal of Biogeography* 29: 1–26.
Swift, D. 1982. 'The Future of African Hunter-gatherer and Pastoral Peoples', *Development and Change* 13: 159–81.
Turner, B., et al. 2003. 'A Framework for Vulnerability Analysis in Sustainability Science', *Proceedings of the Natural Academy of Sciences* 100: 8074–79.
Turton, D. 2005. 'The Meaning of Place in a World of Movement: Lessons from Long-term Field Research in Southern Ethiopia', *Journal of Refugee Studies* 18: 258–80.
Vetter, S. 2005. 'Rangelands at Equilibrium and Non-equilibrium: Recent Developments in the Debate', *Journal of Arid Environments* 62: 321–41.
Ward, D., et al. 1998. 'Land Degradation Is Not a Necessary Outcome of Communal Pastoralism in Arid Namibia', *Journal of Arid Environments* 40: 357–71.
Watson, W. 1958. *Tribal Cohesion in a Money Economy: A Study of the Mambwe People of Zambia.* Manchester: Manchester University Press.
Watts, M., and H. Bohle. 1993. 'The Space of Vulnerability: The Causal Structure of Hunger and Famine', *Progress in Human Geography* 17: 43–67.
Watts, M., E. Sheppard and T.J. Barnes. 2000. *Political Ecology.* Malden, MA: Blackwell.
Werner, W. 2004. 'Landenteignung, Reservate und die Debatte zur Landreform in Namibia', in L. Förster, D. Henrichsen and M. Bollig (eds), *Namibia – Deutschland: Eine Geteilte Geschichte.* Cologne: Ethnologica, pp.292–303.
Wiens, J. 1984. 'On Understanding a Non-equilibrium World: Myth and Reality in Community Patterns and Processes', in D.R. Strong (ed.), *Ecological Communities: Conceptual Issues and the Evidence.* Princeton, NJ: Princeton University Press, pp.439–57.

Part III

Violence, Trade, Conservation and Pastoralism in Africa

Chapter 12

Pastoralism, Conflict and the State in Contemporary Eastern Chad

The Case of Zaghawa–Tama Relationships

Babett Jánszky and Grit Jungstand

Conflicts between pastoralists and farmers occur regularly in Sahelian Africa. In most cases, farmers emerge victorious from such encounters due to their closer relations with state institutions. Pastoralists often rely more on non-state violence to defend their interests and are thereby prone to get into conflict with the state. The case described here is especially interesting as it presents a counter-example. Zaghawa pastoralists are in command of the state as they have held the presidency in Chad since 1990 and have mastered the infiltration of state and economic institutions. In this chapter we trace the process of political, economic and geographical dimensions of the pastoral expansion of the Zaghawa with regard to the economic and political impacts of this process on other groups, taking the example of the Tama.

Throughout history, the whole border region between Chad and Sudan, the settlement area of both Zaghawa and Tama, has been marked by the fight for power and influence between the empires of Wadai and Darfur. In the late nineteenth century the French reached the far east of Chad, while the English had conquered Darfur shortly beforehand. In pursuit of proper economic or political interests – or just for the protection of their own territory and population – ethnic groups inhabiting this area preferred to ally themselves with the strongest possible partners and changed their alliances according to fluctuations of power. The pattern of changing alliances persisted with the demarcation of the border between the two nation-states, and the choice of alliances also led to tensions and fractures within ethnic groups.

Nowadays, the semi-nomadic Zaghawa migrate seasonally with their flocks into Dar Tama, the region to the south of their own territory that formerly had a quite monoethnic population. Over the past two decades, Zaghawa have also increasingly settled in this area. These migration processes have led to a rise in conflicts between Tama farmers and Zaghawa pastoralists, the latter taking advantage of their powerful position within the state to impose their rights and, going beyond that, engaging in cattle raids and the destruction of Tama fields, which they have taken over and used as pasture for their camels. Backed by state authorities, the Zaghawa have no fear of compensation claims, and the fact that they are better equipped with weapons means that they are in most cases protected from Tama vigilante groups. While the Tama relied mainly on strategies of avoidance in the past, they have increasingly engaged themselves in militant opposition movements as a reaction to Zaghawa encroachment, and economic and power imbalances between the two groups have led to a state of generalised insecurity in the region.

The Zaghawa and the State

When Idriss Déby, a Zaghawa of the Bideyat sub-group, came to power in 1990, he promised the people of Chad, 'Je ne vous apporte ni or ni argent, mais la liberté' ('I bring you neither gold nor money, but freedom') (ICG 2008: 9). As a former ally of and army chief under Hissène Habré, the Chadian president from 1982 until 1990,[1] people had low expectations of Déby regarding democratic change and justice. Nevertheless, in 1993 a national democracy conference was organised by Déby's transitional government. As a result, a multiparty system was introduced, and presidential as well as legislative elections were scheduled for the following years. Since then, Chad has held elections more or less regularly, always with outcomes that favour Déby and his party, the Mouvement Patriotique pour le Salut (MPS). Actual voter participation in elections is usually low, and opposition parties, NGOs and international observers often criticise the results as irregular and manipulated. In 2005, the constitution was amended (possibly due to the MPS majority in Parliament) to enable Déby to serve a third term as president. Subsequently, he won the election one year later, which was boycotted by the opposition.

Over the past twenty-two years, the Déby regime has succeeded in clinging to power in Chad, even though it relies on a very narrow base of support while facing extreme unpopularity among the population.[2] The strongest pillar of its relative stability is its ethnically based network, which dominates the military and other relevant sectors, such as the public treasury, the customs and important branches of the public administration. These clientelistic structures and the ill- functioning state nourished since their establishment a continiously growing dissatisfaction of the Chadian population. Many people feel that they are not represented by the state and have the impression of being cheated of their interests and rights as citizens.

However, due to the highly fragmented and weakened nature of the political and military opposition, Déby has so far been able to combat threats to his power.[3] Even though Chad has several registered political parties, they usually rely on narrow regional and ethnic support, and politics is often personalised and lacks programmatic content. In recent years, large parts of the public finances, which have increased significantly due to oil exports since October 2003, have been spent on military equipment, which further lowers the chances of a rebel movement ousting the current government.

External Support for Déby's Regime

The stability of Déby's presidency is also a result of continuous support by and collaboration with external actors. France maintains close links with its former colony, closing its eyes to undemocratic elections and human rights violations. France has a military accord with Chad and has some 1,000 troops stationed in the country. In recent years the French troops have actively participated in military operations against rebel movements attempting to overthrow Déby's government in N'Djaména.[4] The placement of 3,700 European Union Force (EUFOR) troops in Chad in March 2008, subsequently replaced by UN troops in March 2009, was basically a result of pressure from France on the international community.[5] The EUFOR mission in particular was considered by many Chadians as an extension of the French presence in the country, believed to further serve Déby in stabilising his power through securing Chad's eastern border against rebels coming from bases in Sudan.[6]

As well as French influence, the US also has a strong interest in keeping Déby in power. Two of the three members of the oil consortium,[7] Chevron and Exxon, are American companies, and the US government fears that a different government might try to exclude the American companies in favour of Chinese competitors who have also cast their sights on Chadian oil. Furthermore, the US considers Déby's regime an ally in its 'war on terror'.[8]

The support for, or at least the tolerance of, Déby's presidency by the international community is partly a result of the lack of suitable alternatives. Déby apparently is a reliable partner in maintaining Western interests. The regime's current opponents and the leaders of Chadian rebel movements, however, are supported by Sudan and other Arab countries and not seen too sympathetic to maintaining these interests. It is doubtful that the loose alliances formed by the different Chadian rebel groups to overthrow Déby will last in the wake of a successful takeover. Probably none of the rebel chiefs would renounce their claim to the presidency, irrespective of previous agreements about the distribution of government posts. Therefore the risk of a slide into civil war and further destabilisation is seen as high in the case of a successful coup by rebel forces.

It is for these reasons that Déby is seen by the international community and even by some Chadians themselves as 'the lesser evil' and that he faces relatively low pressure and criticism from abroad, this despite the actual situation inside Chad, which is characterised by human rights violations, corruption and a state of impunity for people belonging to the upper reaches of the circles of power, mainly members of the Zaghawa ethnic group.

Social Cohesion within the Zaghawa Community

Many people among the Chadian population we talked to during research between 2002 and 2007 considered the Zaghawa to be arrogant and potentially aggressive, and interpersonal contact with them is avoided by many Chadians, a stance justified by the perceived 'closed-off' nature of Zaghawa society. While more contacts between Zaghawa and non-Zaghawa exist than generally admitted, it is true that the Zaghawa tend to rely mainly on their own ethnic group, a phenomenon which is illustrated for instance by the strong preference for ethnic endogamy and the almost exclusively Zaghawa-based friendship networks of all age groups.[9] However, outsiders tend to generalise when criticising the Zaghawa and consider them as an entirely homogeneous group. By doing so they ignore or deny the deep divisions among Zaghawa, especially when it comes to political fellowship and support, which mainly follow clan and family divisions.[10] As early as the 1990s, resistance against the Déby regime formed within his own ethnic group, especially among the Kobe sub-group.[11]

While the Kobe are prominent in business and administration, many Bideyat have joined the army, which, after Déby's takeover, has become dominated by this sub-group. Nevertheless, many Zaghawa still rely today on pastoralism as their livelihood. Zaghawa military and pastoralists are responsible for most of the cases of violence and injustice in the confrontation of this group with people of other ethnic origin. According to a member of the Chadian human-rights organisation Felidhob, 99 per cent of human-rights violations in eastern Chad in 2002/3 were committed by Zaghawa, and most of these by members of the Bideyat sub-group (as well as Goran, who are sometimes disguised as Zaghawa). While he said that the juridical procedures (if the offender is known) are the same for anyone, he explained that a Zaghawa will never have to stay in prison for a long time. Prison guards are often bribed or put under pressure to free prisoners.[12] But in many cases, conflicts or crimes involving Zaghawa as suspects are not even brought to court since the victims fear reprisals by the offender's family. Other reliable sources claimed that should a victim try to take legal action against a Zaghawa, judges are bribed or threatened, and they will declare the offender innocent or deliver a light sentence.[13] Meanwhile, members of other ethnic groups report that Zaghawa nowadays often avoid the traditional means of interethnic conflict resolution that are common, especially in rural areas, by directly involving state legal institutions that are dominated by their own kin.[14]

It is worth underlining the importance of social cohesion among the Zaghawa, which is mainly based on strong family ties, the respect of traditions and the principle of mutual support. While we focus in this article on Zaghawa pastoralists, the expansionist and sometimes ruthless behaviour of migration can only be understood in relation to the closed-knit family and clan networks that connect Zaghawa from rural and urban backgrounds. Almost every Zaghawa pastoralist can mobilise a network to obtain support in case of difficulties, such as paying 'blood money' (arabic: *diya*; payments made by the family of the offender to the victim's family in cases of injury or killing in order to avoid revenge) claims or putting pressure on the authorities to act in their favour. A very strong sense of solidarity as well as moral and religious obligations commit any well-off or influential Zaghawa to help relatives in need. Various interviews and household studies by Jungstand revealed the immense pressure placed on functionaries and businessmen to intervene in such cases, while a refusal to do so is punished by the community by social exclusion. In order to fulfil these obligations, Zaghawa sometimes appropriate public funds or make use of their authority in politics, the administration or the military to achieve their aims. In this way, they use their powerful position within the Chadian state as a resource to serve their clientelistic networks.

The Chadian state is (and, since its creation, has always been) a weak state with highly informal politics, a situation exploited by elites for appropriate networks. No serious efforts have been made to institutionalise or strengthen the state. Thus, Chadian politics represent a typical case of what Chabal and Daloz characterise as 'the instrumentalization of disorder' (Chabal and Daloz 1999: 155). This should be kept in mind during the following discussion of economic and political relations between Zaghawa and their neighbours in rural areas.

The Ecological Setting in Dar Zaghawa and Dar Tama

The area occupied by Zaghawa and Tama discussed here lies in north-eastern Chad in the region bordering Sudan's Darfur province. The region is part of the Sahel, with its specific climatic conditions and vegetation. The climate is marked by a short rainy season of about three months between June and September and a long dry season lasting the rest of the year. Agriculture is highly dependent on this seasonal rainfall for watering crops, recharging wells and enabling the regrowth of pasture for grazing cattle. The movement of cattle depends on the quality and quantity of rainfall in different micro regions. Nevertheless, annual precipitation is quite low. Rainfall varies between 200 and 400 millimetres per year (Ben Yahmed 2006: 6–7; Mahamoudi 2006: 51–52), decreasing in a northward direction towards the Sahara. The rate of precipitation normally reaches its highest in July and August. However, the region has consistently been affected by droughts, which heavily endanger local economies. Unreliable precipitation (both in quality and quantity) forces the region's inhabitants to secure their livelihood by diverse means and to adopt suitable strategies to minimise the influence of environmental risks on their way of life. Humidity and rain are major factors in people's lifestyles, ranging from semi-nomadic pastoralism in the drier northern parts to sedentarism in the southern parts of the area.

The climate of Dar Zaghawa and Dar Tama is influenced by the Ouaddai Massif that forms the eastern border of the Chad basin. The area lies at an average altitude of around 1,000 meters, with peaks of around 1,300 to 1,400 metres in the Niéré, Maraone and Kapka plateaus (Ben Yahmed 2006: 6–7; Mahamoudi 2006: 50–51). This landscape of pronounced gradients is shaped by erosion and contains numerous wadis. Due to their elevation, the comparatively high amount of water retained in these mountainous areas supports the charging of natural water repositories. These are most often situated alongside wadi riverbeds, so that water resources can also be found throughout the dry season. The wadi reservoirs even allow for the creation of gardens alongside wadis throughout the dry season. The elevation of the area also produces a quite temperate climate. On the plateau around Iriba and Guéréda, temperatures can even decline below ten degrees Celsius during the cold season between November and February.

The vegetation is typical Sahelian, with *Acacia* species being common. The landscape is characterised by steppes with little tree cover in the western and eastern extremes to arboreal savannah marked by degradation and areas of large-scale erosion in the centre (Mahamoudi 2006: 52). With the droughts of the early 1970s and 1980s and the resulting degradation of the vegetation, as well as excessive hunting activities in the past, the fauna has been, besides various small animals, reduced to very few bigger species, such as jackals, rabbits and bustards.

Zaghawa Pastoralism and Migration

The traditional Zaghawa economy is based on semi-nomadic pastoralism. Cattle dominated their flocks until the middle of the twentieth century,[15] but nowadays camels and small stock have become more important due to climatic change, ecological crisis and the resulting shifts in the availability of natural resources. Erratic rainfall and high seasonal and annual variability in the quality and quantity of pasture are typical phenomena in the Sahel. Over the centuries people have developed coping mechanisms to adjust to the inherent risks and insecurity, with flexibility and mobility the main pillars of their pastoral economy. However, the severe droughts of 1968 to 1970, 1972/73 and 1984, chronic political instability and new options for labour migration abroad have led to significant changes in livelihood systems. Today a broad range of alternative economic strategies have been developed in order to cope with the growing insecurity of the pastoralist sector. But we will focus here on the changing patterns of mobility and migration of Zaghawa pastoralists connected with their search for water and better grazing.

It is interesting to note that all the Zaghawa pastoralists interviewed by Jungstand claimed that the availability of water is the major problem in years of drought, which challenges the general perception that the lack of good pasture is the main reason for increased mobility.[16] Wells often fall dry in the dry season and those still carrying water are used for so many animals that watering a herd may take many hours due to the low water level, and wells may be used even at night. Furthermore, many of the hand-dug wells in Dar Zaghawa, built by the herders themselves, are extremely deep (up to forty metres). Only a few permanent wells exist, most of them lying in wadi beds, and these have to be dug anew following each rainy season.[17] So the availability of water is a critical aspect of the Zaghawa pastoralist economy and often determines the choice of when and where to migrate. Of course, the quality of pasture also plays an important role in these decisions, and information on grazing conditions is collected and discussed whenever pastoralists meet each other.

Due to low rainfall and environmental degradation, the annual transhumance routes of Zaghawa herders are nowadays longer (that is to say, people start out earlier in the year after the end of the rainy season) and cover larger distances than thirty to forty years ago.[18] Especially in years of drought, Zaghawa herders move with their animals far into the neighbouring area of Dar Tama, in some years even beyond this and into the areas south of Abéché (Ouaddai, Salamat), thus covering distances of more than 300 km.[19]

Even though the whole Zaghawa economy has undergone diversification, with trade and migrant labour now the main sources of income (see Meerpohl, this volume), and many Zaghawa no longer rely (exclusively) on pastoralism, the pressure on the natural environment has increased. While reliable statistics on the animal population in Chad do not exist, estimates indicate that, after heavy decimation during the droughts of the 1970s and 1980s, pastoralists' herds have recovered and even increased in size, and the total animal population is continuously growing. All Zaghawa herd owners interviewed by Jungstand between 2002 and 2007 claimed that they now have more camels, sheep and goats per herder, and the herds are larger than in the times of their fathers.[20] In addition to this, Zaghawa businessmen or others with high incomes in the urban sector invest in camels, which are tended by relatives back in their home village.[21] These factors, in combination with climatic change and increased risk of drought, are the major cause for the need to migrate towards more fertile regions in the south.

Settling Down in Dar Tama

What started for some as seasonal migration sometimes became permanent resettlement. Since the 1980s, many Zaghawa of the Bideyat and Borogat sub-groups (initially settling in the northern parts of Zaghawa territory on the southern edges of the Ennedi Mountains) have installed permanently in Dar Tama. The drought of 1984/85 that affected much of the Sahel caused many Bideyat as well as other pastoralists from northern Chad (mainly Goran) to abandon their home region and settle in Dar Tama. Today, many settlements of Bideyat and Borogat exist in Dar Tama, often in the proximity of Tama villages, but usually these are smaller in size and mostly constituted by the members of just one family. Some Bideyat, mainly those who have set up villages in the more central parts of Dar Tama, took advantage of areas that had emptied due to Tama migration to Sudan in the 1970s and 1980s, initiated by the droughts and in the hope of finding land or labour in Sudan. According to Tairo Mirri Erdajeb, the representative of the Bideyat in Guéréda (seat of the Tama sultanate and the administrative centre of Dar Tama), a total of approximately 3,000 Zaghawa (Kobe, Bideyat and Borogat) settlements exist within the borders of Dar Tama, most of them founded in the first years after Idriss Déby took power in N'Djaména.[22] This underlines how widespread this phenomenon is, creating a whole new social reality in this formerly scarcely populated area within the last twenty years.

Part of Jungstand's research was carried out in such a settlement, the hamlet of Hudje, which lies on the border between Dar Zaghawa and Dar Tama, some 25 kilometres south of Iriba (see Figure 12.1).[23] Hudje has ninety-three inhabitants; they rely entirely on pastoralism, maintaining a semi-nomadic lifestyle with herds of several hundred camels, cattle, goats and sheep. Herd sizes vary according to pasture conditions and animals diseases. The inhabitants of Hudje are all descendants of Abdullai Mandi Mounour, of the Gueira clan of the Bideyat. His family originate in the area of Nui, a former settlement on the south-western Ennedi Mountains. Droughts forced him and his family to migrate first to Kedego, approximately 30 kilometres north of Iriba. In 1989 the family left this area to settle permanently on the southern edges of Dar Zaghawa, where they had previously only gone during the dry seasons to use the good pasture lands. Even though they still feel a sense of belonging to their area of origin and maintain contacts with family members who remain there, they do not intend to go back as long as the climatic conditions do not change in favour of that region. The place Abdullai Mandi chose to settle was unoccupied, though Zaghawa as well as Tama villages exist in the area. Hudje and its surroundings lack water sources, but the family remains because of the good pasture and tolerates covering long distances to go to the nearest wells.[24] To guarantee use rights of these distant wells, the family has established kinship relations through marriage with the villages close to these wells. While the story of their arrival and installation presents a typical case of migration to Dar Tama, it is important to note that the Bideyat living in this border area claim to have good relations with their Tama neighbours, which is confirmed by the latter. The problems described below stem most often from nomads passing through or Bideyat living further inside Dar Tama.

Access to land in Chad is mainly based on customary use rights. Even though modern laws regulating land matters exist, they lack any practical implementation in rural areas.[25] The traditional system of land rights puts the land in the proximity of settlements, especially the areas which are used for agricultural production, under the authority of the village chief or a '*chef de terre*' who was assigned this authority by a higher-ranking traditional chief (in the case of the Zaghawa and the Tama, the respective sultan). The village chief has the right to distribute uncultivated land to people of his village or even to strangers.[26] Distribution in this way is independent of kinship ties, although long-established settlers are given priority according to customary law. However, access to pasture and wells is open to everyone, indigenes and settlers alike. Only in situations of water scarcity strangers might be denied the use of wells to which people who dug them always have priority access. Access to pastures cannot be denied strangers, even though frequent use of a certain area establishes a kind of priority access right which is seldom challenged by others.

In cases where nomads passing through want to rest for some days in the vicinity of a village, the local chief has to be visited. He receives some gifts (sugar, tea) and is asked for his agreement, and should direct the nomads towards suitable pastures. Newcomers also have to obtain the local chief's permission if they

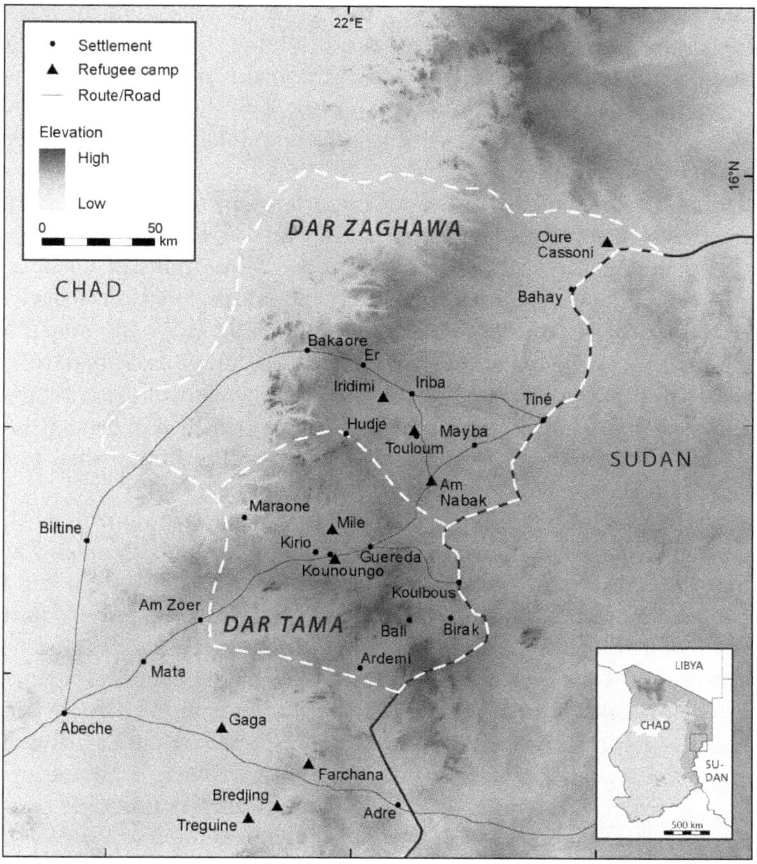

Figure 12.1: Map of Dar Zaghawa and Dar Tama, Eastern Chad.

want to settle permanently on the land under his authority. In the past, as long as land existed in abundance, such requests were usually granted, since the new settlers, like locals, had to pay one-tenth of their harvest to the local chief, thus increasing his prosperity.[27]

Resource Appropriation in Dar Tama

Today, these traditional principles are still in practice but often fail to work as instruments for solving disputes over property and use rights regarding land and water. Resources have become scarce, and in the competition over access to them disparities in power are often more decisive than traditional regulation methods which encourage coexistence. When leaving their own territory for Dar Tama (and beyond), Zaghawa pastoralists nowadays profit from their position within the state. Indeed, since Idriss Déby took power in N'Djaména in 1990, they profit with impunity. This allows them to wander freely beyond the borders of Dar Zaghawa, and even if they violate general rules of cohabitation and resource use they do not fear persecution. Even if the majority of Zaghawa pastoralists might not abuse this situation, the number of conflicts between Tama and Zaghawa is so remarkable that there is justification in talking about a systematic abuse of power by the Zaghawa.[28]

Fear hinders members of other ethnic groups from trying to push back Zaghawa herds from invading their pastures or fields and using their wells. Nowadays, herders are well armed, since access to weapons in Chad is easy, especially for Zaghawa. Decades of civil war and political instability have developed a market for Kalashnikovs and other automatic guns. In comparison to their poorer Tama neighbours, Zaghawa have the financial capacity to buy arms. Many Zaghawa also make use of their links with members of the army to equip themselves with guns and ammunition.[29] As will be explained in detail below, the Tama suffer considerably as a result of the Zaghawa invasion of their territory, especially from camels, which destroy their fields and gardens.

In this situation of mistrust and mutual hostility, the traditional regulation of coexistence no longer functions. In the past, Tama farmers welcomed Zaghawa herds onto their harvested millet fields because the camels fed on the remaining stalks, which reduced the labour required to clean the fields before the next sowing and provided the fields with dung. Nowadays, even these stalks have become a commercial good as they are sold at market as a building material for fences and roofs. So fields which were accessible to Zaghawa herds in the past are closed to them today – a situation which is, in turn, not respected by the Zaghawa. Sometimes problems occur because areas which are heavily frequented and are supposed to remain uncultivated are used by farmers as fields, such as the direct surroundings of wells or transhumance corridors, thus provoking the invasion of animals.

Besides the damage on Tama fields and gardens by Zaghawa herds, the theft of animals – mainly cattle – is also a source of conflict between the two groups. While animal theft has a long tradition among the Zaghawa, the purpose and dimension of this has changed in recent decades. In the past, stealing camels was practised mainly by young Zaghawa in order to obtain camels for bridewealth and, thus, also as proof of their manhood. Today the Bideyat are infamous for their raids into the territory of neighbouring groups, but in the past victims were exclusively other camel-breeding groups, especially Arab groups in (what is nowadays) Sudanese territory (Beck 1996: 78–80). The theft of animals moved to a new dimension when market demand increased, so while they are still occasionally stolen to supplement herds, thefts are mainly done with the aim of an immediate sale.[30]

In every market in the region so-called *dhamins* are supposed to ensure that animals offered at the market have not been stolen by the vendor. Each ethnic group present at an animal market has its own *dhamin*, who has to be a person trusted and respected not only by his own people but also by other groups (Sharif 1998: 148). However, Zaghawa have invented strategies to undermine this institution. The stolen camels are kept in a hidden *ferik* for a period of four weeks – the period a runaway camel has to be kept by someone who finds it before they can sell it in order to give the owner the chance of finding it and claiming it back. However, in this case, the legal owner hardly has a chance of finding his well-hidden camel. After the necessary time has elapsed, the camels are brought to market and the vendor claims that they are strays.

Another strategy is to transfer the camels to distant markets, where their identifying brands are not recognised by the local population. However, since the Tama have never owned large numbers of camels, they have not been the main victims of Zaghawa raids in the past. The situation has changed significantly since the Zaghawa started to dominate the Chadian state. Before, cattle were not stolen because they move slowly and thieves feared that the owners of the animals might catch them on the run. Nowadays, even Tama cattle are taken by Zaghawa because the thieves are well-armed and, even if their owners catch up with them, the Zaghawa do not fear a confrontation. Death occurs occasionally in such fights, with the victims mostly being Tama.

Injustice and Impunity

Mostly, victims of animal theft or aggression will not even try to bring their case to court, since they fear that this would complicate the situation for them, with family members of the perpetrators threatening or harassing them in order to press them to withdraw the complaint. As long as manslaughter is not involved in conflicts between Zaghawa and Tama, there are two ways to deal with problems. The first is simply to do nothing, a tactic usually adopted if the victim is Tama, for the above mentioned reasons. The second possibility is appealing to traditional authorities to handle the case rather than going to the public legal system. In such a case the involved parties address the respective village chiefs first. If they cannot solve the dispute, it will be handed over to the responsible *chef de canton*; if necessary, the next step is to apply to sultan's courts. In cases of manslaughter, the state legal system in the shape of the *Préfet* has to become involved, who will refer the case to the next court. In this context it is interesting to note that during the periods of our fieldwork (2002 to 2007), the position of the local representative of the state administration in Guéréda (2002/3: *Sous-Préfet*, 2003–7: *Préfet*[31]) was occupied most of the time by a Zaghawa. Indeed, from 2006 the official was an illiterate former military person who got this position as a reward for loyalty towards the president. This situation does not encourage Tama to count on him for the solution of disputes and conflicts in which Zaghawa are involved.

To sum up, the major factors enabling Zaghawa pastoralists to move with their herds and flocks into Dar Tama and even to settle there, without being obliged to respect traditional practices and regulations of peaceful cohabitation are: first, the low functionality of both traditional and modern land rights in cases of inter-ethnic encounters;[32] second, the impunity which the Zaghawa enjoy due to their dominance of the political system and the military in a weak state without an independent judiciary; and third, easy access to weapons and ammunition and a high level of generalised violence due to long-standing violent conflicts and political disorder in the whole region. In what follows, the economic impact of Zaghawa pastoralists in Dar Tama will be discussed in detail.

The Tama Local Economy

To understand the current influence of Zaghawa pastoralists on Tama economy and everyday life, one should first concentrate on some details of the Tama traditional economy. The Tama are farmers with a focus on millet cultivation complemented by other drought-hardened crops such as peanuts, sesame and beans. However, considering their difficult living conditions, such as unreliable natural resources and environmental risks, the Tama have developed additional sources of subsistence to ensure their economic status.

Tama households employ a diversity of economic strategies to improve their situation, including animal husbandry, local trade and business, and seasonal migration for wage labour (the latter mostly involving younger and mostly male household members). Gathering wild fruits and herbs is a traditionally important economic activity and is still a common means of contributing to the household subsistence, especially in crisis situations. For example, a plant called *mirrit* is retained as an alternative to millet in case of consecutive failed harvests. People who resisted fleeing Dar Tama during the drought of 1984 reported having survived mainly because of their stock of *mirrit*.[33]

Tama livestock farming mainly concerns the breeding of goats, sheep and cattle. Nevertheless, camel breeding was quite common throughout Dar Tama in the past, the camels being used mainly as transport and pack animals. Furthermore, the region was known for its horses and horse breeding. Early travellers, such as El-Tounsy and Barth, reported on large numbers of horses of good stock (El-Tounsy 1851: 450–57; Barth 1857: 503).

Seasonal nomadism was common in Dar Tama. Most households sent one or two household members into seasonal *ferik* (arab term for seasonal camps of herders outside their homesteads) that were installed in regions of sufficient pasture in Dar Tama and even outside traditional Tama territory. Some families based their household economy exclusively on cattle breeding and gave up their fields and their entitlement to farmland.

However, the household economy of those who entirely concentrated on small livestock farming mostly collapsed during the droughts of the early 1970s and 1980s. During this time a majority of the livestock perished due to insufficient pasture or epizootics. As in such disaster situations, prices for cattle and meat rapidly declined on local markets. Therefore, the sale of remaining cattle normally did not render enough cash to enable households to quickly rebuild new herds after the return of normal natural conditions. Those in possession of camel herds were, to a certain extent, a little more advantageous in facing the drought as camels are more resistant and can cope with larger distances to find suitable pasture or be transferred to distant markets.

Cattle breeding in Dar Tama has never really recovered from these losses and up to today has not reached the same extent as before. Nomadism is still practised only in the northern cantons; in other regions pasture is restricted to a small radius on the perimeter of each village. However, the latter phenomenon has to be related to the current situation of insecurity that we will subsequently return to.

Another main pillar of household income and subsistence is the instigation of small local trade businesses. At least one member of each Tama household is normally present at weekly local markets to sell crops cultivated in their fields or gardens and to sell livestock or meat. In addition, some households also invest in a business in which one or more members of the household are involved. Trade normally starts at a local level, whereby traders buy large quantities of convenience goods such as sugar, tea, batteries and soap at regional main markets to resell in small quantities within their own and neighbouring villages, and sometimes at different local markets. Small-scale trade can contribute to household income and profits may also be reinvested in the business. In this manner some Tama have succeeded in the building up (supra)regional businesses. Comparing local coping strategies today, business and trade can be considered one of the most successful means of accumulating capital goods. However, the fragile security situation threatens trade activities and prevents households from further investments in trade.

Another contribution to household income comes from mostly young male household members, who migrate to urban areas where they engage in wage positions, for example as construction workers or in services. The areas of biggest interest for migration are urban centres in Chad, such as Abéché and N'Djaména, but above all economic hotspots in Sudan. Since the outbreak of the Darfur conflict in 2003 and until Libya's breakdown in 2011, most young men decided to go to Libya, where they normally stayed for one to two years once they found a job. During migration periods they regularly send money to their families back in Dar Tama. Nevertheless, part of a migrant's wage is saved in preparation for his marriage and the bridewealth that is required for it.

Zaghawa Migration, Resource Conflicts and the Tama Local Economy

Prior to the increase in numbers of Zaghawa and other nomads from the north into Dar Tama, and with this the growth in insecurity, the kinds of economic diversification outlined above were apparently successful enough to ensure household subsistence.[34] However, the analysis of today's economic situation shows that Tama cannot cope with the influx of migrants that has increased after the droughts.

In Tama perception, there is a break in local history between the periods before and after the arrival of Goran and Zaghawa in Dar Tama. But what in detail leads to that perception? The presence of Zaghawa and other nomads from the north in Dar Tama had a major impact on the economic adaptation strategies that the Tama had implemented in the past and which we outlined above. Aided by the political position of the Zaghawa within Déby's Chad, nomads coming from northern parts of the country started to compete for natural resources, leading to a fight for material resources.

From the Tama perspective, the Zaghawa increasingly came to and settled in Dar Tama without asking local chiefs for permission. But serious problems with the Zaghawa – or, rather, the perception of them – emerged only a couple of years after the election of President Déby. Tama think that conditions worsened after the massive theft of camels by Zaghawa beginning in the 1980s. Many Tama today report that their families held camels as pack animals or even larger camel herds which they traded in Sudan. Also, the population of horses seems to have been affected by theft. Unfortunately, official figures which would provide the opportunity to compare the number of animals over a longer period of time are not available.

However, the phenomenon of cattle raids can still be observed in contemporary Dar Tama today. As today's camel population is close to zero, theft nowadays is focused on bovines. The raids apparently take place in an organised manner, with gangs of armed men attacking grazing cattle herds in Dar Tama at taking them away. Tama then try to quickly organise themselves into groups to pursue the thieves and recapture the stolen herd, but often they fail. Most Tama cannot afford a gun, and if they could they were scared to wear them in public, and tried to hide them to prevent confiscation. The possession of firearms is officially forbidden in Chad, and in Dar Tama especially government forces often frisk people in search of firearms. This is often not the case if a person is of Zaghawa or Goran descent. As shown above, Zaghawa are backed by a broad network of fellow tribesmen who hold important positions in Chad's police and military, or in other key positions from which they can intervene on behalf of their ethnic brothers. The latter fact is the most important reason why Zaghawa are at an advantage and have been able to create a climate of fear and repression among the Tama.

Because of cattle raids, the Tama have largely given up breeding camels, horses and bovines and today concentrate only on small livestock such as goats and sheep in predominantly small herds.[35] Those households who still invest in cattle breeding today have to spare one or more workers for the surveillance of the herd during the day. Previously this practice was unusual as herds were left to graze freely during daytime. Since herding livestock is normally the task of children, those in charge of it are not able to attend school. Furthermore, even though small livestock, in particular goats, are less susceptible to natural crises, they do not offer the same qualities in terms of capital reserve, especially if only small numbers can be held. That is one main reason why many households today bemoan a lack of cash to buy items besides their daily needs.

The decrease in herd sizes following the droughts and the state of insecurity has also had consequences for bridewealth, which seems to have fallen in quantity since the 1980s. Some Tama reported that before 1984 one had to give six or seven cattle as bridewealth, but today's payments are normally fixed at around three bovines.[36]

Yet livestock raiding is not the only harassment perceived by the Tama vis-à-vis the Zaghawa. With the growing dominance of Zaghawa within the state and their immunity from trial, Zaghawa have increasingly violated long-standing agreements concerning the grazing of harvested fields and gardens. Menacing the basis of the Tama economy, the camel herds of the nomads, passing Dar Tama in search for adequate pastures, enter Tama farmers' fields and gardens, causing massive damage to their crops. The sultan of Dar Tama claims that court cases of destroyed fields and crops have increased dramatically in the recent past.[37] Nowadays fields and gardens are both fenced in by high wooden railings to hinder the camels from entering. This fence construction demands a lot of materials, which are becoming increasingly scarce in Dar Tama. Meanwhile the conflict has gone so far that some Tama have even given up fencing their fields after being threatened by herd owners, who have claimed that their camels would be injured by the wooden fences. In such a situation, field owners are then made responsible for any injury and asked for compensation. Some men interviewed said that often Zaghawa would drive ill or old and collapsing camels onto their fences in order to make the Tama pay for them. It seems incredible, but such cases were repeatedly recorded in Dar Tama.[38] During the harvest season, most households deploy guardians on the fields and gardens to prevent animals from destroying the crops. Nevertheless, Tama often report guarding their fields and gardens to prevent camels from entering so that they cannot be injured or die there. However, the supervision of fields and gardens means additional time and effort for this branch of the economy.

Yet, even more important for the Tama economy is the abandonment of cultivating certain crops. In order to prevent camels from entering their fields and gardens, many households have given up the cultivation of, for example, sorrel and melons, as they are the preferred forage of camels. When the markets of Dar Tama are normally bloated with watermelons in December and January, during the winter season of 2006/7 they were completely empty of them, as Zaghawa attacks on Tama villages had increased throughout 2006 and the whole of Dar Tama was put into a state of emergency.

Insecurity throughout the last decades has also had a significant impact on other important parts of the Tama economy. Local trade has been especially affected by the Zaghawa and Goran presence in Dar Tama. Almost all the households interviewed by Jánszky in 2006 had given up local trade activities involving selling agricultural surpluses as cases of aggression against traders had increased massively.[39] Thus harassment extends to the level of the transfer of goods between local markets and villages. Villagers are often attacked on their way back from markets by bands of men riding camels, who ask them to hand over the cash obtained from their market sales. Often Tama are injured or killed in such encounters, with cases being reported almost on a weekly basis.

The Tama now try to avoid confrontation by relocating trading activities to larger regional markets such as those in Guéréda and Abéché. However, trading at these locations requires enough capital for renting storehouses and higher living expenses. Most rural households do not have sufficient capital reserves for this, so the reaction to the ongoing threat means people commonly give up trading activity.

Finally, the presence of Zaghawa nomads in Dar Tama also hampers another important part of the Tama local economy: gathering activities. Apart from pasture, the immigrants also compete with locals for other natural resources, particularly wild fruits, herbs, wild honey and gum Arabic. Tama declare that nowadays it is not worth searching for these natural materials any longer as they are all taken by passing Zaghawa pastoralists.[40] Furthermore, Tama fear being menaced by criminal bands or armed nomads while out searching. These circumstances represent an additional risk to Tama survival strategies as gathered fruits and materials serve as substitute foodstuffs, especially in crisis situations.

As the analysis of several Tama economic activities in the context of the Zaghawa presence shows, the process of economic diversification to enable survival in the risky environment of Sahelian eastern Chad has become a struggle. All efforts made to adapt economic strategies to current environmental situations and risk are influenced by the threat the Zaghawa represent to the Tama and are, therefore, rather unsuccessful in contributing to better living conditions. Furthermore, the need for the constant surveillance of fields, gardens and herds means a diversion of labour away from other activities that contribute to the meeting of subsistence needs.

However, it is very difficult to figure out who exactly is responsible for raids and violent offences against Tama. In the perception of Tama, all the grievances of everyday life are associated with the Zaghawa. While Tama once referred to a common indigenous history with Zaghawa,[41] they sometimes deny it nowadays.[42] In indigenous discourse, 'men on camel-back' has even become a synonym for 'Zaghawa', which makes it even more difficult to differentiate cases of conflict from an external perspective. Moreover, only a few Tama recognize that 'Zaghawa' attackers are most probably of the Bideyat sub-group, indicating that there are no major problems with relations with the Kobe sub-group living in the sultanate of Iriba.[43]

As mentioned above with regard to the village of Hudje, our research indicates that Zaghawa and Tama do in some places live in close proximity without the tensions and conflicts that we have highlighted, such as in the border region between Dar Zaghawa and Dar Tama. Tama explain this by saying in a situation where people know each other they are likely to be recognised if they participate in a raid, which would risk their status in their community as theft is not approved of. Ancient Zaghawa *ferik*s often have been installed in the neighbourhoods of Tama villages generations ago, which indicates that long-lasting relationships exist. Nevertheless, it often remains unclear who exactly is responsible for organised raids and aggression as the offenders are backed by ethnically defined political or judicial networks and the victims often are scared of the consequences of trying to claim their rights or offering resistance.

Tama Resistance Strategies

As we have shown, the Chadian social, economic and political system is predominantly based on connections with influential key figures, ethnic ties being a major precondition of obtaining political posts and leading positions in the economic sector.[44] Those who most profit from this system are normally those ethnic groups from among which representatives hold state power. This was the case for the Goran nomads when Hissène Habré, himself of Goran origin, was head of state between 1982 and 1990, and it is also true today under the presidency of Idriss Déby, which favours members of his own ethnic group, the Zaghawa. Political participation at a national level for much of the Tama community, however, is hindered by their lack of access to influential circles, but also by a lack of educational attainment. The ethnic ties that give people access to key positions within the state make it more difficult for the Tama to become part of the official sphere and to profit from ethnic networks.

Facing this situation, Tama now feel powerless to act and to push for their rights. Trusting in their position of power within the state, Zaghawa increasingly disregard methods of traditional conflict resolution and mediation in interethnic relations with the Tama. Tama officials report that in cases involving both Zaghawa and Tama, the Zaghawa take the issue directly to tribunals that are controlled by members of Zaghawa clans instead turning to traditional authorities. Hence, Zaghawa do not have to fear being penalised for whatever conflict they are involved in.

To escape this desperate situation, in which their economy is being undermined and they lack recourse to the legal system, Tama have turned to organised resistance since the early 1990s.[45] The first nationally significant Tama rebel movement, the Alliance Nationale pour la Résistance (ANR), was a coalition of several groups that was founded in 1995 under the command of the former chief of staff of the Chadian army, Mohamat Garfa. The armed group was involved in several attacks on government troops in the south and the east of Chad during the following years. In 2003 the ANR signed a peace agreement with President Déby, after which Mohamat Garfa became a minister in the Chadian government. Some members of the ANR continued to resist, but the movement fragmented because of internal discord. Apparently backed by the Sudanese government, in late 2005 Mohamat Nour Abdelkerim, a Tama born and raised in the Guéréda region and a former member of the ANR, founded the Rassemblement pour la Démocratie et les Libertés (RDL). The RDL then allied itself with the Front Uni pour le Changement (FUC) a couple of months later. The FUC was a coalition of several opposition groups, formed in late December 2005 in eastern Chad. As is characteristic for such alliances in Chad, the coherence of the FUC was quite loose, and soon several splinter groups left it (HRW 2007a: 2). From December 2005 onwards, the FUC made several military attacks on government forces, most importantly the attack on Chad's capital N'Djaména on 13 April 2006.

The fact that Tama participated in several major rebel attacks caused increasing mistrust against the Tama community and more and more interethnic tension in Dar Tama. After military confrontations between the government and rebel movements, including the RDL and FUC, cases of Tama being arrested under suspicion increased.

Affected by economic crises and political repression, numerous Tama joined the rebellion. Young men in particular joined the resistance because of lack of employment and general economic hopelessness and because of the political injustice experienced in everyday life, especially in cases of family members or friends injured or killed by Zaghawa or Goran. Rebellion was thus seen as the only way to contribute to an overthrow of the current situation and improve people's living conditions.

In Dar Tama the situation of interethnic tension dramatically worsened throughout 2006 following the attack on N'Djaména in April of that year, and Zaghawa civilians systematically attacked Tama villages, especially in the southern cantons of Dar Tama, stealing cattle and pillaging (ibid.: 25–34). The attacks left several dozen Tama dead and around 2,000 internally displaced as they fled their homes because of insecurity (HRW 2007b: 16–17).[46] The FUC reacted with several counter-attacks, and tried to regain control over Tama territory and push back Zaghawa militias. For this reason several units of the FUC militia were stationed in different parts of Dar Tama and the eastern border region.

However, the situation temporarily changed when Mohamat Nour Abdelkerim surprisingly signed a peace agreement with President Déby in December 2006, after Guéréda, the capital of Dar Tama and seat of its prefecture, had been occupied by FUC fighters in the middle of the same month. The peace agreement contained a plan to disarm FUC soldiers and integrate them into the national army.[47] According to unofficial information, Déby also agreed to leave the administration of the whole of Dar Tama to Mohamat Nour Abdelkerim for a transitional period to 'call to order' the local population and to induce internally displaced Tama to move back to their places of origin.[48] President Deby thereby authorised the rallied FUC battalions to impose order and security in Dar Tama. It is said that Nour set the protection of Dar Tama as one major precondition before signing the agreement and returning from rebellion into the government's ranks.[49] After the signing of the peace agreement, all Tama fighters officially returned to Dar Tama. The sudden presence of FUC soldiers and heavy military equipment in Guéréda and its surroundings gave an idea of how many men and women had joined the rebellion throughout the last decade.

The following weeks and months were marked by ongoing violent ethnic clashes between Tama and Zaghawa. If in some regions, former FUC fighters indeed succeeded in re-establishing security just by their presence, in others they caused new insecurity by leading aggressions against Zaghawa settlements in Dar Tama to take revenge for earlier bloody attacks. Such attacks resulted in a series of ongoing counter aggressions between the two ethnic groups. Also in these riots the identification of participating groups was extremely difficult. For example, rumours said that Zaghawa civilians were supported by members of the national Army in repulsing FUC fighters or in attacking bases of the former FUC, other attacks on FUC encampments were said to have been led by entities of the *Platform for Change, Unity and Democracy* (Socle pour le Changement, l'Unité et la Démocratie / SCUD), another rebel unit led by Zaghawa, and so on (ibid.: 17–18).[50]

Following the 2006 events, insecurity remained a feature of life in Dar Tama. When a Chadian army unit was charged with the demobilisation of FUC fighters in Guéréda in October 2007, parts of the FUC left their bases to move back towards the Sudanese border and further violent clashes broke open. Many former FUC fighters joined other Chadian rebel movements later on. The general lack of security caused the declaration of another state of emergency in the Dar Tama region in October 2007.[51] As a consequence, President Deby dismissed his Tama defence minister, Mohamat Nour Abdelkerim in at the beginning of December 2007.[52] Following these events, the sultan of Dar Tama was also arrested by government forces and kept imprisoned for several months. Although a legal judicial process held in 2009 by the Chadian Supreme Court had found the Sultan not guilty, President Deby has revoked the judicial decree and the sultan was imprisoned anew. The immediate nomination of a new sultan by the Chadian government without a legal judicial process against the precedent is a violation of public and traditional law, and led to a deep split among the Tama community. It also contributed to widening the gap between Tama and Zaghawa as the latter are held responsible for this political act due to their position within the state.

At the time of writing, the relationship between the two groups is still marked by interethnic tensions. Nevertheless, since the signing of the latest peace agreement between Chad and Sudan in January 2010 security situation in the Dar Tama area considerably improved correspondently to the overall security situation in eastern Chad. The peace treaty between the two states entailed amongst other things the withdrawal of mutual support for rebel groups of the two neighbour states.

Conclusion

Pastoralism is declining in many Sahelian states as well as in other arid zones of Africa (e.g., Markakis 1993). The main reasons for this are the political marginalisation of pastoralists, their involvement in endemic conflicts and general impoverishment, as well as the privatisation of land that was formerly used as pasture. Contrary to this trend, pastoralism in Chad is an important economic sector and animals are (with oil and cotton) among the major country's export products. In northern Chad, pastoralism is the most important subsistence strategy, despite the absence of official state support for pastoralist communities. The so-called 'new paradigm' in pastoralist development, encouraging mobility and empowering local communities to manage their own natural resources (Niamir-Fuller 1999: 4), only very recently became a focus of development projects that to date have very limited radii and results. Despite this, Chadian pastoralists have managed to maintain their livelihoods and even commercialise it.

But a closer look at the situation in Chad shows that not all groups have succeeded in maintaining their traditional survival strategies in the same way. This imbalance can only be understood with regard to the specific political setting in Chad, which clearly enables and even favours the Zaghawa with regard to maintaining a profitable pastoralist economy. On the other hand, the Tama, in reaction to frequent raids, have given up their pastoralist activities to a large extent. Furthermore, they have also been forced to adapt all their economic strategies to help them mitigate the consequences of Zaghawa migration into Dar Tama and the increasing insecurity that has ensued.

The power imbalance between the two groups has led to a breakdown of traditional mechanisms of cohabitation between Zaghawa and Tama. While some Zaghawa pastoralists do not see the necessity of collaboration when they can achieve their goals in other (illegal) ways, Tama peasants are forced to react with strategies of avoidance to cope with field damage and raids from invading Zaghawa pastoralists. To a large extent, the Tama's adaptation strategies have failed, leading to serious economic decline and finally to increasing violence through the formation of a militant opposition movement.

As a consequence, relations between the two groups today are marked by ongoing mistrust, violent attacks by militias on both Tama and Zaghawa residents, and a fragile state of security in the whole of Dar Tama as well as its border areas.[53] In addition to this, the region has been the scene of violent clashes between rebel movements based in the border area of Sudan and the Chadian army until recently. Without further clear government attempts to sustainably stabilise the region, and as long as it follows an ethnic-based political agenda and allows judicial corruption, the situation is likely to re-deteriorate within the near future.

In this chapter we have shown how the political, spatial and economic expansion of Zaghawa pastoralists has affected the Tama economic system as well as the political and security situation in Dar Tama. The 'capture' of the state by an ethnically based – pastoralist – network was the precondition for the pastoralist expansion in Chad, which has enabled Zaghawa pastoralists to gain and maintain power and influence, and exercise control over resources at the cost of the rest of the Chadian population.

Notes

1. Hissène Habré's presidency was characterised by a one-party state and the aggressive suppression of any opposition movement. The regime was also especially infamous for the use of torture, political detention and the killing of political opponents.
2. No other president of Chad has managed to maintain power for a comparable period of time.

3. For a detailed description of opposition groups and rebel movements, see May and Massey (2007: 38–43).
4. In March 2006, French intelligence uncovered a plot to shoot down Déby's plane. In April 2006, the rebel coalition the Front Uni pour le Changement (FUC), which seized the towns of Am Timan, Goz Beida and Mongo and then attacked N'Djaména, was defeated by the Chadian army, backed by help from the French (May and Massey 2007: 34). During rebel attacks on N'Djaména in February 2008, the French are said to have given logistical support to the Chadian military, therewith securing Chadian government forces the ability to launch counter-attacks on rebel positions.
5. The UN troops were part of the Mission des Nations Unies en République Centrafricaine et au Tchad (MINURCAT). This is a UN mission established by the Security Council and has the mandate to improve security and protect refugees and civilians in the conflict-ridden regions of eastern Chad and north-eastern Central African Republic.
6. Rebel attacks most often emanate from Sudanese territory as important branches of the Chadian armed opposition are backed by the Sudanese government. The mutual support of rebel groups by the two neighbour states is a strategy in the ongoing proxy war between the two states, with Sudan supporting rebels fighting Déby while Chad assists rebel groups in Darfur.
7. In 2003, Chad became an oil-exporting country, drilling oil in the south of Chad and transporting it via a pipeline to the port of Kribi in Cameroon. The ownership of the Chad Oil an Exploitation Project is held by a consortium of three companies (Exxon/Mobil, Petronas Malaysia and Chevron) and the governments of Chad and Cameroon.
8. Special units of the Chadian army, among them many young men from the pastoralist north, are trained by US Special Forces within the framework of the Trans-Sahara Counterterrorism Initiative. The special training that these troops receive is thought to be one reason for the defeat of rebel attacks in recent years by the Chadian national army (May and Massey 2007: 46).
9. Regarding marriage, especially for Zaghawa women, weddings with 'outsiders' are rare, since such a woman is considered 'lost' to her kin group in this patrilineal society, and the bridewealth demanded by Zaghawa is so high that other groups are hardly able or willing to pay it.
10. Marchal (2006: 20–24) presents a good overview and analysis of schisms within the Zaghawa ethnic group and the political implications of these.
11. Zaghawa are divided into three major sub-groups: Kobe, Bideyat and Borogat (see Haggar 1999). Haggar, himself a Kobe, gives a detailed account of the formation of a Zaghawa opposition party and movement against Déby.
12. Interview with K.M., 14 April 2003.
13. Interview with V.B., 25 October 2004.
14. Traditional law blends with Islamic shariah law and is executed by local office holders such as the sultan and village chiefs, often assisted by members of Islamic committees or an imam.
15. While the Bideyat have always kept large flocks of camels, the Kobe specialised in cattle breeding (see Tubiana 1985).
16. Similar observations have been made among the Himba of arid Namibia (Behnke 1999: 185–216).
17. Due to its remoteness, Dar Zaghawa has so far received very little assistance from development projects. Consequently, there are very few permanent, cemented wells.

18. Until now, the degree of environmental degradation has been relatively low compared to other Sahelian regions, probably because of the relatively low population density, the absence of intensive farming (which ensues with the extension of farmland and large irrigation schemes), and the lack of development initiatives aimed at settling pastoralists, which create conditions of overgrazing and overstocking. However, since the influx of tens of thousands of Sudanese refugees into eastern Chad because of the conflict in Darfur, some micro-regions are extremely burdened by the increased pressure on resources.
19. Before the outbreak of violent conflict in Darfur in 2003, many Zaghawa pastoralists also crossed the border into Sudan in the search of better pastures. This trans-border pastoralism has almost stopped in recent years due to fear of attacks and raids.
20. Sharif Harir, who did research among the Bideyat in Chad, confirms the observation of increased herd size (Sharif 1996: 163).
21. Little has observed similar developments in Kenya (Little 1985).
22. Interview with Tairo Mirri Erdajeb, 5 July 2003. This estimate seems quite high, but considering the fact that most of these settlements consist only of the members of one extended family, it is probably realistic.
23. For a detailed account of this case study, with a special focus on seasonal migration movements and the impact of droughts and political conflict, see Jungstand (2007).
24. The well where drinking water is fetched and small stock watered is 4 kilometres from Hudje; the wells where camels and cattle are watered are up to 20 kilometres away.
25. For details of laws relating to land, see CEFOD (2004).
26. According to customary law, land already cultivated by a person (who got it from the local chief) becomes that person's quasi-property, and can be inherited by their descendants.
27. Despite paying this 'harvest tax' to a local chief, migrants remained under the political and moral authority of their own chiefs, even if they did not settle in the new place but remained in their home region.
28. The phenomenon is not confined to relations with the Tama, being also known between Zaghawa and other ethnic groups, depending on where Zaghawa pastoralists move.
29. For similar examples of increases in violence among pastoralist societies in East and North-east Africa, see Bollig and Österle (2003).
30. Bollig and Österle discuss the problem that 'it is nearly impossible to inventorise the gains of militias and raiders' (Bollig and Österle 2003: 121). This is also the case of the Chadian Zaghawa, as no detailed data on such gains could be obtained since those involved hardly ever talk about raiding activities to an outsider.
31. In 2002, a decentralisation process was started in Chad, including the restructuring of the administrative units. What previously had been a *Sous-Préfecture* (with a *Sous-Préfet*) became in most cases a *Département* (with a *Préfet*).
32. As long as conflicts involving land occur between members of the same ethnic group, the traditional regulation and settlement methods usually function well.
33. Interview with A.B., 15 January 2007.
34. Literature on the Tama is extremely scarce. Information or details about single aspects of their economic or cultural life are basically non existent, especially from a historical perspective. The assumption that subsistence was successful or more successful in the past is based on interviews and fieldwork conducted by Jánszky between 2005 and 2007.
35. Small livestock is no also no longer safe from theft. Since the arrival of tens of thousands of Darfur refugees in the area, such cases are nowadays reported on a nearly daily basis. Apparently the theft of goats and sheep is seen by refugees as a means of obtaining meat, which is not distributed by humanitarian agencies. Alternatively, the animals can be sold to obtain cash.

36. Interview between A.M. and Jánszky, 11 March 2006.
37. Interview by Jánszky, 14 March 2006.
38. Interviews by Jánszky, 2005 to 2008.
39. These interviews were part of a survey undertaken of forty-two households in the village of Kirio in central Dar Tama.
40. While Bideyat migrating to Dar Tama still gather wild cereals, the Kobe living directly north of Dar Tama ceased to engage in this tedious activity due to their relative prosperity, allowing them to buy millet in case of harvest shortages. In comparison, Tama women have been observed by Jungstand coming to Dar Zaghawa in order to gather wild grains, with the Zaghawa having no objection to this, since they themselves did not have any need of this supplementary food.
41. Interview with H.M., 17 April 2006.
42. The issue of shared or common history also came out in interviews by Jánszky in 2006. For example, there have been arrangements concerning exemption from paying blood money (*diya*) between some Zaghawa clans and the Tama. Such arrangements are losing their importance since Zaghawa demand horrendous *diya* payments in cases today.
43. Evidence here is extremely difficult to come by as there is no documentation of tribunals. Court hearings normally do not lead to satisfactory information since most of those questioned are of Zaghawa descent and thus part of the 'protection network'.
44. Of course, there are exceptions. For instance, Mohamat Garfa, one of the former leaders of the opposition Alliance Nationale pour la Resistance (ANR), held a ministerial post between 2003 and 2008. Also, Mohamat Nour Abdelkerim, the former leader of the Front Uni pour le Changement (FUC), became Minister of Defence in 2007 after having signed a peace deal with President Déby. On these two cases, see below.
45. Resistance is not confined to the Tama and has involved other ethnic groups in Chad. For example, the Dadjo have rebelled in significant numbers since 2001.
46. This is additionally born out by interviews with OCHA (Office for the Coordination of Humanitarian Affairs) by Jánszky, January 2007.
47. Details of the circumstances and conditions of the accord are unavailable. The only official information available is 'Le texte de l'Accord de paix entre la République du Tchad et le Front Uni pour le Changement Démocratique (FUC)', published by the Chadian government on 26 December 2006. Retrieved 12 March 2007 from: http://www.reliefweb.int/rw/rwb.nsf/db900sid/KHII-6WV5SB?OpenDocument. The process of integrating former FUC fighters into the Chadian army was still ongoing in late 2009.
48. 'Situation Trouble Généralisée dans le Dar Tama', *Tchadactuel*, 3 March 2007.
49. Interviews by Jánszky in Dar Tama, January and February 2007.
50. Interviews by Jánszky, January and February 2007.
51. 'Etat d'Urgence à Sens Unique', *Tchadactuel*, 18 October 2007.
52. 'Déby Limonge son Ministre de la Défense, l'ex Rebelle Mahama Nour', *Tchadactuel*, 1 December 2007.
53. 'Les Brèves de N'djaména: Combats dans le Dar Tama', *Tchadactuel*, 19 August 2008.

References

Barth, H. 1857. *Reisen und Entdeckungen in Nord- und Central-Afrika in den Jahren 1849 bis 1855*. Gotha: Justus Pertes.

Beck, K. 1996. 'Nomads of Northern Kordofan and the State: From Violence to Pacification', *Nomadic Peoples* 38: 73–98.

Behnke, R.H. 1999. 'Stock Movement and Range Management in a Himba Community in North-western Namibia', in M. Niamir-Fuller (ed.), *Managing Mobility in African Rangelands*. London: Intermediate Technology Publications, pp.184–216.

Ben Yahmed, D. (ed.). 2006. *Atlas du Tchad.* Paris: Editions J.A.

Bollig, M., and M. Österle. 2003. 'Continuities and Discontinuities of Warfare in Pastoral Societies', *Entwicklungsethnologie* 12(1/2): 109–42.

CEFOD. 2004. 'Recueil de Textes sur le Droit Foncier au Tchad'. N'Djaména: Centre d'Etude et de Formation pour le Développement.

Chabal, P., and J.-P. Daloz. 1999. *Africa Works: Disorder as Political Instrument.* Oxford: James Currey.

El-Tounsy, M. Ibn-O. 1851. *Voyage au Ouaday.* Paris: Duprat.

Haggar, B. B. 1999. *Wirtschaft und Gesellschaft der Beri - Ein ostsaharisches Feldbauer-Hirtenvolk im soziokulturellen Wandel.* Gießen: Gießener Schriften zur Wirtschafts- und Regionalsoziologie.

HRW. 2007a. 'They Came Here to Kill Us: Militia Attacks and Ethnic Targeting of Civilians in Eastern Chad', Human Rights Watch Report 19(1A).

——— 2007b. 'Trop Jeunes Pour la Guerre: Les Enfants Soldats dans le Conflit Tchadien', Human Rights Watch Report 19(9A).

ICG. 2008. 'Tchad : Un Nouveau Cadre de Resolution du Conflit', International Crisis Group Report 144.

Jánszky, B., and T. Jánszky. 2008: 'Tchad/Soudan: Des Alliances Changeantes', *Outre-Terre: Revue Française de Géopolitique* 20: 289–300.

Jungstand, G. 2007. '"Going to Greet the Relatives": Household Economy, Mobility and Politics in Dar Zaghawa, NE Chad', in O. Bubenzer, A. Bolten and F. Darius (eds), *Atlas of Cultural and Environmental Change in Arid Africa.* Cologne: Heinrich-Barth-Institute, pp.174–77.

Little, P.D. 1985. 'Absentee Herd Owners and Part-time Pastoralists: The Political Economy of Resource Use in Northern Kenya', *Human Ecology* 13(2): 131–51.

Mahamoudi, A.M. 2006. 'Le Biltine', in N. Karaki (ed.), *A la Découverte du Tchad. Lac et Biltine.* N'Djamena: Centre Al-Mouna, pp.46–60.

Marchal, R. 2006. 'Tchad/Darfour: Vers un Système de Conflits', *Politique Africaine* 102: 9–28.

Markakis, J. (ed.). 1993. *Conflict and the Decline of Pastoralism in the Horn of Africa.* Houndsmill: Macmillan.

May, R., and S. Massey. 2007. *Chad: Politics and Security*. Geneva: WriteNet/UNHCR.

Niamir-Fuller, M. 1999. 'Introduction', in M. Niamir-Fuller (ed.), *Managing Mobility in African Rangelands.* London: Intermediate Technology Publications, pp.1–9.

Sharif, H. 1996. 'Management of Scarce Resources: Dryland Pastoralism among the Zaghawa of Chad and the Crisis of the Eighties', in A.G.M. Ahmed and H.A.A. Ati (eds), *Managing Scarcity: Human Adaptation in East African Drylands.* Addis Ababa: Organization for Social Science Research in Eastern and Southern Africa, pp.101–9.

——— 1998. 'Emissaries for Conflict Management among the Zaghawa on the Sudan–Chad Border', in M.A.M. Salih and J. Markakis (eds), *Ethnicity and the State in Eastern Africa.* Uppsala: Nordiska Afrikainstitutet, pp.147–55.

Tubiana, M.-J. 1985. *Des Troupeaux et des Femmes.* Paris: L'Harmattan.

Chapter 13

Unofficial Trade When States Are Weak

The Case of Cross-border Livestock Trade in the Horn of Africa

Peter D. Little

Introduction

In June 2001, Abdi and Dahir sit around a small hotel room in Garissa, a growing town of about 50,000 in north-eastern Kenya, discussing the 'pros and cons' of the current cattle trade between stateless Somalia and Kenya. They lament the fact that they have to sell their animals quickly because there are not sufficient areas nearby to allow them to graze, and it is expensive to purchase fodder to sustain them. Corruption by local officials in Kenya is another issue and 'everybody is looking to be paid a little *chai* [tea, bribe] or they will make problems'. Abdi and Dahir arrived yesterday from a small settlement in southern Somalia about eight hours by vehicle over poorly maintained roads. Their herd of about sixty cattle is nearby, having been trekked overland by herders from the Lower Jubba area of conflict-ridden Somalia about eleven days earlier. They hope to sell all of them at the weekly market on the following day, and then take their earnings to Nairobi to purchase consumer goods to take back to Somalia or remit the money back via informal wire transfers. Abdi and Dahir are one of a growing number of Somalis who had shifted their livestock business to cross-border trade after the Somali state and its southern export markets collapsed in 1991.

Transborder livestock trade in the Horn of Africa,[1] including along the Kenya–Somalia border, represents a particularly important and challenging unofficial activity. On the one hand, it epitomises the essence of informal or 'shadow' trade, operating along remote borders in a vast region, where government presence is particularly weak or, in some cases (Somalia), absent. In many instances it represents the only type of exchange in these border areas, since extremely poor regional infrastructure and communications impede official trade between neighbouring states. Illustrative of this reality is the fact that official annual exports of cattle from Ethiopia, the most populous country in the region, were less than

2,000 in 1999, but unofficial trade was estimated to be more than twenty-five times this amount in the same year (see Teka, Azeze and Gebremariam 1999). For some observers, transborder trade in the Horn of Africa and elsewhere on the continent represents a normal market response to cumbersome export regulations and regional comparative advantages, and should be encouraged as a means to meet local demand that is not being met by national production and markets, and to strengthen food security among pastoralists (see Zaal and Polderman 2000). These same supporters often argue that the practice predates colonial and postcolonial state boundaries and, thus, reflects long-standing indigenous patterns that are more efficient than formal trade channels (see Meagher 1997). For some governments, however, cross-border trade reflects a potential loss of foreign exchange, an illegal activity (smuggling) that creates unfair competition for official traders and food producers. Their position argues for increased regulations and taxes, policing, and/or forcing the trade into formal market channels.

In this chapter I discuss cross-border livestock trade in the Horn of Africa, with a particular emphasis on cattle trade along Somalia's borders, perhaps the most significant form of clean commerce in the region.[2] It draws on my own research on the topic dating back to the mid 1980s and a recent collaborative project on cross-border trade and food security that I co-directed (see Little 1992, 2003; Little, Teka and Azeze 2001). The chapter reflects the situation of cross-border livestock trade in the region up to 2008. The first part of the chapter discusses the macro context of transborder trade in the region and provides a snapshot of its importance and scale. The second part of the discussion addresses the special case of Somalia and its voluminous transborder livestock commerce with neighbouring states. Somalia has been without a recognised government since 1991, when the Siad Barre regime was overthrown and the country was thrown into a long period of political turbulence that continues to exist. It represents the longest period in modern history without a formal state and effectively with a stateless economy. It will be shown that even without a government Somali commerce is more accurately viewed as an extreme entry along a continuum of formal-to-informal (or legal-to-illegal) trade rather than a distinctly unique case. It is a question of degree, not difference, and there are other parts of Africa that have internationally recognised governments but where large parts of the economy and territories operate outside the legal realm. In the third section of the chapter the significance of particular local institutions in transborder trade are assessed, while the final part summarises key points and suggests that African governments might be better off by recognising rather than discouraging cross-border livestock trade.

The 'Macro' Picture

International boundaries throughout the Horn of Africa have important economic and ecological characteristics that generally distinguish the region from other parts of Africa. The incredible vastness of the region's borders makes administrative presence and controls very expensive relative to most other African regions. Somalia's border with Kenya and Ethiopia alone is more than 1,000 kilometres long. Most of the borderlands are also characterised by arid and semi-arid environments and livestock-based economies, where trade in livestock and livestock products assume considerable importance. As a commodity, livestock has features that make it amenable to cross-border trade even in situations of widespread insecurity. It is a mobile and high-value commodity that can be transported overland rather than on roads, and can easily be moved across borders. With market infrastructure, veterinary certification facilities and transport infrastructure so poorly developed, official export trade in the commodity between neighbouring countries is very limited.

For the livestock herders or pastoralists who dominate the area's border regions, market alternatives to cross-border trade are minimal. Often markets in neighbouring countries are considerably closer and more lucrative for herders than those found in their own countries. Overall, the pastoral groups of the borderlands are weakly integrated with most areas of their own countries, and domestic channels in many countries hardly provide adequate outlets for the sale of their livestock and livestock products. It is simply easier for many to sell their animals at nearby markets, which are often across an international border. Moreover, huge numbers of cattle are often found in or near border zones and, in the case of southern Somalia (near the Kenya border), cattle numbers in the border zones are among the highest in the Horn of Africa, with estimates as high as 700,000 head (see Little 2003). These animals represent an important attraction for traders who operate in the border areas.

Generally, the development of transborder livestock trade stems from ecological, geographic, political and social factors that advantage certain zones (Teka, Azeze and Gebremariam 1999: 1). In some countries, like Ethiopia, unfavourable domestic market policies, and distances to key market towns (such as Addis Ababa), poor infrastructure and low domestic purchasing power favour informal exports to neighbouring countries where prices are better (ibid.). Throughout the region social relations based on clan affiliation, religion (Islam), kinship and friendship shape the existence of cross-border trade, and the presence of the same ethnic groups on both sides of the border further strengthens it. Pastoral trading partners across international boundaries can often draw on common languages and identities in conducting business. Consequently, cross-border trade in the Horn of Africa has always assumed considerable importance in the economies and societies of the region, 'even when governments have attempted to discourage it' (ibid.: 1). In fact, despite its importance some states usually appreciate only the illegal dimension of cross-border trade and, when they act, their normal response is to try to control it.

In terms of livestock trade in the study region, two important markets – one regional and the other external – shape its direction. In the southern border areas of the region the importance of Nairobi, a city of about 3.5 million people and the key regional market in the Horn, cannot be underestimated. Cattle prices in Kenya are often 20 to 25 per cent higher than in neighbouring countries, which explains why cross-border cattle trade, including from Tanzania, Ethiopia, and Somalia, is mainly unidirectional: it flows from neighbouring countries to Kenya. In the early 1990s Kenya liberalised its meat trade and numerous small-scale slaughter facilities emerged. It is estimated that Nairobi purchases about 450,000 cattle annually (2002), with about 410,000 for slaughter and 40,000 for restocking or fattening on nearby commercial ranches (Little 2003). Of the total number of animals, Somalia supplies approximately 75,000 (16 per cent) and Ethiopia about 40,000 (8 per cent). Not surprisingly, the vast growth in informal exports to Kenya kept urban meat prices in Nairobi relatively low and stable during 1996 to 2003.

In terms of the region's principal external market, the Middle East, the bulk of the livestock for the Somali portion of this trade comes out of northern Somalia (Somaliland and Puntland). Approximately 50 to 60 per cent of the animals for this trade are informally sourced across the border in eastern Ethiopia. The export trade in small stock from Somaliland, the breakaway autonomous state in the north of Somalia, and Puntland, the regional state in the northwest, was larger in the late 1990s than before 1991, when the country had a government (FAO et al. 2004). However, it has dropped by as much as 40 per cent following a series of import bans imposed by Saudi Arabia due to fears of Rift Valley fever, a mosquito-transmitted disease that mainly affects livestock but can be lethal to humans (see Little 2003). In 1998 Somalia (including Somaliland) accounted for 850,000 goats and 1.25 million sheep, or more than 95 per cent of all goat exports and 52 per cent of sheep exports from the entire East Africa region (Zaal and Poderman 2000: 17–18). These market achievements also made stateless Somalia one of the largest exporters of live animals in the world, and much of this came from informal cross-border trade. A modest estimate would be that this cross-border trade generates about US$25 million for the economy of eastern Ethiopia.

In addition to livestock, other goods are exchanged informally across borders, including the most lucrative – the mild narcotic *khat*.[3] Several of these products are critical foods for local consumption and local food security. The movement of traders from southern Somalia to key Kenyan markets following the collapse of the Somalia state has facilitated cross-border trade because these merchants often maintained their linkages to suppliers in Somalia. Not only cattle but food aid (especially cooking oil and wheat), pasta and electronics from the other side of the border find their way into north-eastern Kenya's retail sector. During the occasional border closures of the late 1990s and 2000s, merchants made it very clear to me that their businesses were strongly dependent on transborder trade with stateless Somalia (see also FAO 2003). Analyses of prices for critical foods, like wheat flour and maize, revealed large price spikes when unofficial cross-border trade was negatively affected either through trade embargoes or floods (Little 2003). Local shortages of key foodstuffs are not uncommon in the border regions when commerce is slowed. It should be noted that in the region the Ethiopian government has been most vigilant in trying to control cross-border trade, including in livestock, and has effectively classified it as contraband (illegal) trade and confiscated animals on numerous occasions (Little 2005). On the other hand, Kenya effectively allows cross-border trade in livestock, and traders and government officials openly discuss it, although in theory it is still illegal according to Kenyan law. In a visit to Garissa market near the Kenya–Somalia border during September 2008, market officials and traders identified the numerous animals in the market from Somalia and spoke openly about the importance of transborder trade, an openness that was also prevalent during 1996 to 1998.

Figure 13.1: Cattle market in Garissa, Kenya, September 2008 (photograph by Peter Little).

Figure 13.2: Cattle being loaded in Garissa, Kenya (photograph by Peter Little).

In many parts of Africa it often is difficult to distinguish between formal or official and unofficial or informal trade activities, even in cases where states exist and controls are in place (see Hibou 1999). For example, those livestock that are trekked across the Somalia border to be sold are officially taxed at different market centres in Kenya and, eventually, sold through formal market channels in urban centres. Does this constitute official (that is, state-recognised) or unofficial trade? The answer to that would depend on the level of the market chain one examines. Is it at the point when the animal crosses the border illegally or further up the chain when it is sold through licensed traders and butchers? Contrary to common perceptions, unofficial transborder livestock trade also generates significant amounts of local taxes and permit revenues at higher levels of the market chain.

One needs only to visit Eastleigh, Nairobi, a centre of Somali business life and a key part of the cross-border trade story, to see how the formal (legal) and informal (illegal) merge in complex ways. The shopping area itself, which is widely noted to have the busiest commercial avenue in Nairobi, attracts middle- and upper-class shoppers from around the city and other parts of the country. They come to shop for bargains and, in some cases, to purchase counterfeit identity cards and the like. Upscale brands of fashion, electronics and other consumer items imported through 'duty-free' Somalia can be purchased at 20 to 30 per cent below prices elsewhere in town, and services, such as the internet and telephones, can be obtained at a fraction of normal costs. Unregistered (illegal) phone services stand in close proximity to official phone booths, and fuel stations selling contraband diesel are found within a few blocks of a licensed petrol station. That other openly informal urban-based economies dominated by diaspora trading groups like Somalis have taken hold in many African cities, including parts of Johannesburg, South Africa (Rogerson and Rogerson 1996; Peberdy and Rogerson 2000), demonstrate their prevalence on the continent. It is a growing phenomenon that has largely gone unexplained and, in many cases, undocumented (Murray n.d.). In this sense Eastleigh is 'openly informal', neither hidden from authorities nor entirely consistent with an official, public place of business. At the same time, it is integral to the service economy of Nairobi and its 3.5 million residents, who depend on it for low-cost products and services.

Similar to the cross-border commerce in northern Kenya, the prevalence of illegal trade and its tight integration into daily practice serves to legitimise it. Somali livestock that is imported illegally into Kenya but openly traded, discussed and inscribed in daily practice assumes almost official status in the border regions, but is not reflected in most trade statistics. In 2001 some local Kenyan officials with whom I spoke did not think cattle were contraband commodities or the trade in them illegal, when in fact on paper it really is. Other smuggled imports, such as drugs or weapons, are secretively traded but treated as 'more illegal' or 'dirty' in status than animals. If discovered by Kenyan border officials, it is likely that they will be confiscated.

Livestock Trade in the Kenya–Somalia Borderlands

With the collapse of the Somali state in 1991, there was a spectacular 'boom' in cross-border livestock trade between Kenya and Somalia (see Figure 13.3). In the case of Garissa, Kenya, a key market town near the Somalia–Kenya border, the aggregate value of cattle sales has grown by an astounding 500 per cent since 1991, and 700 per cent since 1989 (see Figure 13.4). About 70 per cent of the animals end up at the terminal market of Nairobi, with the remainder mainly directed to Mombasa. Analyses of interview data from two different periods of research, 1986 to 1988 and 1996 to 2001, demonstrate that many Somali traders have benefited from the growth in cross-border trade to neighbouring countries. As one merchant told me in 1998, Somalia is now one large 'duty-free' zone, a dream situation for an entrepreneur seeking profits. Another merchant remarked that, 'Since the fall of the government we have been doing well in livestock trade'.[4] It should be noted that while livestock enters Kenya without formal custom duties or taxes, when the Somali trader sells an animal at a Kenyan market s/he must pay a market tax to local officials, often to the particular county or municipal council body with tax authority in the area. Depending on the destination of the animal, the trader or buyer must also pay a fee to a veterinary department official to have the animal cleared for movement outside the area. And, of course, at the animal's final destination of sale (often Nairobi or Mombasa) other taxes are paid. In short, while cross-border trade is informal and does not incur state taxes, market transactions on the Kenyan side, which can be as many as two or three additional times for each animal, is official, with taxes and other fees being collected.

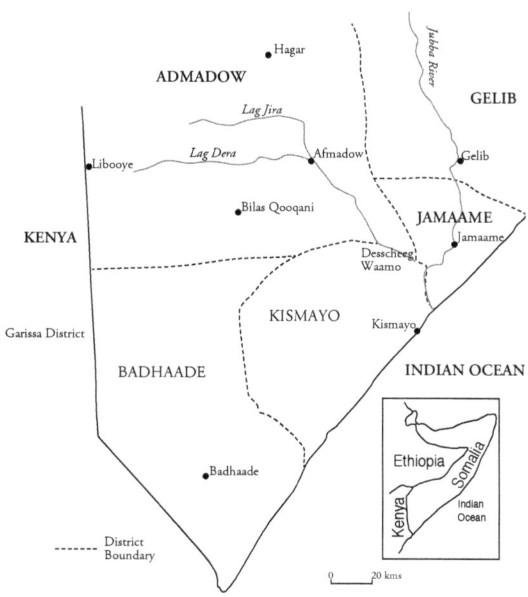

Figure 13.3: Lower Jubba Region: Southern Somali Borderlands.

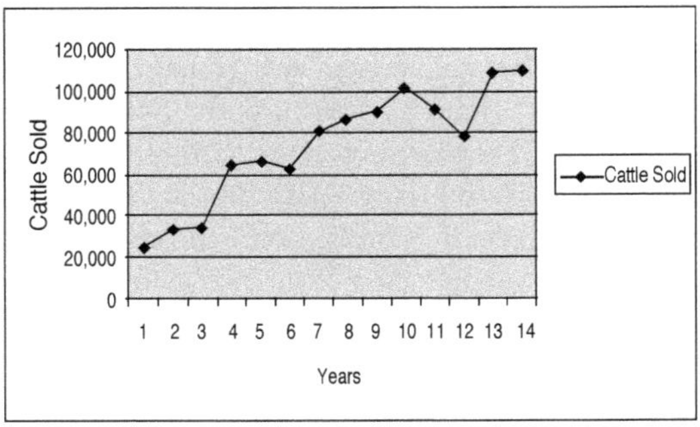

Figure 13.4: Cattle sales at Garissa, Kenya, 1989–2002.

Current cross-border livestock trade has some parallels with the pre-1991 situation. It is an activity that was significant during the late 1980s, but grew considerably during the 1990s (see Figure 13.3). Similar to the current situation, the activity was theoretically illegal in the 1980s but was openly discussed by traders and even some government officials. Many of the traders, who engaged in the trade in the pre-war period, are still involved despite the political changes. During 1998 to 2001, the legality of the cattle trade was challenged only once by the Kenyan government, when it closed the border and blocked trade during short periods during 1999 and 2000. Along the Somalia–Kenya border little official effort is made to halt livestock movements since the trade is so important for the country's livestock sector and economy generally.

Effects of War and Statelessness

The prolonged conflict in southern Somalia has had a significant impact on the country's livestock markets. While four market channels – overseas export trade, a national market centred on Mogadishu (the former capital of Somalia), regional domestic trade centred on Kismayo (the former capital of the Lower Jubba Region), and informal transborder trade – existed prior to 1991, the cross-border commerce is the only channel that escaped large-scale devastation. In fact, it has grown considerably as a result of the volatile conditions in Somalia. The regional domestic market in the Lower Jubba still exists, but herders increasingly avoid Kismayo market because of conflict. This isolation originally stemmed from divisive politics and clan relations among the main clan-based factions in the area, especially tensions in the 1990s between Harti members who reside in Kismayo town and the surrounding area which is controlled by the Ogaden clan, and between the Marehan and the Harti in the early 2000s (see Little 2005). These strained relations were manipulated and aggravated by regional faction heads (warlords). Because they have to traverse Ogaden-controlled lands, Harti-owned cattle and products from Kismayo were rarely transported to lucrative Kenyan markets after 1991 despite the obvious economic benefits of doing so. Less than 2 per cent of cattle sold during March to November 2000 at the largest Kenyan border market were from Kismayo,[5] and virtually no Harti traders were found at the market.

Since 2007 much of the conflict around Kismayo has resulted from disputes between factions of the Union of Islamic Courts, especially the al Shabaab group, and supporters of the recently formed and UN-backed transitional federal government.[6] In short, hostilities have decreased Kismayo's role in regional livestock trade and created extreme market and price distortions in and around Kismayo town, at one time the major market for the entire Jubba Valley, southern Somalia.

Currently, prices for cattle in Somalia increase vis-à-vis distance to the Kenyan border rather than toward the country's largest city, Mogadishu, as was the case in the 1980s. On the Kenyan side, regional prices generally reflect distance to Nairobi and Mombasa, the main terminal markets. In Kenya prices move in the opposite direction: they are lowest near the border and increase as one moves south and west, where the large urban markets are located. When the animal reaches Garissa, the major transit market in Kenya and probably the largest cattle market in East Africa, cattle prices are about 45 per cent higher than in southern Somalia. In the final stages of the market chain, the animals are transported from Garissa to Nairobi and sold at prices up to 200 per cent higher than in southern Somalia. Based simply on the economics of these facts, it is hardly astonishing that flows of cattle from Somalia to Kenya increased in the 1990s. What is surprising, however, is that price differences between Kenya and Somalia are not greater under the current volatile conditions.

There is little question that the national cattle market in Mogadishu has been badly affected by post-1991 events and conflict. While prices rose during 1987 to 1998 at market centres involved in cross-border trade (Afmadow and Liboi), they actually declined in US dollar terms for Mogadishu. Prior to the collapse of the national government, cattle prices at Somali markets were always highest in Mogadishu. For the border town of Liboi, prices doubled between 1987/88 and 1996, a reflection of the strong growth in cross-border trade with Kenya. Not surprisingly, the market catchment for the cross-border trade has expanded greatly toward Mogadishu since the 1980s, as Kenyan-based traders find it profitable to source cheap cattle well into the interior of Somalia and almost up to the hinterland of Mogadishu.

What does it mean for economic agents, like traders and middlemen, in a context where the formal economic and political institutions in a country have collapsed? Based on the Somali example, the impacts are not as significant as one might envision. For the economic agent, port fees for importing goods and moving them safely must still be paid, but instead of going to representatives of a central state they are going to warlords and militias. The entrepreneur still has little access to formal banking and financial institutions, as was the case in the pre-1991 era, although small-trader access to informal financial facilities has actually improved in the past seventeen years. The feeble nature of economic institutions and financial systems in countries like Somalia has meant that their collapse has been relatively insignificant for much of the population.

Social Relations

Whether it is state-sanctioned (legal) or not, trade requires norms and some level of mutual trust to function. To trust somebody assumes that agreed outcomes, behaviour and prices are followed and that parties do not have to carefully monitor each other (Gambetta 1988; Ensminger 1992). Trust and fair practice are critical in the current Somali economy, where legal enforcement is virtually impossible and money and cash advances are often made over very long distances. It is not unusual for a Somali trader to advance the equivalent in Kenya shillings of several thousand US dollars to a middle person to purchase animals for them. In these cases, there are important trust relations based on kinship and other social relations that serve market and finance functions and help to reinforce these cash transactions. Where trust-based relationships exist, market transaction costs are reduced because informal credit and market contracts are more easily extended without formal contracts or other agreements (Ensminger 1992: 104–5). In times of conflict and uncertainty, these relationships become even more important and can lead to highly exclusionary practices, whereby traders of certain clans and groups exclude, in some cases violently, others from participating in the trade. This may help to explain the predominance of strong ethnic and clan-based trading coalitions in the transborder trade between the conflict-ridden Ogaden region of eastern Ethiopia and neighbouring Somalia and Somaliland (see Umar and Baluch 2007). These networks are infused with the strong presence of dominant clans in trading activities in Somalia, as well as throughout northern Kenya and eastern Ethiopia. These discriminatory practices are often reinforced by local politicians and militia heads who enforce rigid territorial boundaries between different clans and sub-clans, which serve to only further instigate deep-seated hostilities among different groups.

In the Somali case, trust has been strongly challenged by recent acts of seemingly random violence, but it was also compromised earlier by the brutalities of the Barre government and its representatives. In some cases recent events have reinforced the importance of trust within certain social groups, such as clans, while unfortunately heightening conflict and distrust between certain social groups. In the words of Anna Simons, 'Now more than ever, knowing genealogy does chart who can and who cannot be trusted' (Simons 1998: 70). The segmentary kinship principle of political alliances and trust has always been present in Somali politics (see Little 2003), but has assumed more significance recently as faction leaders have deliberately used clanship ('genealogy') as a political weapon. Through the misdirected actions of warlords and militia leaders, clans and their territories have become forcibly isolated from each other and interactions restricted by armed factions. In these volatile, risky environments, where rival identities can be a death sentence, Somali traders often rely on their extensive kinship ties and on members of their clan and sub-clan rather than others.

The expansive clan system of the Somalis has abetted traders in gaining access to markets, finance and information, both during periods of boom and bust. It has also aided the global trade and remittance networks that have helped to keep the Somali economy afloat.[7] As noted earlier, while my evidence is not as strong, there is little doubt that common religious bonds based on Islam have also facilitated the formation of market networks, the flow of information and finances between different agents in the market chain, and the fostering of trust-based relations in the cross-border trade. Evidence from elsewhere in north-eastern Kenya (Ensminger 1992) and West Africa suggests that Islam can play an important role in livestock trade, although there is little indication that institutions, such as the Muslim brotherhoods which are so important in the Senegalese trading diaspora, have much of a role in Somali transborder commerce. Nonetheless, with the collapse of the centralised state, there is little doubt that Islamic identity and religion is generally more important among Somalis now than prior to 1991 (see also McGown 1999).

Somali traders and transporters are found throughout East Africa and well into central Africa and the Middle East. These networks have grown in significance with the implosion of the Somali state and massive exodus of Somalis into the diaspora. The segmentary kinship and clan idioms strengthen them by enhancing trust and guarding against deception in business dealings. 'Greater Somalia' based trade and finance networks have expanded beyond the Horn of Africa to encompass parts of the Middle East and to link numerous international cities to Somalia, including Dubai and London. While a significant trade diaspora existed prior to 1991, the collapse of the state and the ensuing problems accelerated its growth, with livestock being just one of its important commodities.

As noted earlier, in the Horn region, similar ethnic groups with common languages straddle international borders. This facilitates cross-border trade and channels the trade through certain centres and merchants. For example, in the border region of north-eastern Kenya and southern Somalia, Somali pastoralists are found on both sides of the border but there are marked differences in clan composition that influence the nature of cross-border trade. Because of the rigour required and hardships endured, many traders in the business are young (less than forty years old).

Who Benefits?

Historically, in Africa and elsewhere, informal sector activities such as petty trade have been associated with the poor, those who are cut off from salaried employment, financial institutions and lucrative licensed trade. In many cases, informal trade is highly gendered with high levels of female participation, especially women who pursue petty trade as a survival strategy to feed their families (Little and Lundin de Coloane 1992). While there are some poor, small-scale traders involved with transborder cattle trade, many of the actors are not poor by local and regional standards and very few are women (see below). Indeed, there is considerable wealth and market concentration among cross-border traders, with about 20 per cent of traders accounting for more than 60 per cent of market transactions and 50 per cent dealing with less than 15 per cent of the trade. The high standard deviation and range in volume of sales in trader samples from the Somali borderlands support this pattern of inequity. For instance, in the sites in Kenya and southern Somalia where cattle trade dominates, 34 per cent of cattle traders sell less than 300 cattle per year (Table 13.1). By contrast, a small minority (13 per cent in 1998) had annual sales in excess of 1,200 animals. A similar trend is also observed in the small-stock (goats and sheep) and camel trades, where a few traders have annual sales three to four times higher than most merchants.

As noted earlier, the Somali civil war instigated an increase in cross-border cattle trade and the merchants, both large- and small-scale, associated with it have benefited. Those traders who were involved in informal, transborder trade fared much better after the state collapsed than merchants associated with other kinds of trade, especially the official export business. The boom in cross-border trade encouraged a relatively large percentage (42 per cent) of the 84 traders interviewed in 1996 and 1998 to start their businesses during the previous five years. The growth in the cross-border trade, coupled with the lack of viable alternative employment options in the region, accounts for relatively large increases in the number of cattle traders in recent years. More than 92 per cent of traders interviewed were Somali, and the bulk of these were of the Ogaden clan.[8] And in a sample of 84 traders, 100 per cent were male; in a random observation of more than 150 sellers on three market days in Garissa in 2001, only one female livestock trader was observed who had any involvement with the cross-border cattle trade. However, women traders are more involved with the cross-border livestock (sheep and goats) trade between Ethiopia and Somaliland (Umar and Balauch 2007). Involvement of non-Somalis in the trade only occurs on the Kenyan side at higher levels of the market chain (such as in Nairobi), although they were more active at smaller markets in the late 1980s. As I have pointed out elsewhere (Little 2003), trade has become increasingly clan-based as a result of the instability of the 1990s and the blatantly clan-based tactics of different militia heads.

Table 13.1: Annual sales of cattle by traders, Somalia/Kenya border.

Range of Annual Sales	Percentage of Total	
	1996	1998
1 to 300	17	34
301 to 600	18	26
601 to 900	20	5
901 to 1200	28	22
1200+	17	13
TOTAL	100	100
Average Sales (Mean):	943	855
Standard Deviation:	760	883
n=94 traders.		

Source: Little (2003: 117).

It is possible to calculate economic returns for traders at different levels in the market chain, and to determine where profits are highest. Table 13.2 illustrates the kinds of costs and profits that can be accrued along a common cross-border route, the market channel between Afmadow in Somalia through Garissa to Nairobi. The table assumes that there are two traders involved in the market chain, one based in Garissa who buys from a Somalia-based middleman, and one based in Nairobi who purchases from the Garissa trader. The estimates include other assumptions that will vary from year to year. In very dry years, when animal sales are slow, trader costs allocated to fodder and water can be as much as five to six times the figures in Table 13.2.

As would be expected, the highest risks and costs in the cattle trade involve the initial purchase and transport. The net return for the trader (Trader 1 in Table 13.2) who buys directly from Somalia and sells at Garissa is 15 per cent. The return is almost the same at 16 per cent for the trader who purchases at Garissa and then resells in Nairobi (Trader 2). The high cost of transport between Garissa and Nairobi (about 12 per cent of the Garissa purchase price) accounts for most of Trader 2's costs. If Trader 2 purchases an animal directly in Somalia rather than at the Garissa market, then the net return per animal can be as high as 31 per cent. Thus, while the risks can be excessive in some years, the returns from the cross-border trade also are very high – well above profits in other types of agricultural trade and higher than returns that I observed in 1987/88.

The Importance of Local Institutions and Practices

Like any form of risky, long-distance trade, cross-border cattle trade involves a myriad of institutions, actors and relationships. The stateless nature of transborder trade in Somalia requires that traders rely on a range of local institutions and practices to facilitate their businesses. Most of the important institutional innovations are centred on informal finance, broker services and transport activities that are increasingly shaped by clan and sub-clan relations.

Informal Finance Services

Very few traders (less than 10 per cent of the total) have access to formal sources of finance. Informal finance supplements the lack of formal market access and, as noted above, trust-based relationships play an important role in these transactions (see also Teka, Azeze and Gebremariam 1999: 54). In the case of the Somali border areas, informal finance services minimise risks associated with carrying large amounts of cash in an unstable environment. Somali border traders can take their earnings to Nairobi, convert them into dollars and then 'wire' them back to money houses in Somalia, where they can be picked up by associates. This informal practice, called the *hawala* or *hawilaad* system (meaning 'transfer' in Arabic),

Table 13.2: Trader returns in the cross-border trade, 1998–1999.

	Amount US$	Net Return
ITEM	Per cattle+	
Initial Purchase Price from Herder	108.00	
Purchase Price from Middlemen, Afmadow	128.00	
Transport cost (Afmadow-Garissa)	3.00	
Hired Herd Labour	1.60	
Security/Transit Fees	0.40	
Water (50 days @ .08)	4.00	
Medicine/dips	1.82	
Fodder (Garissa market) (dry sea. only)	0.60	
Risk from loss (theft, drought, etc.) (6%)	7.68	
Broker Fee (Afmadow)	1.25	
Broker Fee (Garissa)	1.67	
Council Tax (Kenya)	1.33	
Currency transaction/conversion fees	5.28	
Trader 1 Costs	156.63	
Sale Price, Garissa	176.00	
TRADER 1 RETURN		15%
Transport Cost (Garissa-Nairobi)	20.15	
Movement Permit/Fees	1.33	
Hired Labour	0.33	
Water	1.00	
Fodder (Garissa and Nairobi)	0.60	
Market/Municipal Tax—Nairobi	1.33	
Broker Fee (Garissa)	1.67	
Broker Fee (Nairobi)	2.50	
Trader 2 Costs	204.91	
Sale Price, Nairobi	233.00	
TRADER 2/MIDDLEMAN RETURN		16%
GROSS DIFFERENCE BETWEEN ORIGINAL PURCHASE AND FINAL PRICE	105 (82%)	
TRADER RETURN IF ONE TRADER ONLY AND PURCHASES FROM SOMALIA AND SELLS IN NAIROBI		31%

Source: Based on Little (2003: 111).

increasingly involves mobile phones and it avoids the need to carry large amounts of cash across the border (see also Mahmoud 2003). In other cases, traders will convert part of their earnings in Kenya into tradable goods, which they will arrange with a wholesaler to be picked up at the border to avoid the risk of travelling in north-eastern Kenya with excess money.[9] These transfer services are mediated through informal money houses and middlemen, who assume special importance in most forms of long-distance trade, including livestock. While much of the livestock trade is calculated in Somali shillings, and final payments are made in Somali shillings or US dollars, the actual handling of cash in large transactions is minimal.

As noted earlier, many of the individuals who work together in these networks are often bound by kinship and common religion and/or ethnicity. In times of conflict and heightened uncertainty, such as is currently the case in Somalia, these networks assume even more significance as actors turn 'inwards' and favour transactions with those whom they know well. Umar and Baluch (2007), for example, document how trader-based networks affect financial and trade flows along the border of eastern Ethiopia and Somalia. They argue that the highly uncertain trade environment, due to political instability, conflict and random product confiscations by government officials, has created tightly structured clan-based trading corridors (networks) where flows of cattle and finances are limited to a particular corridor. These corridors 'serve to protect but also to limit the volume and value of trade … [E]ach corridor is dominated by two or three large clans and managed by different sets of traders, guarantors and credit suppliers whose ties are clan-related and whose operations are founded on trust' (ibid.: 8). As pointed out earlier, on the positive side, relations of trust based on clan and/or other social relations can serve important market and finance functions (see Mahmoud 2003).

In the region under discussion, many of the important informal finance businesses that traders use are based in Nairobi. The enterprises usually charge fees of 3 to 6 per cent to send funds from Kenya to locations in Somalia. Different forms of transfer and credit minimise risk and reduce the physical handling of worn Somali notes, which slows their deterioration and helps to keep them in circulation. As noted earlier, the system sometimes functions in the idiom of the clan system, and knowledge of sub-clan and family relations can be used as cross-checks on identity. A report commissioned by the United Nations Development Programme (UNDP) describes how informal money houses operate, in this case in relation to the remittance business:

> A major financial innovation, since the collapse of the State (Somalia), has been the rise of *hawilaad* or remittance companies. There are several major remittance companies and innumerable small ones. The large companies handle remittances of funds sent by the Somali Diaspora to relatives in country, charging a commission of 3 per cent to 7 per cent (depending on the level of competition and the size of the remittance). Using a mix of telephone, fax, and HF radio, and relying on a worldwide network of agents, the *hawilaad* companies can instantly transfer money from a Somali in Canada to

> his family in Bosasso. Increasingly, *hawilaad* companies also transfer money between businessmen both within Somalia and internationally, allowing merchants to make fast purchases of commodities growing scarce in a local market. (UNDP 1998: 15)

The scale of the informal banking system is extraordinary, and at least one of the larger Somalia-based enterprises has branches in many cities around the world. As a point of comparison, the US-based company Western Union charges a 12 per cent fee to send cash from the USA to the Horn of Africa. Unfortunately for many Somali families, the largest of these *hawala* banks, al Barakaat, abruptly closed in November 2001 over concerns that it laundered funds and generated profits for the terrorist network al-Qaida (IRIN 2001; Vesley 2002). Based on a US-led initiative, the indictment also claimed that at the highest levels the bank was controlled by key al-Qaida supporters. Thus, a financial institution that had evolved as a viable operation in such a short time following the collapse of Somalia's finance sector, and was depended upon by literally thousands of Somali families in the region and around the world, is now labelled a terrorist-linked organisation. There has been an active lobby by development agencies, such as the UNDP, to re-open al Barakat, but this had not yet happened. However, other informal money companies in Somalia have mainly filled the void left by the departure of al Barakat, and the financial practice of international money transfers remains strong.

Market Brokers

Poor access to market information is a major risk associated with cross-border trade, although mobile phones have improved the situation. Since official information about livestock prices is virtually non-existent, traders rely on informal means of obtaining market information.[10] This complicates the search for markets by individual traders and herders, who often rely on local brokers (*dilaal*) for assistance. The broker's role is to match the buyer with a seller, the latter often having travelled 100 kilometres or more to market, and to insure the legitimacy of the sale. It is a common institution throughout the Middle East and Islamic regions of West and East Africa (see Cohen 1969). Once again, the notion of trust is inherent to the institution of *dilaal* and relevant to understanding 'unofficial' activities like cross-border trade.

In all of the markets, including Nairobi, there are *dilaal* to match sellers and buyers and negotiate prices on their behalf.[11] A *dilaal* works in the market on behalf of both buyers and sellers, and can be compensated by both. The fee is usually around 1 to 2 per cent of the price of the animal, anywhere between US$1.25 to US$2.50 per animal. In some cases, the fee is cut in half, with both buyers and sellers paying part of the fee; in others buyers and sellers may be working with different brokers and pay them separately. For sellers, these arrangements remove the burden of finding a buyer and negotiating a price. An older, trustworthy broker may work for several parties simultaneously.

Brokers show an uncanny ability to gauge animal weights, quality and prices accurately. This is where an experienced trader holds an advantage over younger merchants and explains why many of the major cattle brokers in the region have been in the business for several years. Early attempts by both the Kenyan and Somali governments to 'rationalise' the system by introducing scales for weighing animals proved futile, as brokers had little use for such expensive devices. In the Nairobi marketplace, for example, Somali traders almost always deal with Somali rather than non-Somali brokers. In Somalia the role of *dilaal* in the 1980s was so significant that they were officially registered and licensed by the local administration, and required to pay a small tax. In Kenya, *dilaal* recently organised themselves into an association in at least one important border market town, Moyale. The group has their own offices and has registered with the local administration (Mahmoud 2003).

The roles of and fees charged by brokers changed very little between 1988 and 2001, a surprising outcome given the increased insecurity in the border region following the collapse of the Somali state. If theft and insecurity were so rampant, one would have expected broker fees to have increased considerably as a result of the added risks of doing business and possibly from increased cattle theft. Since a broker is paid to insure the legitimacy of a transaction, if a stolen animal is sold the *dilaal* is responsible should the original owner seek compensation or the return of the animal. Yet, brokers' fees have not grown very much and are still around 1 to 2.5 per cent of the value of the transaction, which is what they were in the late 1980s. This lack of change would imply that risks of transactions in stolen animals are not especially high.

Brokers can reduce stubborn market bottlenecks that hurt sellers, whether they are caused by herders or traders. By finding buyers in a timely fashion, brokers can help sellers to avoid the added costs of maintaining animals until they are sold. Recall that sellers often have transported their animals over very long distances before they reach the market. For example, at the key border market of Garissa, merchants must buy fodder and water for their animals when they cannot find a buyer and the herd must remain in the area. A quick sale through a broker minimises outlays on fodder, water and hired labour. However, even *dilaal* have little control over external events that can greatly affect the market and the demand for animals. For example, when there is an unexpected glut of animals on the Nairobi market from neighbouring countries, such as Tanzania and Ethiopia, Somalia-based traders may have to wait several days in Garissa before transporting cattle down country. When this happens, the trader often has little recourse but to purchase fodder or to move animals into surrounding range lands.

Transport

A third informal practice that facilitates cross-border trade is long-distance trekking. Indeed, one of the most significant and intriguing aspects of the cross-border trade involves the trekking of cattle over several hundred kilometres, which can take up to several weeks. Trekking is an age-old profession that dates back to the 1800s and the early caravan trade (Dalleo 1975). It takes place along most of the Kenya–Somalia border and throughout the Horn of Africa generally. In most cases, cattle are moved overland with three trekkers and an armed security person for approximately every 100 cattle. There is a designated 'head' trekker who is directly responsible to the trader and who may be employed on a fairly regular basis. In southern Somalia, this individual must make payments to pass through territories controlled by different factions and sub-clans, and like the position of *Abbaan* (protector) in the days of the nineteenth-century caravan trade (see Lewis 1994: 115), they are responsible for the safe passage of the herd. The head trekker or protector usually comes from one of the main lineages or sub-clans whose territory the market animals must traverse.

Table 13.3: Transport costs in cross-border cattle trade, 1996 and 1998.

Market Route	Estimated Distance (km)	Number of days[3]	Cost Head (US $) 1996/1998	Cost per km 1996/1998
Garissa-Mombasa[1]	470	1–2	16.15/27.83	0.34/0.059
Garissa-Nairobi[1]	420	1	13.29/20.15	0.032/0.048
Mandera-Isiolo[2]	620	3	ND/16.67	ND/0.0269
Afmadow-Garissa	240	14	2.32/3.00	0.010/0.013
Baidoa-Garissa	550	28	4.59/5.92	0.010/0.010
Dinsor-Garissa	490	23	4.03/2.08	0.010/.004
Dhobley-Garissa	160	8	2.03/ND	0.012/ND
Kismayo-Garissa	350	ND	ND/2.92	0.010
Libooye-Garissa	150	6	1.60/2.50	0.011/.017
Mogadishu.-Afmadow	300	15	3.27/ND	0.011
Mogadishu-Garissa	540	28	4.15/4.92	0.0100/.010
Wajir-Garissa	250	ND	1.90/11.671	0.008/0.047
Isiolo-Nairobi[1]	250	1	ND/14.17	ND/0.057
Mandera-Isiolo	600	36	ND/2.63	ND/0.004
Dinsor-Moyale	540	32	ND/6.00	0.011
Sakow-Mariakani (Mombasa)[4]	730	30	ND/8.33	ND/0.011

ND=No data.
Source: Author's data.

Notes:
1. These prices reflect the use of motorised transport (truck).
2. This price was mainly for motorised transport but included some trekking on foot.
3. There is considerable variation in this category depending on environmental and security conditions at the time, the particular motivations of the trader, and the skills of the herders. These numbers represent estimated averages based on trader interviews.
4. The last part of this trek (100 km) often involves motorised transport.

The trader is likely to have a young relative accompany the animals on the trek, because of the physical rigour involved and to safeguard their property. In some cases it can take almost a month to reach a key market from interior locations in Somalia, and substantial animal weight loss can occur.[12] One of the trekkers is usually responsible for procuring rice, flour, sugar and tea for the trip and serving as the cook. The trader normally covers the cost of food, and on long treks an entire sack of flour or rice (50 kilograms or more) will be purchased and transported on a pack animal (usually a donkey).

Based on data in Table 13.3 that show the different trekking routes, there appears to be a functioning market for animal trekking services based on distance and cost. As would be expected, transport costs per animal generally correspond with distance to market and type of transportation involved, except where insecurity is particularly problematic. On the Kenyan side, the movement of animals between Garissa and the terminal market of Nairobi is by lorry, especially since the completion of a new tarmac road in 2002. The Baidoa (Somalia) to Garissa and Dinsoor (Somalia) to Garissa routes are examples of high security risks and, not unexpectedly, they experience relatively high transport costs. It was assumed that transport costs in the region would have risen exorbitantly throughout the past decade because of insecurity, but this has not been the case for many of the key routes. Movement costs per animal rarely exceed US$0.01 to US$0.18 per kilometre, and only modest increases have taken place since the Somali government's collapse in 1991. There were also minor differences in transport costs between the Kenya- and Somali-based circuits.

Concluding Remarks

The materials presented here show the extent to which cross-border livestock trade drives the economies of the Horn of Africa's border regions and areas well into the interiors of Ethiopia, Somalia and Kenya. They provide telling evidence of the importance of cross-border commerce for inhabitants of the border region, and even for consumers and producers located several hundred kilometres from the borders. That the activity is effectively illegal does not detract from its significance or from the fact that some regional governments do little to discourage it. Examples from this and other studies (see Little, Teka and Azeze 2001) show what happens to local incomes, food security, welfare and local and regional markets when cross-border commerce is disrupted. In the case of the Ethiopia–Kenya border, it was found that revenues from cross-border livestock trade financed informal grain imports from the surplus-producing highlands of southern Ethiopia to areas of northern Kenya (Teka, Azeze and Gebremariam 1999: 50).

But are the results presented here applicable to areas outside the Horn of Africa? It has been my contention that while cross-border trade along Somalia's borderlands is an extreme case of what I have called elsewhere an 'openly illegal activity' (Little 2003), studies from other parts of Africa show that unofficial cross-border trade is absolutely critical to local and regional economies, including in Mozambique, Sierra Leone, Djibouti, Liberia, the Democratic Republic of the Congo (DRC), Benin, Angola and several other African states where governments exist (see, e.g., Nugent and Asiwaju 1996; WFP 2006; Raeymaekers 2008). The same also holds true for large parts of Central Asia and the former USSR (see Burawoy and Verdery 1999; Humphrey 1999). Some of these countries engage in unofficial cross-border trade involving high-value commodities, such as diamonds and iron ore, and attract investment from legitimate transnational corporations (see Richards 1996; Ellis and MacGaffey 1997; Ellis 1999). The effective collapse of national monetary systems and financial and regulatory institutions in African countries like the DRC and Angola also means that most of the domestic economy is unofficial and transborder trade is mediated through private entrepreneurs not government bodies. As this chapter has shown, in strict terms this practice is considered smuggling, but the term is increasingly meaningless in large parts of Africa where the difference between what is officially legal and illegal is increasingly blurred (Hibou 1999).

To conclude, while some of the governments of the region in the late 1990s and early 2000s made positive remarks about the cross-border trade in livestock,[13] these have diminished in the past few years. New regional concerns about security in the post-9/11 era have come to influence recent policy initiatives concerning international borders, and, unfortunately, and these concerns have the potential to negatively affect all types of informal transborder commerce. Kenya closed down much of the 'clean' trade between Kenya and Somalia under international pressure during parts of 2003 and 2006/7, although it relaxed its stance in late 2007. Ethiopia, in turn, has voiced recent concerns about 'contraband' trade along its borders and does not seem to distinguish between trade in 'clean' (and essential) products from other kinds of border trade, even though as recent as 2000 it relaxed barriers on a lot of petty trade in foodstuffs along its borders (Umar and Balauch 2007). Nonetheless, the reality is that unofficial transborder trade in livestock and other critical commodities plays an important role in local and national development and will continue to do so until alternative markets and livelihoods are found for these areas and their inhabitants. Governments in the region should recognise this and take positive steps to support rather than discourage the activity for the benefit of their economies and citizens.

Notes

The research reported in this chapter was supported by a Research and Writing Grant from the John D. and Catherine T. MacArthur Foundation. Additional funding was provided under the BASIS Collaborative Research Support Program (Grant No. LAG-A-00-96-90016-00) and the Global Livestock Collaborative Research Support Program (Grant No. PCE-G-00-98-0036-00). I wish to acknowledge the strong support of my colleagues, Hussein A. Mahmoud, Tegene Teka, Abdel Ghaffar M. Ahmed and Alemayehu Azeze, with whom I participated in a study of cross-border trade and food security in the Horn of Africa between 1998 and 2001.

It goes without saying that none of the above institutions and individuals is responsible for the contents of this chapter. An earlier version of this chapter was presented at the EGDI-WIDER conference on 'Unlocking Human Potential: Linking the Informal and Formal Sectors', 17–18 September 2004, Helsinki, Finland. Parts of this chapter are based on Little (2003).

1. The region normally includes the Sudan, Ethiopia, Eritrea, Djibouti, Somalia and the north-eastern part of Kenya.
2. I use the term 'clean' commerce for trade in relatively benign commodities like cattle or grains in order to distinguish it from commerce in 'dirty' or generally harmful goods, like narcotic drugs and weapons.
3. This is the mildly narcotic plant that is widely chewed by Somalis and is grown mainly in Meru District, Kenya, and in parts of the Ethiopian highlands. It is very difficult to gain data on this trade but it clearly represents the most valuable commodity trade, including livestock, in Somalia and its border areas. For a good analysis of the trade, see Anderson et al. (2007).
4. Trader interview, 24 May 1996.
5. Stefano Tempia (personal communication).
6. For background details on recent politics in southern Somalia, which can change on an almost daily basis, see Menkhaus (2007).
7. It is estimated that Somalis outside the country may remit as much as US$1 billion in cash per annum to relatives and friends living in Somalia (World Bank 2007).
8. For more details, see Little (2003).
9. Hussein Mahmoud (2003) records a practice whereby transborder traders who sell animals in Nairobi transfer their cash earnings to a wholesaler from the border who is in Nairobi. The wholesaler, in turn, buys goods in Nairobi with the trader-supplied money and transports the products to border towns for sale. The person then orders a business associate or partner at the border to repay the livestock trader or his partner. The livestock trader may have a partner at the border who receives the cash and then re-initiates procuring animals for sale in Nairobi. This important informal practice allows both the livestock trader and the wholesaler to conduct business without personally transferring large amounts of cash across vast areas of insecure territory. Bandits in northern Kenya are less likely to attack a lorry if it is only transporting goods.
10. Some NGOs and government organisations operating in the area collect prices on different types of animals, grain and fuel wood. However, the dissemination of this information is limited and its market coverage is limited. Moreover, even when it is available herders and livestock traders rarely use this information.
11. For similar cases of brokers in livestock markets elsewhere in Africa, see Cohen (1969) and Manger (1984).
12. It was virtually impossible to estimate weight loss from trekking, although herders pointed to it as a problem. The amount of weight loss will depend on range conditions, the number of trekking days, the herding skills of the trekkers and other factors.
13. See 'Livestock to Move Freely across Borders', *Daily Nation* (Nairobi), 9 November 2000; see also Teka and Azeze (2002).

References

Anderson, D.M., et al. 2007. *The Khat Controversy: Stimulating the Debate on Drugs.* New York: Berg.

Burawoy, M., and K. Verdery (eds). 1999. *Uncertain Transitions: Ethnographies of Change in the Postsocialist World.* Boulder, CO: Rowman and Littlefield.

Cohen, A. 1969. *Custom and Politics in Urban Africa.* Berkeley: University of California Press.

Dalleo, P.T. 1975. 'Trade and Pastoralism: Economic Factors in the History of the Somali of North-eastern Kenya, 1892–1948', PhD dissertation. Syracuse, NY: Syracuse University.

Ellis, S. 1999. *The Mask of Anarchy: The Destruction of Liberia and the Religious Dimension of an African Civil War.* New York: New York University Press.

Ellis, S., and J. MacGaffey. 1997. 'Le Commerce International Informel en Afrique Sub-saharienne', *Cahiers d'Etdues Africaines* 145: 11–37.

Ensminger, J. 1992. *Making a Market: The Institutional Transformation of an African Society.* Cambridge: Cambridge University Press.

FAO. 2003. 'Closure of Garissa (Kenya) Cattle Market: Food Security Implications in the Garissa Marketshed in Kenya and Somalia', unpublished paper. Nairobi, Kenya: Food Security Analysis Unit of the Food and Agriculture Organization.

FAO, World Bank and European Union. 2004. 'Towards a Livestock Sector Strategy: Final Report', unpublished report, No. 04/001 IC-SOM. Rome: Food and Agriculture Organization (FAO).

Gambetta, D. (ed.). 1988. *Trust: Making and Breaking Cooperative Relations.* Oxford: Blackwell.

Hibou, B. 1999. 'The "Social Capital" of the State as an Agent of Deception, or the Ruses of Economic Intelligence', in J.-F. Bayart, S. Ellis and B. Hibou (eds), *The Criminalization of the State in Africa.* Oxford: James Currey, pp.69–113.

Humphrey, C. 1999. 'Traders, "Disorder," and Citizenship Regimes in Provincial Russia', in M. Burawoy and K. Verdery (eds), *Uncertain Transitions: Ethnographies of Change in the Postsocialist World.* Boulder, CO: Rowman and Littlefield, pp.19–52.

IRIN. 2001. 'US Shuts Down Money Transfer Group', *Integrated Regional Information Network of the United Nations.* Retrieved 8 November 2010 from: http://www.irinnews.org/report.asp?ReportD=13593.

Lewis, I.M. 1994. *Blood and Bone: The Call of Kinship in Somali Society.* Lawrenceville, NJ: Red Sea Press.

Little, P.D. 1992. 'Traders, Brokers, and Market "Crisis" in Southern Somalia', *Africa* 62(1): 94–124.

——— 2003. *Somalia: Economy Without State.* Oxford: James Currey.

——— 2005. 'Pastoralism in a Stateless Environment: The Case of the Southern Somali Borderlands', *Geography Research Forum* 25: 128–47.

Little, P.D., and I.B. Lundin de Coloane. 1992. 'Petty Trade and Household Survival Strategies: A Case Study of Food and Vegetable Traders in the Peri-urban Area of Maputo, Mozambique', Working Paper No. 70. Binghamton, NY: Institute for Development Anthropology.

Little, P.D., T. Teka and A. Azeze. 2001. 'Cross-border Livestock Trade and Food Security in the Horn of Africa: An Overview', unpublished research report. Madison: Land Tenure Center, University of Wisconsin.

McGown, R.P. 1999. *Muslims in the Diaspora: The Somali Communities of London and Toronto.* Toronto: University of Toronto Press.

Mahmoud, H. 2003. 'The Dynamics of Cattle Trading in Northern Kenya and Southern Ethiopia: The Role of Trust and Social Relations in Market Networks', PhD dissertation. Lexington: University of Kentucky.

Manger, L.O. (ed.). 1984. *Trade and Traders in the Sudan*. Bergen: Department of Social Anthropology, University of Bergen.

Meagher, K. 1997. 'Shifting the Imbalance: The Impact of Structural Adjustment on Rural–Urban Population Movements in Northern Nigeria', *Journal of Asian and African Studies* 32(1/2): 81–92.

Menkhaus, K. 2007. 'The Crisis in Somalia: Tragedy in Five Acts', *African Affairs* 106 (424): 357–90.

Murray, M. n.d. 'Diasporic Peoples and Transnational Communities: The Changing Dynamics of the New Urban Migration', unpublished manuscript.

Nugent, P., and A.I. Asiwaju (eds). 1996. *African Boundaries: Barriers, Conduits, and Opportunities*. London: Pinter.

Peberdy, S., and C. Rogerson. 2000. 'Transnationalism and Non-South African Entrepreneurs in South Africa's Small, Medium, and Micro-enterprise (SMME) Economy', *Canadian Journal of African Studies* 34(1): 20–40.

Raeymaekers, T. 2008. 'Conflict and Food Security in Beni-Lubero: Back to the Future', in L. Alinovi, G. Heimrich and L. Russo (eds), *Beyond Relief: Food Security in Protracted Crises*. Warwickshire: ITDG Publications/Practical Action Publishing, pp.169–95.

Richards, P. 1996. *Fighting for the Rain Forest: War, Youth and Resources in Sierra Leone*. Oxford: James Currey.

Rogerson, C.M., and J.M. Rogerson. 1996. 'The Metropolis as Incubator: Small-scale Enterprise Development in Johannesburg', *Geojournal* 39(1): 33–40.

Simons, A. 1998. 'Somalia: The Structure of Dissolution', in L.A. Villalon and P. Huxtable (eds), *The African State at a Critical Juncture: Between Disintegration and Reconfiguration*. Boulder, CO: Lynne Reinner, pp.57–73.

Teka, T., and A. Azeze. 2002. 'Cross-border Trade and Food Security in the Ethiopia–Djibouti and Ethiopia–Somalia Borderlands', Research Report Series No. 4. Addis Ababa: Organization for Social Science Research in Eastern and Southern Africa.

Teka, T., A. Azeze and A. Gebremariam. 1999. 'Cross-border Livestock Trade and Food Security in Southern and Southeastern Borderlands of Ethiopia', Research Report Series No. 1. Addis Ababa: Organization for Social Science Research in Eastern and Southern Africa.

Umar, A., and B. Baluch. 2007. 'Risk Taking for a Living: Trade and Marketing in the Somali Region, Ethiopia'. Addis Ababa: UN-OCHA/Pastoral Communication Initiative Project.

UNDP. 1998. 'Human Development Report: Somalia'. Nairobi: United Nations Development Programme.

Vesley, M. 2002. 'Somalia: US Cuts Net Lifeline', *African Business* 44 (January).

WFP. 2006. 'Informal Cross Border Food Trade in Southern Africa'. Nairobi: World Food Programme Famine Early Warning Systems Network.

World Bank. 2007. 'Remittance Trends 2007', Migration and Development Brief No. 3. Washington, DC: World Bank.

Zaal, F., and A. Polderman. 2000. 'High Risk, High Benefits: International Livestock Trade and the Commercial System in Eastern and Southern Africa', unpublished paper presented at the Organization for Social Science Research in Eastern and Southern Africa conference, 'Globalisation, Democracy and Development in Africa: Future Prospects', 24-28 April 2000, Dar es Salaam, Tanzania.

Chapter 14

Pastoralism and Trans-Saharan Trade

The Transformation of a Historical Trade Route between Eastern Chad and Libya

Meike Meerpohl

Introduction

This chapter explores the relationship between pastoralism and trans-Saharan trade between Chad and Libya, with a focus on the importance of livestock husbandry for nomadic and semi-nomadic groups in the sub-Saharan regions of north-eastern Chad, with specific attention to the Zaghawa. Furthermore, it will outline how the Zaghawa have transformed from being semi-nomadic pastoralists to traders. This transformation has come about due to migration and wage labour stimulated by the affects of drought on their homeland with resulting extensive interregional networks used for trade links. Additionally, savings accumulated from wage labour have enabled many to send remittances back to the families that remained behind. This money seems to have been mostly reinvested in camel husbandry instead of in cattle, resulting in large camel herds used in a profitable camel trade. Based on a description of the historical developments over the last two hundred years, this chapter will analyse the transformation of the former Wadai to Benghazi trade route into modern-day trade links between Chad and Libya. This transformation will shed light on all aspects of trade: the goods traded, the routes followed, the people involved, the management of the journey, as well as the organisation and control of trade. The recent livestock trade between Chad and Libya exemplifies the re-establishment of trade links between the two countries, which had broken down during the Chadian civil war (1966 to 1978) and the Chadian–Libyan war (1978 to 1987). Throughout history, former trade routes have been continuously influenced by external factors, and currently the Darfur conflict and technical innovations impact on these trade routes as well. The chapter concludes that several extraneous influences and impacts have stimulated the present-day trans-Saharan livestock trade, wherein the camel has been transformed from a pack animal, transporting commodities, into a trade good itself.

The information presented here is based on twenty months of ethnographic fieldwork conducted in north-eastern Chad and Libya between 2003 and 2007. This main research base was the border town of Tiné, located at the north-eastern corner of Chad on the border with Sudan. Data was collected through semi-structured and open-ended interviews at Tiné market with authorities, inhabitants and key persons. A further survey was done of the marketplace and its shops. Besides conducting research in Tiné, data was collected through interviews with randomly selected camel traders, middlemen and herders at camel markets in Abéché and Biltine and in north-eastern Chad with pastoralists, herders and camel breeders. Data was additionally collected during two trips with camel herders in 2003, covering a distance of about 300 kilometres between Abéché and Tiné, and during another trip in 2006 of about 1,000 kilometres between Tiné and Kufra, Libya, which aimed at gaining an understanding of the conduction of today's trans-Saharan camel trade, its organisation, management, logistics and risks.

Regional Background

The focus of this chapter is the area between north-eastern Chad and south-eastern Libya, geographically situated on the southern fringes of the Sahara and the sub-Saharan zone. South-east Libya is a vast desert, which is devoid of rain and subsequently devoid of vegetation apart from that found at larger oases such as Kufra and Jalu, where trade, agricultural projects and, in recent years, oil exploitation form the most important economic activities. North-eastern Chad has an arid to semi-arid environment as a result of highly unpredictable rainfall. The region of Dar Zaghawa lies within northern Chad, which is of specific interest. In Arabic, *dār* describes a house or homeland, and in this case defines the area traditionally inhabited by the semi-nomadic Zaghawa. Dar Zaghawa is divided by the international border between Chad and Sudan (see Figure 14.1).

Dar Zaghawa is situated on a plateau lying approximately 600 metres above sea level. It includes part of the Ennedi Highlands, which rise up to 1,450 metres (Harrir 1986: 3).[1] In Dar Zaghawa, the mean annual rainfall is between 250 and 300 millimetres. The rainy season begins in June and lasts until August. The higher parts of the Ennedi receive more rain on average, while north of the Ennedi, in the Sahara, rainfall is low and the area is nearly devoid of vegetation. Due to the relatively high amount of the seasonal rainfall, the Ennedi Highlands are dense with vegetation consisting of grassland, scrub and woodland. In general, rainfall is sporadic but consistent enough to support grassland pastures and intermittent rivers. Seasonal rainwater flows eastwards into the Wadi Tiné and Wadi Howar. Agriculture is confined to wadis where millet, tomatoes, onions, okra and melons are cultivated.

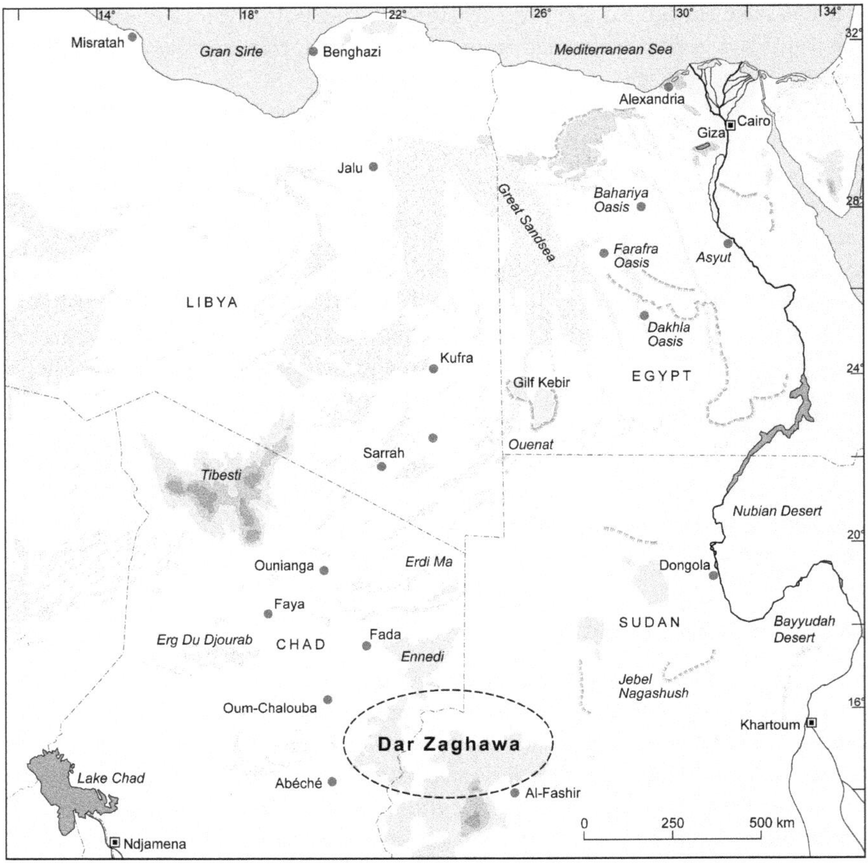

Figure 14.1: Map of Libya, Egypt, northern Chad and northern Sudan, showing Dar Zaghawa.

Years with minimal rainfall have a significant impact on the vegetation, grazing possibilities, and subsequently on the inhabitants and their economic activities. Over the last few decades, the population of the region has increased, with an annual population growth rate of 2.7 per cent (Ngadoum and Moriconi-Ebrard 2006: 26); livestock numbers have also increased. As a result of droughts in 1968, 1972/73 and 1984/85, and overgrazing, pasture and water have become scarce, which has forced the inhabitants to move to areas further south and to encroach into the homelands of other ethnic groups, resulting in ethnic tensions and violent conflicts (see Jánszky and Jungstand, this volume). The pressure on grazing land and water resources has also been an impetus for investment in camels, which are better adapted to the harsh environment than cattle and can make use of remoter grazing areas (Ibrahim 1984: 156).

The Historical Background of the Trans-Saharan Trade

In the region under discussion, trade between pastoralists and inhabitants of oases has long taken place, and long-distance trade routes have been existence for centuries in the broader region. Over time, various trans-Saharan trade routes arose between what are now Chad, Sudan and Libya. In the following description, the historical Wadai–Benghazi trade route is of particular interest, located between the former empires of Darfur and Kanem-Bornu (see Figure 14.2).[2]

Historically, the oases of the Sahara have been strategic points along trade routes and meeting places for the inhabitants of the fringes of the Sahara desert. Oases in south-eastern Libya, such as Kufra and Jalu, were used as transit points for caravans following the different trans-Saharan routes, coming from sub-Saharan regions such as the Wadai sultanate. The sultanate of Wadai, founded at the beginning of the seventeenth century by Abd al-Karim Ibn Djamé, was located in today's eastern Chad (Tubiana 1960: 49). The Wadai empire controlled a large

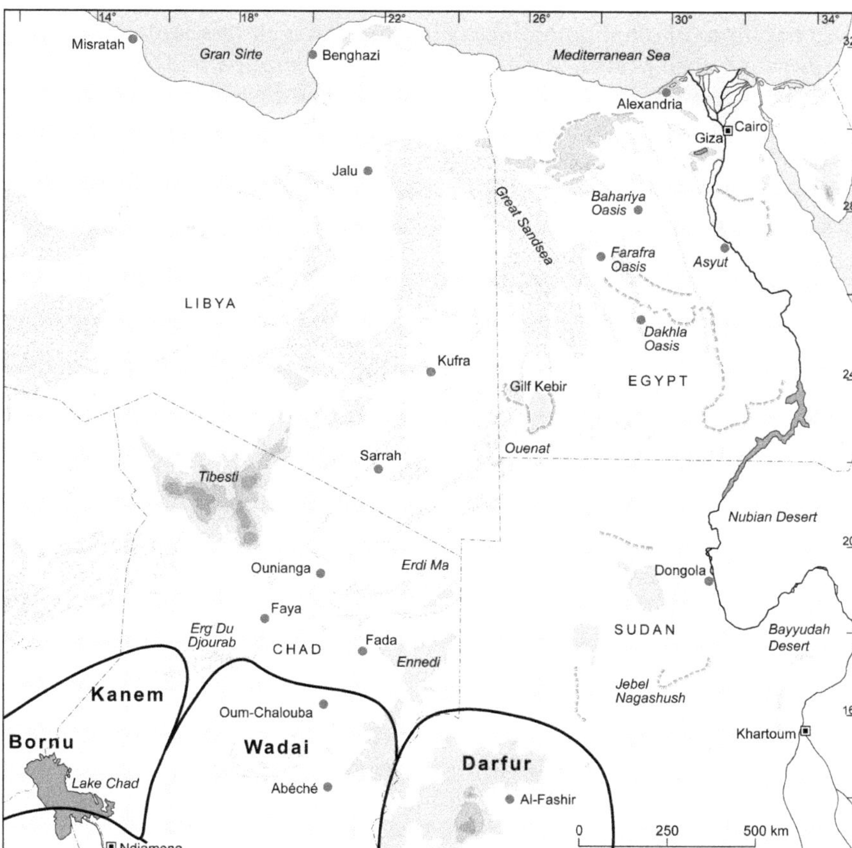

Figure 14.2: Map of the historical empires of Wadai, Darfur and Kanem-Bornu.

area but was positioned between the two more powerful empires of Darfur and Kanem-Bornu. The power and influence of the empires of Darfur and Kanem-Bornu meant they dominated trade in the region. As a result, the Wadai empire only had a few trade links. Its neighbouring empires profited by controlling the existing trans-Saharan trade routes to the north. The empire of Kanem-Bornu controlled the trade route west of Wadai, starting from Lake Chad and ending in Tripoli. Darfur, to the east of Wadai, controlled the trade on the *darb al-arbaʿīn*, Arabic for 'the forty-day route'. The *darb al-arbaʿīn* connected the Nile region with Darfur. It was Egypt's most important trade route into the African interior, due to the profitable slave trade in the eighteenth and nineteenth centuries. Before it became a slave-trade route, it was used by small traders and ethnic groups in Egypt and as well as by traders of the empire of Darfur. However, by the end of the seventeenth century, the slave trade between Darfur and Egypt was well established (O'Fahey 1978: 31). In the war between the empires of Darfur and Wadai from 1700 to 1750, many prisoners from Wadai were traded north as slaves and exchanged for goods. Under the Keira dynasty of Darfur, the slave trade was the most significant source of income, and raids (*ghazwa*) in the southern provinces such as Bahr al-Ghazal supplied the empire's traders with slaves. From 1790 to 1874, caravans arrived in Egypt almost every year and provided Egypt not only with slaves but also with ostrich feathers and ivory. Besides slaves, about four-fifths of the camels arriving with the caravans were sold in Egypt as well, since fewer goods were carried on return journeys and water carried for the slaves was not necessary. It is estimated that about 4,400 camels were sold in Egypt annually in the eighteenth century (Walz 1986: 34).

During the nineteenth century, the *darb al-arbaʿīn* remained the most important trade route located to the east of Wadai. Following the Anglo-Egyptian annexation of Darfur in 1874, the slave trade became illegal. Traders, offering slaves in Egypt's markets were prosecuted. This led to the near abandonment of the trade route. Only a few caravans with salt arrived, but the export boom from Darfur came to an end, and the trade route was completely abandoned by the end of the nineteenth century. Subsequent to the collapse of the trade route, the Darfur empire came to an end during the Mahdiya period.[3] Ali Dinar, the last sultan of the Fur, re-established the Darfur empire after the Mahdiya period ended in 1898. However, Ali Dinar could not revive the *darb al-arbaʿīn* trade route because trade between the Mediterranean and sub-Saharan regions had been diverted to the Wadai–Benghazi trade route under the control of the sultan of Wadai by that time (La Rue 1983: 656).

Trade links between Wadai and the Mediterranean began at in the beginning of the nineteenth century. The sultan of Wadai at the time, Abdel Karim Sabun Ibn Saleh (sultan from 1803 to 1813), placed great importance on participation in trade with the Mediterranean region after observing the flourishing economies of the neighbouring sultanates from their well-established trade in slaves, feathers, ivory and gold, exchanged for sugar, cloth, tea and weapons. Due to the neighbouring empires' profitable trade with the Arabic world, close relationships

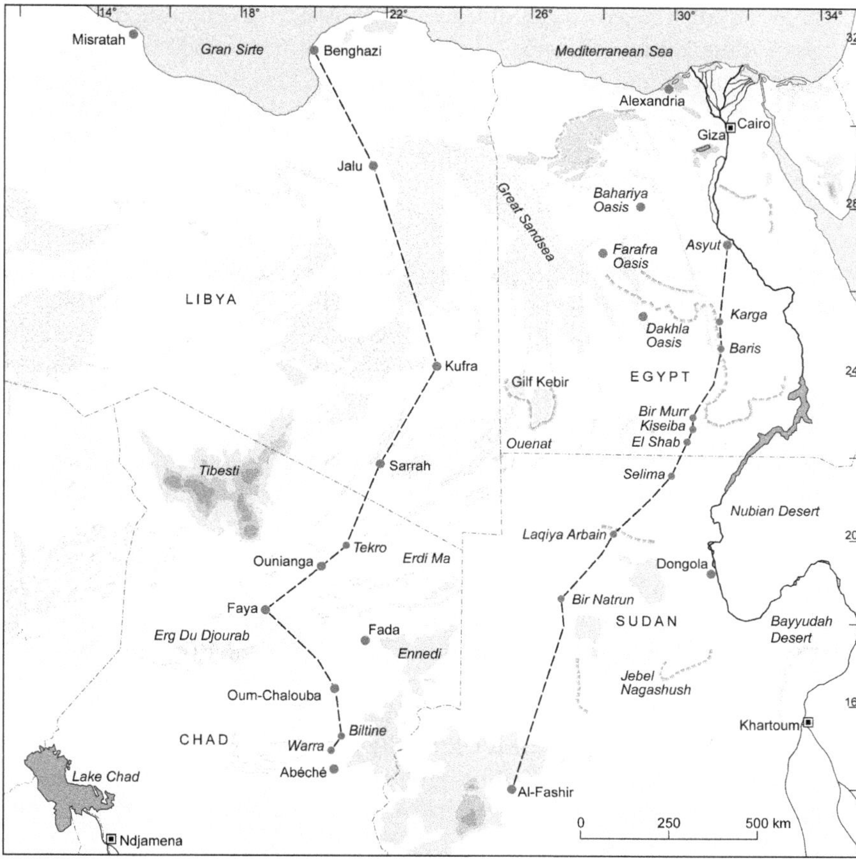

Figure 14.3: The *darb al-arba'in* and Wadai–Benghazi route.

had been established and weapons were received in exchange for slaves. The well-equipped army of the Darfur empire triumphed in rivalries with neighbouring empires and gained power over them (Doutoum 2006: 16). Around 1809, a trader from Jalu, located in north-eastern Libya, made his way to the sultan of Wadai in Warra, the capital of Wadai. This contact resulted in the establishment of a trade route between Wadai and Benghazi via the oases of Kufra and Jalu, situated in the Libyan Desert. The route was precarious and disrupted several times due to ethnic conflicts in the region. Some ethnic groups – such as Tubu, Goran and Bideyat (a Zaghawa sub-group), who inhabited the northern reaches of Wadai – were known for raiding caravans and made travel in the northern regions uncertain and dangerous. These groups were well acquainted with the area and controlled the water wells and routes. These disruptive elements made trade difficult for many years, consequently limiting the number of trade caravans using this route.

Yet in the late nineteenth century, after many years of disruption, the Wadai–Benghazi route became the most important long-distance trade route between the Mediterranean and the sub-Saharan regions of East Africa: 'between 1890 and 1900 an average of 200 to 300 camels entered Gialo [Jalu] from Kufra each week. This trade through Kufra continued until 1912' (Ziadeh 1958: 59). The extension and growing importance of this route was attributed to the abandonment of the *darb al-arba'īn* between Darfur and Egypt and other trade routes in Africa, as well as the control and influence of the Sanūsīya in the region. The Sanūsīya brotherhood was established in 1837 as a reformist movement seeking a return to the simple faith of early Islam (Vandewalle 2006: 18–20). The Muslim brotherhood, founded by Sayyid Muhammed ibn Ali al-Sanūsī (1787–1860) from Morocco, governed from their headquarters in the Cyrenaica region of the east coast of Libya from 1843 until 1896, when they moved the headquarters to Kufra (Wright 1989: 85). As a missionary order, it sought to reform the lives of the Bedouins and convert the non-Muslim peoples of the Sahara and Central Africa to Islam (Vandewalle 2006: 18–20). The son of the founder and chief of the brotherhood, Sidi Muhammed al-Mahdī bin Sayyid Muhammed al-Sanūsī (1845–1902) proclaimed that all inhabitants of the southern regions – including oases in Borkou[4] and Ennedi, the homelands of the Goran, Bideyat, Gaeda, the Arabs of Wadai and the Tuareg of the Ahoggar of southern Algeria – should swear loyalty to the brotherhood. By the twentieth century, the order was well-established among most of the inhabitants of the oases of the northern Libyan Desert and central Sahara. Sidi Muhammed al-Mahdī himself travelled to Wadai, Darfur and further areas within the Sudan to spread his doctrine. His plan was to pacify the different ethnic groups and to create a flourishing flow of trade goods. Most of the traders, who had an alliance with the Sanūsīya, did not have to pay taxes for passing through the land; they simply paid for protection for themselves and their caravans through the Sanūsīya leader. For this service, traders supplied the Sanūsīya leader with presents, such as tea, soap, candles and cloth. They also conducted trade in his name and exchanged ivory, ostrich feathers and slaves in Benghazi and Cairo.

In 1900, Muhammed al-Mahdī travelled to Wadai where he met the sultan of Wadai and made him the chief of the brotherhood in the region. He also attempted to unite the Libyan traders in Wadai against the French colonial power with the sultan's assistance. He supplied the sultan with tea, sugar and cloth, to which the sultan reciprocated with three horses as present to the Sanūsīya leader from Kufra (Djian 1991: 120). These trade links were established and flourished under the protection of the Sanūsīya. With the Sanūsīya's establishment of schools and religious centres for the dissemination of Islam, trade increased and became more profitable over the years. Their influence resulted in the settlement of many conflicts, which brought social and political stability to the region, making travel possible without fear of raids (Vikør 1995: 211).

The capital of the Wadai empire was moved from Warra to Abéché in 1850 due to a lack of water. Nimro, situated close to Warra, was the economic centre of Wadai. The main traders were the Jallaba, who had a strong influence on trade connections in Darfur and had settled at Nimro to take advantage of the trade opportunities there. They operated in the name of the sultan of Wadai and made use of his connection to the Sanūsīya. The Jallaba are known as an urbanised trading class of Sudanese Arabs from the Nile region who inhabited many areas of Sudan and neighbouring countries where they expanded their economic and political influence (Walz 1986: 72–74). After Abéché had become the capital of Wadai, the Jallaba began to establish markets there, increasing its importance as not only a political but also an economic centre. The Jallaba not only represented the trader elite and controlled trade, they also brought Arabic culture to the Wadai region, which persists up to today. As Ferrandi states:

> assis sur des nattes des hommes blancs et graves boivent lentement du thé dans des verres minuscules, un esclave noir surveille l'eau bouillante, le maître casse des pains de sucre à gros morceaux, tous les convives se livrent dans le recueillement à un des grands plaisirs de l'Afrique arabe: boire un sirop de thé, assis à terre, les jambes croisées et le dos appuyé à un coffre de bois peint.[5] (Ferrandi 1912: 361)

In the trans-Saharan trade between the Mediterranean and the Wadai sultanate, slave trading and the exchange of commodities, such as ostrich feathers, ivory, tea, sugar and a small amount livestock, was extremely profitable for many years. But with the fall of Wadai to the French in 1909 and the invasion of Libya by the Italians between 1912 and 1914, the Wadai–Benghazi trade route was abandoned. The colonial powers banned the slave trade, subjugated various ethnic groups and sultanates, destroyed traditional structures, established new borders and transformed the different autonomous empires into nation-states. Years of insecurity and uncertainty followed, and civil wars and ethnic conflicts erupted within and between countries. Some of these internal tensions and hostilities continue today.

In Wadai, the end of the trans-Saharan trade was followed by a drought between 1912 and 1914, which initiated widespread migration out of the Wadai region. Jallaba traders returned to their native homelands in the Nile area north of Khartoum, while trade in Wadai came under the control of the Hausa and other local African traders, but became restricted to a regional scale.[6] With the establishment of Fort Lamy (present-day N'Djaména) by the French colonial power in 1900 and their concentration on cotton exports, the eastern economic centre of Abéché lost its importance (Arditi 1993: 184). Trade in the west fell into the hands of the colonial power, while in the north, Libyan traders established trade links between the oases of Kufra and Faya, a town in northern Chad. Arabic merchants at that time dominated economic activities in the northern regions of Chad, commercialising cereals, dates, salt, cloth and tea, while local ethnic groups had difficulties in taking part in trade between the northern regions and Libya. When Chad gained independence in 1960, many of the Libyan traders left Chad with only a few remaining. The resulting trade vacuum was filled by members of the French and Chadian military (Arditi 2005: 24).[7] Eventually, with the withdrawal of the French military in 1965 and Libyan traders in the 1970s, commercial relations between Chad and Libya were interrupted several times (Arditi 1995: 851). During the Chadian civil war (1966 to 1978) and the Chadian–Libyan war (1978 to 1987), trade connections broke down completely. Conflicts, wars, crises and instability in the region made commerce between Chad and Libya impossible for many years.

Pastoralism in Eastern Chad

Since the ethnic groups of northern Chad had difficulty in taking part in long-distance trade interactions between Chad and Libya, raising livestock was one of the major economic activities in the sub-Saharan region and it became, and still is today, one of the main sources of livelihood for a large number of Chad's population. Camels, donkeys, goats and sheep, and in more southern regions cattle, graze the scrub and grasslands. Camels are especially valuable, because of their ability to survive under conditions of scarce resources. In the northern regions, the camel therefore became the prevailing livestock animal, while in southern regions cattle, sheep and goats dominate pastoralists' herds. The size of Chad's camel herds is difficult to estimate, since census data are almost non-existent. A marked decline occurred between the 1970s and 1980s, when armed conflicts restricted access to many areas. The number of 355,000 head given by Chapelle and based on data from 1970 (Chapelle 1986: 91) is certainly an underestimation of the camel population. A more realistic number is that of Clanet, who estimated 1,500,000 head (Clanet 2006: 34).[8]

The pastoralists' preoccupation with their herd, its maintenance and its prosperity dominates their daily lives. Livestock is an indispensable source for milk, which provides the core of the herders' nutritional needs in remote areas. In addition, livestock provides meat, wool and leather and serves as the basis for other material needs. Chapelle indicates that studies often focus only on the importance of livestock in relation to the meat industry, while the importance of livestock for the daily necessities of the population is often overlooked (Chapelle 1986: 88–89). In order to maintain their herds and to successfully undertake livestock husbandry, pastoralists are spatially mobile. Seasonal rainfall and the availability of water are the bases of choosing pastureland for grazing. Pastoralists' mobility is adjusted to the short humid period of the rainy season and the long dry period of the rest of the year. During the rainy season, the higher northern ranges of pasture are used. In some wet years, the northern *jizu* pastures bloom,[9] which is preferred by the northern camel breeders because livestock can graze for several months there without water and will give a lot of milk. During dry periods, the herds are moved south to the lower areas, where water is still available and the vegetation persists longer in wadi beds and other fertile areas.

Semi-nomadic groups of the northern regions of Chad, such as the Zaghawa, prefer to graze their livestock within their tribal homeland, while nomadic groups, such as the Awlad Rashid or Beni Halba, move north to the edge of the Ennedi Highlands to provide their herds with grass and water. After the rainy season, they return south, staying close to the wadis. The semi-nomadic groups also move southwards, but stay within their tribal homelands as long as grass is available. However, they will move beyond their homelands because of droughts, overgrazing or the intrusion of herders from other ethnic groups. Gaining access to water and grassland is often accomplished with violence. Certain ethnic groups have influential connections with those in political office, the army and the police, and this can prove advantageous for them when they move south and pass through the homelands of others. Tribal and family networks, and political connections, allow them to make use of wells, fields and grasslands for their herds with little interference (see Jánszky and Jungstand, this volume). Pastoralist livelihoods are thus not only dependent on environmental and geographic conditions but also on the ability to mobilise political connections and social networks.

Zaghawa Adaptations to a Changing Environment

Between 1968 and 1985, the sub-Saharan regions were severely affected by three major droughts: in 1968, 1972/73 and 1984/85, resulting in a significant reduction of livestock herds, especially of cattle. Most of the livestock either perished or was sold. This rush to market caused animal prices to plummet, while on the other hand prices for crops rose sharply as a consequence of the drought and reduced harvests. The imbalance between prices for millet and livestock resulted in a widespread famine in the whole sub-Saharan region (De Waal 1989: 137). The situation in Dar Zaghawa was also desperate, compelling the inhabitants to adapt. Since 1969, the inhabitants have had to evolve risk-minimising strategies to cope with changes resulting from drought and to secure their livelihoods and minimise their losses. As De Waal points out, because of the droughts:

> pasture [was] seriously degraded. Many wells [were] dry. From 1979 to 1984 Wadi Hawa flowed only once, and the *jizu* pastures never bloomed. The Zaghawa rejected the notion that their land was overpopulated or overstocked. 'The rain cheated us', [was] their complaint. Nevertheless during the fifteen years to 1984 the Zaghawa effectively adapted to a changing environment. When the famine of 1984/5 struck, they were well-prepared. (ibid.: 93–94)

The first significant adaptation to these drastic changes was the restocking of herds with camels and goats, which are better adapted to survive on the scarce resources of the region and thus more resistant to unpredictable rainfall than cattle and sheep. This was made possible by primarily using remittances from family members who had migrated. In the 1950s, Zaghawa herds were mainly composed of cattle. Speaking about this period, Tubiana stated that, 'the Zaghawa are chiefly breeders of cattle, sheep and goats; next come camels, whose usefulness lies in the fact that they provide good pack animals for long trade journeys'

(Tubiana and Tubiana 1977: 40–41). Even until the 1960s, the number of cattle was two to three times higher than that of camels within Dar Zaghawa, but in the 1970s the number of cattle began to decline. This corresponded to an increase in goats and sheep but also in the camel population. For the period between 1965 and 1975, Ibrahim stated that for the Sudanese part of Dar Zaghawa the number of cattle decreased by 66 per cent while camels increased by 370 per cent, sheep by 26 per cent and goats by 660 per cent (Ibrahim 1984: 156), a change which applies to the Chadian part of Dar Zaghawa as well.

Ali, a 51-year-old Zaghawa camel breeder and trader living in Chad, has been breeding camels for many years. He grew up as a herder by supervising his father's herds within Dar Zaghawa, and he later inherited three herds of camels, two herds of cattle and two flocks of sheep from his father. During the droughts of 1968, the 1970s and 1980s, which struck Dar Zaghawa badly, he lost all his cattle and sheep. However, instead of reinvesting in cattle and sheep, he decided to concentrate on camel husbandry exclusively. Today he is a successful owner of five camel herds of approximately 500 head and is sure that his decision to exclusively keep camels, made due to the changing environment, was the right one considering the last few years and the size of his current camel herd.

While cattle are vulnerable to drought, camels can live without water for a month and are able to consume scarce vegetation over wide distances. The increase in goat numbers is due to the fact that goats can be nourished with almost any kind of bush and tree, even the poisonous *Calotropis procera*, a bush which other livestock leave untouched. While goats are raised primarily for milk consumption, which in drought years is the primary nourishment, sheep are reared for market in order to fulfil monetary needs. The increase of livestock numbers has continued, despite its negative impact on the fragile environment. Even though the number of animals is incomplete or estimated, overstocking and overgrazing has obviously resulted in desertification (Ibrahim 1980: 123–27).

Through abundant remittances from family members abroad, Zaghawa individuals became prosperous and managed to benefit from changes caused by the droughts. Individuals were able to buy livestock at markets at modest prices, restocking their herds with camels and goats, resulting in large herds after the droughts. This allowed them to prosper from the boom in the following livestock trade. In contrast, less prosperous households that suffered large herd losses had difficulties in developing a new herd due to fact that they had few reproductively active animals left in their herds. Normally only male animals are sold, but in times of need, such as droughts, female animals are also sold, which significantly reduces the potential for recuperation after droughts (De Waal 1989: 160).

A second adaptation by the Zaghawa to the droughts was migration to urban areas in search of wage labour. This migration had already started after dry years in the mid 1940s and reached its peak in the 1970s and 1980s. In the 1980s, approximately two-thirds of the Zaghawa population left Dar Zaghawa, migrating to urban centres or to other Arabic countries in search of wage labour. This labour migration was fuelled simultaneously by the oil boom in neighbouring Arabic countries, which became the favoured destinations of the numerous Zaghawa who took advantage of these opportunities (Ruppert 1988b: 64). Wage labour abroad enabled numerous men to send remittances to family members in the drought-affected areas, which were often used to restock and restructure herds in order to ensure the livelihoods of those remaining in Dar Zaghawa. After several years of profitable work abroad, many Zaghawa returned to Chad and Sudan and settled in the capitals or other market towns. The majority used their savings to start commercial endeavours and establish businesses. The growth and expansion of these businesses eventually allowed them to control markets in Darfur, as well as dominating the markets in Omdurman, to the west of Khartoum (Ibrahim, Ruppert and Yath 1991: 94), and some years later markets in Chad. The result of these significant shifts in the economy of the region is that the Zaghawa made 'a remarkable transition … from an agro-pastoral economy, which has been subjected to recurrent drought disasters, to a diversified economy with trade as a supporting pillar' (Ibrahim 1997: 275).

For the new business, the Zaghawa relied on broad networks established by family members living in towns and villages outside Dar Zaghawa, such as Khartoum, N'Djaména and elsewhere. With the new concentration on trade, many Zaghawa gave up their semi-nomadic activities, but not necessarily their practice of livestock husbandry. Many Zaghawa continued raising livestock with the support of other family members or paid herders (Ruppert 1988a: 214). Some men invested directly in their own herds, which were herded under the supervision of close family members. A town-based elite thereby started to own and to dominate large parts of grazing herds in remote areas. This phenomenon, a new form of pastoralism, has been observed in other areas of Africa as well, such as Kenya, where Hogg has described it as 'new pastoralism' (Hogg 1986; see also Little 1985). The importance of livestock husbandry in the region declined from being the core component of Zaghawa livelihood and economy to a secondary component. Trade and business, meanwhile, became their main source of income, and involved broad family networks resulting from labour migration and resettlement in different trading towns and capitals. Close family members in other towns became important trading partners. Subsequently, close trading links were established between the urban centres of Chad and Sudan and the periphery of Dar Zaghawa, profitably serving rural and in turn urban needs.

Thus, over recent decades the Zaghawa reacted to a changing environment by altering their livelihood strategies and economic activities. As Ibrahim describes it: 'Since the early 1970s … the Zaghawa have made an economic and social breakthrough. They have managed to overcome their physical and economic situation. They have achieved this by diversifying their activities and particularly by going into the trade business' (Ibrahim 1997: 275). In the 1970s and 1980s Zaghawa traders concentrated their trade links mainly on the Sudanese part of Dar Zaghawa due to the political instability in Chad at the time, but with the coming to power of Idriss Déby in Chad in 1990, trade was re-established in Chad and across the border with Libya.

In Dar Zaghawa, pastoralism was the most significant economic activity and for many decades the core livelihood strategy. With the changing conditions of the last few decades, pastoralism gave way to trade, particularly livestock trade, as the main livelihood strategy. The capability of the Zaghawa to remain flexible and react by migrating and diversifying their livelihood strategies by concentrating on trade and investing in livestock husbandry seem to be the key reasons for their ability to maintain their livestock herds, although with a shift in the kinds of animals herded, and to sustain their livelihoods after the droughts of the 1970s and 1980s. Many Zaghawa succeeded in moving into the highly profitable trade economy, not only regionally but also through trans-Saharan trade interactions, which had been in the hands of Libyan and Jallaba traders before the 1960s (Abdul-Jalil 1988: 31).

The Re-emergence of Trade Links between Chad and Libya in the 1990s

Since the beginning of the 1990s, interactions and trade relations between Libya, particularly the oasis of Kufra, and the sub-Saharan region of Chad have re-emerged (Bennafla 2004: 91). The resumption of Libyan–Chadian relations began in 1994 with the decision by the International Court of Justice to declare the Aouzou Strip a part of Chad and to order the withdrawal of Libyan troops from Chadian territory (Bennafla 2000: 68).[10] The improvement in relations had wide-ranging effects on trade and trade routes, particularly in the northern regions where in 1996 70 per cent of commodities found in northern economic centres of Chad originated from Libya. The prices of Libyan commodities were lower than goods transported from Chad's capital N'Djaména (Arditi 2003: 189).

During the presidency of Hissène Habré (1982 to 1990) and even more so with the presidency of Idriss Déby (1990 to the present), corruption and currency counterfeiting became commonplace in Chad. As a result of the presidency of Idriss Déby, a Zaghawa himself, many Zaghawa today occupy numerous positions within the government and military. These positions of power are used to take part in and profit from unofficial and illegal trade practices over the borders with Sudan and Libya. Customs offices became monopolised by the leading political elites of the Zaghawa and Goran. Often taxes from custom payments are not paid into the national treasury but end up in the hands of private persons (Arditi 2003: 194). Cross-border trade also became a way to skip official tax payments and to gain profits by taking advantage of various currencies and diverse nationalities. In border areas, such as Dar Zaghawa, profit making through unofficial international trade is a common practice. This includes trade interactions to the north, via trans-Saharan trade routes to Libya.

The connection between the military and business is widely acknowledged and accepted, especially in corrupt countries and fringe areas (Roitman 1998: 315–16). Dar Zaghawa, located in the north-eastern corner of Chad, is only one example of this phenomenon. Profiting through other illegal practices is frequent in these areas, including stealing livestock and, more recently, banditry, particularly in regions with minimal government presence and control, such as Dar Zaghawa.[11]

Today's Trans-Saharan Livestock Trade

As shown above, the region of north-eastern Chad is rich in livestock, such as camels, goats and sheep and in more southerly regions cattle, which are important export commodities for the country. In contrast, the country is short of industrial goods, such as sugar, textiles, tea and other essentials. Libya on the other hand, with its Mediterranean ports, its numerous trade links and its position as a transit country, is able to supply sub-Saharan countries with many of these commodities. Due to the large amount of desert in the country, there is insufficient camel husbandry in Libya to sustain the consumption of camel meat. The ability of Chad and Libya to provide each other with needed essential commodities was the most important reason for the re-establishment of trade links between the two countries in the 1990s.

The present day trans-Saharan livestock trade represents the transformation of the historic Wadai–Benghazi trade route. From past to present, the types of commodities have changed, as has the mode of transportation. While the camel replaced the donkey as the mode of transportation in the first millennium BC, in the last few decades trucks have replaced the camel. The camel's importance in trade has shifted from a pack animal for carrying trade goods to a trade commodity itself. Between the sub-Saharan regions and Kufra, camels are now traded by the thousands each month. Large camel herds of up to 500 head are grazed in scrubland areas in order to supply families with daily needs. Camel husbandry is an important part of people's subsistence in the sub-Saharan regions of Chad, and it provides the Chadian livestock market with the required stock for trade to the north. Camels are generally only sold at market when daily life commodities are insufficient or when other monetary needs arise. Herewith the regional camel trade begins at the local markets, when pastoralists sell their camels because they need cash or other necessities.

Mahmoud, a 24-year-old camel herder of the nomadic Arab-Misria, made a six-day trip from Batha, south of Abéché, to the Abéché livestock market to sell one grown male camel. During a five-month period he came to the market two times to barter a camel in order to supply his family with essential needs. After selling his camel to a wholesale trader, he purchases primarily sugar and tea, dates and cloth. After two days he returns to his family by camel with the acquired goods.

The primary camel markets in Chad are located in Abéché, Tiné, Biltine and Kalait. One of the larger market centres is Abéché, the capital of Wadai since 1850.[12] The Abéché camel market is open daily throughout the year. Following the rainy season, when nomadic groups use the Wadai region to graze their herds, up to 400 camels a day can come to the market. In contrast, only a few dozen are sold during the dry season, when the nomads move with their camels to fertile wadis in more southerly regions. The camels sold at the market are primarily young males of four to five years, which are removed from the herd to prevent fighting and breeding with the mares. Older males that can no longer be used for transportation or riding might also be found at the market, sold as slaughter animals to camel traders. Females are rarely seen in the markets as they are the basis for maintaining the fertility of the herds.

Another camel market is located further north in Tiné, which is located on the Chad–Sudan border. The Tiné market is considered the main camel market in the north. It is the second biggest camel market after Abéché and the largest economic centre of the northern region. The camel market takes place three days a week, while other shops are open daily. In contrast to the market in Abéché, where most of the animals sold come from Arab nomads, the Tiné camels originate from semi-nomadic pastoralists of the more northern regions of Chad, such as Zaghawa and Goran. The remoteness of these areas causes many to travel long distances, often travelling ten to fifteen days to reach the marketplace in order to sell camels and purchase daily needs.

The traders in the Tiné and Abéché markets are primarily Zaghawa and secondly Goran. In Kufra, local oasis inhabitants of the Zuwaya tribe function as brokers and middlemen between traders from northern Libya and the traders from Chad. On market days in Chad, between two to twenty traders vie to buy camels from suppliers to expand their export herds. They barter for camels to assemble herds of approximately 60 to 250 camels to take to the market at Kufra, Libya. These camels are held close to the market centre by a few herders until a sufficient number is acquired. Some traders will combine their herds with others for the journey north, while more prosperous traders assemble their own herds. Once the herds are assembled and ready to be taken to Kufra, traders will contact a partner there to inform him about their upcoming arrival. The herd is sent on foot through the desert north to Kufra.

From Abéché two routes can be taken (see Figure 14.4). The first passes Tiné or Iriba, and heads north to Bao, the Mourdi Depression, the Erdi Ma region and then on to Kufra. Alternatively, from Tiné some groups choose a route east of the Ennedi Highlands via Qarab at-Tom. This route allows the herders to bypass Chadian authorities, something which they do because of lack of official papers such as proof of tax payments, or when attempting to smuggle female camels across the border. The other possibility from Abéché is the way via Biltine and Kalait, onwards to Archei, Demi and north through the desert to Kufra. The distance between Tiné and Kufra is approximately 1,000 kilometres and 1,300 kilometres for that between Abéché and Kufra.

The Organisation and Management of Trans-Saharan Trade

For the long-distance desert crossing to Kufra from Tiné or Abéché, the trader supplies the herders accompanying the camels with canisters of water, millet, dried tomatoes, dried okra, onions, tea, sugar and shoes. The trader himself stays at the market. A typical herd of approximately one hundred camels usu-

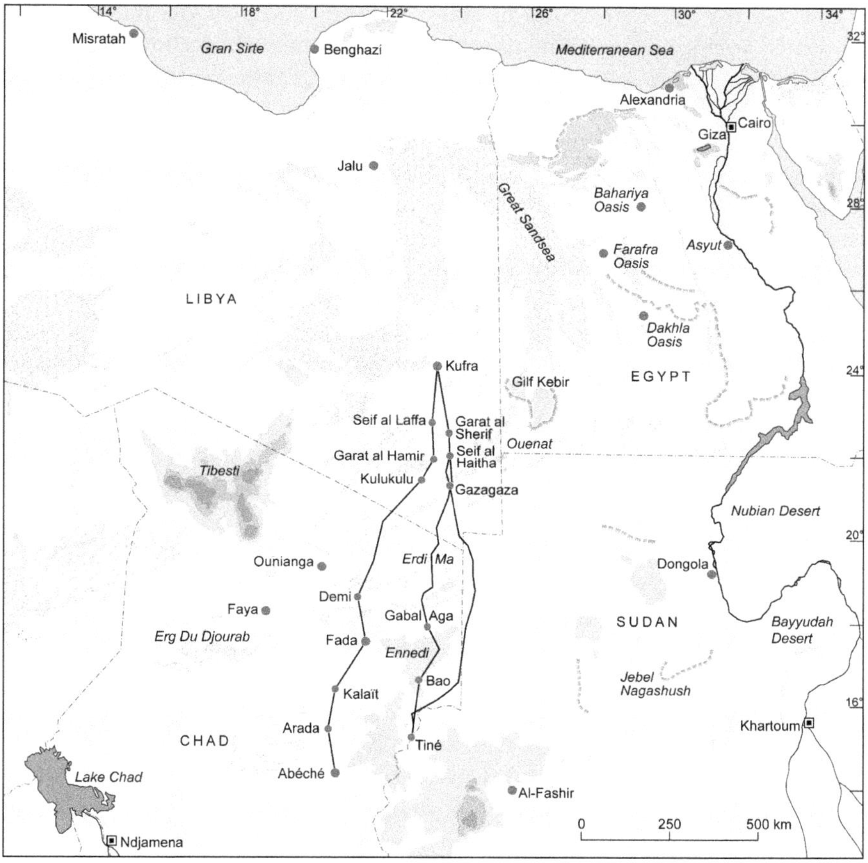

Figure 14.4: Routes taken by camel herders from Abéché and Tiné to Kufra.

ally requires four to five herders to accomplish the journey. The herds consist of mostly young males with a few grown camels, used for riding and transporting supplies. Despite laws forbidding the cross-border trading of female camels, some traders send mares as well. In instances where a trader assembles more then one herd to take to market, he will separate the camels and group the younger animals together with any mares, while the second group will be made up of the older males in order to keep them from being distracted by the mares.

In the past, the camels in caravans were tied to one another; today the camels are handled as a herd and allowed to walk freely, being driven forward by the herders in much the same way as sheep and cattle. The herders keep the animals in a close group to prevent individual camels from going astray. By keeping the camels in a close group, the herders take advantage of the camel being a gregarious animal that prefers to stay together in a group. However, the herders are aware of the downside as well: if one camel starts to run, the rest of the herd will follow and the herders may not be able to keep up and may lose the herd. The herdsmen attempt to maintain a speed that is slow enough to allow the herders following on foot to keep up, not so fast that the camels start to run, and fast enough to avoid individual camels from breaking away from the herd. The herders walk and run behind the camels to keep them together in a group as close as possible. Because of the extreme conditions, endurance, strength and determination are the basic characteristics required for herders to survive the harsh environment and to succeed in a desert crossing with livestock.

Each camel group is led by a *khabīr*, the guide of the group and usually a close relative of the trader. The *khabīr* is responsible for the well-being of the assemblage. He is the leader and determines the route, the speed and stops made by the group, the structure and organisation of the herd, the rationing of water, tea and sugar, and he has the final word in any decisions. His words and orders are obeyed without objection or question. He is chosen by the trader to lead the herd across the desert based on his knowledge of the region and the route. Surprisingly, the *khabīr* is often young, usually in his early twenties, and has accumulated most of his experience in trans-Saharan trade over several years under the leadership of an experienced *khabīr*. They typically start as a herder, driving camels through the desert. After multiple journeys over the years, they gain the necessary knowledge of crossing the desert and the organisation of trans-Saharan crossings. After participating in five to ten desert crossings and acquiring the essential skills, a herder can be promoted to a *khabīr* after a few years and will be able to guide a herd himself. Regardless of training and experience, there are some *khubarā* (plural of *khabīr*) who, due to lack of knowledge and/or ability, or because of circumstances beyond their control, lose their way in the desert and perish together with their herds. While caravan leaders once acquired knowledge from their fathers and ancestors, within the Zaghawa community today a *khabīr* learns the skills of survival in the desert from other family members or herders. Because of the instability in Chad prior to the 1990s and the lack of long distance trade, many fathers of today's *khubarā* did not work as *khabīr* themselves. Although they were experienced in raising camels, they did not have the skills and knowledge necessary to enable their sons to become a *khabīr*.

Rarely is a single camel herd driven north on its own. When the *khabīr* leaves the market centre with his camel herd, normally one or two other herds follow, up to two kilometres apart, with the herds walking one behind the other. This gives the opportunity to the *khubarā* to exchange information and decide on the route together. They can also exchange rations, such as sugar, tea or water, if one group runs short.

From Abéché, they travel slowly northward through the wastelands and grass pasture, along wadis, with vegetation composed mostly of bushes and acacia trees, until they reach the Ennedi Highlands. The slow pace allows the camels to graze and browse. In areas with dense vegetation, the herders may stop the camels for a short period in order to allow them to fill their stomachs. The slow speed also enables the camels to grab leaves whilst passing bushes and trees. Wadis and wells located on the way are used to supply the camels with water, as well as to refill the herder's water canisters.

A typical day of the route's first segment is characterised by nearly continuous walking from sunrise to sunset, which is generally from six in the morning to five or six in the evening. In the middle of the day, a break of about two hours is usually taken to prepare sweet tea, to cook millet with a sauce (made of dried tomatoes, okra and meat, and known as *asīda*) and to rest. At this time, the camels browse in the surroundings and rest as well. At dusk, the herders stop the camels to prepare the herd for the night. They fetter at least one leg of all the camels with an *'aqāl*, a cord made from old cloth, and a piece of wood. This limits the camels' mobility and prevents them from wandering off during the night. Some camels have both front legs tied up to prevent fighting. These are usually

Figure 14.5: Camel caravan on its way to Kufra.

Figure 14.6: Camel caravan on its way to Kufra.

Figure 14.7: Camel caravan on its way to Kufra.

big male camels, which were sold from herds with females where they may have been the leader or held a particular position within a herd's hierarchy. By combining these grown male camels from different herds into a new one, many of them will fight for dominance or leadership positions while travelling, especially during the night. Fighting can result in injuries to legs, the neck and hump from biting. On long desert trips, these injuries may weaken a camel and eventually cause it to die. Each camel represents capital for the trader and every lost camel is a loss of investment, so the *khabīr* and herdsmen make every effort to ensure each camel survives the journey. As such, the herders quickly identify the camels that are prone to fight and ensure these are tied up with both legs during the night. The same problems appear during the daytime, when the herders drive their camels north through the desert, the same camels continue their fights when walking in order to emphasise who is the strongest in the group. Similar complications occur when passing grazing herds consisting of female camels. The males become distracted by the mares and try to leave the herd. This is common in the sub-Saharan regions south of the Ennedi, where many camel herds are grazed and bred on green pastures after the rainy season. The herder's task becomes difficult in these areas by trying to keep sometimes up to 250 camels together, often having to chase them back together or run after strays. However, once they arrive in the desert areas, the camels usually cease fighting and stay close together to conserve energy because of the cold wind, lack of food and water, and exhaustion.

Upon reaching the Ennedi region, the *khabīr* increases the pace of the herd. Due to the lack of vegetation and scarcity of water, the camels are driven as fast as possible through the harsh environment devoid of vegetation. This accelerated speed increases the risk for the herders, who become tired faster and risk complete exhaustion. Based on the herders' strength and endurance, the camels are now driven as long as possible on continuous walks, often starting out before sunset at about five in the morning and continuing long into the night, sometimes after midnight. The herders typically average two to five hours of sleep a night over a period of about two weeks. The herd travels roughly 80 kilometres a day at an average speed of 5 kilometres per hour. This means roughly sixteen to seventeen hours of constant walking daily, with only a one hour midday break for tea and food. This continuous trekking throughout the day, even when alternating between walking and riding every two hours, drains the strength and endurance of the herders, slowly reducing them to complete exhaustion. In hot conditions, the camels are driven during the night, when temperatures are low and less energy is required for travel, and then rest in the middle of the day, when the heat is unbearable. In the winter, when temperatures at night can reach freezing point, the herders will take a break during the middle of the night as the cold makes walking impossible. At very low temperatures, the whole camel herd will refuse to move. A cold wind blowing from the front will stop the camels in the first row, causing them to attempt to turn their backs against the wind, and consequently the whole herd will stop. This prevents the herd from being driven forward, resulting in an inevitable stop for the night. At this point, the camels are rounded up close together to keep them warm by resting next to each other. With empty stomachs, the cold may harm the camels and will hinder them from standing up in the morning to continue the journey.

The herders make their sleeping place in the middle of the herd, using it as a windbreak to protect themselves from the cold and wind. A small fire is used to prepare tea and millet gruel. Due to the lack of firewood in the desert, fire is rarely used to warm the frozen body but reserved instead for cooking. Firewood is essential for food preparation and is gathered in the sub-Saharan zones and loaded onto the pack camels. Once out of sub-Saharan regions, little or no firewood can be found. A miscalculation of the required amount of firewood can be disastrous since the herders' food supplies need to be cooked before being consumed. The cold desert nights drain the herders' strength and sleep is nearly impossible. The herders' inappropriate equipment, such as thin blankets and clothing, provide little protection against the cold temperatures. A herder's income typically does not exceed €150 to €200 per trip, paid by the trader, while the *khabīr* receives a camel or about €300, which is used exclusively to provide for their families or to accumulate bridewealth items, but not for equipment.

Hassan, a 24-year-old Zaghawa and *khabīr* of the group and nephew of the trader, went about six times to Kufra before guiding camels to the Libyan market. When questioned about his earning, he answers that he will receive one camel from his uncle for every trip he undertakes to stock his own small camel herd, which he needs to do as he soon wants to get married to his cousin. For the bridewealth, he will receive further camels from his broader family and will not only use his own camels. His own small growing herd of about eighteen camels will be the foundation and capital basis for his future and newly established family.

In the desert zones, the camels walk side by side in rows of about twenty, with the others following behind. The camels at the front, which during the colder time of the year walk against the cold northerly winds, are usually in the second row the next day, due to the exhaustion from marching against the cold wind. The camels suffering from severe exhaustion, worn feet or sore wounds are always last in line and are continually pushed forward by the herders. Camels that refuse to continue walking or stand up, even after intensive beatings, are left behind in the desert to die. It is common for some camels on trips through the desert to be lost due to physical fatigue. The many carcasses along the route testify to the difficulties of the journey and the unforgiving nature of this harsh environment.

The duration of the journey from the markets in Chad to Kufra varies depending on the starting point, the route taken, the *khabīr*'s management abilities and many other factors. The journey from Tiné to Kufra takes approximately seventeen to twenty days, and starting in Abéché adds an additional five to six days. In former times it required up to seventy days, as Cordell describes: 'From Kufra caravans reached Wadai generally by the way of Ouinanga and Oum Chalouba, stopping only overnight. Travel time was between sixty to seventy days, and the journey ended at Arada or Nimro outside Abéché' (Cordell 1977: 26). The historical Wadai–Benghazi route passed to the west of the routes taken today: from Kalait westwards towards Faya and the Ounianga lakes, Tekro, Sarra, Buseima and then Kufra, using natural oases for water supplies on the way. The reduction in time is due to two main factors. The first is that the camels traded today do not carry any trade goods on the desert crossing, with the exception of the few camels within the herd used for riding and the transportation of the herders' possessions, water and food. Not having to carry burdens slows their exhaustion, requires less water and allows the camels to walk faster and on a more direct route through the desert. Secondly and more importantly is a new form of management resulting from technical innovations which allow the use of an even more direct route through the desert without need of natural oases.

Natural and Mobile Oases: External Influences through Technical Innovations

Traditionally, natural oases supplied caravans with water; however, today trucks loaded with water and food are used to sustain the herd during its desert crossing, essentially acting as mobile oases. Additionally, satellite phones have been used in the desert region since 2004 and enable *khubarā* to stay in contact during the trip with the traders at the home market, as well as with the trading centre in Kufra and its market prices. This constant contact with the traders facilitates water management for the camels. Traders in Chadian market centres and their counterparts in Kufra organise the supply and the delivery locations prior to the departure of the herd. Trucks loaded with water and straw leave Kufra in order to meet the camel herd in the desert approximately 350 kilometres south of Kufra. The meeting points are known landmarks, such as stretches of sand dunes or outlying mountains. Based on the size of the herd, up to five trucks will meet the herders and their livestock at the agreed location.

Usually, the feeding and watering of the exhausted camels takes approximately a whole day and night. This rest is necessary to allow the herders, and especially the camels, to regain their strength and continue the march north. The herders are also resupplied with water, firewood, provisions and batteries for their satellite phones and torches. The herders typically rest and enjoy a good night's sleep, as do the camels. The herd is then driven north for another five days before reaching Kufra, although at a much slower pace due to exhaustion. The water supply brought by truck shortens the route substantially, allowing the herders to drive the camels straight north through the desert rather then following the historic routes situated along natural oases. The use of trucks combined with satellite communications allows the desert to be crossed regardless of natural water supplies (Meerpohl 2007).

In 2006 some of the Libyan truck drivers were already familiar with Global Positioning System (GPS) navigation. Although GPS is still rarely used, some truck drivers began to mark the meeting points for the trucks and the camel herds into GPS systems in order to assist inexperienced drivers in locating them. Chadian *khubarā* still rely exclusively on their knowledge and experience in crossing the desert. Some have begun to use satellite phones, but the majority are still not in possession of these expensive appliances. Nevertheless, even relying on these technical innovations, the long distance trek to Kufra through the harsh environment of the Sahara tests the herder's and camels' strength and endurance. Sandstorms, poor visibility, cold wind and lack of water, food and knowledge weaken the entire group. Some camels become too weak to continue and are left behind to perish in the desert. Herders may also lose their way and die in the desert because of a *khabīr*'s lack of knowledge, a miscalculation of the water and food supplies needed, or loss of direction or disorientation.

Marketing in Kufra

Upon the herd's arrival in Kufra, it is stopped and inspected by Libyan authorities, which monitor arriving camel groups in the vast desert to ensure that import taxes are paid properly before entering the market. The assemblage is therefore brought to a designated waiting area to keep them from bypassing customs and to undergo a veterinary check before the camels can be brought to market. They will remain under the supervision of a veterinarian and Libyan customs for approximately five days, before the required official papers are provided. The most common maladies encountered at this time are worn feet, swollen ankles, raw skin and exhaustion. The camels are fed and watered at the Chadian trader's cost for a period of one to two weeks in the market to rest and recover. This allows the camels to regain the weight that was lost during the more than 1,000 kilometre trek through the desert. At the marketplace, camels are sold to Libyan traders, primarily from northern Libya. Traders from Kufra function as brokers and middlemen. Traders buy hundreds of camels at a time, with the younger ones being the most preferred. In contrast to earlier times, the march of the camels ends in Kufra; they do not walk the entire way to Jalu or Benghazi as the caravans once did coming from the former empire of Wadai. Once in Kufra, the majority of the camels are traded and transported to Misratah, the second biggest camel market in Libya, located on the Mediterranean coast. The camels are loaded onto trucks carrying twenty head at a time and transported by road 1,000 kilometres northwards to be sold to slaughterhouses. On average, traders make about €100 profit per camel at the market in Kufra. Through import and export agencies in Kufra, profits from camel sales are primarily reinvested in sugar, which is transported to Chad to supply the population's daily sugar needs. Once all the camels are sold, the herders accompany the trucks and trade goods back to Chad and return to their starting point.

The Darfur Conflict as an External Impact

As mentioned earlier, when it was in operation, the Wadai–Benghazi trade route was interrupted conflicts, wars and instability in the region. Bideyat and Goran in particular raided caravans and profited from banditry. Additionally, the Chadian civil war and the Libyan–Chadian war interrupted trade links between Libya and sub-Saharan regions. Over the years, instability and tension have been constant variables, forcing traders to adapt to the changing trade conditions. Presently, the Darfur conflict has the greatest and most wide-reaching impact on cross-border trade in the region. The most notable impact has been the closure of the border between Sudan and Libya and between Sudan and Chad, reducing the trade triangle that had existed between the three countries. A solution to this became necessary in order to maintain trade links and ensure people's livelihoods and economic survival. New trade networks were rapidly established to ensure that markets remained supplied with the essential goods for daily life. Before the conflict, close trade networks existed between markets in Sudan and Libya, including livestock trade. Before the conflict most camels sent to Kufra were from Sudanese markets, such as Mellit. This practice changed with the beginning of the Darfur conflict, when herds were raided and stolen by insurgents. Livestock trade between Darfur, Sudan and Egypt, which included large numbers of camels from Chad, was also abandoned. Due to instability in the region and border closures, camels were no longer brought from Darfur to Kufra. Pastoralists have left Darfur and have sought stability for themselves and their livestock in Chad or other areas within Sudan. The camel market in Tiné, because of its border location, lacked livestock and traders in the first years of the conflict, that is 2003/4. The Tiné markets were supplied with livestock from both Chad and

Sudan before the crisis, but with the border between the two countries closed, an influx of refugees from Darfur, and the resulting cessation of livestock traffic out of the area, the camel trade became particularly reliant upon the limited livestock raised in Chad. As a result, many traders withdrew from the markets in the early years of the conflict. They left for other market centres in Chad or Sudan in search of more profitable opportunities. Other trading networks, especially with areas in Sudan, broke down entirely.

Not to be deterred, three years after the beginning of the Darfur conflict, the inhabitants have adapted to the changes and the new circumstances. A significant portion of the camel trade was diverted into Chad because of the abandonment of trade links between Darfur and Libya as well as Darfur and Egypt. Before the conflict began, approximately 80 per cent of camels traded at Kufra market came from Sudan while at present about 95 per cent of the camels are delivered from Chad, a change that has resulted in only a small decline in total camel numbers sold at Kufra. While Sudanese commodities rarely appear in Chadian markets at present because of the conflict, Sudanese pastoralists graze their camels on Chadian pastures and continue to bring their camels to Chadian instead of Sudanese markets. As a result, livestock trade in Chad has boomed while retail business has declined, due to a lack of customers.

Conclusion

Between the Mediterranean Sea and the sub-Saharan zone, trans-Saharan trade remains an important activity for the inhabitants of the edges of the desert, enabling them to exchange commodities and sustain their livelihoods. The shortages of specific goods and products reinforced the necessity for long-distance trade routes in the past. While shortages still occur up to the present day – with sugar, tea, textiles and industrial goods in short supply in northern Chad and camels in Libya – the historic Wadai–Benghazi trade route and the system of trans-Saharan trade has been transformed in response to external and environmental pressures to ensure the supply of people's daily needs.

Over the last two hundred years, trade links between Chad and Libya have changed in many respects. In the late nineteenth century, due to the Sanūsīya, the Wadai–Benghazi route was the most important long-distance trade route between the Mediterranean and the sub-Saharan regions of East Africa. With the fall of the Sanūsīya, the rise of colonial powers, civil wars and international conflicts, the trans-Saharan trade routes were abandoned. However, after years of war and instability in Chad, resulting in many trade disruptions, trans-Saharan trade has re-emerged in a new form. Within the newly established trade links, the control of trade has changed as well. Historically, Jallaba traders from the Nile region, Wadaian traders under the authority of the sultan and traders linked with the Sanūsīya brotherhood controlled the trade. Control has today shifted to Zaghawa and Goran traders. Migration, stimulated by several droughts, was responsible for a remarkable transition of Zaghawa from semi-nomadic pastoralists to prosperous traders with wide ranging networks. Additionally, savings made through wage labour and trade were often invested in camel herds, resulting in absentee herd owners who concentrated on the herd's reproduction instead of subsistence. Subsequently, large camel herds have formed the basis of a flourishing trans-Saharan camel trade since the 1990s. This trade is a response to the large demand for camel meat in Libya, and profits from this trade have been used to buy industrial goods for resale in Chad. The position of Zaghawa and Goran in this trade has been facilitated by their fundamental knowledge of Chad's northern regions, but also by their ability to mobilise connections in the customs services, police and army. As a result, Zaghawa and Goran comprise the majority of the camel traders in Chad today.

Throughout the years, external pressures were a constant yet unpredictable component, continually influencing the long-distance trade routes. While in earlier years, existing trade routes such as the *darb al-arba'īn* and other more western trade routes were abandoned because of external causes, the Wadai-Benghazi route remained and extended, allowing trade from other areas to be diverted to this trade link. Later, instability in Chad had an impact on this route as well, causing interruptions leading to the near abandonment of this trade route. Presently, the Darfur conflict is the primary external pressure on trade routes in the region and the reason for the border closures and the abandonment of trade links between Sudan and Libya. Adjusting to the new circumstances, trans-Saharan trade to Libya expanded into Chad on a large scale.

While trans-Saharan trade was once almost impossible without the donkey or the camel as a mode of transportation, there have been several changes in the late twentieth century. Within the long-distance trade, the transportation of goods has been mostly taken over by motorised vehicles, with only regional trade continuing to make use of camels for carrying commodities. Along with this change in the means of transportation, trade goods have also undergone a transformation, specifically in regards to the camel's role in the trans-Saharan trade: the camel, now replaced by trucks, has lost its importance as a means of transportation and has instead become a trade good itself. Today, more than 10,000 camels annually are driven north to Kufra to be sold in markets for meat. The structure of these caravans has changed as well, with camels no longer being tied one behind the other as they once were in traditional caravans; nowadays, they are driven through the desert as a herd. Also, taking advantage of technical innovations, today's herders are able to use straighter, shorter routes to bring camels to market by using satellite phones to coordinate travel and to order trucks to meet the herd at agreed locations and that act as mobile oases loaded with food and water. This has reduced the travel time from sixty to seventy days in the past to twenty to twenty-five days today. To sum up, the trade routes used historically have been transformed, whereby camels, once used for transportation, have become the core trade good today, facilitating extensive trans-Saharan trade to re-emerge and expand across the region.

Notes

The fieldwork on which this chapter is based was carried out in north-eastern Chad and Libya between 2003 and 2007. The research was funded by the German Research Foundation (Deutsche Forschungsgesellschaft). The names of all Zaghawa persons mentioned in the text have been changed.

1. The Ennedi Highlands are a sandstone mountain range, surrounded by sand on all sides. The plateau is rich with flora and fauna. Deep valleys provide water throughout the year, and even species of Nile crocodiles are found.
2. Darfur is Arabic for 'Land of the Fur', referring to the Fur sultanate in western Sudan between the fourteenth and twentieth century. Founded by the Tunjur dynasty, it was at its height under the Keira dynasty between the seventeenth and twentieth centuries. The Keira dynasty ended in 1916 when the British annexed Darfur to Sudan (O'Fahey and Spaulding 1974). Kanem-Bornu was a kingdom of the Sef dynasty, probably founded around the mid nineteenth century, north-east of Lake Chad. The empire controlled the area until the nineteenth century. Its territory at various times included what is nowadays southern Chad, northern Cameroon, north-eastern Nigeria, eastern Niger and southern Libya. Because of its location it served as a point of contact in trade between North Africa, the Nile Valley and the sub-Saharan region (Taher 2006).
3. In 1881, a religious leader in Sudan named Muhammed Ibn Abdalla proclaimed himself the *Mahdī* ('the guided one') and started a war in order to unify the peoples of western and central Sudan. His followers took the name *Ansar* ('the followers'). Taking advantage of conditions resulting from Ottoman-Egyptian exploitation, the *Mahdī* led a nationalist revolt resulting in the fall of Khartoum in 1885 (Westphal 2002).
4. Borkou is a region of Central Africa, primarily located in northern Chad, forming part of the transitional zone between the Sahara and sub-Saharan zones. It is bounded to the north by the Tibesti Mountains, the south and east by the plains of Wadai and Darfur, and to the south-west by Lake Chad.
5. 'Sitting on mats, the white, grave men slowly drink their tea in tiny tea-glasses, a black slave keeps an eye on the boiling hot water, the head of household breaks the sugar loaf into big pieces, the guests enjoy one of the big pleasures of the Arab African world: drinking sweet tea, sitting cross-legged on the floor and leaning back against a painted wooden box.'
6. The Hausa are a Sahelian people, originating from Niger and Nigeria. During pilgrimages to Mecca, many Hausa settled down in other African countries, starting agriculture and trade.
7. Chad's northern region Borkou-Ennedi-Tibesti (BET) remained under French military control until 1965.
8. This figure is based on an official number of 1,169,803 for the year 2000, given by Chad's Ministère de l'Elevage. The ministry's official livestock data estimates an annual growth rate of 3 per cent for camels (RdT 2004: 27). This growth rate seems to be very low when compared to Lancaster and Lancaster (1990: 184), who give an annual growth rate of 35 per cent, but who on the other hand also leaves external factors such as drought, theft, loss, disease, infertility and stillbirths unconsidered.
9. The word *jizu* is a Sudanese term for ephemeral succulent vegetation, which occurs sporadically in the dry winter months in the Wadi Howar region (Wilson 1978: 327) and consists of plants such as *Indigofera berhautina, Neurada procumbens, Tribulus longipetallus, Cyperus conglomeratus Stipagrostis acutiflora* and *Fagonia bruguieri* (Baumann 2008: 18).
10. The Aouzou Strip is an area in northern Chad which lies along the border with Libya, extending south about 100 kilometres into Chad's Borkou-Ennedi-Tibesti region. A

dispute over the control of this area and its annexation by Libya in 1976 led to war between the two countries from 1978 to 1987.

11. International organisations, operating several camps for Darfur refugees in eastern Chad, have lost over forty vehicles within the last two years through banditry, mainly committed by Zaghawa and Chadian military members, who use them for private purposes, disposal or for military or rebel activities.
12. Chad's eastern region is named Wadai, after the former sultanate.

References

Abdul-Jalil, M.A. 1988. 'Some Political Aspects of Zaghawa Migration and Resettlement', in F.I. Ibrahim and H. Ruppert (eds), *Rural–Urban Migration and Identity Change: Case Studies from the Sudan*. Bayreuth: Bayreuther Geowissenschaftliche Arbeiten, pp.13–36.

Arditi, C. 1993. 'Commerce, Islam and Etat au Tchad (1900–1990)', in E.P.L. Gregoire (ed.), *Grand Commerçants d'Afrique de l'Ouest*. Paris: Karthala, pp.174–220.

——— 1995. 'Le Commerce de Dattes du Borku (Tchad)', *Cahiers des Sciences Humaines* 31(4): 849–82.

——— 2003. 'Le Tchad et le Monde Arabe: Essai d'Analyse de Relations Commerciales de la Période Précoloniale à Aujourd'hui', *Afrique Contemporaine* 207(3): 185–98.

——— 2005. 'Le Rôle Etrangers dans la Constitution d'une Classe de Commerçants au Tchad au XX. Siècle', *Bulletin Méga-Tchad* 2005(1): 21–26.

Bennafla, K. 2000. 'Tchad: l'Appel des Sirènes Arabo-islamique', *Autrepart* 15: 67–86.

——— 2004. 'La Réactivation des Echanges Transsahariens: l'Example Tchado-libyen', in W. Marfaing and S. Wippel (eds), *Les Relations Transsahariennes à l'Epoque Contemporaine: Un Espace en Constante Mutation*. Paris: Karthala, pp.89–110.

Baumann, K. 2008. *Intra- und Interspezifisches Verhalten von Equiden und Boviden in Gemeinschaftshaltung im Zoo Hannover*. Diplomarbeit im Fach Zoologie. Göttingen: Georg-August-Universität zu Göttingen.

Chapelle, J. 1986. *Le Peuple Tchadien: Ses Racines et Sa Vie Quotidienne*. Paris: L'Harmattan.

Clanet, J.C. 2006. 'Elevage et Pêche', in D.B. Yahmed (ed.), *Atlas du Tchad*. Paris: Les Editions J.A., pp.34–37.

Cordell, D.D. 1977. 'Eastern Libya, Wadai and the Sanusiya: A Tariqa and a Trade Route', *Journal of African History* 18: 21–36.

De Waal, A. 1989. *Famine that Kills. Darfur, Sudan*. Oxford: Oxford University Press.

Djian, G. 1991. 'Étude Sur les Senoussistes et Leur Action dans le Centrafricain', *Islam et Sociétés au Sud du Sahara* 5: 107–39.

Doutoum, M.A. 2006. 'Histoire', in D.B. Yahmed (ed.), *Atlas du Tchad*. Paris: Les Éditions J.A., pp.16–17.

Ferrandi, J. 1912. 'Abéché', *Renseignements Coloniaux* 10: 349–70.

Harrir, S. 1986. *The Politics of 'Numbers': Mediatory Leadership and the Political Process among the Beri 'Zaghawa' of the Sudan*. Bergen: University of Bergen, Department of Social Anthropology.

Hogg, R. 1986. 'The New Pastoralism: Poverty and Dependency in Northern Kenya', *Africa* 56: 319–33.

Ibrahim, F.H. 1980. Desertifikation in Nord-Darfur: Untersuchungen zur Gefährdung des Naturpotenzials durch nicht Aagepaßte Landnutzungsmethoden in der Sahelzone der Republik Sudan. Hamburg: Universität Hamburg.

——— 1984. *Ecological Imbalance in the Republic of the Sudan, with Reference to Desertification in Darfur*. Bayreuth: Bayreuther Verlagsgesellschaft.

——— 1997. 'Observation on the Mobility of the Zaghawa of Northern Darfur, Sudan', in F.H. Ibrahim an H. Ruppert (eds), *Coping with Resource Scarcity: Case Studies from Tazania and the Sudan.* Bayreuth: Bayreuther Geowissenschaftliche Arbeiten, pp.274–90.

Ibrahim, F.H., H. Ruppert and Y.A. Yath. 1991. 'Identitätswandel bei den Seminomadischen Migranten in der Hauptstadt des Sudan: Dargestellt am Beispiel der Zaghawa, der Dinka und der Nuer', in F. Scholz (ed.), *Nomaden, Mobile Tierhaltung: Zur Gegenwärtigen Lage von Nomaden und zu den Problemen und Chancen Mobiler Tierhaltung.* Berlin: Verlag Hans Schiler, pp.1–112.

La Rue, G.M. 1983. 'The Export Trade of Dar Fur, ca. 1785–1875', in G. Liesegang, H. Pasch and A. Jones (eds), *Figuring African Trade.* Berlin: Reimer, pp.635–68.

Lancaster, W., and F. Lancaster. 1990. 'Desert Devices: The Pastoral System of the Rwala Bedu', in J.G. Galaty and D.L. Johnson (eds), *The World of Pastoralism: Herding Systems in Comparative Perspective.* New York: Guilford, pp.177–94.

Little, P. 1985. 'Absentee Herd Owners and Part-time Pastoralists: The Political Economy of Resource Use in Northern Kenya', *Human Ecology* 13(2): 131–51.

Meerpohl, M. 2007. 'Mobile Oases in the Libyan Desert', in O. Bubenzer, A. Bolten and F. Darius (eds), *Atlas of Cultural and Environmental Change in Arid Africa.* Cologne: Heinrich-Barth-Institute, pp.138–41.

Ngadoum, R., and F. Moriconi-Ebrard. 2006. 'Population', in D.B. Yahme (ed.), *Atlas du Tchad.* Paris: Les Editions J.A., pp.26–29.

O'Fahey, R.S. 1978. 'Slavery and the Slave Trade in Dar Fur', *Journal of African History* 14(1): 29–42.

O'Fahey, R.S., and J.L. Spaulding. 1974. *Kingdoms of the Sudan.* London: Methuen.

RdT. 2004. 'Rapport Annual des Statistique'. N'Djaména: Ministère de l'Elevage de la République du Tchad.

Roitman, J. 1998. 'The Garrison-entrepot', *Cahiers d'Etudes Africaines* 38(2–4): 297–329.

Ruppert, H. 1988a. 'Konflikte bei der Wanderung von Nomaden in den Städtischen Raum: Das Beispiel der Wanderung der Zaghawa nach Omdurman und der Hadandawa nach Port Sudan', *Die Erde* 119: 211–17.

——— 1988b. 'The Migration of the Zaghawa to Khartum/Omdurman Conurbation', *Bayreuther Geowissenschaftliche Arbeiten* 11: 63–78.

Taher, M.A. 2006. 'Kanem, Creuset des Peoples: Cahiers d'Histoire', *A la Découverte du Tchad* (Centre Al-Mouna, N'Djaména) 2: 47–62.

Tubiana, M.-J. 1960. 'Un Document Inédit sur les Sultans du Waddāy', *Cahiers d'Etudes Africaines* 1: 49–112.

Tubiana, M.-J., and J. Tubiana. 1977. *The Zaghawa from an Ecological Perspective: Foodgathering, the Pastoral System, Tradition and Development of the Zaghawa of the Sudan and the Chad.* Rotterdam: Balkema.

Vandewalle, D. 2006. *A History of Modern Libya.* Cambridge: Cambridge University Press.

Vikør, K. 1995. *Sufi and Scholar on the Desert Edge.* London: Hurst.

Walz, T. 1986. *Trade between Egypt and Bilād as-Sūdān (1700–1820).* Cairo: Institut Français d'Archéologie Oriental du Caire.

Westphal, W. 2002. *Sturm über dem Nil: Der Mahdi Aufstand.* Cologne: Parkland.

Wilson, R.T. 1978. 'The "Gizu": Winter Grazing in the South Libyan Desert', *Journal of Arid Environments* 1: 327–44.

Wright, J. 1989. *Libya, Chad and the Central Sahara.* London: Hurst.

Ziadeh, N.A. 1958. *Sanusiyah: A Study of a Revivalist Movement in Islam.* Leiden: Brill.

Chapter 15
Pastoralism and Nature Conservation in Southern Africa

Susanne Berzborn and Martin Solich

Introduction

Pastoralism in Southern Africa has been massively affected by various approaches to nature conservation that in their manifold organisational forms have created new chances but mainly taken away options for the Nama of the South African Richtersveld, the herders of Lesotho, for pastoralists in northern and southern Namibia as well as in western Botswana. While the impact of livestock keeping on the Southern African environment is well documented (e.g., Shackleton, von Maltitz and Evans 1998; Jürgens, Gotzmann and Cowling 1999; Hoffman and Ashwell 2001), the influence of nature conservation on livestock keepers plays a minor role in scientific discourse. In this chapter we strive to fill this gap in research on livestock keepers and nature conservation in Southern Africa. Our focus is on 'official' contemporary state- or NGO-driven initiatives of nature protection and not on the conservation practices of pastoralists.

Environmental initiatives have become one of the most important forms of external interventions in Africa (Neumann 2000; Schroeder 2000). As current research demonstrates, the impact of political contexts and approaches on nature conservation is enormous (e.g., Singh and van Houtum 2002; Brosius 2004). A political ecology approach is often applied to analyse views on contemporary nature preservation, depicting it as a global programme of socio-ecological regulation (Bryant 1998; Stott and Sullivan 2000; Walker 2005; Biersack and Greenberg 2006). This perspective 'points us toward understanding how local patterns of land use are related to broader social, political and economic conditions, and how the environment serves as a locus for the enactment and perpetuation of patterns of inequality' (Brosius and Russell 2003: 51). It deals with the struggles among diverse actors over access to and control over natural resources (Wilshusen 2003: 42). This chapter will make use of the concepts of political ecology to place trends in conservation in the broader context of respective political agendas and to trace how powerful individuals and organisations shape conservation practices in Southern Africa.

The case of Southern Africa is remarkable because of the long-term European influence on environmental management as well as on the allocation of land. Also, the recent past of the South African Apartheid regime has had a massive impact on land issues beyond its borders, especially in Namibia. From the advent of colonialism, much pastoral rangeland in Africa was expropriated for wildlife conservation on the premise that pastoralists overgrazed and damaged the land (Homewood and Rodgers 1987). From the 1980s onwards, a new discourse emerged, and thinking shifted towards a more participative way of dealing with pastoralists in the various countries of Southern Africa. It was perceived that the integration of local knowledge into nature conservation practices and the creation of (economic) incentives to protect the environment would enhance prospects for the sustainable management of resources. It was acknowledged that Africans have for centuries shaped and maintained the allegedly 'wild' and pristine landscapes which were now deemed to be in need of protection. Recently, another trend in conservation has emerged, especially in Southern Africa, which challenges the participative approaches of the 1980s and early 1990s. Grand projects stretching beyond national boundaries are meant to improve game conservation and to increase tourism.

The chapter is structured chronologically, and after introducing the three main trends in conservation (colonialism and fortress conservation, participative approaches, transfrontier initiatives), case studies will illustrate the different phases of nature protection and their social and economic impact on livestock keepers. The contribution is based on our own empirical work in the Richtersveld, South Africa, and in the Lesotho highlands, supplemented by other case studies in Southern Africa. The historical and political similarities and differences of the Southern African states allow for a comparative approach and contribute to an understanding of the dynamics of pastoralism and nature conservation. We will examine the following questions: What are the general conditions? How did approaches to nature conservation change? Who are the main actors? What is the impact of conservation initiatives on pastoralist societies and ways of life? How do local people react to the imposed initiatives and to what degree were and are they able to influence the shape of protected areas?

Fortress Protection of African Wildlife

Western ideas of nature and its conservation were transported to Africa through the expansion of colonialism (Neumann 1998; Luig 1999). Not surprisingly, the first international convention on the conservation of African wildlife was signed in 1900 in London, the capital city of the main colonial power. Grove (1990) identifies three reasons for the attention paid by the colonial regimes to nature conservation. First, having control over natural resources for the needs of the colonial state and private capital. Second, protecting the colonial state's

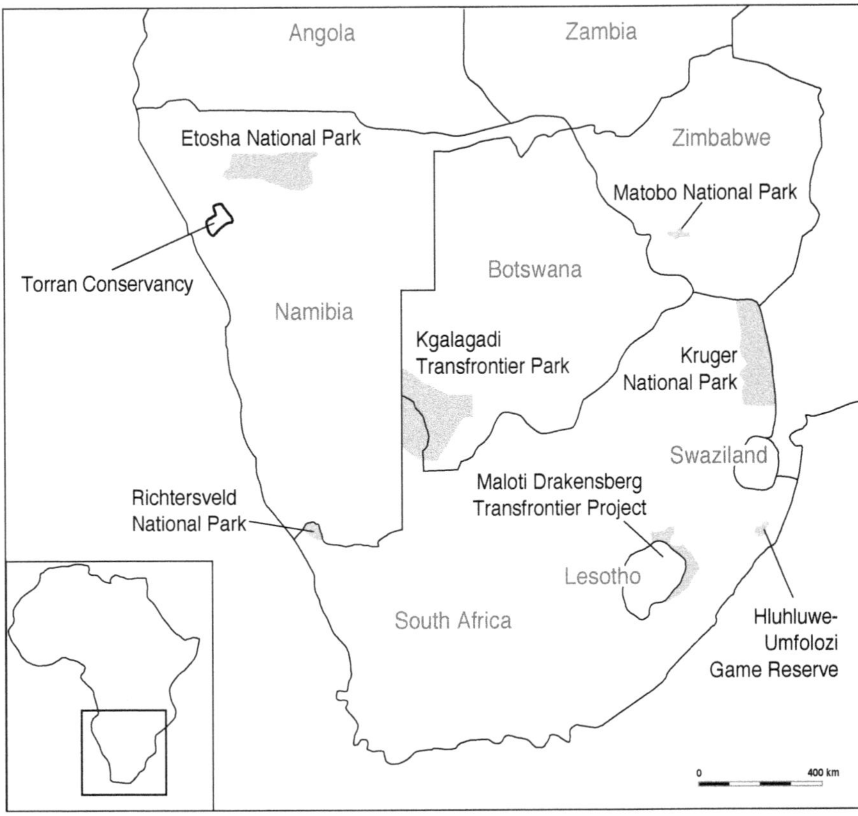

Figure 15.1: The location of the case studies in Southern Africa.

own economic interests by the early recognition and localisation of environmental degradation and climatic change. Third, maintaining aesthetic values. Most remarkable was the demarcation of wilderness areas, which were preferably free of people. What was conceptualised as the 'African wilderness', however, was mainly used as a hunting area for colonial officers and settlers where they could act out chauvinistic ideals of manhood by hunting large mammals and other spectacular species. The Africans were instead made responsible for degradation and nature destruction (Hulme and Murphree 2001; Borgerhoff Mulder and Coppolillo 2005). The division of land into areas adequate for the use of settlers, for intensive use by African farmers and wilderness areas was constantly enforced by colonial administrations.

Scientists supported the ideas of early conservationists. They saw the need for protecting non-arable land based on experiences gained in other colonies where severe degradation had taken place. British scientists saw forests as 'environmental catalysts' that could protect the surrounding lands (Grove 1990: 21). So it was that the colonial administration of Great Britain developed the first forest reserves and exported this nature conservation concept to its colonies as one way to fight degradation. In the Transvaal, Natal, Southern Rhodesia and Nyasaland, game reservation policies were enacted in 1858 and 1888 and justified with similar arguments as those for the establishment of forest reserves (ibid.: 37). In 1898, the first large protected area was established in Southern Africa – the Sabi Game Reserve, which later became the Kruger National Park. Other parks followed, and all these African parks adopted the North American model of Yellowstone National Park by ignoring traditional land use by local people. This mostly meant the exclusion of African farmers, pastoralists and hunters from the land where new reserves were established (ibid.). Other strategies to fight degradation were developed by European colonial administrations, which had a strong impact on the land use practices of pastoralists. In South Africa, for example, grass-burning prevention regulations were introduced (ibid.).

The Establishment of Protected Areas

With the advancing modernisation of society, parks were depicted as a peaceful and primitive retreat for white people, as 'nostalgic spaces for alienated urbanites' who were stressed by modern urban life (Brooks 2005: 223–24). Especially the Kruger Park was for the white population 'linked to sentimentality and nostalgia for a romantic and rural past which was endangered by urbanization and increased industrialization' (Carruthers 1995: 99). 'Wild' hunters and gatherers, but also 'native' mobile stock farmers, 'noble savages' all, could be accommodated in national parks as they were perceived to be part of 'nature', being picturesque, living in harmony with the environment and far from modern civilisation (Neumann 1998: 18, 125). But as soon as local people were too 'modern' and could not serve as an antithesis to the civilised world of the white population, they were no longer welcome in protected areas. Stock farming was considerably affected by nature conservation policy, and mobile pastoralists in particular were evicted as they were seen to have a detrimental effect on natural resources, especially on wildlife. They used seemingly empty, peripheral territories, which could be easily declared as protected areas by the authorities, such as Etosha National Park in the dry lands of central Namibia. Game reserves were preferably established on marginal lands not suitable for agriculture or, as in the case of South Africa, on segregated ethnic 'homelands' where the population was less able to resist land alienation (Beinart 1989). Both the colonial and the Apartheid regimes had in common that they concentrated their nature conservation efforts mostly on the protection of nature in areas where Africans were seen responsible for degradation and for threatening wildlife. The enormous impact of European settlers was deliberately ignored. They were, however, from a contemporary perspective, responsible for much of the decline in wildlife (Carruthers 1995: 16–19, 91; Luig 1999: 18). The establishment of protected areas threatened the livelihoods of pastoralists and farmers threefold: they no longer had access to their land, they could not supplement their diet with hunting, and their livestock was killed and crops were destroyed by wild animals.

An exemplary case of the period is that of the Hluhluwe and Umfolozi game reserves, Zululand, South Africa. Africans were first tolerated but later forcibly removed from their land under the pretext that their livestock impeded anti-disease campaigns. Proclaimed as game reserves in 1895, the colonial authorities would have liked to re-proclaim the area a national park, following the Kruger model (Brooks 2005: 223). However, this endeavour was hindered, among other things, by the fact that Zululand did not only have valuable big game like the white rhino, but also commercial farmers. The game hosted tsetse fly that transmitted *nagana*, a livestock disease, and sleeping sickness. There were hence conflicts between the preservation of wild animals for tourists and the farming of white settlers and African livestock keepers. In 1929, a strict *nagana* control programme was initiated in the course of which most of the wild game in the reserve was shot (ibid.: 226). The local rural people were first seen as being part of the valued 'primitive' nature. From 1939 onwards, however, fences were erected and the conservation area was consolidated through incorporating a 'corridor' between the Hluhluwe and Umfolozi reserves. This corridor was inhabited by Africans that were initially persuaded, or rather pushed, to leave, being finally forcibly removed in 1944. The reasoning behind the fencing as well as the removal of people was the necessity of game control in order to put the neighbouring commercial farmers at ease with the expanding conservation area. The African stock farmers in the corridor area were levied with a new livestock tax, as it was argued that the domestic stock in the native areas was the reason for the *nagana* problem. In 1942, all stock had to be removed from the corridor (ibid.: 232). According to oral histories recorded by Brooks, these removals affected hundreds of households and were depicted as a temporary measure and as part of the anti-*nagana* campaign. Therefore, many people agreed to move with their stock. The underlying reason was, however, that Africans and their livestock disrupted the image of a pristine Africa and a 'pure' space of nature. They continued to act as tourist guides and game guards but were no longer allowed to live in the area with their livestock (ibid.: 235).[1] This had devastating effects on the life of the people. They became impoverished, their social institutions were eroded and the resettlements led to more pressure on natural resources in the new area (Fabricius and de Wet 2002).

A similar experience affected the transhumant pastoralist Ndebele of Southern Rhodesia, who were blamed for fire setting. Their land was eventually annexed for conservation as the Matopos National Park, formally established in 1926, and for European farmers, who were, however, responsible for many fires themselves (Grove 1990). While the occupancy of the park by Africans was not seen as problematic in the beginning, from the late 1940s onwards they were blamed for damaging the ecological system through the use of fire. The park administration depopulated the protected area and allowed only a small number of Ndebele to reside in the park (Ranger 1999).

While pastoralists where evicted from many nature reserves, another solution to the perceived conflict between pastoral land use and conservation was followed in the Etosha game reserve in northern central Namibia. Etosha, which was established in 1907, included the territory of San as well as pastoralist Himba and Herero. The resident Hai||om were for almost fifty years considered a part of nature and allowed to reside, hunt and gather inside the conservation area (Dieckmann 2007: 186). A lot of them possessed some livestock. However, in the late 1940s, policies changed and the legislation became more rigorous. Hai||om were now allowed to hunt only two species of game and the permitted number of livestock was reduced (ibid.: 186–87). It was further decided that it was not worthwhile preserving the Hai||om as 'bushmen' anymore (ibid.: 189). In 1954, the majority of Hai||om were evicted from Etosha. Officially, the extermination of the game population by the 'bushmen' was the reason for this. Remarkably, the extermination of game by Hai||om had not beforehand been brought up as an argument. On the contrary, archival records over the years stated that meat consumption by Hai||om did not threaten the game population within the reserve. The need for a labour force on white-owned farms seemed to be a more substantial basis for the decision to evict them (ibid.: 190), even if the argument was framed in terms of nature conservation. In contrast to the usual procedure, the pastoralist Himba and Herero living in Etosha were not evicted. The 16,000 square kilometres of land used by them was cut out of the Etosha National Park and de-proclaimed in 1970 (Holmes 1992: 211). Nevertheless, this separation of land used by pastoralists from an originally larger protected area is an exception.

Even if the concept of national parks is rooted in colonial times, the heyday of the parks came in the 1970s and 1980s. The number of national parks increased significantly, which can also be ascribed to the protected area movement that gained popularity all over the world. Africa was always a focus of conservationists, and their activities were supported and popularised by the frequent presence of African spectacular species in the mass media. Thereby, the beauty of those landscapes was introduced to a wide audience in Western countries as wilderness (Adams and Hulme 2001). The conservation movement, supported by Western governments and donors, had the power to influence the new political elites in African countries to expand protected areas. However, in most of the cases, the establishment and expansion of protected areas still ignored the needs of the local people affected (Wells 1996).

Range Management Strategies to Fight Degradation

Efforts to change grazing practices as a means of fighting land degradation and desertification – already identified in colonial times as a problem with economic and social impacts – were, alongside the establishment of protected areas, a focus of conservations. Pastoralists were seen as responsible for overgrazing (Homewood and Rodgers 1987). In general, their communal land-use practices were regarded as threatening to the environment. The succession of droughts in Africa beginning in the 1970s convinced governments in Southern Africa to establish broader environmental programmes based on current theories of ecosystems (Niamir-Fuller and Turner 1999; Rohde et al. 2006). The concepts of range management and, later, natural resource management, developed in North America, played a strong role in Southern Africa. Referring to scientific theories of carrying capacity, international organisations and governments started to influence the grazing practices of pastoralists through such things as the Lesotho Range Management Programme. A similar programme was also introduced in Botswana with the Tribal Grazing Lands Policy in 1975, aiming at the privatisation of communal rangeland (Rohde et al. 2006; Taylor 2007). The following detailed case study from the Lesotho highlands will illustrate the approach to nature conservation during this phase and trace its impact on local livestock keepers.

The Basotho started to populate the highlands of Lesotho in the early nineteenth century under King Moshoeshoe I, the founder of the Basotho nation. The mountains, representing two-thirds of Lesotho's territory, today mark the eastern frontier of the country, which is surrounded by the Republic of South Africa. In 1870, the kingdom of Lesotho was integrated into the administration of the British Empire to provide protection for the Basotho against the Boers. The Basotho used the foothills of the Maloti Mountains as a sanctuary during the *lifaqane* wars and as pasture land for their big herds. Under the rule of Moshoeshoe, the highlands were reserved for the herds of the chiefs of the lowlands. Since the 1830s, population growth and outside pressure exerted by land loss to Boers and British settlers led to the slow but constant settlement of higher mountain valleys, which were previously inhabited by groups of San people. To this day upper areas in the mountains are used as summer grazing pastures where mostly young herders spend the warmer snow-free months with livestock. However, the number of livestock from the lowlands grazing in the mountains has decreased due to stock theft and land disputes. Nevertheless, chiefs situated far away from the pastures currently claim grazing rights in the highlands of Lesotho. Cattle as well as small stock (sheep and goats) have always played an economically important role in Lesotho. Colonial officers strongly supported the shift to market-oriented livestock rearing – mostly by the introduction of new species such as angora goats (see Quinlan 1995). Furthermore, a system called *mafisa* allowed the political elite of chiefs to bind followers through the loan of livestock, which shows the important social and symbolic function of livestock among the Basotho.

Since the reign of Moshoeshoe, Lesotho has been an important exporter of crops and meat. This changed when productivity started decreasing in the 1930s. Colonial officers saw Basotho land-use management as unsustainable and responsible for this decrease. New methods of crop production such as terracing were introduced but with questionable results. Additionally, overgrazing was seen to be responsible for rangeland degradation and therefore a reduction of livestock was supported by the improvement of formal markets along with the allocation of communal pastures to local villages as an element of stronger grazing control.

However, traditional concepts of grazing control, called *maboella*, were ignored by the British administration. Authors such as Meena Singh suggest that the changes in farming under the colonial regime are probably the cause for later productivity problems (Singh 2000; Showers 2005).

In the 1970s, rangeland degradation was still identified as one of the biggest problems of the newly independent state of Lesotho, and debates about carrying capacity had profound impacts on the pastoralists of the highlands of Lesotho. Lesotho was seen by Western scientists as significantly overstocked, and poor livestock management was attributed by experts from the United States Agency for International Development (USAID) to Basotho livestock-rearing practices. Following the tradition of the colonial administration, the government of independent Lesotho implemented further regulations for grazing control supported by USAID (Turner 2003).

The Range Management and Grazing Control Regulations of 1980 led to the Range Management Area Programme (RMAP), initiated in 1982, which forced livestock owners to apply for grazing permits to the Ministry of Agriculture. The Range Management Division was established to implement the RMAP. In some regions, livestock owners had also to organise themselves into Grazing Associations (GAs) which were officially responsible for the management of territorially defined Range Management Areas (RMAs). Today, GAs unite local livestock owners who pay obligatory fees to the association and in return have access to the land of the RMA connected to the specific association. A further aim of GAs is to improve the quality of their livestock by cooperation. Besides other common activities, such as animal health services, the actions of GAs concentrate on restricting grazing rights and establishing rotational grazing to fight the problem of overgrazing. The central Range Management Division provides range management officers especially trained to deal with grazing management. In local communities they are supported by locals, called 'range raiders', whose primary function is to establish if the grazing plans are being followed by pastoralists. 'Range raiders' are therefore mostly elected from among the membership of GAs. These imported concepts of range management, originally from North America, were strongly supported by USAID, which was responsible for the Sehlabathebe RMA pilot project implemented in 1982. The later established RMAs were based on the technical and informative findings of this project. The whole programme rests upon ecological concepts of equilibrium and carrying capacity models which dominated the discourse on degradation in the 1970s.

Nowadays a number of authors clearly think that this perspective may be dangerously simplistic (Behnke and Scoones 1993; Warren 1995; Vetter 2005). Some critics (Quinlan 1995; Chakela 1999; Turner 2003) list a number of further controversies connected to this kind of grazing control, which the Basotho people were induced to adopt. The exclusion of those who are not members of GAs and unclear definitions of rights and power over the grazing areas are leading to persistent disputes in the community. Originally, traditional authorities, such as chiefs and headmen, had the power to make decisions over issues of natural

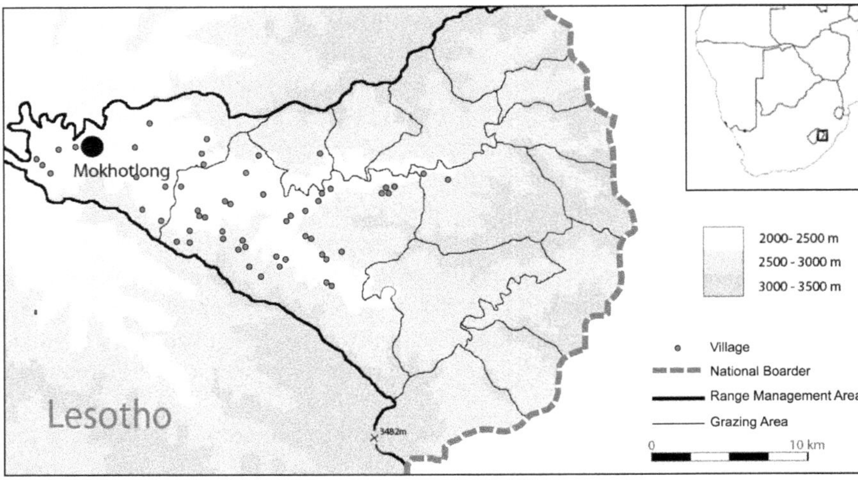

Figure 15.2: The Mokhotlong RMA as an example of a Range Management Area in eastern Lesotho.

resources, but now their role in natural resource management is no longer clearly defined. Furthermore, the high membership fees and the costs related to the required pasture rotation seem to favour farmers with higher numbers of livestock. In contrast, local and participative control through formally democratic GAs is being presented as a positive effect of establishing RMAs (Artz 1993).

In the context of the traditional concepts of *mafisa* and *maboella*, changes in range management have not only had impacts on the environment and livestock farming, traditional structures of authority were also confronted with a shift of power inside communities. Even now the division of authority is still unclear and causes conflicts. It is also important to realise that these externally introduced changes meant a shift from traditional pastoralist practices to market-oriented management. For example, the traditional systems of land distribution clearly changed. Land which was formally distributed by chiefs is increasingly inherited within families. Whether the establishment of RMAs and GAs has affected the environment significantly is not easy to answer. But the fact is that the government and NGOs still identify the very same problems of degradation and environmental destruction as they did previously (Quinlan 1993).

New Kinds of Conservation: Sustainability and Equity

In South Africa following the introduction of majority rule in 1994, and in Namibia following independence in 1990, hitherto existing conservation policies were challenged. As a result of political changes, the majority of people expected an improvement in their livelihoods. The redistribution of land as well as the achievement of secure tenure were high on the political agenda, and in South Africa there were many land claims pertaining to protected areas. The new governments were no longer able to appropriate land for conservation and exclude the local population from the politics and economics of conservation. Conservation – especially in the form of huge national parks – is still strongly associated with Apartheid among the black population in Namibia and South Africa, and with authoritarian rule in other countries (Wells 1996; Magome and Murombedzi 2003; Borgerhoff Mulder and Coppolillo 2005). On the other hand, South Africa and Namibia were at that time already signatories to a number of international treaties that demanded nature conservation, even though they had to struggle with the high maintenance costs of huge protected areas. However, the search for alternative ways of conservation also affected pastoralists in Southern Africa.

Table 15.1: Southern African signatories to international conventions on conservation.

International Conventions on Conservation	
African Convention on the Conservation of Nature and Natural Resources*	1968
Convention on Wetlands of International Importance Especially as Waterfowl Habitat**	1971
Convention for the Protection of the World Cultural and Natural Heritage	1972
Convention on International Trade in Endangered Species of Wild Fauna and Flora	1973
Convention on the Conservation of Migratory Species of Wild Animals***	1991
Convention on Biological Diversity	1993

* without Namibia, South Africa and Zimbabwe
** without Swaziland and Zimbabwe
*** only South Africa

One popular way to link conservation and rural development was the introduction of Community-based Natural Resource Management (CBNRM) in Namibia and Integrated Conservation and Development Projects (ICDP) in South Africa. In the 1980s, mainstream conservation discourse fundamentally changed. The new point of view was that protected areas should no longer be managed exclusively by the state but that local people should be included in conservation projects. The idea was that the integration of local knowledge into nature conservation practices and the creation of (economic) incentives to protect the environment enhance the prospects for sustainable management of resources. The involvement of the local community was based on a threefold argument: more community involvement in decision making processes would, first, decentralise power to the local level and contribute to the democratisation of rural societies and an allocation of rights of administration, and second, the indirect use (revenues from tourists, leasing, and so on) would create economic benefits for local communities. Both – increased participation in decision making and benefit sharing – would, third, increase local interest in and thereby contribute to conservation as well as to the enhancement of rural livelihoods. In CBNRM areas, local communities should manage the environment by themselves and generate monetary benefits through its commercialisation. The intension is that communities reconcile the consumption of natural resources with the conservation of biodiversity. While the collection of valuable wild plants and the sustainable farming of livestock are examples of the consumption of natural resources, ecotourism is seen as a positive way of commercialising areas of high biodiversity which leads to the conservation of natural resources (Hutton, Adams and Murombedzi 2005). While the land of pastoralists was massively reduced due to colonialism and the establishment of national parks, community-based conservation allowed for pastoralist land use and acted as a retreat for livestock owners.

The idea of community involvement in conservation is at odds with the protectionism of former times and reflects the shift towards democratisation in Africa in the 1990s. It became nearly impossible to ignore people in or close to protected areas (Borgerhoff Mulder and Coppolillo 2005). Community-based Conservation (CBC) became a popular concept, which was seen as an ideal way of promoting local participation, economic development and nature conservation at the same time.[2] It has to be stressed, however, that community conservation approaches do not replace earlier ones but are rather integrated with other forms of conservation.

The well known Zimbabwean Communal Areas Management Programme for Indigenous Resources (CAMPFIRE) that strived to improve the livelihood of rural people through the use of wildlife was a role model for many other community-based initiatives. In Botswana, a formal CBNRM programme started in 1992 (Taylor 2007). Community Management Areas (CMAs) were planned as multiple-use areas and a Natural Resource Management Project (NRMP) was designed. Communities were allocated the right to manage the resources within their area, and they had to establish legal entities and register as a trust, with most of the revenue returned to the communities. In Zambia, the Administrative Management Design for Game Management Areas (ADMADE) started in 1988 with assistance from the World Wide Fund For Nature (WWF) and USAID. It was an integrated wildlife conservation and community development programme aimed at preserving wildlife by giving local communities tangible benefits from wildlife management.

To involve local people was not a new idea. Already in the late 1960s, UNESCO's Man and the Biosphere Programme propounded the integration of human land-use practices in nature reserves. The importance of community involvement had also been mentioned during the World National Park Congress in 1982 in Bali. Another benchmark for the new view of conservation was the United Nations Conference on Environment and Development (UNCED), held in Rio de Janeiro in 1994, where changes in protected-area categories were made by the International Union for the Conservation of Nature and Natural Resources (IUCN) (Borgerhoff Mulder and Coppolillo 2005). In nearly all the categories a limited possibility of livestock rearing within protected areas was accepted:

'Indigenous and traditional communities inhabit large areas of land with low population density, and practise low-intensity activities like self-subsistence hunting and gathering, or even traditional, migratory pastoralism with limited impact on ecosystems and wildlife' (Bishop et al. 2004: 135).

In the context of community involvement, the concept 'sustainability' plays a central role. The concept underlies almost every recent approach to nature conservation. The most common definition, 'development that meets the needs of the present without compromising the ability of future generations to meet their own needs' (WCED 1987: 43), was developed by the World Commission on Environment and Development in the name of the United Nations. This independent commission was founded by the United Nations in 1983 to prepare a report concerning perspectives for worldwide development without depleting natural resources or harming the environment. Their report, commonly known as the Brundtland Report (WCED 1987), popularised the notion of sustainable development so successfully that it has since been taken up by almost every international institution, agency and NGO. USAID, for instance, initiated a huge and influential Natural Resources Management programme that aimed to 'promote the wise stewardship of natural resources and sustain them for future generations', and the 'conservation of natural resources while supporting the economic growth of communities who depend on these resources for their livelihoods'.[3] In Southern Africa alone, fifty-two projects were funded by USAID, and some of them are still ongoing. The thinking that conservation and development should be integrated found its way into the mainstream with strong support from the most important international conservation organisations and their manifesto 'Caring for the Earth' (IUCN 1991). The manifesto resulted in conservation being integrated into development projects.

Communal Conservancies in Namibia: The Case of Torra

In Namibia, new legislation in 1996 provided for the establishment of communal conservancies that bear the potential for the local population to benefit from wildlife-based enterprises. By allocating limited rights to wildlife and access to economic benefits derived from tourism, incentives were created to manage wildlife on a sustainable basis (Long and Jones 2004). The conditions for being able to register as a conservancy include forming an accountable, representative organisation with individual membership, a legally recognised constitution, clear-cut boundaries with neighbouring communities, and a land-management plan with multiple-use zones (Hohmann 2003). Many conservancies suffer from conflicts between pastoral land use or agriculture and wildlife as farmers lose stock to predators and crops are destroyed by wild animals (Murphy et al. 2004). Further, during the establishment of conservancies, the applicants have to cope with the many pitfalls that can undermine their efforts to register a conservancy, including the definition of membership and of boundaries, internal conflicts or uncertain land tenure (Jones 1999; Hohmann 2003; Sullivan 2003). Some of the conservancies are, however, economically very successful and have received non-financial benefits such as capacity building, recognition from neighbours and outside authorities or increased pride through control over resources and their own livelihoods (Jones 1999; Long 2004b).

The Torra conservancy in north-western Namibia was one of the first conservancies registered in 1998. It is situated in a wildlife-rich, semi-arid area characterised by gravel plains and rugged mountains and covers 3,520 square kilometres. It is sparsely populated (less than 1,000 inhabitants) and has an ethnically mixed community whose livelihoods are based on farming, mainly with small stock, small-scale gardening and limited wage labour (Long 2004a: 17). Some 87 per cent of households have access to livestock. To protect them from predators, owners usually herd their stock together in large herds (Suich 2003).

The success of the Torra conservancy applies to conservation as well as to economic development. A wildlife population that was almost wiped out has flourished and game censuses estimate unprecedented numbers of wildlife such as black rhino, elephants and predators (Nott, Davis and Roman 2004). In 2000, Torra became the first financially self-sustaining conservancy. Benefits derive mainly from lucrative tourism joint ventures (Long 2004a: 19). Considerable income is generated by a luxury lodge that is staffed entirely by local people, and by trophy hunting and sales of live game. The conservancy can pay out dividends to its members, who in addition benefit from jobs, game meat as well as from livestock-protection measures and preventive actions against human–wildlife conflicts. The livestock keepers are integrated and respected and did not lose out as in other conservation areas. The Torra conservancy has, for example, planned to allocate some of its income to stock farmers for veterinary drugs and to improve livestock health (Long and Jones 2004: 155).

While increasing wildlife numbers are beneficial for trophy hunting, they have negative effects on local stock farmers as conflicts escalate between humans/ livestock and wildlife. Stock losses and crop damage can be extensive and do not only cause financial loss but threaten the food security of local people and increase their vulnerability (Murphy et al. 2004). Torra therefore participates in a pilot compensation scheme that provides financial compensation for cases of stock loss from certain wildlife species and funeral costs in the event of the death of a conservancy member (IRDNC 2003). The conflict between humans/ livestock and wildlife is taken seriously, but there is still uncertainty about who is responsible and has the authority to deal with those conflicts – the conservancy or the Ministry of Environment and Tourism (MET). In times of drought, for instance, wildlife compete with livestock for grazing, and the community would like to reduce not only livestock numbers but also those of wildlife to secure grazing areas. However, they are only entitled to do so after the MET has approved a licence, which is a lengthy process. People are allowed to kill protected wildlife if human life or the lives of stock animals are actually being threatened, but not if wildlife are only damaging crops (Murphy et al. 2004: 113).

Another problem, with regards to stock farming, is access to (emergency) pasture by people living adjacent to a conservancy. In an area characterised by low and erratic rainfall such as the Torra conservancy, it is necessary for stock farmers to have access to grazing across conservancy boundaries. Formerly, traditional authorities regulated access to land, but nowadays a conservancy is a new stakeholder. It has, however, no power to exclude outsiders because it does not have full ownership rights over the land (Long and Jones 2004: 148–49). For instance, the Torra conservancy committee zoned land for wildlife and tourism use. When livestock farmers from outside did not adhere to the rules that regulated grazing access, the conservancy could not do anything about it. Therefore, Long and Jones (ibid.: 157) argue that there is a need for tenure security to protect access rights and make it possible for the conservancy to exclude outsiders, even though this exclusivity of land access might reduce the possibility of conservancy members grazing outside their own area in times of need.

In total, the Torra conservancy demonstrates that livestock keepers can be accommodated in conservation initiatives, and economic development can coexist with the preservation of natural resources. In 2004, this conservancy won the Equator Initiative Prize awarded by the United Nations Development Programme for outstanding community projects that reduce poverty through the sustainable use of biodiversity.

The Co-management of National Parks

There has also been a shift towards more community participation and involvement among state-led national parks, traditionally committed to the 'fortress' conservation ideal. ICDPs are implemented usually in connection with protected areas and address mostly the local people living around them, who are often pastoralists. Cooperation between the park management and local people should lead to a better understanding of the goals of nature conservation and to the development of income opportunities in similar ways to the case of CBNRM. These projects are the result of the recognition that protected areas fail if they exist in opposition to neighbouring communities (Wells 1996; Wilshusen et al. 2002). In South Africa, as in other countries around the world, co-managed or contractual national parks have emerged as new models of conservation areas in which responsibilities and benefits are shared between rural landowners and conservation agencies (Reid and Turner 2004: 223). In 1983 and 1986, the South African National Parks Act of 1976 was amended to allow national park expansion through contractual agreements. The main difference to conventional national parks is the fact that the land is owned by a local community or private and corporate landowners that decide to dedicate their land to conservation and therefore enter co-management arrangements and/or a contract with conservation authorities. This can have advantages for both sides: conservation authorities can extend protected areas into areas where it would not be possible to own the land. It further gives them the possibility of addressing land claims by people evicted from national parks without 'losing' the land for conservation purposes. For local people, the co-management of national parks might enhance their livelihoods in comparison with other land-use options and contribute to social and institutional empowerment (ibid.: 224–25).

In the following, we portray a community-based contractual park but want to point out that fully co-managed national parks remain in the minority. The Richtersveld National Park in South Africa, established in 1991, was the first communal contractual park. The whole park is co-managed by the local population and South African National Parks (SANParks) (Berzborn 2006: 360–65). Some local communities, such as the Makuleke who were removed with their stock from their lands in 1969 to make space for an expansion of the Kruger National Park, lately received the restitution of rights to land inside national parks under the South African land-reform programme. They developed slightly different models of co-management. In these cases, only a certain part of a national park is under co-management.

Richtersveld National Park

The Richtersveld is a communal area situated in the north-western corner of South Africa, immediately south of the !Gariep (Orange) River which forms the border with Namibia.[4] A former 'coloured reserve', Richtersveld is home to about 3,500 inhabitants who live in four villages. Many of them identify themselves as Nama, others as Bosluis Baster, and a minority as Xhosa. Pastoralism was and still is very prominent in the area. Even if, for most of today's population, stock farming is of relatively little significance with regard to income generation, it plays a major role in terms of enhancing the resilience of households through the diversification of the economic base (Berzborn 2007). Further, livestock have remained a very important element of the identity of the people. Usually, livestock are herded by a family member or a hired herder who lives on a stock post in the veld while the owners of the animals live in the village.

The semi-arid to arid environment of the Richtersveld is characterised by a high biodiversity and many endemic plants (Williamson 2000). Partly, the vegetation belongs to the Succulent Karoo (Jürgens 1991) which is one of the eight biodiversity hotspots in Africa as defined by Conservation International.[5] From the 1970s onwards, international scientists and the South African National Parks Board (NPB) began to think about creating a national park to conserve this biodiversity (Boonzaier 1996). Previously, the region was seen as a scenic but not very interesting mountainous desert. As described above, the dominant paradigm at the time was that human impact, in particular the communal grazing practices of pastoralists, had a damaging effect on natural resources. Hence, in 1989 a wilderness park was planned, which meant that local pastoralists would be kept out. With the help of external actors, the plan to dispossess the local population of grazing rights and exclude them from decisions relating to the planned national park area was stopped. After tenacious negotiations, the first South African National Park

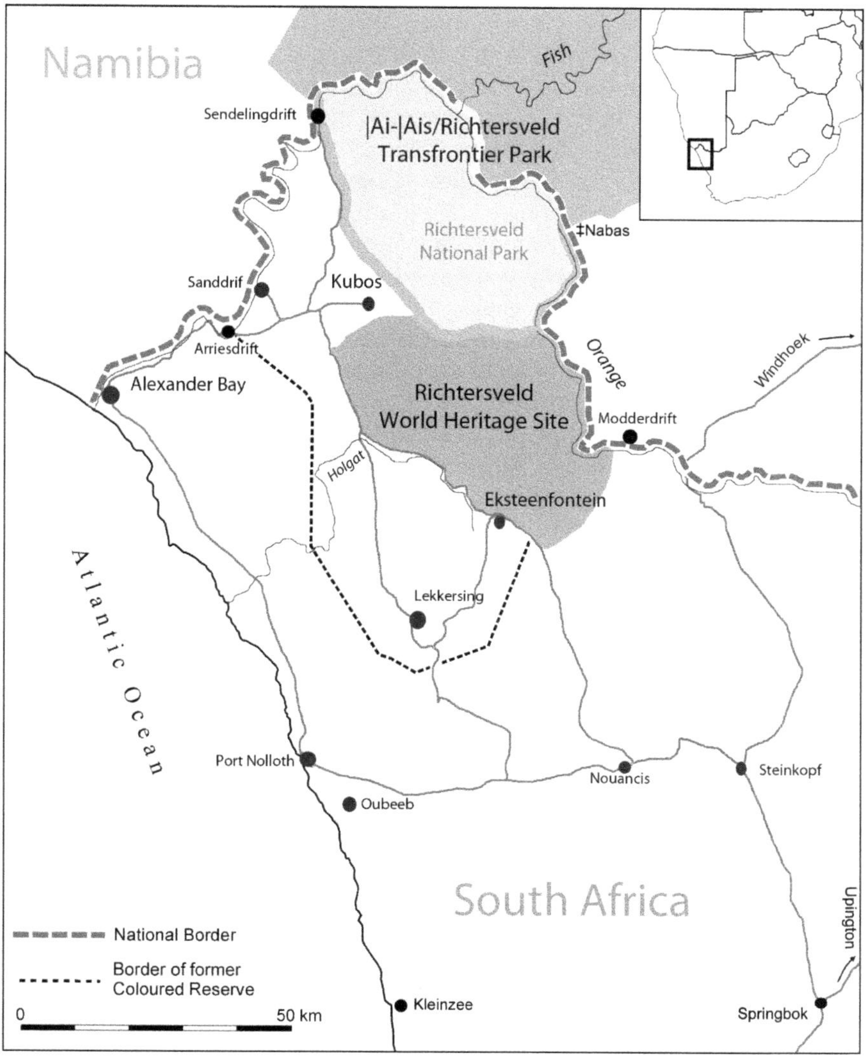

Figure 15.3: Richtersveld National Park, Transfrontier Park and the Richtersveld World Heritage Site.

under communal co-management was established in 1991 and covered some 1,620 square kilometres. The local community remained owners of the land and leased the park area to NPB (today SANParks).[6] Income from the park is paid into a community trust for development and community projects. The four villages and stock farmers are represented on the management board of the park, and stock farming as well as the collection of such things as medicinal plants is allowed inside the park under certain conditions.[7] Stock numbers inside the park are limited to 6,600 small-stock units, and only those stock farmers that grazed their stock in the national park area previously are allowed. Besides the rent received for the land and a limited number of jobs in the national park, the community can profit from ecotourism options and intangible benefits such as increasing pride in Nama culture, a revival of the Nama language, and projects involving arts, crafts and oral history that were initiated in conjunction with the national park. It has to be stressed, however, that the benefits for local people result from the fact that they have not been deprived of their grazing lands rather than from financial gains generated by the park. Opportunities for generating income from ecotourism are fewer than in the Makuleke region of the Kruger National Park.

During negotiations, local people and their supporters argued that the traditional land-use practices of the semi-nomadic Nama pastoralists shaped the landscape over centuries as a result of their sustainable transhumant grazing practices. NGOs and the (inter)national scientists linked to them were instrumental in establishing this picture of the sustainable traditional land-use practices of the resident Nama (Bollig and Berzborn 2004). Ethnographic research was conducted to assess Nama herding practices and it was found that there are elaborate rules for herd management (Archer, Boonzaier and Smith 1989). Archaeological studies also showed that pastoralists have lived in the area for 2,000 years (Webley 1990, 1997), and botanists established that traditional herding is not destroying the environment.

In the case of Richtersveld National Park (RNP), the move towards increased participation was pushed by a bottom-up initiative and cannot be regarded as a voluntary decision by the conservation agency. But the spirit of the late 1980s made it possible for the local community to successfully resist the alienation of their land. On the one hand, local people could take advantage of the paradigmatic shift in the assessment of traditional herding practices and their environmental consequences, whereby indigenous knowledge was valued and traditional grazing practices no longer demonised. This shift has to be interpreted as another boost to Nama identity. While being Nama was suppressed and had a negative connotation during Apartheid, it now provided a great opportunity. On the other hand, the success of the Richtersveld people must be seen in the context of the transformations that have been taking place in South Africa since 1990. There were signs that the end of the Apartheid era was on its way and that the disenfranchisement of the black population would come to an end. This political environment, together with changes in conservation thinking and external support, facilitated the empowerment of the local population. Those directly affected by the establishment of a national park could now play a significant role in shaping the form of conservation in their area.

Despite the success of RNP, there are many critical voices at the local level as well as at the level of SANParks. The main bone of contention among the local community is that their representatives lack real power on the management board and that decisions are not made in a participatory manner. In addition, with regards to job creation, the tangible benefits for the community amount to less than was promised. Some of the stock farmers that did not qualify as park farmers complain that they have lost access to occasional grazing and that there are now two classes of stock farmers. For SANParks, the community is often perceived as obstructive. A further complaint is that there is insufficient grazing management. Trust in productive cooperation was also undermined by the fact that the management plan was only signed in 2002, more than eleven years after the establishment of the park, as it was difficult to come to an agreement on the exact modalities of stock farming.

The Richtersveld Communal Conservancy and World Heritage Site

The idea of establishing a communal conservancy in the southern Richtersveld evolved as a result of discontent with developments in the RNP (see Figure 15.3). Supported by EcoAfrica, an NGO that has been working for many years in the Richtersveld, a conservation area was planned from 1998 onwards in which the local community would have the power to actively shape the preservation of nature. This initiative was accepted as one project to be included in the Integrated Development Planning (IDP) in 2000.[8] In particular, people saw the need to control tourists who drive off-road, damage petroglyphs and illegally collect plants. In addition, the self-management of natural resources was desired as well as capacity building. Another concern was the development of ecotourism as a viable and sustainable alternative to other land-use practices in a form that allows the local community to directly benefit from profits. After a three-year participatory process, a conservancy management plan was completed. Representatives from very different groups and institutions in the Richtersveld formed the management committee.

The communal conservancy reached legal status when the Richtersveld Cultural and Botanical Landscape was declared a World Heritage Site (WHS) in June 2007. Traditional pastoralism features highly in this WHS. In the application (Thornton et al. 2006), as well as in the South African government notice of the WHS (RSA 2007), the communal grazing practices of the Nama are depicted as being in harmonious interaction with nature and not having a negative impact on the highly valued natural resources. The documents read almost as a return to the 'noble savage' image and stress a traditional lifestyle and continuity over more than 2,000 years.[9] The application further justifies the qualification for WHS status by indicating that there are no pastoral sites in Africa with WHS status and that pastoralists are currently only represented in one WHS by the Laponian area of Sweden.[10]

The declaration of the WHS led to divided reactions. On the one hand, it is celebrated as a great success for a local community that wants to manage their own resources on a sustainable basis and strives to protect the environment. It is further perceived as a boost with regards to tourism, access to funding and job creation. Hence, the declaration of part of the Richtersveld as a WHS provides hope for the economic development of the whole community. On the other hand, there are concerns that the process seems to have been pushed through too fast by donors as well as by members of the conservancy committee. Additionally, there are a number of other concerns. The first concerns the borders of the area, which has a large buffer zone where land use will be restricted. The second hinges on the fact that, while part of the Richtersveld has been declared a WHS, and therefore falls under the authority of the provincial government, land ownership of the larger Richtersveld area has not been clarified. There is a land claim still pending under the Transformation of Certain Rural Areas Act (1998), and therefore a final decision over land ownership has not yet been accomplished. This could lead to a conflict, however, between having full authority over the land and having WHS status. At the moment, the government assigns the management of the WHS to the community, but the community will always be dependent on the goodwill of the government with regards to authority over the land. The third concern is the fact pastoralist land-use and a pastoralist lifestyle are depicted in very traditional terms in the WHS documents. One of the main reasons for obtaining WHS status was the traditional and sustainable pastoralist land-use in the area. But the way that pastoralism was portrayed does not always correspond with the reality of contemporary life. For example, great emphasis was placed on traditional, portable shelters (*matjies huise, ||haru oms*) made of reed mats. But there is virtually no stock post at the moment that uses just reed mats because they are often replaced by hessian, plastic or other materials. In addition, the impermanence of the structure is stressed in the application as well as in the official government proclamation of the WHS. There are, however, quite a few farmers who nowadays prefer to live in caravans or corrugated-iron houses at their stock posts and do not change them as often as was traditionally done. It is thus feasible that they might be forced to change their livestock management practice to make it more traditional than it is today.

Large-scale Projects and Transboundary Protected Areas

The CBNRM-related euphoria of the late 1980s and 1990s declined towards the end of the century (see, e.g., Oates 1999; Terborgh 1999). The community-involvement approach has been criticised in various quarters for having unrealistic assumptions about being able to improve the sustainability of livelihoods (Patel 1998). Further, CBNRM programmes have been perceived as failing to create representative bodies for the communities involved (Sullivan 2002; see also Taylor 2007), and the success of CBNRM initiatives with regard to conservation has come under attack by researchers as well (e.g., Kangwana 2001). CBNRM projects are increasingly perceived to fail; hence, the main goal of CBNRM – the integration of conservation and sustainable rural development – is sometimes not achieved, and calls to return to 'fortress' conservation have become louder. Still present is the discussion about who owns 'nature', the state or local communities (Wilshusen et al. 2002).

Following criticism and growing scepticism with regards to community-based conservation, another approach became fashionable in the 1990s: transboundary conservation. Transboundary conservation projects have gained popularity amongst conservationists, governments and donors. Supporters refer to the concept of ecosystems and the conclusion that human-made national frontiers ignore the existence of natural ecosystems. They see a need for transnational cooperation in conservation, with a focus on identified regional ecosystems in borderlands (Westing 1993; van der Linde et al. 2001; Hanks 2003; Swatuk 2005). These projects differ in form and scale and are labelled in many different ways. The term Transboundary Natural Resource Management (TBNRM) covers all kinds of initiatives. The most popular form of TBNRM are Transfrontier Conservation Areas (TFCAs), which cover a defined territory and contain mostly one or more protected areas close to international borders, supplemented by private and/or communal lands (Griffin et al. 1999).

Scientists have been talking about the need for transboundary conservation for some time, but until the 1990s it had not played a big role in Southern Africa (Magome and Murombedzi 2003). During the last few years, TBNRM has been welcomed as a universal solution to environmental, social and economic problems. The first aim of this conservation strategy is to protect cross-border eco-regions and enable the free migration of large mammals such as elephants. Second, non-governmental and governmental organisations perceive TFCA projects as conducive to economic development as they are seen to attract lots of tourists, and thereby generate job opportunities for the local population. Third, there is the hope that the management of TFCAs will decrease conflict potential between countries involved in them (Westing 1998). They are hence often called Peace Parks.

The most important non-governmental supporters of the establishment of TFCAs in the Southern African region are the IUCN and the Peace Park Foundation (PPF), who have negotiated quite successfully with governments and donors (Budowski 2003; Draper, Spierenburg and Wels 2004). Whereas the IUCN is responsible for conceptualising the theoretical background for TBNRM globally, the PPF is an NGO with massive international donor funding and is active in project planning and the implementation of most TFCAs in Southern Africa. Both organisations aim for nature protection and are financed by donors in Western industrial countries. Supported by governments and these NGOs, a number of TFCAs are currently being developed in Southern Africa. The Kgalagadi Transfrontier Park (straddling Namibia, Botswana and South Africa) became the first TFCA in 2000, and the |Ai|Ais/Richtersveld Transfrontier Park was established in 2003. But lately, the huge Gaza/Kruger/Gonarezhou TFCA has received the most attention (Magome and Murombedzi 2003). Scientific and other literature usually refers to this TFCA (e.g., Wolmer 2003; Hughes 2005), which includes the Great Limpopo Transfrontier Park (established in late 2000), of which Kruger National Park is a part.

This new trend of establishing TFCAs is not unproblematic. The whole discourse is highly politicised and it is not easy to overlook the multitude of interests (Duffy 1997; Ramutsindela 2003). With up-scale planning, these outside interventions have a big impact on the local population. The enormous scale of the projects automatically limits possibilities of participation (Fall 1999). As transnational cooperation requires respecting the sovereignty of respective states, cross-border management is complicated (Agrawal 2000). The difficult coordination of natural resource management at a regional level leads to the danger of weakening the influence of local actors. This can be interpreted as a step back to the top-down approach in conservation that was the basis of colonial 'fortress' conservation' (Draper, Spierenburg and Wels 2004). Some scholars claim that national governments extend their power into regions that were formerly on the periphery of state control by establishing TFCAs (e.g., Dzingirai 2004). While national parks were often situated in border regions and not easy to control and monitor, they have now moved to the centre of interest and lost their peripheral connotation. Besides the common notion of 'peace parks', TBNRM has the potential of leading to new conflicts and brings unachievable hopes for unproved future economic development through tourism. Problems also involve the practicality of establishing new TFCAs in Southern Africa. It is also questionable whether or not TFCAs are more successful with regards to nature conservation than other strategies (Wolmer 2003; Hughes 2005).

As a matter of fact, pastoralists and their need for extensive grazing areas are especially affected by TBNRM projects. Along with the danger of land absorption, pastoralists are challenged by the loss of possible participation in large-scale planning and by the spread of epidemics across state borders due to wildlife migration. The case of the Maloti-Drakensberg project serves as an example of the possible impact of TBNRM on pastoralists in Southern Africa.

The Maloti-Drakensberg Transfrontier Conservation and Development Area

In November 2001 the Maloti-Drakensberg Transfrontier Conservation and Development Project (MDTP) was implemented with the main aim of founding a future TFCA in a 300-kilometre-long border region between eastern Lesotho and the Republic of South Africa. The project was funded by the Global Environmental Facility (GEF) and the World Bank with around US$16 million. Based on a bilateral agreement, the MDTP was split into two national divisions to ensure the national sovereignty of both countries. The Lesotho division was organised under the patronage of the country's Ministry of Tourism, Environment and Culture and had its headquarters in Lesotho's capital, Maseru. Like other TFCAs, the project's goals were biodiversity conservation and secondly the promotion of nature-based tourism for the economic development of the region. The protection of biodiversity was seen as especially important as the Drakensberg (or Maloti Mountains) is the only high altitude mountain range in Southern Africa where nature is considered to be unique. The *Ukhahlamba-Drakensberg* Park on the South African side was even declared a UNESCO World Heritage Site (World Bank 2001). The project area on the Lesotho side of the border instead covers mostly the communal rangelands used by locals keeping livestock, mostly cattle, sheep and goats, for subsistence and economic reasons.

For our present purpose, it is interesting to examine what future meaning a TFCA between Lesotho and the Republic of South Africa will have for the pastoralists of the highlands of Lesotho. The project implementation plan states that, 'the main issues in Lesotho are (i) grazing pressure on rangelands containing globally significant biodiversity, (ii) lack of a protected areas system, (iii) lack of conservation management capacity, and (iv) poor utilization of the potential for nature based tourism' (ibid.: 7). Attempts to reduce grazing pressure would have a significant impact on all local pastoralists, whereas the generation of income through nature-based tourism will probably play a role for a minority of local people over the next few years. The establishment of new protected areas is promoted not only as a basis for the further development of nature-based tourism. In Lesotho, where less than 1 per cent of the country is protected, it is seen as an important end in itself. But in a small, overpopulated and overstocked country, only relatively small peripheral areas can be declared protected without a negligible impact on the pastoralists. Those protected areas currently being planned are small when compared to the South Africa. In Lesotho nearly all higher lying valleys apart from villages and fields are used by pastoralists, and hence conserving biodiversity whilst ignoring them is impossible.

To involve pastoralists in protective actions outside protected areas, the MDTP has attempted to improve the Range Management Area (RMA) concept in Lesotho (ibid.). Managed Resource Associations (MRAs) should in future replace established Grazing Associations (GAs) and involve other natural resource users other than livestock owners, such as traditional healers (Büscher 2005). This idea is based on among other things the argument that former exclusion of other resource users was responsible for the poor outcome of the range management programme. In this context, the RMAs will be renamed MRAs. The MDTP identified focal regions where pilot projects under the new MRA concept were to be run, but until now there has been no impact on the pastoralists in these regions. A need for better protection of, for example, wetlands and their high biodiversity value is primarily promoted through environmental education inside the MRAs. In the opinion of the project, improvements in environmental knowledge and the inclusion of other parties other than the stockowners should lead to environmental protection. Recognising that the RMA concept had not really been successful (Quinlan 1995; Büscher 2005) and that no extra financial and/or personal support was planned in 2006 by the state authorities for the management of the MRAs in the future, achieving the MRA goal is in doubt. In spring 2006 in the high-altitude Mokhotlong district, home to some 130,000 people, there was only one range management officer and one officer of the Ministry for Tourism, Environment and Culture. This situation is further complicated by the unresolved problem of authority over natural resource issues between local governments, chiefs and GAs.[11]

In global conservation discourse, the participation of people has since the very beginning been an important issue. In the MDTP, 'community involvement' was one of eight components and had its own budget. In participative regional workshops organised by small consultants, stakeholders had a forum in which to discuss and formulate advice, but their real influence on decision making-processes remained unclear. The matter was complicated by the fact that the English language was used in all official and discussion papers, whereas the English skills of many participants are low. The difficulty of translating English terms regarding conservation into Sesotho – the mother tongue of most Basotho – is a general problem. On the other hand, most ordinary farmers in the focus areas were not even aware of the work of the MDTP. Although there was a lot of effort made by the project to inform local people in large meetings and events, the language problem and the fact that they did not receive direct benefits led to low interest among those pastoralists who operated away from the tourist routes. A staff member of the MDTP interpreted this low concern with the 'criminal interests' of local people, who are involved in drug trafficking and stock theft, issues which are also seen as central problems of transboundary cooperation. Most staff members of the MDTP, Basotho from the lowlands, were additionally confronted with the situation of being strangers among pastoralists from the highlands. The establishment of a TFCA in the Maloti-Drakensberg region fell short of expectations of conservational and economic improvements expressed by the involved parties. Biodiversity conservation without significant improvements in people's livelihoods is impossible in Lesotho because of the political priorities in this relatively poor country. Hence the actions of the authorities will still concentrate mostly on tourism which is seen to have the potential to create new jobs and to gain financial benefits. However, the relatively small tourism sector of Lesotho stands in clear competition to the already highly developed South African tourism industry.

Conclusion

This contribution has showed how the handling of the environment is embedded in particular political contexts. The approach to nature conservation in the whole of Southern Africa was affected by varying global trends and has accordingly changed over time. Our comparative analysis at a regional scale has further revealed that political contexts cannot be reduced to national politics. General trends in nature conservation are connected to globally changing political contexts concerning colonialism, democratisation and transnational integration. These global trends can, however, be traced back to changing discourses and politics in the Western world. The living conditions for pastoralist societies have differed according to varying approaches to nature conservation. Attempts to conserve natural resources and biodiversity have had, and still have, enormous impact on livestock keepers and their access to grazing land. Coming back to the questions addressed at the beginning of the chapter, we draw the following conclusions.

General Conditions in Southern Africa

The establishment of settler colonies in Southern Africa in colonial times resulted in a highly skewed distribution of land. For centuries, the great majority of people suffered oppression, discrimination and forced removal through colonial and

Apartheid legislation. The removal and resettlement of people created space and a labour force for white-settler-dominated freehold farmland. A specific approach to nature conservation built upon the establishment of uninhabited protected areas led to further land dispossession. These processes resulted in the emergence of highly fragmented societies in Southern Africa. With the advent of independence and changes in political rule, governments have started giving priority to a more equal distribution of land and have reconsidered conservation practices. In the light of global discourses on human rights, justice and the preservation of biodiversity, the participation of local people in conservation became a necessity. Nowadays, prestigious large-scale projects are supported by the regional economic integration of the Southern African countries. They further reflect a trend in the global community to enhance cooperation with regard to identified eco-regions that stretch beyond national boundaries.

The Main Actors in Nature Conservation in Southern Africa

During the colonial and Apartheid eras, a small white minority defined the shape of conservation with the help of scientists from their home countries. From the 1980s onwards, (inter)national NGOs entered the scene and broadened the spectrum. Under their influence, local people gained a voice. CBNRM projects were relatively clearly arranged and the affected communities had the chance to become actors in the shaping of conservation efforts. Large-scale TBNRM projects are much more government driven than the smaller CBRNM approach, and involve a multitude of interests and a variety of actors. A high number of international NGOs and state agencies in Southern Africa are focused on nature conservation and now concentrate on prestigious TBNRM projects. This decreases the potential for local people to influence the shape of conservation areas and subordinates the needs of pastoralists once again to the demands of conservation. Nevertheless, the new trend of TBNRM is not to be compared with the repressive nature of colonial- and Apartheid-era conservation practices.

The Impact of Conservation Initiatives on Pastoralists

Due to colonialism and Apartheid, pastoralism in Southern Africa ceased to be practised on a wide scale. The territories of livestock keepers were annexed for exclusive use by European colonists or for nature conservation. Examples where pastoralists were evicted from conservation areas include the Matopos National Park and Hluhluwe and Umfolozi game reserves. While pastoralists were often tolerated in the beginning as being 'part of nature', they were almost always finally evicted from conservation areas. The case of Etosha, where pastoralists' land was de-proclaimed and cut out the national park, remains an exception. Restrictions on access to grazing land fundamentally threatened the livelihoods of pastoralists. With the advent of CBNRM projects, they got the chance to regain access to certain lands and take part in decision making. As the example of the

Torra conservancy shows, the establishment of a conservation area can also lead to improvements in the living conditions of livestock keepers, even when there are serious conflicts between wildlife and livestock. Recently, stock farmers have once again come to compete with wildlife that is highly valued as a magnet for tourists in transfrontier conservation areas, but also often interferes with the land use of local people. Evictions in the name of conservation can actually no longer take place under the postcolonial and post-Apartheid governments, but democratisation and the devolution of rights to the local level is less of a reality than it should be when it comes to the transboundary management of natural resources.

The Power of Pastoralists in Nature Conservation Politics

The ability of pastoralists to engage in nature conservation politics varies according to the different approaches of how to handle the preservation of the environment. While room for manoeuvre was rather small under the repressive colonial and Apartheid regimes, livestock keepers were able participate in the shaping of conservation areas when global thinking shifted and CBNRM programmes were launched in Southern Africa. The case of the Richtersveld in South Africa clearly shows that a local population can prevent a national park not accommodating them and thereby defend their access to grazing lands. Recent changes at the Richtersveld park, the establishment of the communal conservancy as well as the World Heritage Site status show that local people have the opportunity to participate in the shaping of conservation areas. Usually, local people are supported in their efforts by external actors such as NGOs or consultants who play a major role in empowering people.

In the future, nature conservation across borders will play a central role in the whole Southern African region, where countries are connected through economic, social and political ties. We consider it essential to integrate pastoralists and other local land users into these conservation efforts to ensure their stability and long-term acceptance. To provide a wide base for analysis, there is still a need for further empirical studies that shed light on the dynamics of TBNRM and livestock keepers. We still do not know enough about the impact of large-scale conservation areas on livestock keepers and the role pastoralists can play in conserving biodiversity across national boundaries.

Notes

1. The corridor was successfully returned to Hlabisa-Mpukunyoni under the South African land reform programme. The claimant community has agreed that they will not physically occupy the property, and the conservation authority will remain managers of the land for conservation purposes.
2. For an overview, see, e.g., Hulme and Murphree (2001), Fabricius et al. (2004) and Brosius, Tsing and Zerner (2005).
3. Quoted from: http://www.nric.net/. Retrieved 23 October 2007.

4. Berzborn conducted twenty-four months of ethnographic fieldwork in the Richtersveld between 1999 and 2007. The main findings are published in Berzborn (2004, 2006).
5. See: http://www.biodiversityhotspots.org. Retrieved 26 September 2007.
6. Technically, ownership rights in the land are still vested in the Minister of Land Affairs as the case involving the Transformation of Certain Rural Areas Act (1998) is not yet complete. But it is more than likely that the land will be transferred to the Richtersveld Communal Property Association in the near future.
7. For an overview of the 1989 and 1991 conditions, see Berzborn (2006: 362). Even diamond mining is allowed inside the national park.
8. IDP is a government-driven process that requires local communities to identify future development needs and land-use practices.
9. There is a basal date for herding sheep in the Richtersveld of around 2000 BP, and more recent dates of 1300 BP, 450 BP and 330 BP (Webley, Archer and Brink 1993; Webley 1997). However, archaeological research shows no clear evidence that the Richtersveld was occupied continuously from 2000 BP until the present. This might be explained by the fact that there have not been enough excavations and/or the density of the population was too small, but it might also indicate that there was no continuity in pastoral land use (L. Webley, personal communication).
10. On this matter, see Thornton et al. (2006: 75).
11. Interview with Phallang Lebesa, MDTP district conservation officer, 4 May 2006.

References

Adams, W.M., and D. Hulme. 2001. 'Conservation and Community: Changing Narratives and Practices in African Conservation', in D. Hulme and M. Murphree (eds), *African Wildlife and Livelihoods: The Promise and Performance of Community Conservation*. Cape Town: David Philip, pp.9–23.

Agrawal, A. 2000. 'Adaptive Management in Transboundary Protected Areas: The Bialowieza National Park and Biosphere Reserve as a Case Study', *Environmental Conservation* 27(4): 326–33.

Archer, F.M., E. Boonzaier and A.B. Smith. 1989. 'Communal Grazing Patterns in the Northern Richtersveld'. Cape Town: Unpublished report submitted to the National Parks Board.

Artz, N.E. 1993. 'Local Participation, Equity, and Popular Support in Lesotho's Range Management Area Programme', *African Journal of Range and Forage Science* 10(1): 54–62.

Behnke, R.H., and I. Scoones. 1993. 'Rethinking Range Ecology: Implications for Rangeland Management in Africa', in R.H. Behnke, I. Scoones and C. Kerven (eds), *Range Ecology at Disequilibrium: New Models of Natural Variability and Pastoral Adaptation in African Savannas*. London: Overseas Development Initiatives, pp.1–30.

Beinart, W. 1989. 'Introduction', *Journal of Southern African Studies* 15(2): 143–62.

Berzborn, S. 2004. 'Haushaltsökonomie, Soziale Netzwerke und Identität: Risikominimierende Strategien von Pastoralisten und Lohnarbeitern im Richtersveld, Südafrika', PhD dissertation. Cologne: Universität zu Köln.

——— 2006. *Ziegen und Diamanten: Sicherung des Lebensunterhalts im Ländlichen Südafrika*. Berlin: LIT Verlag.

——— 2007. 'The Household Economy of Pastoralists and Wage Labourers in the Richtersveld, South Africa', *Journal of Arid Environments* 70(4): 672–85.

Biersack, A., and J.B. Greenberg (eds). 2006. *Reimagining Political Ecology*. Durham, NC: Duke University Press.

Bishop, K., et al. 2004. *Speaking a Common Language*. Cardiff: International Union for the Conservation of Nature.
Bollig, M., and S. Berzborn. 2004. 'The Making of Local Traditions in a Global Setting: Indigenous Peoples' Organisations and Their Effects at the Local Level in Southern Africa', in P. Probst and G. Spittler (eds), *Between Resistance and Expansion: Explorations of Local Vitality in Africa*. Münster: LIT Verlag, pp.297–329.
Boonzaier, E. 1996. 'Negotiating the Development of Tourism in the Richtersveld, South Africa', in M.F. Price (ed.), *People and Tourism in Fragile Environments*. New York: Wiley, pp.123–37.
Borgerhoff Mulder, M., and P. Coppolillo. 2005. *Conservation*. Princeton, NJ: Princeton University Press.
Brooks, S. 2005. 'Images of "Wild Africa": Nature Tourism and the (Re)Creation of Hluhluwe Game Reserve, 1930–1945', *Journal of Historical Geography* 31: 220–40.
Brosius, J.P. 2004. 'Indigenous People and Protected Areas at the World Parks Congress', *Conservation Biology* 18(5): 609–12.
Brosius, J.P., A. Lowenhaupt Tsing and C. Zerner (eds). 2005. *Communities and Conservation: Histories and Politics of Community-based Natural Resource Management, Globalization and the Environment*. Walnut Creek, CA: AltaMira Press.
Brosius, J.P., and D. Russell. 2003. 'Conservation from Above: An Anthropological Perspective on Transboundary Protected Areas and Ecoregional Planning', *Journal of Sustainable Forestry* 17(1/2): 39–65.
Bryant, R.L. 1998. 'Power, Knowledge and Political Ecology in the Third World: A Review', *Progress in Physical Geography* 22(1): 79–94.
Budowski, G. 2003. 'Transboundary Protected Areas as a Vehicle for Peaceful Co-operation', unpublished paper presented at the 5th World Parks Congress, 12–13 September 2003, Durban, South Africa.
Büscher, B. 2005. 'Land and Resources in a Transfrontier Setting: The Case of the Maloti-Drakensberg Transfrontier Conservation and Development Project', PLAAS Policy Brief 19. Cape Town: Programme for Land and Agrarian Studies, School of Government, University of the Western Cape.
Carruthers, J. 1995. *The Kruger National Park: A Social and Political History*. Pietermaritzburg: University of Natal Press.
Chakela, Q.K. (ed.). 1999. *State of the Environment in Lesotho 1997*. Maseru: National Environment Secretariat.
Dieckmann, U. 2007. *Hai//om in the Etosha Region: A History of Colonial Settlement, Ethnicity and Nature Conservation*. Basel: Basler Afrika Bibliographien.
Draper, M., M. Spierenburg and H. Wels. 2004. 'African Dreams of Cohesion: Elite Pacting and Community Development in Transfrontier Conservation Areas in Southern Africa', *Culture and Organisation* 10(4): 341–53.
Duffy, R. 1997. 'The Environmental Challenge to the Nation-state: Superparks and the National Parks Policy in Zimbabwe', *Journal of Southern African Studies* 23(3): 441–51.
Dzingirai, V. 2004. *Disenfranchisement at Large: Transfrontier Zones, Conservation and Local Livelihoods*. Harare: International Union for Conservation of Nature, Regional Office of Southern Africa (IUCN ROSA).
Fabricius, C., and C. de Wet. 2002. 'The Influence of Forced Removals and Land Restitution on Conservation in South Africa', in D. Chatty and M. Colchester (eds), *Conservation and Mobile Indigenous People: Displacement, Forced Settlement, and Sustainable Conservation*. Oxford: Berghahn, pp.142–57.

Fabricius, C., et al. (eds). 2004. *Rights, Resources and Rural Development: Community-based Natural Resource Management in Southern Africa.* London: Earthscan.
Fall, J. 1999. 'Transboundary Biosphere Reserves: A New Framework for Cooperation', *Environmental Conservation* 26(4): 252–55.
Griffin, J., et al. 1999. *Study on the Development of Transboundary Natural Resource Management Areas in Southern Africa: Main Report.* Washington, DC: Biodiversity Support Program.
Grove, R. 1990. 'Colonial Conservation, Ecological Hegemony and Popular Resistance: Towards a Global Synthesis', in M.J. MacKenzie (ed.), *Imperialism and the Natural World.* Manchester: Manchester University Press, pp.15–50.
Hanks, J. 2003. 'Transfrontier Conservation Areas (TFCAs) in Southern Africa: Their Role in Conserving Biodiversity, Socioeconomic Development and Promoting a Culture of Peace', in U.M. Goodale et al. (eds), *Transboundary Protected Areas: The Viability of Regional Conservation Strategies.* New York: Food Products Press, pp.127–48.
Hoffman, M.T., and A. Ashwell. 2001. *Nature Divided: Land Degradation in South Africa.* Cape Town: University of Cape Town Press.
Hohmann, T. 2003.'"We Are Looking for Life, We Are Looking for the Conservancy": Namibian Conservancies, Nature Conservation and Rural Development: The N|a-Jaqna Conservancy', in T. Hohmann (ed.), *San and the State: Contesting Land, Development, Identity and Representation.* Cologne: Rüdiger Köppe Verlag, pp.209–56.
Holmes, T. 1992. 'Conservation Activities in the Kaokoveld (North-west Namibia)', *Biodiversity and Conservation* 1: 211–13.
Homewood, K., and W.A. Rodgers. 1987. 'Pastoralism, Conservation and the Overgrazing Controversy', in D. Anderson and R. Grove (eds), *Conservation in Africa: People, Policies and Practice.* Cambridge: Cambridge University Press, pp.111–28.
Hughes, D.M. 2005. 'Third Nature: Making Space and Time in the Great Limpopo Conservation Area', *Cultural Anthropology* 20(2): 157–84.
Hulme, D., and M. Murphree (eds). 2001. *African Wildlife and Livelihoods: The Promise and Performance of Community Conservation.* Oxford: James Currey.
Hutton, J., W. Adams and J. Murombedzi. 2005. 'Back to the Barriers? Changing Narratives in Biodiversity Conservation', *Forum for Development Studies* 2: 341–70.
IRDNC. 2003. 'A Quick Guide to the Human–Animal Conflict Compensation Scheme (HACCS) for Kwandu and Mayuni Conservancies'. Windhoek: Integrated Rural Development and Nature Conservation.
IUCN. 1991. 'Caring for the Earth: A Strategy for Sustainable Living'. Gland, Switzerland: The World Conservation Union and United Nations Environment Programme and World Wide Fund For Nature.
Jones, B.T. 1999. *Rights, Revenue and Resources: The Problems and Potential of Conservancies as Community Wildlife Management Institutions in Namibia.* London: International Institute for Environment and Development.
Jürgens, N. 1991. 'A New Approach to the Namib Region, I: Phytogeographic Subdivision', *Vegetatio* 97: 21–28.
Jürgens, N., I. Gotzmann and R.M. Cowling. 1999. 'Remarkable Medium-term Dynamics of Leaf Succulent *Mesembryanthema* (*Aizoaceae*) in the Winter-rainfall Desert of Northwestern Namaqualand, South Africa', *Plant Ecology* 142: 87–96.
Kangwana, K. 2001. 'Can Community Conservation Strategies meet the Conservation Agenda?' in D. Hulme and M. Murphree (eds), *African Wildlife and Livelihoods: The Promise and Performance of Community Conservation.* Oxford: James Currey, pp.256–66.

Long, S.A. 2004a. 'Framework and Methodologies', in S.A. Long (ed.), *Livelihoods and CBNRM in Namibia: The Findings of the WILD Project.* Windhoek: Directorates of Environmental Affairs and Parks and Wildlife Management, the Ministry of Environment and Tourism, pp.13–24.

——— 2004b. 'The Implementation of CBNRM', in S.A. Long (ed.), *Livelihoods and CBNRM in Namibia: The Findings of the WILD Project.* Windhoek: Directorates of Environmental Affairs and Parks and Wildlife Management, the Ministry of Environment and Tourism, pp.41–53.

Long, S.A., and B.T. Jones. 2004. 'Conservancy Institutions and Governance: Implications for Livelihoods', in S.A. Long (ed.), *Livelihoods and CBNRM in Namibia: The Findings of the WILD Project.* Windhoek: Directorates of Environmental Affairs and Parks and Wildlife Management, the Ministry of Environment and Tourism, pp.139–59.

Luig, U. 1999. 'Naturschutz im Widerstreit der Interessen im Südlichen Afrika', in G. Meyer and A. Thimm (ed.), *Naturräume in der Dritten Welt: Ausbeutung, Nachhaltige Entwicklung oder Schutz?* Mainz: Johannes-Gutenberg-Universität, pp.11–35.

Magome, H., and J.C. Murombedzi. 2003. 'Sharing South African National Parks: Community Land and Conservation in a Democratic South Africa', in W.M. Adams and M. Mulligan (eds), *Decolonising Nature: Strategies for Conservation in a Post-colonial Era.* London: Earthscan, pp.108–34.

Murphy, C., et al. 2004. 'The Costs of Living with Wildlife', in S.A. Long (ed.), *Livelihoods and CBNRM in Namibia: The Findings of the WILD Project.* Windhoek: Directorates of Environmental Affairs and Parks and Wildlife Management, the Ministry of Environment and Tourism, pp.105–17.

Neumann, R.P. 1998. *Imposing Wilderness: Struggles Over Livelihood and Nature Preservation in Africa.* Berkeley: University of California Press.

——— 2000. 'Primitive Ideas: Protected Area Buffer Zones and the Politics of Land in Africa', in V. Broch-Due and R.A. Schroeder (eds), *Producing Nature and Poverty in Africa.* Stockholm: Nordisk Afrikainstitutet, pp.220–42.

Niamir-Fuller, M., and M. Turner. 1999. 'A Review of Recent Literature on Pastoralism and Transhumance in Africa', in M. Niamir-Fuller (ed.), *Managing Mobility in African Rangelands.* London: Intermediate Technology Publications, pp.18–46.

Nott, C., A. Davis and B. Roman. 2004. 'The Torra Conservancy in Namibia', in C. Fabricius, et al. (eds), *Rights, Resources and Rural Development: Community-based Natural Resource Management in Southern Africa.* London: Earthscan, pp.200–9.

Oates, J.F. 1999. *Myth and Reality in the Rain Forest: How Conservation Strategies Are Failing in West Africa.* Berkeley: University of California Press.

Patel, H. 1998. *Sustainable Utilisation and African Wildlife Policy: The Case of Zimbabwe's CAMPFIRE – Rhetoric or Reality?* Massachusetts: Indigenous Environmental Policy Centre.

Quinlan, T. 1993. 'Native American Reservations in Africa?' *Cultural Survival Quarterly* 17(2): 42–44.

——— 1995. 'Understanding Development: Different Perspectives on Livestock and "Range" Management in Lesotho', in *People, Rangeland and Development.* Pietermaritzburg: Grassland Society of Southern Africa, pp.34–41.

Ramutsindela, M. 2003. 'Glocalisation and Nature Conservation Strategies in 21st-century Southern Africa', *Tijdschrift voor Economische en Sociale Geografie* 95(1): 61–72.

Ranger, T. 1999. *Voices from the Rocks: Nature, Culture and History in the Matopos Hills of Zimbabwe.* Oxford: James Currey.

Reid, H., and S. Turner. 2004. 'The Richtersveld and Makuleke Contractual Parks in South Africa: Win-Win for Communities and Conservation?' in C. Fabricius et al. (eds), *Rights, Resources and Rural Development: Community-based Natural Resource Management in Southern Africa*. London: Earthscan, pp.223–34.

Rohde, R.F., et al. 2006. 'Dynamics of Grazing Policy and Practice', *Environmental Science and Policy* 9: 302–16.

RSA. 1998. 'Wet op die Transformasie van Sekere Landelike Gebiede, Wet 94 van 1998', *Staatskoerant* 401(19417): 1–15.

——— 2007. 'Department of Environmental Affairs and Tourism: Proclamation of the Richtersveld Cultural and Botanical Landscape as a World Heritage Site', *Government Gazette*, 4 July.

Schroeder, R.A. 2000. '"Re-claiming" Land in the Gambia: Gendered Property Rights and Environmental Interventions', in V. Broch-Due and R.A. Schroeder (eds), *Producing Nature and Poverty in Africa*. Stockholm: Nordisk Afrikainstitutet, pp.268–94.

Shackleton, S., G. von Maltitz and J. Evans. 1998. 'Factors, Conditions and Criteria for the Successful Management of Natural Resources Held under a Common Property Regime: A South African Perspective', Land Reform and Agrarian Change in Southern Africa, Occasional Paper Series 8. Cape Town: Programme for Land and Agrarian Studies, School of Government, University of the Western Cape.

Showers, K.B. 2005. *Imperial Gullies: Soil Erosion and Conservation in Lesotho*. Athens: Ohio University Press.

Singh, J., and H. van Houtum. 2002. 'Post-colonial Nature Conservation in Southern Africa: Same Emperors, New Clothes?', *GeoJournal* 58: 253–63.

Singh, M. 2000. 'Basutoland: A Historical Journey into the Environment', *Environment and History* 6(1): 31–70.

Stott, P., and S. Sullivan (eds). 2000. *Political Ecology: Science, Myth and Power*. London: Arnold.

Suich, H. 2003. 'Summary of Partial Results from the Socio-economic Household Survey Regarding Community-based Natural Resource Management and Livelihoods in Caprivi and Kunene', WILD Project Working Paper 12. Windhoek: Directorates of Environmental Affairs and Parks and Wildlife Management, the Ministry of Environment and Tourism.

Sullivan, S. 2002. 'How Sustainable is the Communalising Discourse of "New" Conservation? The Masking of Difference, Inequality and Aspiration in the Fledgling "Conservancies" of Namibia', in D. Chatty and M. Cholchester (eds), *Conservation and Mobile Indigenous People: Displacement, Forced Settlement and Sustainable Development*. Oxford: Berghahn, pp.158–87.

——— 2003. 'Protest, Conflict and Litigation: Dissent or Libel in Resistance to a Conservancy in North-west Namibia', in D.G. Anderson and E. Berglund (eds), *Ethnographies of Conservation: Environmentalism and the Distribution of Privilege*. Oxford: Berghahn, pp.69–86.

Swatuk, L.A. 2005. *Peace Parks in Southern Africa: Parks for Peace or Peace for Parks? Issues in Practice and Policy*. Washington, DC: Woodrow Wilson International Center for Scholars.

Taylor, M. 2007. 'Community Based Natural Resource Management (CBNRM) and Pastoral Development in Botswana: Implications for San Land Rights', in M. Bollig et al. (eds), *Aridity, Change and Conflict in Africa*. Cologne: Heinrich-Barth-Institut, pp.345–62.

Terborgh, J. 1999. *Requiem for Nature.* Washington, DC: Island Press.

Thornton, M. et al. 2006. 'The Richtersveld Cultural and Botanical Landscape', unpublished paper.

Turner, S. 2003. 'Rangelands in Lesotho: Can Common Property Resources Succeed in the 21st Century', paper delivered at the 7th International Rangelands Congress, Durban, South Africa.

van der Linde, H., et al. 2001. *Beyond Boundaries: Transboundary Natural Resource Management in Sub-Saharan Africa.* Washington, DC: Biodiversity Support Program.

Vetter, S. 2005. 'Rangelands at Equilibrium and Non-equilibrium: Recent Developments in the Debate', *Journal of Arid Environments* 62: 321–41.

Walker, P. 2005. 'Political Ecology: Where is the Ecology?' *Progress in Human Geography* 29(1): 73–82.

Warren, A. 1995. 'Changing Understanding of African Pastoralism and the Nature of Environmental Paradigms', *Transactions of the Institute of British Geographers* 20(2): 193–203.

WCED. 1987. 'Our Common Future' [Brundtland Commission Report]. Oxford: Oxford University Press for the World Commission on Environment and Development.

Webley, L. 1990. 'Prehistoric Land and Resource Use in the Richtersveld', unpublished report. Rondebosch: Interdissiplinêre Navorsingsprogram met die betrekking tot die Richtersveld Nationale Park.

——— 1997. 'Jakkalsberg A and B: The Cultural Material from two Pastoralist Sites in the Richtersveld, Northern Cape', *Southern African Field Archaeology* 6(1): 3–19.

Webley, L., F.M. Archer and J. Brink. 1993. 'Die Toon: A Late Holocene Site in the Richtersveld National Park, Northern Cape', *Koedoe* 36(2): 1–9.

Wells, M. 1996. 'The Social Role of Protected Areas in the New South Africa', *Environmental Conservation* 23(4): 322–31.

Westing, A. 1993. *Transfrontier Reserves for Peace and Nature.* Nairobi: United Nations Environment Programme.

——— 1998. 'Establishment and Management of Transfrontier Reserves for Conflict Prevention and Confidence Building', *Environmental Conservation* 25(2): 91–94.

Wilshusen, P.R. 2003. 'Exploring the Political Contours of Conservation: A Conceptual View of Power in Practice', in S.R. Brechin, et al. (eds), *Contested Nature: Promoting International Biodiversity with Social Justice in the Twenty-first Century.* Albany: State University of New York Press, pp.41–57.

Wilshusen, P.R., et al. 2002. 'Reinventing a Square Wheel', *Society and Natural Resources* 15: 17–40.

Williamson, G. 2000. *Richtersveld: The Enchanted Wilderness.* Hatfield: Umdaus Press.

Wolmer, W. 2003. 'Transboundary Conservation: The Politics of Ecological Integrity in the Great Limpopo Transfrontier Park', *Journal of Southern African Studies* 29(1): 261–78.

World Bank. 2001. *South Africa - Maloti - Drakensberg Transfrontier Conservation and Development Project.* Washington, DC: World Bank.

Part IV

Pastoral Modernities in Africa

Chapter 16

The Indigenisation of Pastoral Modernity

Territoriality, Mobility and Poverty in Dryland Africa

John G. Galaty

> Rather than a refusal of the commodities and relations of the world-system, this more often means … a desire to indigenize them. The project is the indigenization of modernity.
>
> —Marshall Sahlins

Introduction

Pastoralism is one of the world's great livelihood systems, one that allows humans to inhabit the enormous tracts of arid and semi-arid lands that stretch across the globe by living off the products of domestic herbivores that consume grass and shrubs. Up to one half of the world's land is used as rangeland grazing, the proportion rising in the 'arid continent' of Africa. Nomadic or semi-nomadic pastoralists predominate across the dryland Sahelian and Saharan belt that transects West and North Africa and the Horn of Africa, where the belt of aridity turns southward to penetrate East Africa. Much of Southern Africa is also characterised by aridity and thus also by high livestock and pastoral populations.

Of the countries of the 'bottom billion' that contain the poorest 20 per cent of the world's five billion people, many lie in the arid and semi-arid lands of Africa. In these countries, populations are low, people widely scattered, water scarce, towns dispersed and services remote. Many dryland nations are landlocked, and most have few resources, weak governments and are prone to conflict; also – unless they have oil or precious minerals – they are quite poor (Collier 2007). Though aridity is a predictor of poverty, not all dryland inhabitants are poor, though most are at risk from unpredictable rainfall and uncertain livelihoods. Pastoralists in particular may own substantial wealth in their livestock, which simultaneously represent stores of food, capital and value, and serve as the world's oldest currency, used in both social and market exchange (Schneider 1979). Although the international poverty line is conventionally set at US$1 per day per person, poverty is especially complex where much subsistence production does not run through markets, or, as among pastoralists, where there is a disparity between three key measures of economic well-being: capital assets, revenue flows and food security (Sen 1993).

This is especially so for pastoralists who, although they may hold considerable capital assets in livestock and land, depend on highly variable markets to obtain household incomes and experience radical shifts in access to food in relation to climatic and agricultural cycles. Indeed, during the colonial period, officials were quite uncertain whether pastoralists were among the poorest of their citizens, due to their non-materialistic, non-sedentary indigenous lifestyles and vulnerability to drought, or among the richest, due to the value of their livestock holdings. This confusion parallels paradoxical views held today: pastoralists are desperately in need of radical change and 'development' but that when it arrives they tend to suffer from it; they are traditionalists who would benefit from the innovations of modernity but are already very much part of the modern world and must be very selective in which of its elements they embrace. The value of selectivity especially applies to the wrenching impacts on their systems of territoriality that processes of development and modernisation have had, which affect how dryland dwellers hold and use land, especially by diminishing their mobility practices, long their critical adaptation to living in drylands.

This chapter examines how African pastoralists, whose herding strategies have relied on time-tried techniques and technologies and whose social relations are defined by customary institutions, confront and indeed embrace the modern world and bend it to their needs and interests.[1] The notion that pastoralists are anti-modern or 'traditionalistic' is often used to politically discredit these rural communities whose lives are bound up with the husbandry of domestic livestock using natural pastures found in drylands. But pastoralists seize useful opportunities in the contemporary world that appear to arise from global versus local, customary norms and institutions. A framework of global institutions is often identified with modernity; it includes political, economic, infrastructural and cultural elements largely forged in the twentieth century and increasingly shared by most peoples and countries throughout the world, including pastoralists. Yet the requirement that herders extract energy from sparse and arid rangelands makes a framework of local institutions feasible and even necessary, most notably in defining complex systems of pastoral territoriality through which husbandry is practised. Rather than tracing rural practices back to one or another of these grand frameworks, I ask here how the systems of territoriality and patterns of mobility that we observe today in fact reflect multi-levelled institutional interactions between local, state and global influences. Here, I propose, is where we find the indigenisation of modernity that defines contemporary pastoralism.

Today, pastoral existence is essentially framed by the following realities: the structures of the nation-state, including borders, security services, taxation, administrative bodies and social services; the mechanisms of the market, including the production, distribution and valuation of commodities on a larger and larger scale; increasingly rapid dissemination of information, made possible due to new technologies and the spread of common codes and languages; ever-heightened mobility via new forms of transportation and investment in the infrastructure of travel, which has become more affordable through economies of scale and more efficient technologies; and the creation and diffusion of forms of culture that belong to no one source but to all, given their association with the world of information and commodities, including clothing fashions, music, literature, political habits, sports and lifestyle expectations. Pastoralists also share a customary framework with many arid land dwellers and other rural citizens: livelihood practices that include extensive animal husbandry pursued through mobile forms of land use, face-to-face societies, segmentary kinship, gender and age-based social structures, oral communications, and locally salient cultures and languages.

Pastoralists are experts at raising domestic animals; the more they rely on livestock the more they attain livelihood security through extracting products from livestock that they either consume or sell. They use natural pastures for grazing, which for ecological reasons demands herd and household mobility the extent of which, from season to season, is determined by the abundance or scarcity of rainfall, the quantity and distribution of grasses that rain nurtures, and the extent of their political control of territory. Given their scarcity of non-livestock based capital, pastoralists tend to combine innovations that cost money with customary practices that do not: boreholes with dug wells, veterinary medicine with often effective local remedies, upgraded stock with indigenous breeds, purchased foods with home-produced foods based on milk, meat and dryland grains, trucks to ship stock with strategies of trekking, markets with local exchange. Despite the high degree of family and household autonomy necessitated by decentralised, flexible and opportunistic movement, pastoral communities are bound together through complex social ties that emerge out of multi-scalar social structures, by languages reproduced through local patterns of contact and engagement, and by cultures built out of shared systems of signification. Institutions of social differentiation within pastoral societies are often defined by gender, age, generation, lineage or occupational statuses, and by the particularities of locality and ritual life.

Because the two are so intertwined in contemporary life, it seems false to categorically distinguish between customary and modern orientations, yet Maasai – to evoke one example – do see clanship, age-set systems and local knowledge as belonging to themselves, while governments, telephones, and churches belong to everyone. To describe how elements of the modern world are used for indigenous aims and purposes, much as pastoralists use trucks and mobile phones to facilitate pastoral herd movement, Sahlins (1999) speaks of the 'indigenisation of modernity'. But pastoralists also practice the 'modernisation of the indigenous' when drawing from cultural strategies proper to their indigenous lives to pursue modern purposes and aims in ways specific to given locales, such as calling in livestock loans to help pay school fees.

This chapter aims to discuss the 'modernity' of strategies pursued by pastoralists as they respond to the constraints imposed on them by the states in which they live and by the global influences that for both good and ill impinge on them. In comparative context, I will first examine climatic and institutional structures that underpin the patterns of mobility and systems of territoriality practised by four African pastoral communities: the Nuer of the southern Sudan, the Fulbe living along the Niger inland delta in Mali, the Maasai of the Rift Valley region of southern Kenya and northern Tanzania, and the Tswana of eastern Botswana (see Figure 16.1). I will then focus on political events and development initiatives that have affected these groups in recent decades, through which state and global influences have dramatically altered the social contexts in which they practise mobility and territoriality. Dryland herding communities often resist impositions of the state and hold in abeyance wider forces of change that threaten their control over critical resources, but at the same time they pursue innovations of their own that 'indigenise' the novel opportunities that they encounter, inventing the terms by which the modernity of contemporary pastoralism should be defined.

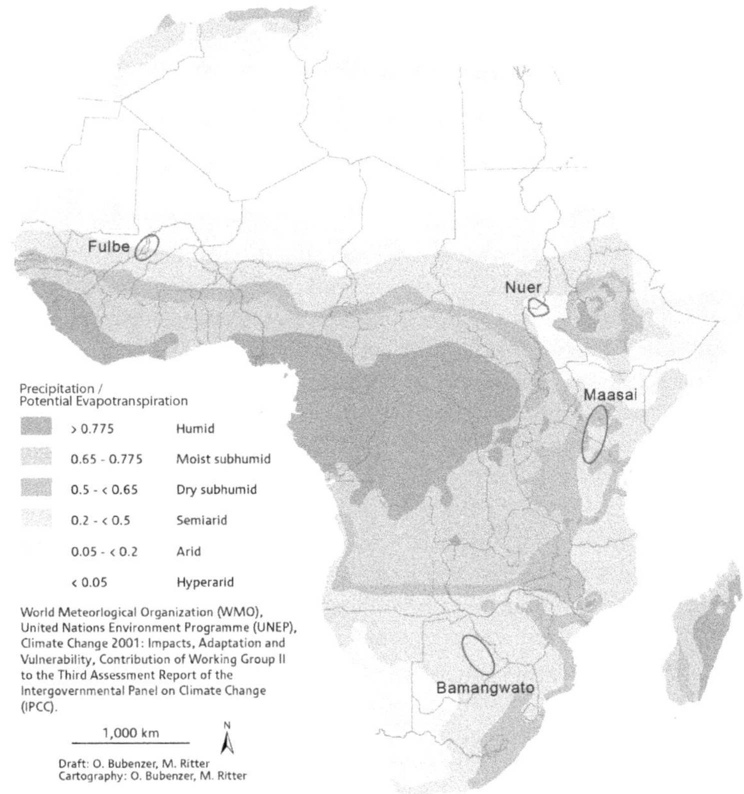

Figure 16.1: Location of the cases discussed in the context of climatic zones in Africa (after Bubenzer et al. 2007: 18).

Pastoral Institutions, Dryland Constraints, and Poverty

North (1990: 126) emphasises the fact that some of the great institutional innovations that shaped the modern world transformed uncertainty into risk, 'risk' being amenable to more precise assessment and avoidance or mitigation. The new institutional economics of North emphasises that institutional innovations tend to be adopted when they increase returns on capital or labour, which better strategies of assessing, avoiding and insuring against loss do accomplish. But in explaining the origins of modern capitalist institutions, he was too dismissive of pre-modern organisation (his examples are tribal organisation, the *suq* and the caravan trade), which he claimed failed to induce institutional modifications, such as increasing the mobility of capital, lowering information costs or spreading risk (ibid.: 125). Clearly there are useful innovations that can enhance pastoral productivity, but too often changes championed as 'development' have achieved the opposite, decreasing productivity and increasing risk for rangeland dwellers. The inhibition of pastoral movement that usually accompanies progressive sedentarisation (whether forced or voluntary) seems a key element in a pastoral development package that encourages the adoption of fixed land husbandry, but these trends too often undermine the key strategy that makes pastoral land use in arid and semi-arid lands viable: each move by pastoralists is calibrated to increase animal productivity and diminish risk.

Focusing on institutions of pastoral territoriality, it is clear that increasing productivity and decreasing risk are continuous preoccupations for pastoralists, achieved in particular through controlling pasture lands and regulating the safe movement of herds, households and people – and at the same time guaranteeing the efficient mobility of both (animal) capital and information. Through territorial systems, rights in and norms for negotiating access to resources, duties towards others regarding land held in common, and powers of holding, disposing and using land, and excluding others from it, whether through force or persuasion, are defined and secured. Increasingly, such societal systems of tenure are undergoing formalisation, with various parties gaining legal rights to land long held by pastoral societies, whether the state, local elites, foreign companies, conservation entities or communities themselves. North wisely observes that the formalisation of property rights, often seen as a prerequisite for increasing economic efficiency, may not be beneficial to those who enjoyed the informal property rights that are subsequently eroded (ibid.: 100). If we generalise this point, the wholesale adoption of new institutions or technologies aimed at achieving higher returns may in fact undermine the very local institutions that underpin livelihood security, diminish risk and make optimal use of rangeland resources. For example, too often technical innovations in improved livestock breeds, prophylactic veterinary medicine, paddock management, privatised tenure or modern ranching have failed to achieve higher yields or returns, although they have displaced productive techniques and forms of pastoral organisation that actually do work.

A key question explored in studies on pastoralism is the extent to which the mobility of herds and households that characterises pastoral practices is a response to environmental or socio-political factors (McCabe 1994: 73). Aridity is a factor of 'stress' that exacerbates environmental threats to human well-being, putting livelihoods in jeopardy through climatic unpredictability (Bollig 2006). Drought can be devastating both for dryland farmers without reliable irrigation, and for livestock herds that suffer from scarcity of pasture. In this regard, pastoral mobility, as the most effective strategy for exploiting scarce pasture and avoiding damage to rangeland environments, is correlated with low rainfall. In Africa, the greatest diversity of pastoral societies is found in arid, semi-arid and dry sub-humid regions that receive from less than 200 up to 800 to 1,000 millimetres of rainfall per year. Here, mobility, settlement and land use are primarily shaped by the quantity and distribution of rainfall (see Figure 16.1). Pastoral mobility varies along a temporal continuum and shifts with variation in both seasonal and yearly rainfall and forage quality and abundance, from very frequent movements in arid, to only seasonal mobility in semi-arid, to almost sedentary husbandry in sub-humid regions.

While it seems clear that micro movements by herders to access pastures or water are essentially environmental responses, mobility may also be employed to avoid insecurity, to elude state control, or to access towns or markets. So nomadic movements may both be adaptive responses to aridity and sparse pasture and strategic initiatives by herders seeking to avoid threats or access opportunities. McCabe (1994: 81) reports that, for four herd owners from the Ngisonyoka Turkana of Kenya, between 56 to 65 per cent of moves were attributed to environmental factors, 20 to 27 per cent to security factors, and up to 9 per cent to social factors. Many herders, however, seek simultaneously to optimise several factors, seeking better pastures, avoiding conflict zones, positioning the household to facilitate a later move, coordinating with neighbours, moving closer to friends, visiting towns, and so on. Improving access to grazing may be a constant factor influencing where herders move, but other factors are never absent and may serve as triggers initiating moves and guiding precisely where destinations will be.

Decisions take place within larger systems of territorial governance, which open and close options and situate the placement of homesteads within a configuration of social affiliations and networks. Systems of territoriality are underpinned by patterns of environmental perception, which give definition to spatial configurations of resources, to attributions of rights and claims to land, water and grazing, and to social and political networks that form an invisible lattice across the countryside, following structures of power and authority (Casimir 1992: 19). Cooperation and territorial organisation vary with each society's core institutions; these institutions provide local frameworks for managing the multitude of risks that states spend the majority of their national budgets in addressing. Consider the institutional safeguards against risk that governments in industrialised societies underwrite: insurance, welfare benefits, health care, police and security forces, controls on foods, agricultural and pharmaceutical products, safe transport, construction codes and regulations, provisions for safe leisure, labour codes, and so on. Pastoral societies face threats to their well-being from diverse sources, whether recurrent drought, periodic livestock epidemics, livestock theft and loss, injuries from violent conflict, demographic growth outpacing resources, depletion, overuse or degradation of pastures, loss of access to resources, loss of land, human illnesses and accidents, food insecurity, and so on. Threats of environmental hazard or political dangers are met by moves to mitigate or avoid these risks, providing buffers against uncertainty (Bollig 2006). Many strategies are embedded in local institutions and social networks, others in bodies of local knowledge and perception.

In the absence of the effective governance that one might wish states would provide, pastoralists have evolved customary territorial practices, such as herd or household mobility, to increase security in the face of risk, by avoiding political or environmental danger, protecting human and livestock communities, accessing services while avoiding disturbances, sometimes using, sometimes avoiding towns and trading centres, and consolidating protective social ties. Some of these situations are universal but some are specific to the particular physical and social environments that pastoralists inhabit in the African drylands, and have changed over time. This will be seen below as we investigate how, in four societies, pastoralists respond to the exigencies of managing herds in varied environments through territorial practices – involving rights in and control of land and mobility – that purposefully engage with influences of contemporary modernity that limit and constrain their use of land under new and challenging conditions.

Four Cases of Pastoral Mobility and Territoriality

To establish a comparative framework that captures aspects of pastoral social and environmental diversity in Africa, this section examines four systems of pastoral territoriality. The discussion will emphasise the geographical and climatic factors that constrain and shape pastoral movements and the multi-scale institutional framework that has customarily defined their social, political and territorial relations. The presentation emphasises variation in pastoral resources, core institutions and patterns of mobility. The Nuer, Fulbe, Maasai and Tswana cases were chosen because they represent much of the geographical diversity that occurs across the African drylands, illustrate distinct systems of ecology and territoriality, and are well-known and even prototypical exemplars of African pastoralists. The Nuer practise flood-plain herding around Nile tributaries, the Fulbe practise movement across climatic gradients north and south of the Niger-Bani river system, the Maasai practise variations on vertical movement between highland and lowlands, and the Tswana practise cyclical centre/periphery movements around central water points; their core institutions involve Nuer segmentary lineages, Fulbe patrilineal clans, family-based herding groups and a regional ethnic hierarchy, the Maasai a territorially segmented age organisation, and the Tswana a stratified system of chiefdoms. Figure 16.1 demonstrates that these four cattle-keeping communities straddle and seasonally move across arid, semi-arid and dry sub-humid zones, which on average receive from 250 to 800 millimetres of rainfall per year. In wetter areas, domestic livestock become an adjunct of cultivation, while in drier areas camels are herded and cattle thrive only in the wet season.

The section asks how customary territorial systems have been shaped by a combination of local environmental conditions and social and political institutions. A later section will explore how territoriality – including mobility, land use and systems of rights – has been affected by upheavals and interventions generated by state and global forces in recent years. Moreover, how have indigenous pastoral communities responded to currents of modernity as they have sought to secure their livelihoods and their hold on land under difficult contemporary circumstances?

The Nuer of Southern Sudan

In the classic case described by Evan-Pritchard (1940), the Nuer along with the Dinka and other ethnic groups occupied land in the basin of the Nile and its tributaries in southern Sudan (see Figure 16.2). A segmentary lineage system provided a framework for political relations, with territorial sections being organised in terms of a dominant lineage related to other lineages (and thus sections) in terms of a hierarchically organised genealogical model. Similar to other systems found among the Bedouin and other Arabic speaking tribes and the Islamic-influenced Somali, the Nuer established political order through the threat of feud, by which injuries to an individual are rectified by negotiated compensation payable upon threat of retaliation, according to a segmentary logic, by which individuals, families and lineages relate to each other as members of the highest order genealogical unit that divides them.

Along with other members of their section (or 'tribe', as Evans-Prichard called territorial fractions), Nuer benefited from the early rains in May to allow their cattle to graze on young, succulent grasses. As the rains progressed, lower ground became flooded so Nuer occupied ridges where their villages were located and planted crops (mainly sorghum and millet). As water receded and grasses dry, they burned the tough lowland grasses and moved cattle to graze on new grass that grows after fire. Then, as the dry season advanced and as crops ripened around November, they retreated with cattle to the banks of Nile tributaries and swamplands to find riverine grazing and to fish near their camps, while a second harvest was reaped near the villages (see Figure 16.3). Evans-Pritchard succinctly described the rotation between villages and camps: 'Nuer are forced into villages from protection against floods and mosquitoes and to engage in horticulture, and are forced out of villages into camps by drought and the barrenness of vegetation and to engage in fishing' (ibid.: 63).

The Fulbe of Doukoloma, Mali

This Sudanic pulse-like pattern of movement from rivers to hillocks, according to a rhythm of rainfall and flooding, can be contrasted to the bidirectional north–south spatial displacement that takes place in the West African Sahel,

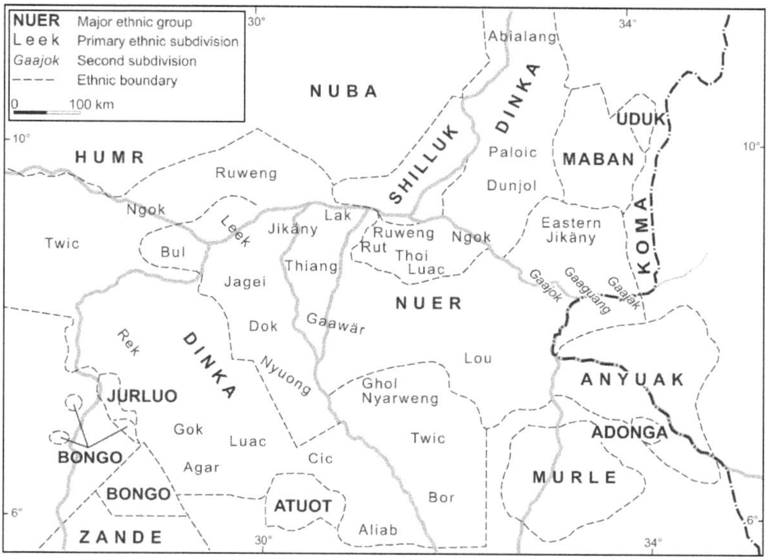

Figure 16.2: Approximate territories of principal ethnic groups in the Upper Nile Region of the Sudan.

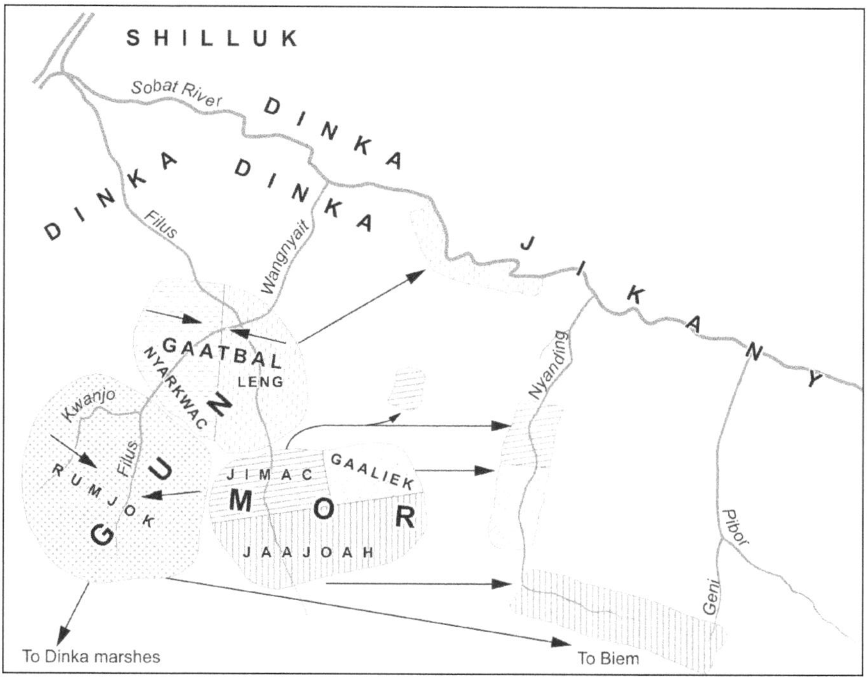

Figure 16.3: Movement of Lou Nuer tribal sections from villages to dry-season camps (after Evans-Pritchard 1940: 56).

which is determined by a combination of rainfall patterns, the rising and falling of the Niger and Bani rivers, and the agricultural cycle. The Sahel experiences heat and rain from June to mid October, then gradual cooling and the cessation of rain from mid October to December, with a dry and cold period from mid December to mid March, and then a hot and dry season from mid March to June. The cattle-keeping Fulbe, who live in the region of Doukoloma, north of the Bani River in Mali (see Figure 16.4), are divided into patrilineal clans (*lenyol*) which were previously critical in controlling grazing lands, especially in the rich Niger delta. Highly autonomous patrilocal households, often made up of a Fulbe man and his married sons, live as a minority among their neighbours, the Bamana. Fulbe practise limited cultivation of subsistence crops (mainly millet and maize), but unlike the Bamana do not grow cash crops. Fulbe are most dedicated to cattle husbandry, which includes transhumant movement across the Bani and Niger rivers on an annual basis, the timing being determined both through autonomous household decisions and in coordination with the entire group (Grayzel 1990: 48).

In October, herds are moved into nearby forests to avoid fields as the millet ripens, but after the harvest in November they are brought into the fields to graze

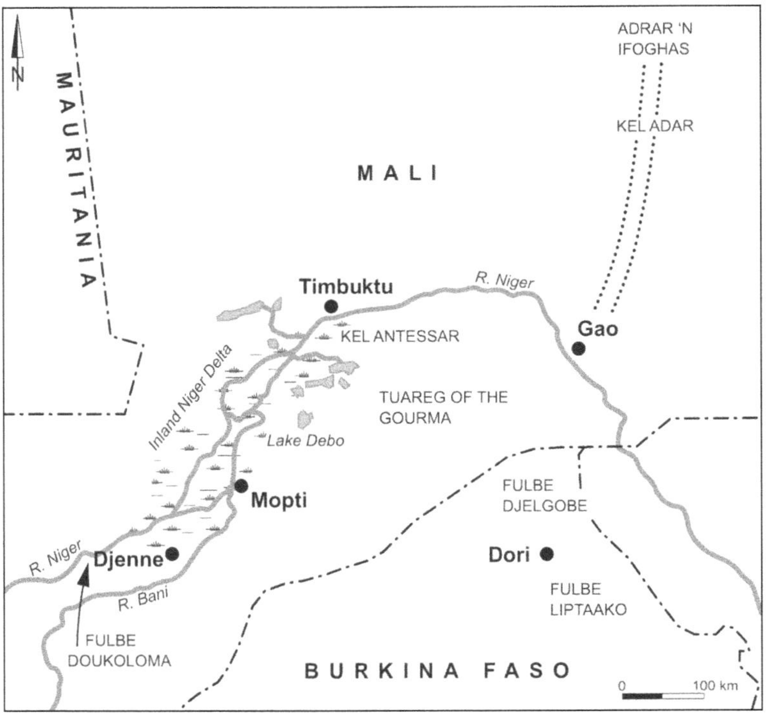

Figure 16.4: Sahara-Sahelian pastoralism around the Niger Inland Delta (after Homewood 2008: 21).

on the stubble and fertilise the land. Waiting for the harvest to run its course south of the river in December, herders depend for local grazing more on good relations with the local Bamana, whom they provide with milk and plough oxen in return for grazing access. The transhumance begins with the crossing of the

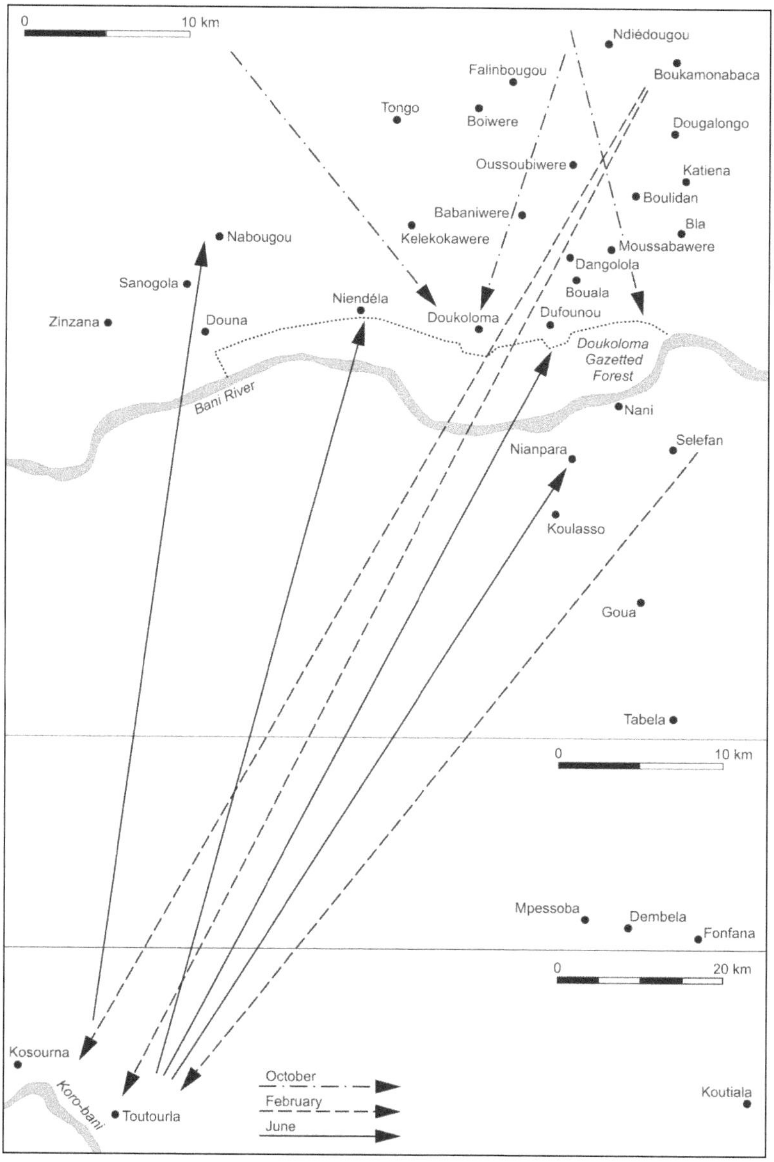

Figure 16.5: Doukoloma Fulbe herd movements in the Bani River region of the Niger Inland Delta (after Grayzel 1990: 50–53).

river, after which herds and herders operate independently from their homes, and continues until May, as herds move southward to avoid being overtaken by subsequent moves of cattle. In June, when rains begin to fall in the south, herds turn back northward to avoid mud and mosquitoes, and in July they re-cross the

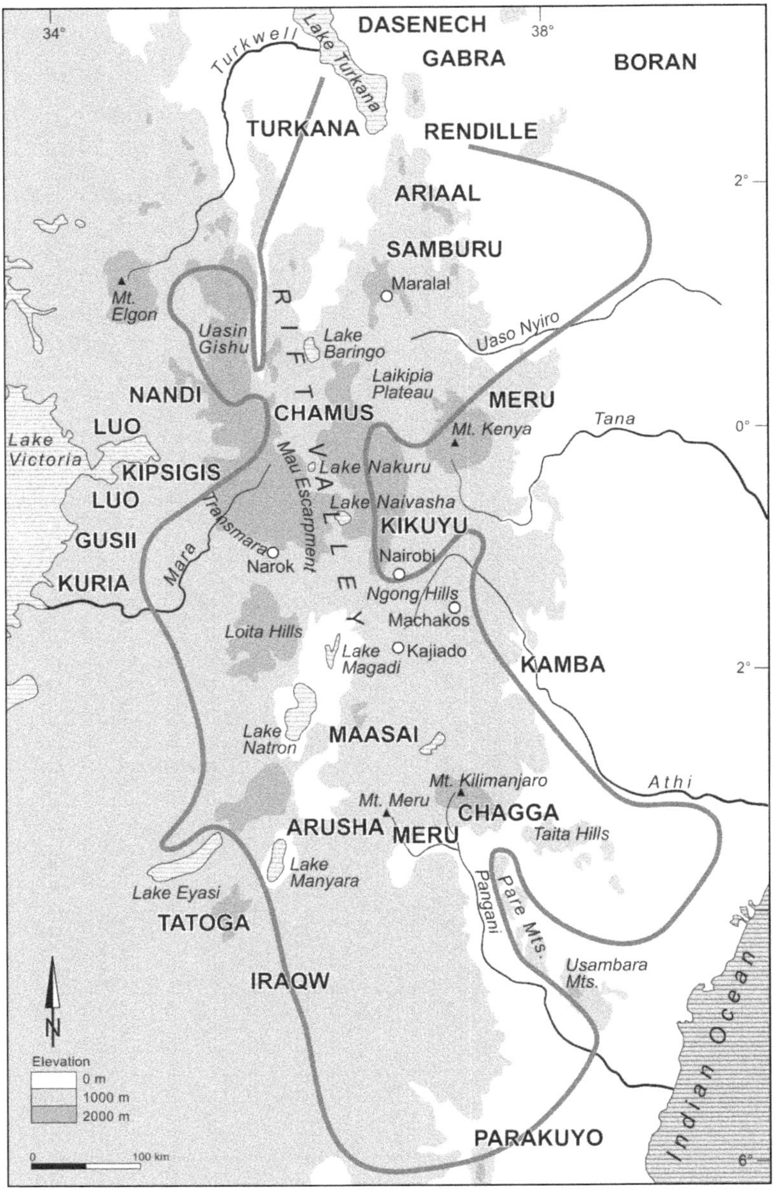

Figure 16.6: Maa-speaking people and their neighbours (after Spear and Waller 1993: 3).

river, the first to return to Doukoloma being the oxen used to plough the fields (see Figure 16.5). All herds hasten to cross as the river rises, mid July being a time of celebration as households are reunited; in August and September, livestock enjoy rich local pastures and are kept in corrals near homesteads to keep them out of planted fields (ibid.: 49–54).

Essentially, the dry season pushes cattle southward and then the wet season pushes them back north, with Tuareg camels descending from the Sahara in their wake to make use of the remaining shrubs and stubble. It is noteworthy that land is serially used by several producers in turn, who retain seasonal rights: farmers or agro-pastoralists to cultivate, cattle herders to bring animals to graze stubble, camel herders to profit from residues and browse.

The Maasai of Kenya and Tanzania

More complex, multi-directional territorial patterns can be seen among the Maasai and other closely related Maa speakers (Samburu, Chamus, Arusha and Parakuyo) of East Africa, who have long occupied the Central Rift Valley region (see Figure 16.6). Maasai territoriality involved a multi-scalar system of political 'sections' (*il-oshon*) organised through interaction between clanship and age organisation, with the latter providing the framework for the political order, the former a cross-cutting set of ties. At each scale, a spatial 'order' of social relations was identified in terms of which pastoral functions were and even now are carried out. Maasai sections were politically autonomous units – roughly comparable to the 'tribes' described by Evans-Pritchard for the Nuer – that occupied and defended territory that included both wet- and dry-season grazing. Each section included an effective army, organised through the warrior class of an age set with its leadership, which was built up out of local battalions formed within geographical sub-sections or neighbourhoods, within which most pastoral movements occurred and even now occur (see Figure 16.7).

Although sections identified strongly with territory, and in some cases were either named after territories or territories after them (Kaputie after the Kaputie Plains, for instance, Laikipiak and Uasin Gishu after the plateaux where they originated), they tested the limits of their regions by seasonally expanding, and in some cases routinely mixing with, members of other sections in wet-season pastures or around permanent water that both claimed some rights to. Within neighbourhoods, pastoral land use was coordinated across grazing areas and seasonality through meetings of male elders, often convened by age-set leaders. A multi-family homestead (*enkang*) usually participated in joint pastoral grazing, often with labour sharing, while within a homestead each constituent polygamous family (*ol-marei*, sing., 'family', or *enk-ishomi*, 'family gate post') both owned and managed the livestock herds on which its members subsisted.

Livestock productivity benefits from a bimodal pattern of rainfall, which yields new grass twice a year that can nourish calves and stimulate milk yields throughout the year. Long rains begin around the end of March and run into

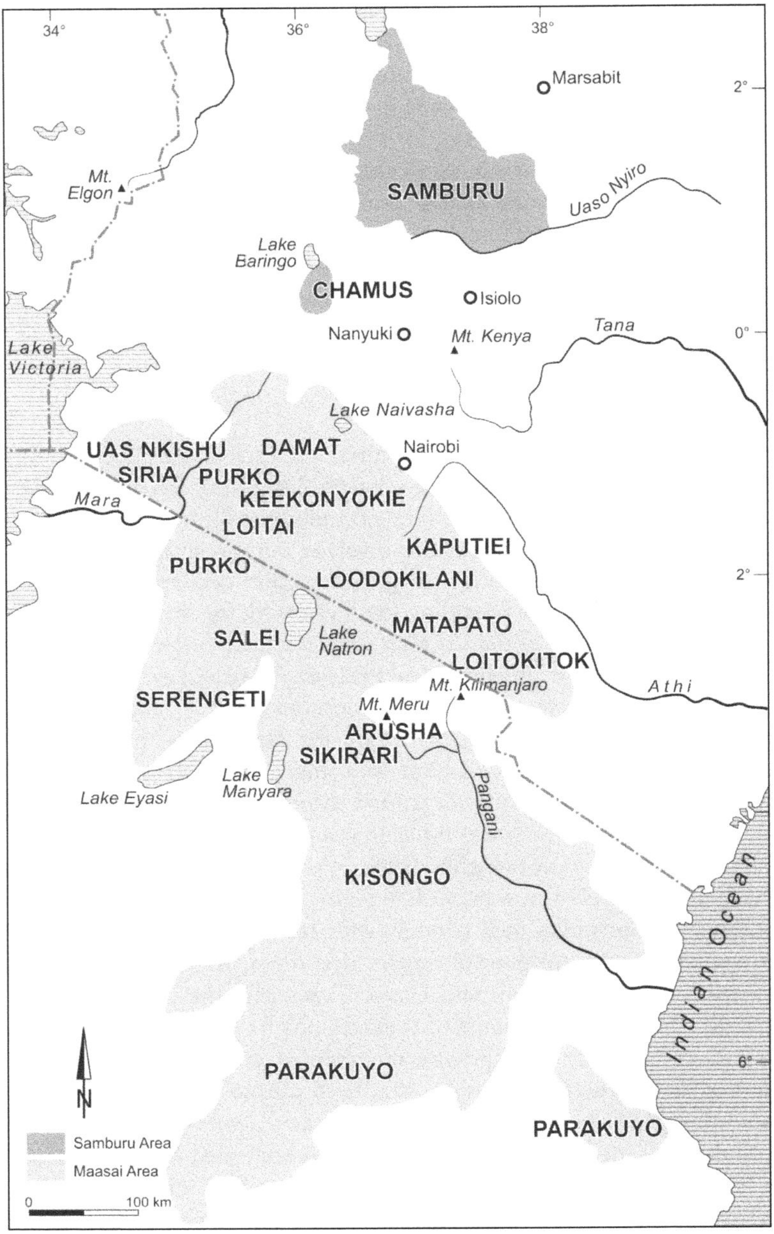

Figure 16.7. Maasai sections and other Maa-speaking groups (after Galaty 1993: 71).

June, followed by a cold, dry season, while the short rains begin in late October and run to December, followed by a hot, dry season. As new grass is established, livestock are taken to wet-season grazing areas, regions without permanent water that can only be exploited when there is standing water in pools or artificially built dams. As water dries up and seasonal pastures are expended, herders gradually retreat to areas with permanent water near lakes, rivers or boreholes. As dry seasons progress, the reduction of tsetse flies bearing animal trypanosomiasis in bushlands near permanent water, such as along the lower Uaso Ng'iro, allows livestock to use pastures near water, especially in swamp and riverine areas where succulent grasses remains abundant even in drought.

In the nineteenth century, Maasai political sections controlled adjacent regions of valley and plateau pastures, the latter often near highland forests that provided fresh water and cool grazing in the dry season. Thus Maasai land use involved vertical movement up escarpments to the wetter hills and mountains in the dry season, or towards areas of low-lying swampy seepage near river beds or at the foot of escarpments, and down to the drier and hotter valley floor grazing in the wet season (Sperling and Galaty 1990). The Keekonyokie (KK), for instance, moved from the Kedong' Valley up to the Kinangop Plateau along the Aberdare Escarpment and the higher regions north and west of Nairragie Enkare along the Mau Escarpment, and made use of the fresh waters and swamp grazing of Lake Naivasha and the Ewuaso Kedong' River on the valley floor. With the rains, they moved south of Mount Suswa (Oldoinyo Oonyokie) to exploit seasonal pastures that lacked permanent water. Purko (PK) and Damat (DT) moved from the Naivasha and Nakuru basins up to the middle and northern Mau, Lodokilani (LD) from the hot, dry plains near Magadi to the swamps and riverine pastures of the lower Ewuaso Ng'iro, up to the Nguruman Escarpment, Matapato (MT) towards the Namanga Hills, the Kisongo (KS) of Loitokitok between the slopes of Kilimanjaro and the swamps of Kimana and Oltukai (now Amboseli Park), the Kisongo of Ngorongoro from the vast wet-season grazing towards the Serengeti to the dry-season grazing of the crater, the Loita (LT) from the Loita Plains up to Loita highland forest, and so forth (see Figure 16.8).

Historically important alliances of territorial sections were built up on the basis of social and dialectical commonalities through negotiations between leaders of age-set battalions within a roughly segmentary structure, and more recently have proven effective in land conflicts. At the most encompassing level, Maasai recognised a cultural affinity within *ol-Maa*, the community of Maa speakers, but since Maa speakers are one another's most intense competitors this rarely added up to extensive cooperation (Galaty 1993). My point is that the highly individualistic and opportunistic herd movements, organised at the family, household and neighbourhood levels, that make pastoral land use so effective cannot be understood apart from this complex, multi-scalar system of territoriality, organised through age-set structures. Progressively through the colonial and postcolonial periods, boundaries of Maasai territory and sectional boundaries became increasingly well defined, but the institutional structure remained in place, and even today exists side by side with the governance structures of the modern state.

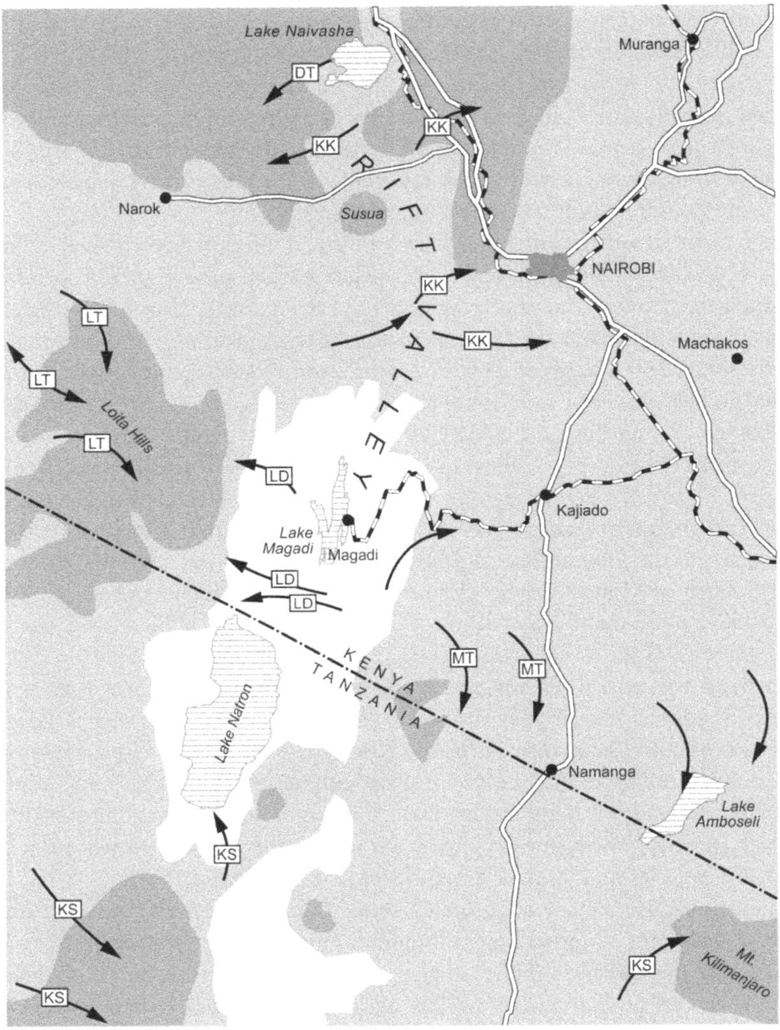

Figure 16.8: Dry-season herd movements by Maasai sections in the central Rift Valley (KK = Keekonyokie; KS = Kisongo; LD = Loodokilani; LT = Loita; MT = Matapato).

A Digression on Pastoralism and Egalitarianism

Theorists have proposed that pastoralism engenders egalitarianism. This is due to the autonomy enjoyed by pastoral families that follows from their high mobility, which itself grows out of the process of livestock management. Also, it is virtually impossible for an ownership class to monopolise livestock wealth, since livestock is a form of wealth that increases with herd reproduction and is moveable (Schneider 1979). Seeming to confirm the notion of egalitarianism, many pastoralists are intensely proud, share an ethos of fierce independence and recognise the reciprocal rights of other pastoralists to make their own decisions and to participate in local governance (Salzman 2004). However, this argument fails to address the heterogeneity of political forms found in pastoralist societies. True, the Nuer, the Fulbe, the Maasai and many other pastoralist societies practice their husbandry through highly autonomous cattle-owning families, live in homesteads where decisions are largely made independently, and recognise the right of each man to participate in political life and to own livestock, which as objects of subsistence, material wealth and cultural value are the basis for cultural esteem and material independence. But pastoralism is also central to numerous hierarchically organised societies in Africa, including highly stratified chiefdoms and states where livestock and land is owned or controlled by kings, chiefs or a dominant class. Tuareg of the Sahara benefit from the labour of non-livestock owning slaves, claim the harvests of land-tied agricultural labourers, and enjoy the fruits of a regional ideology of pastoral superiority (Bernus 1981). Ownership of livestock by the Tutsi king was a token of control over the complex, stratified Rwandan kingdom during the precolonial period (Bonte 1991), and livestock keeping was a key component in the organisation of the precolonial Tswana state. The egalitarian nature of pastoralism might be better stated as follows: that the nature of livestock wealth and the highly decentralised process of holding and herding domestic livestock, including mobility, lends livestock owners autonomy and a sense of equal status with their fellows, but that these pastoral institutions may be embedded within more complex and hierarchical political structures in which the high prestige of pastoralists is in effect purchased at the cost of subordinating non-pastoralists, whose low prestige resulted from their serving as labourers within a stratified system.

The Bakgatla and Bamangwato Tswana of Botswana

The precolonial Tswana state, or *morafe*, was a multi-ethnic but centralised polity that bound together people of diverse ranks, including royal or aristocratic families, commoners, strangers or aliens, and servants, around the person of the chief or king (Peters 1994: 22–25). The state was spatially organised around a town in which all citizens resided, and was ritually centred on the chief. Towns ranged in size up to 10,000 to 15,000 people, with smaller mini-states or wards (*khotla*) mirroring the structure of the state itself, with a leader around whom patrilineal descent groups constituted their family compounds. Within the principal town, which was the 'centre of authority and order for the state', ranked groups were arrayed in and around the chief's compound, and when the town periodically moved – to seek water, pasture and land to cultivate – the social map changed to recognise adjustments in rank. Ideologically, the political order represented concentric circles around the central *kgotla*, the 'great enclosure' that served as the chief's court, ritual ground and a public meeting place (ibid.: 23). Peters proposes that the hierarchy of ranks was ideologically thought of less as a ladder than as circles where those of lesser rank were more distant or 'behind' those of greater rank (ibid.: 26–27).

Cattle were considered the most important form of wealth and the embodiment of social relations and wealth, and their allocation was one of the chief's major prerogatives. Loans of *mafisa* cattle by chiefs provided supporters with the products of domestic animals in exchange for providing care. Chiefs could also allocate privileged access to well-watered pastures and water points to favoured followers (ibid.: 29–30). The expansion of the state westward towards the Kalahari involved the creation of territorial officers or 'governors' who supervised the chief's cattle posts and the people living or hunting in the territory, and so doing concentrated authority and power as well as livestock (ibid.: 36–37). In line with the concentric configuration of power, cattle were moved from town centres to the 'bush' at the distant periphery in accord with the seasons, with hunters serving as servile herders at the farthest remove from the chiefly towns.

In the Central District of Botswana, the dominant Bamangwato live in large, town-like sites made up of numerous houses and stone walls in a centralised settlement system typical of the southern Tswana. Land use is defined by zones, with the main settlement at the centre, surrounded by arable lands where fields are located, beyond which are grazing areas and more remote regions used for hunting (Hitchcock 1990: 219). At the time of the first rains, in October and November, families move out of the main town into the fields, in which they live until March during the rainy season, while they are planting and tending their crops, before returning to the main settlement after the harvest in April as the cold and dry season approaches. Cattle posts, located in the grazing areas, are occupied year round by male family members and hired help or servants, who occasionally go further afield to hunt game. Land use was strictly defined, with arable and grazing lands segregated under the control of the chief (ibid.: 224).

As is the case more generally, four classes composed this socially and politically stratified system: the chief, his family and royal relatives; the common people or (literally) 'servants', absorbed by conquest or submission; foreigners (literally 'refugees') who lived outside the royal capital; and the lowest class of serfs, made up of ethnic groups of low status (ibid.: 221–22). In fact, the Bamangwato who mainly inhabited the capital comprised only 18 per cent of the district population, while others lived in the more peripheral areas. In the Western Sandveld, the click-speaking Basarwa – considered 'serfs' – made up three-fourths of the total population. The Bamangwato kept cattle posts as much as 150 miles away from their main settlements (ibid.: 225); the overseer (*modisa*) was responsible for allocating grazing lands, with cattle posts located not only according to water and grazing availability but also by the decision of the authorities. New sections were created when the chief established a new large cattle post for his heir, and at the same time established 'a new ward of commoners to assist in its management'. As the Bamangwato expanded, they extracted tribute from low-status groups who became serfs, who provided goods to the ward heads and provided labour in herding, in return for which they received clothing, food and drink (ibid.: 227–31).

Cattle posts were placed in remote areas as much for political as economic reasons, since they served to legitimise the claims of the Bamangwato and the Bakgatla to border areas and extend their administrative control (ibid.: 235). Posts were set up near permanent springs and with the establishment of new posts herd owners gained control of the labour of the hunting peoples of the area (such as San or click-speaking Basarwa, Bakgalagadi), since Tswana water rights superseded the rights of local groups (ibid.: 238). Movement of livestock around cattle posts, towards fresh pasture during the rains and back to permanent water in the dry season, occurred within a larger multi-ethnic regional system, with cattle, water and labour being controlled on the basis of this stratified order of classes (see Figure 16.9). At the very least, one must conclude that the autonomy and self-reliance that may emanate from the herding process does not preclude its insertion into wider systems of political hierarchy and inequality. Indeed, where egalitarianism is marked, it is often among free pastoralists in contradistinction to their relations with subordinate and subjugated classes, whose perceived inferior-

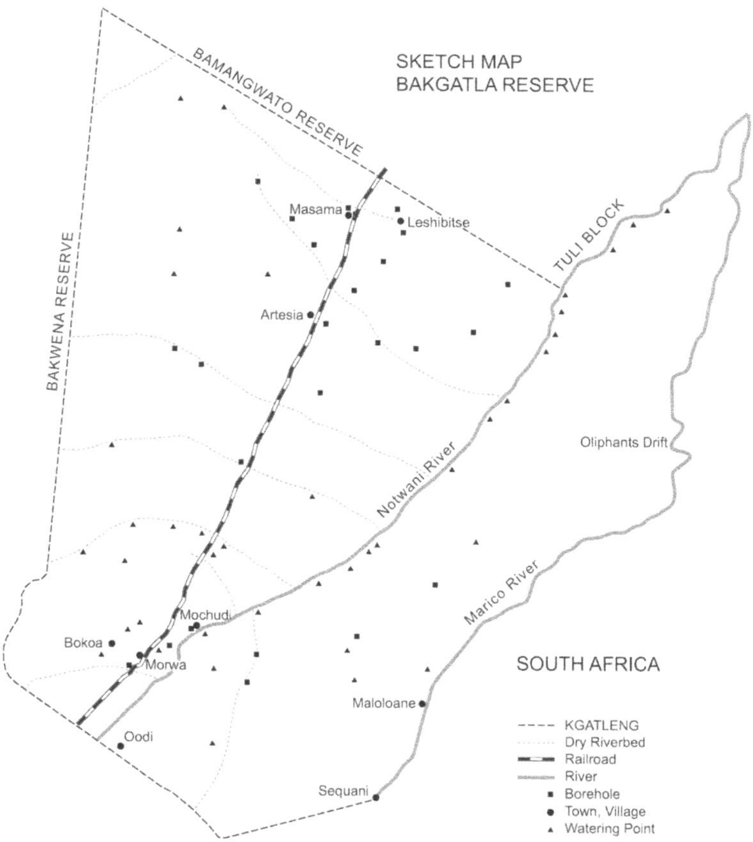

Figure 16.9: The distribution of boreholes in Kgatleng, 1939 (after Peters 1994: 60).

ity is the coin that purchases the sense of pastoralist superiority. Clearly we need a more powerful theory of egalitarianism and inequality than the simple deducing of egalitarianism from the constraints of pastoralist practice and autonomy. In contexts of scarce water and an ideology of hierarchy, livestock can indeed be monopolised, becoming both a privilege and an underpinning of superior status, and the basis for subordination of lower status groups.

Social Pathways and Diversity in Pastoralist Land Use

These four cases – which illustrate societies similar in their intensive pursuit of pastoral livelihoods – represent several forms of social and environmental diversity that occur in widely varying geographical and climatic systems found in Africa. Patterns of movement are, on the one hand, entrained by the seasonal distribution of grazing resources and rainfall. The Nuer move towards, and then away from, the Nile and its tributaries, responding to seasonal flood patterns and the rhythms of grazing, agriculture and fishing. The Fulbe cross the Niger and Bani rivers southward when they are low, as the dry season progresses, and northward as they rise, with livestock congregated in villages and dispersed in northern pastures in the agricultural season. To simplify, Fulbe are attracted northward by rain and southward by drought. In contrast, the Maasai move in more complex, multi-directional patterns, that befit the bimodal rainfall pattern of the region, seeking more remote, waterless rain-fed pastures during the wet season and higher dry-season pastures and perennial streams in the dry season, moving from grasses that spring up only in rain to more resilient, permanent pastures or highland forest grazing. The Tswana move outwards towards cultivated fields and towns towards a drier periphery most suitable for grazing where cattle posts have been created around wells and boreholes; cattle graze – often untended – in broad concentric circles, periodically drawn back to water.

All four of these societies practice semi-nomadic movements that help them articulate settlements with access to pastures and availability of water. Do they have a sense of 'home' to which – after their peregrinations – they seek to return? Evans-Pritchard is clear that Nuer dry-season villages where they cultivate are sites of permanence, dry-season camps where they fish being temporary abodes. Fulbe depart from and return to villages near the river, where they cultivate or trade with those who do, and create temporary encampments in more distant pastures to the south (in the dry season) and north (in the wet season). Maasai too are most comfortable in wet-season pastures, where they relax, since dry-season moves are often carried out by smaller groups of young men and more frequently. Tswana towns and villages are quite permanent, with cattle posts being frequented by workers. The customary oscillations of all four groups have a certain regularity in normal years: 'pulse like' according to seasonality. But the spatial configuration that defines their primary movements are quite different: the Nuer from hillock villages to riverside cattle camps, back and forth; the Fulbe pursue long distance linear transhumance back and forth across major rivers, riverside forests and fields serving as their hub; the Maasai move more vertically, from lowland wet-season to highland dry-season pastures; and the Tswana follow a concentric pattern between centres of social life and water to distant pastures.

Patterns of movement are not only shaped by the ecological distribution and seasonal availability of resources but also by the social construction of herders' relations to those resources – in short, property relations – and to other social groups. Institutionally, the Nuer and the Fulbe strongly depend on lineage systems for their core system of social organisation, the Maasai on age-set systems and the Tswana on a social order organised according to stratified chiefdoms. Rights in resources derive both from group affiliations and the establishment of 'presence' as herders make claims-through-use (Galaty 2005a). As opposed to a generalised notion of 'common property' open to all that pastoralists are too often wrongly associated with, it is affiliation with a Nuer 'tribe' such as the Lou or the Leek, a Maasai *olosho* section such as the Keekonyokie or Lodokilani, a Fulbe local group such as the Doukoloma, or a Tswana *morafe* chiefdom, such as the Bamangwato, that carries with it rights to live in and use the resources of a certain territory. But more particular claims are established by other processes, such as allocation of land or a cattle post by a Tswana chief, establishing residence in a Maasai dry-season neighbourhood through formal request, negotiating the return use of a farmer's stubble by Fulbe, or joining a wet-season village for farming by a Nuer. Let us not forget that pastoralists are customary experts at establishing claims to water or pasture through simply showing up, perhaps with friendly greetings, but often with an implicit threat of force. At the first appearance of herders, their use of pastures is seen as a gift from locals; subsequent appearances represent claims that over time become rights. If the formalisation of rights through the state has been intended to make rights claims through negotiation and pressure irrelevant, in many cases property continues to be a social as opposed to a simply legal process.

There is a dynamic to herding which cannot be captured through analysis of formal rights or rules, or even by patterns of decision making, for herding involves an often- surreptitious search for grass that is often 'opportunistic', audacious and forceful (Scoones 1995). Equally, pastoral movements are undertaken within the sort of political context alluded to under the discussion of the Maasai above, which includes relations between pastoral sections and between herders and farmers, who over time both compete and cooperate. 'Governance' of territory thus occurs not only within institutional systems – of kinship, clanship, lineages, sections, and age sets, as well as networks of alliances or partnerships – but also of intra- and inter-ethnic dynamics between territorial groups. There is a path dependence to pastoralists' movements that biases where moves can most profitably be made, due to attraction to or affinity with (or antagonism with) groups and communities who make up the social topography of a region. Research on the Nuer speaks of their important relations with the Dinka, who – though 'foreigners' to Nuer – are both friends and antagonists whom Nuer raid and by whom they are subjected to raids for cattle (Evans-Pritchard 1940:

122–23). If the Bamangwato are a minority in their country, it is because they are surrounded by San-speaking Basarwa and Bakgalagadi whom they have incorporated as bound labour (once termed 'slaves' or 'serfs') who provide crucial herding labour in the pastoral process (Wilmsen 1989; Hitchcock 1990: 231). Fulbe existence is closely inter-related with the more intensively farming Bamana (or elsewhere to Hausa, Dogon or Wolof), who, forming the vast majority in Doukoloma, sell their labour to Fulbe and negotiate with Fulbe over access to forests and fields and trade during the latter's transhumance (Grayzel 1990: 39–41). From northern Kenya to central Tanzania, Maa speakers have had close relations with other pastoralists and to neighbouring hunter-foragers and agriculturalists, most notably to the Kikuyu of central Kenya, with whom economic complementarity long obtained until the onset of postcolonial land conflicts exacerbated over the last fifteen years by electoral politics (Galaty 2008). Close relations with other pastoralists, including both exchange and conflict, occurs between Samburu and Ariaal and the Borana, Turkana and Pokot of northern Kenya, with Kipsigis and Nandi of western Kenya, and with Barabaig in northern Tanzania.

One can contrast situations where pastoralists represent majorities in their territories and interact with other pastoralists or farmers at their peripheries (for example, Maasai and Nuer), and situations where herders are a minority among farmers with whom they live seasonally in intimate though often uneasy proximity, recognising or contending over complementary access to resources (for example, the Fulbe and Bamana, Kisongo Maasai and Arusha, and the Parakuyo Maasai and numerous cultivating communities in central Tanzania). Clearly, the ethnic politics of accessing resources through mobility is more complex where groups are intertwined. In every case, it is difficult to separate out the environmental and political factors involved in pastoral mobility, and it is imperative to examine questions of mobility and semi-nomadic land use from the perspectives both of the relatively autonomous herder and of the complex institutional and social structures, both constraining and enabling pastoral movement, which form invisible social pathways across the African topography.

Postcolonial Territoriality and Mobility

The customary nature of territorial organisation does not take away from its continuing viability under contemporary conditions of pastoral husbandry, but each level of pastoral scale takes on a new meaning in the contemporary situation. In the postcolonial period (that is, since independence was achieved by most African nations in the early 1960s), several important transformations have profoundly affected systems of pastoral territoriality, most specifically patterns of mobility, settlement and property, but also the exercise of political authority, governance and provision of local security.

Firstly, growing population and increased need for land has led to farmers encroaching on rangelands, often encouraged by the state; this has circumscribed grazing areas and led to increased conflict between herders and farmers (Campbell et al. 2000). Secondly, increases in pastoral populations have led to urban migration and the growth of small towns and trading centres, drawing pastoralists towards settled areas in order to access employment, health services, schools and markets. While supporting economic diversification and absorbing population growth, this has curtailed the orbits of pastoral movement (Fratkin and Roth 2005). Thirdly, class formation and social stratification have strengthened the hand of the new entrepreneurial and political elites who wish to appropriate rangeland resources, whether for mineral extraction, tourism, conservation or commercial development (Galaty 2005b). Fourthly, the development of an international market in automatic weapons coupled with a sense of pastoralist alienation and grievance has given rise to cases of armed resistance by pastoralist populations, whether through increased violence within and between pastoralist populations, endemic banditry or insurgencies against forces of the state, all of which have profoundly disrupted dryland societies and economies (Eaton 2008). Fifthly, states have joined with international donors to initiate programs of land reform in range areas, which although aimed at development of the livestock sector have often undermined the land rights of pastoralists. In short, decreasing control by pastoralists over rangeland territories has greatly complicated the practice of mobile grazing, access to critical resources, food security and land governance while increasing land fragmentation at the cost of livelihoods and conservation (Galvin et al. 2008).

In the sections that follow, I will examine major events that have impacted on pastoralist land use and mobility, in effect putting the systems of territoriality of the four cases we have already examined into their wider political and cultural contexts, including changes in rangeland property discussed in most detail with respect to the Maasai of Kenya and Tanzania. The modernity of contemporary pastoralism emerges out of the forces and events enumerated in the five trends just presented for the postcolonial period, but it should be noted that population growth, the development of small towns, markets and other sedentary institutions, growing social stratification, and shifts in claims to land are long-term processes that began in the colonial and in some cases precolonial periods.

The Nuer

The Nuer were swept up in the Anyanya rebellion that broke out in 1961, as the south refused to be controlled by what they saw as the Arab-dominated government that took power at Sudan's independence. With international mediation, the south was guaranteed autonomy according to the 1972 Addis Ababa Accords, but – aggravated by the Jonglei Canal project that would have changed the south's Nile basin ecology by diverting its water northward for irrigated agriculture, unilateral imposition of shariah law and the abrogation of the unity of the south, carried out after President Numeiri seized power in Sudan – the south turned again to armed rebellion (Hutchinson 1996). The Nuer and the

Dinka together formed the major fractions of the Sudan Peoples Liberation Army (SPLA), led by the Dinka, John Garang (Deng 1995). Northern armies ravaged southern grazing lands periodically from 1989, and, stimulated by the Popular Defence Forces Act that empowered the enemies of southern communities by making them agents of the Sudanese state, the pastoralist Bagghara Arabs, allied with northern government forces, raided Nuer communities for livestock and slaves. With agriculture disrupted and livestock and people being seized, the south entered a period of destitution, ameliorated only by a pipeline of international food aid entering the southern Sudan through northern Kenya.

In 1991, as Ethiopian support for the SPLA ceased with the overthrow of the socialist Dergue government in Ethiopia, a faction led by the Nuer, Riek Machar, tried to overthrow Garang and then split off from the main SPLA, initiating Nuer–Dinka strife in the south (Hutchinson 1996).[2] There is some evidence that the US began providing arms to the SPLA around 1996, to counter the government of Bashir and perhaps pressure him to expel Osama bin Laden (which occurred that year). International mediation and negotiations carried out between 2002 and 2004 led to the Comprehensive Peace Agreement (CPA) between the north and the once-again united south in 2005, followed by the formation of a unity government with the SPLA chairman serving as president of the autonomous region of Southern Sudan and vice-president of Sudan. A continuing source of aggravation is the as yet unsettled status of the oil-rich Abyei Province, from which Nuer have been largely displaced as oil companies have established oil-producing enclaves in the blocks allocated by the Sudan government to oil companies from China, Malaysia, India and, until recently, Canada, from which oil was exported both during the war and continues today (see Figure 16.10). The Government of the Southern Sudan (GOSS) gained the right of self-determination. A referendum on secession was successful with 98.83% support, which led to the attainment of independence on 9 July 2011. Continuing conflict between the two states has occurred over land and control of oil-rich border regions, potentially putting current oil contracts with foreign companies into question. The rights of the Nuer to the pastures they were denied by decades of warfare are of great weight, since their ancestral lands sit on considerable oil wealth. In the meantime, the war that lasted from 1983 to 2005 profoundly disrupted Nuer livelihoods and access to pastures due to armed raids, loss of livestock, loss of pastures through their expulsion from key areas, and loss of mobility due to generalised insecurity. The Nuer movement pattern from hillock villages to riverside camps has been profoundly disrupted by insecurity, massive loss of livestock and the concentration of populations in towns where famine relief and extensive services are provided by the UN, through the structures of the new GOSS. Fighting over the modern spoils of nationhood, the societies of Southern Sudan transcended their particular identities in the context of a civil war in which global alliances played a key role. But tensions between Nuer and Dinka over authority, land, and access to patronage have undergone a resurgence within the context of southern governance.

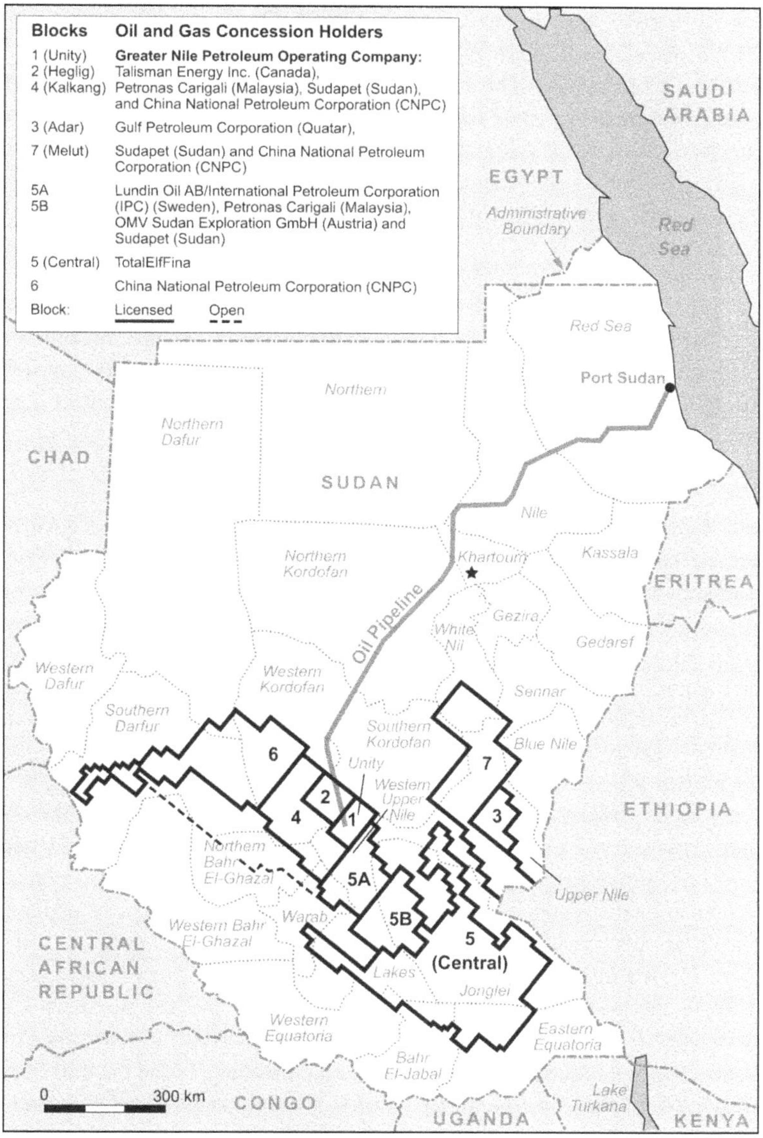

Figure 16.10: Oil concessions in central and southern Sudan.

The Fulbe

The factors that have influenced the evolution of Fulbe herding practices in Mali include the growing role of the state, the spread of agriculture and wage labour, and the growth of agricultural populations, which has increased land-use conflicts with herders. Fulbe scattered after a revolt against French colonisers was suppressed in 1892, and began to concentrate in Doukoloma in response to the colonial confiscation of livestock during the Second World War, since the area was wilder and less settled. Though it was seen by Fulbe as a 'good place to hide', it was rendered healthier after the war when the French sought to reduce predators and tsetse flies, improving the value of grazing resources for Fulbe (Grayzel 1990: 63). Through providing security and environmental control, colonialism did open up new grazing territories and, by creating towns and centres, expanded the commercial market for livestock and the demand for wage employment. Given their general distaste for agriculture, Fulbe were quick to use the ox plough since they possessed livestock to use and found it easier in cultivating their millet, maize and cow peas (ibid.: 44). Given their ethic of autonomy, Fulbe evade the condescension with which they view labouring for other people by working as herders for relatives (paid in livestock, not money), for non-kin at a greater distance, or as specialised herders or livestock agents for non-Fulbe (ibid.: 45).

Present government policy has in principle rendered all uncultivated land open to grazing by anyone, thus allocating land rights to farmers but not to herders. By thus undermining even local use rights of herders, the government strengthens farmers' over herders' rights, thus encouraging cultivating drylands, which effectively excludes pastoralists from critical resources and undermines the complementary use of land according to seasons by both parties that has long characterised pastoral farming relations (ibid.: 64). Fulbe access to riverine areas for grazing and transhumance has in this way been severely limited, exacerbating their relations with farmers who deny them their traditional rights, diminishing the grazing lands available to herders and undermining their seasonal claims, thus making transhumance more and more difficult even as drier conditions – seen at the extreme in the great Sahelian drought of the early 1970s – make it more and more necessary. In contemporary Mali, violent conflict has arisen between Fulbe and Dogon over attempts by the latter to prevent herders from accessing pastures near fields, so now only Fulbe women can enter towns in order to sell milk, as their men would be at risk. Showing remarkable flexibility and characteristic opportunism, Fulfulde-speaking herders have spread eastward across Africa, from furthest West Africa to Sudan, Ethiopia and the northern Congo. But faced with governments that have limited their rights, and with conditions of insecurity in Chad and the Central African Republic, Fulbe have been forced to retreat into Cameroon, where they enjoy a legally fragile existence but press their claims to indigenous and aboriginal rights based on new forms of identity (Pelican 2009). Due to their inclusion in modern states across West Africa (as far east as the Sudan and western Ethiopia), Fulbe have had to reformulate the stance of ethnic honour that they have long used to assert their position within a regional status hierarchy towards cultural strategies that have indigenised their assertion of rights as modern citizens to land based their 'autochthonous' history of pastoralist land use.

The Maasai

In Tanzania, Ethiopia, the Sudan and elsewhere, state ownership of land has led to pastoralists' territories being taken for ranching, agricultural or tourist enterprises, whether operated by the state or – in the neoliberal period – by private companies. In Kenya, the richest Maasai lands were expropriated for European settlement in the early part of the twentieth century, with Maasai moved into a formal 'reserve' in what is now Kajiado, Narok and Trans-Mara districts (Hughes 2006). European settlements were placed under private title, as were highland farms held by Africans during the late colonial period, and as a matter of course Maasai and Samburu districts underwent a process of rangeland enclosure including comprehensive land adjudication and titling beginning in the 1970s, under which individual and group ranches were created (Galaty 1994a). Demarcation and privatisation of these pastoralist lands has been accompanied by far-reaching changes in patterns of settlement and land use. The group ranch programme was conceived in the mid 1960s; at the same time comparable experiments in pastoral development were being pursued elsewhere in Africa, such as in Botswana (the TGLP to be discussed below), Tanzania (the Maasai range and livestock programme) and Niger. But nowhere was transformation of pastoralist land holding carried out so unremittingly as in Kenyan Maasailand, perhaps because the region lay sufficiently near to the capital city of Nairobi and in sufficient proximity to highland areas that it was deemed attractive to others as well (Mwangi 2007). It is not by chance that the major wildlife reserves and national parks of Kenya and Tanzania were created in pastoralist regions, most specifically Maasai and Maa-speaking Samburu areas, because these pastoralists have abjured consumption of game meat and have considered wildlife and livestock able to live in complementarity on the East African grasslands. But the demarcation of pastoralist lands created a contradiction between the goals of sedentarising pastoralists, with the ill-conceived aim of making them into dryland farmers, and of conserving great herds of wildlife, which require not just the concentrations of water and swamp resources around which parks and reserves have been founded but also more extensive pastures into which they can disperse and through which they can migrate during wet seasons (Thompson and Homewood 2002).

While group ranches were originally created with the aim of securing land for collective pastoralist use (within particular locations of Maasai sections), their viability was soon undermined through three processes: the extraction of parcels from group holdings through the granting of individual titles to local elites; the corrupt inclusion of outsiders on the list of registered members; and the eventual subdivision of group ranches among their registered members, both through outside political pressure and internal dynamics (Galaty 1994b; Lesorogol 2008). The latter occurred as ordinary group members saw their group holdings shrink as more extractions occurred, their land being used as reserve grazing by elites who owned their own individual ranches, and their numbers growing as new generations of young men were added to the register. In effect, a 'run' on the group ranch occurred, as members increasingly concluded that they would be better off claiming a portion of land now rather than risk receiving less or even nothing in the future (Galaty 1999). But when many group ranches were subdivided, some members found that, due to corruption, allotments were made to outsiders, some rich and politically influential, while they received little or nothing. And many who received land were quickly pressed to sell portions to eager outsiders, and many found that the need to make a living under the strictures of diminished land with fewer livestock found them cash poor, so that they sold land simply to live (Mwangi 2007). In this way, titling land to make it secure, which is the classic argument in favour of formalising land tenure, has in fact led to a drastic undermining of pastoralist land holding.

This process has been paralleled by several other factors of social change, including: increased educational participation, requiring school fees and taking children out of the pastoralist labour pool (Archambault 2011); greater commercialisation of livestock, to feed the national market and provide cash incomes (McPeak and Little 2006); and increased sedentarisation, as families have settled, in part to access social services, in part to claim land (Fratkin and Roth 2005). Livelihoods have also been diversified, as some pastoralists have adopted agriculture and others have leased their land for commercial agricultural production, especially for wheat production in Narok and Samburu districts (Thompson and Homewood 2002). The expansion of agriculture into the hills and swamps that have long provided the means for pastoral survival in dry seasons has also created conflicts both with herders and wildlife, and has compromised the pursuit of wildlife conservation in key areas surrounding parks and reserves (Campbell et al. 2000). The loss of land in the interest of conservation, to the expansion of agriculture and through a fraudulent titling process, has left Maasai pastoralists disillusioned, and this has certainly contributed to the violence that has surrounded the transition to democratic electoral politics (Galaty 2008).

The creation of group ranches has lent local customary politics an additional institutional platform, along with traditional age-set leadership that has functioned side by side with the provincial administration. In principle, group ranches were supposed to decrease pastoral mobility, as were individual ranches, with livestock numbers being limited to the 'carrying capacity' of each parcel of land. The inapplicability of this static ranching model to actual rangeland conditions in Africa has been demonstrated by studies of range ecology (Scoones 1995). But most importantly, the dynamic mobility that underpins the success of rangeland pastoralism has been severely compromised by demarcating and enclosing pastoral lands. How have Maasai pastoralists responded in terms of new strategies of territoriality and mobility as they have charted evolving forms of land use?

The loss of lands in the central Rift Valley to colonial settlement between 1904 and 1913 severed the link between several Maasai sections and highland pastures on the Mau Escarpment, above Lakes Naivasha and Nakuru, and on the Kinangop Plateau to the south-west side of the Aberdare Mountains. The progressive demarcation of districts, sectional or 'locational' boundaries, group ranches and finally individual ranch boundaries have inhibited herd and household mobility, disrupted coordinated lowland/highland movements, and increased conflict arising from the transfer of resources such as swamps, boreholes, wells and trekking paths from community to privatised holdings. Still, like water flowing downhill, livestock movements must occur for the health of animals and pastures, but they now tax the ingenuity of herders.

Under greater restrictions at home, most pastoralists with their livestock spill over their own boundaries in times of drought, seeking pasture by moving into forests, national parks, common lands by the side of roads and around the international airport, and into the capital city; they have sought access to large under-utilised farms, have negotiated access to grazing lands of friends where some grazing remains, or have invaded commercial ranches in areas once owned by pastoralists. In short, the new mobility makes use of the interstices of modern life where grass is valued less than by pastoralists. But, more importantly, the unrealistic restrictions that formalised land tenure places on pastoral land use are consistently evaded, ignored or bargained over, as herd owners use time-tried strategies of flexibility and negotiation. These are precisely the attributes by which Berry (1993) characterises customary tenure and forms of local resistance that arise in response to overly formalised property systems, ways of bending rules that in fact make such systems work. At the same time, pastoralists have used new technologies to exchange information and make their strategic moves more efficient. The mobile telephone is now used by herd owners in instructing their workers where to move herds, in negotiating land access and in accessing prices at markets without having to actually appear (Galaty 2007). During the drought of 2008–2009), none of the seven families from Oltepesi in Keekonyokie section interviewed about mobility strategies practised over the previous year remained on their individualised ranch parcels; due to lack of local grazing, all had sought pasture elsewhere, in Keekonyokie or in other sections, using kinship or friendship networks. By necessity, they crossed other individual ranch lands, usually without permission and in some cases by cutting fences, to reach sites that had slightly more grass than their home areas. But, following the precept of 'it is your turn today, the next turn will be mine' (*naailookino mootiok kima*), they did not refuse other herders permission to use their ranch areas when they received some rain.

Access to land is no longer just a local affair, nor just a legal one, as herders submit their demands to political leaders who press their claims for emergency access to national lands, such as parks, reserves, forests and roads. Pastoralists also use the political channels of government and civil society to influence legislative and constitutional debates over property since they see their land rights as issues of high policy. Their claims are increasingly articulated in terms of indigenous rights, with their national status strengthened by being projected onto the international forums of the UN and organisations of civil society (Galaty 2008).

The Tswana

During the colonial period, Botswana functioned as a livestock-keeping appendage of South Africa, providing meat in exchange for other goods. Colonial authorities interpreted Tswana customary law as holding that producers could graze an indefinite number of cattle but could neither fence, sell nor lease grazing lands (Worby 1988: 160; Hitchcock 1990: 217). Following the post-independence drought of 1965, cattle numbers tripled within a decade (to over 3 million head), but were increasingly held by fewer, richer people (Worby 1988: 160–61). Although traditionally anyone had unrestricted access to natural water sources, manufactured water sources – such as wells and boreholes – belonged exclusively to whomever invested labour and capital in constructing them (ibid.: 162). While chiefs retained the right to allocate land in the post-independence government, their actual control of land diminished as water sources were developed, which in arid Botswana entailed monopoly control of the grazing that surrounded a water point and cattle post. Hitchcock points out: 'although it was impossible for tribespeople to gain *de jure* rights over areas of land, the development of water sources, in effect, gave them control over areas surrounding the water point. In this sense, water was crucial to the land tenure system in the grazing areas of Botswana' (ibid.: 217).

The Tribal Grazing Land Policy (TGLP), initiated in 1975, aimed to remove large-scale livestock owners from communal areas, which had become overgrazed, by allocating them new grazing areas in the sand veld, where individuals and groups gained exclusive leasehold rights in exchange for rent (ibid.: 248). The programme thus radically redefined tenure in the communally zoned (or 'tribal land') areas, where most people and livestock lived, by creating designated commercial farming areas where large stock holders and syndicates formed leasehold ranches, 'communal' grazing areas where enclosure was prohibited, and 'reserve' areas to meet future grazing needs, a category soon abolished given the high demand for land (Peters 1994). Wealthy cattle owners were the major beneficiaries, who, faced with the possibility of gaining exclusive title and the ability to fence land, undertook a rapid land grab launched by the indiscriminate digging of boreholes (Worby 1988: 165). Poorer herders were either excluded or forced to pay grazing fees. Overgrazing on the commercial ranches, exacerbated by the establishment of watering points in very dry regions, soon forced some herders back to the communal lands, reversing the expected flow of people and livestock. Rather than increasing their security, the poor have experienced loss of land and economic opportunities. And the provision of individual title has not resulted in control on livestock numbers or protection against overgrazing (Hitchcock 1990: 248–49). One outcome anticipated by planners was the emergence of an entrepreneurial class of livestock keepers, who benefited from close ties to government to secure water rights and thus land and from special agreements with the British government to export commercially produced meat to Europe. In the meantime, poorer Tswana herders seem to have fewer prospects and diminished access to the water and grazing resources they need. Furthermore, San-speaking Basarwa have lost hunting territories to herders and often live on wages earned for herding services as they increasingly become a rural proletariat (Wilmsen 1989), though their assertion of indigenous rights in the international forum has lent publicity to their struggle to gain recognition of their rights to territories now being claimed by the government.

Conclusion

This chapter has examined territoriality and mobility practices in four sites of dryland pastoralism in the context of two frameworks: the first concerns the interaction between seasonality, local ecology and socio-political institutions; the second, the impact of state and global dynamics on localities, including development initiatives that have affected pastoralist property and land rights and experiences of conflict.

Pastoralism has been described as a continuous search for protein across the African landscape. Across the Sahel, a hot, rainy season begins in May and extends into October, with a long and coolish dry season occurring until mid March, becoming warmer but still dry through May or June. Nuer retreat to their agricultural villages as the rains increase, and only move onto the riverine flats of the Nile system with the dry season, for grazing and fishing. With the rain, the Fulbe concentrate their livestock near villages north of the Niger River during the agricultural cycle, or in some cases take them to more distant pastures northward into the desert, and then retreat to riverside grazing as the dry season progresses, before moving across the river to seek grazing further south. River and rain establish the tempo of people's movements, their links with farmers the precise direction of their moves. In East Africa, Maasai pastoralists enjoy life in their main residences during the two annual rainfall seasons in November to December and April to June, though also exploiting distant pastures which are only accessible when rain pools form; as the hot and cold dry seasons in January to March and July to October progress, they move more temporarily to areas of perennial grass, whether swamps, riverine forests or highlands, often separating herds from homesteads. In Botswana, rains from October to March allow livestock and people to depart villages and spread out, people to agricultural fields, domestic animals to more distant pastures. After the harvest, from May to September, people retreat to villages and towns for the long, dry season, while most livestock retreats to cattle posts, clustering around sources of permanent water.

Clearly, no matter how important the distribution of edible protein may be, the ecology of the African grasslands is not sufficient to predict or to explain livestock movements, which are also intimately tied to a hierarchy of social and economic institutions and other livelihood dynamics: the larger agricultural cycle, the social relations of herders, the availability of labour throughout the year, their own involvement in diverse forms of food production and markets, ties between herders and farmers, social networks, and systems of rights in land

and grazing resources. But the quest to find nourishment for domestic stock is incessant, with the diverse social and economic factors mentioned above being modalities, opportunities and constraints that shape how this overriding requirement is served.

The second framework, emanating from state and global influences, provides yet another set of opportunities and constraints that have altered the political ecology of pastoralism. Five factors of change were identified above that are profoundly altering the political terrain of pastoralism: first, population growth and encroachment by non-pastoralists on grazing lands; second, the growth of small towns and increased sedentarisation; third, heightened social stratification and the role of elites in appropriating pastoral resources; fourth, the spread of arms and increased conflict among pastoralists and between pastoralists and the state; and fifth, the role of development and land reform in fragmenting rangelands and dispossessing pastoralists. These factors are at work throughout Africa, although their weight differs between regions and countries.

Encroachment by non-herders on pastoral lands is more striking where higher rainfall makes agriculture possible, even in drylands, in particular in localised oasis, riverine, swamp, escarpment, forest or highland regions where water seepage or even irrigation make cultivation possible, precisely the sites valued as dry-season grazing that significantly enhance the productivity of the wider and drier reaches of pasture. Fulbe increasingly experience friction with cultivators such as the Bamana, who seek to exclude them from the banks of great rivers, as do the Maasai who have been losing lands to Kikuyu and other cultivators in the highland Mau forests, along the Nguruman Escarpment, throughout the Ngong Hills and in the foothills of Kilimanjaro. This is less an issue when herders like the Nuer are also farmers, or where pasturelands are too dry for cultivation, such as in the Kalahari and Saharan fringes, the former utilised by Tswana or Herero herders, the latter by northern Fulbe and Tuareg.

The concentration of people near sites of trade and social services is widespread; this trend in fact helps pastoralism to thrive by absorbing population growth in villages, towns, cattle posts and market areas, providing jobs and other forms of economic diversification. The phenomenon has also led pastoralists to settle in some areas as a strategy of claiming land, which when linked to non-mobile husbandry can threaten the well-being of the grassland ecology by concentrating livestock. Settlement is often associated with modernity and a higher quality of life, since settled families often access health services, schools, markets and communication networks. But when herders settle they often experience lower levels of nutrition and health, and where animal husbandry is curtailed they can suffer from loss of wealth and well-being (Fratkin and Roth 2005). All of the groups reviewed here have found an alternative strategy of combining settlement with continuing livestock mobility, carried out by cadres of young men or hired help.

With the spread of education, the growth of rangeland towns, the development of a political and bureaucratic class, and the increase in the importance of commercial trade, pastoralist communities have experienced social stratification based less on livestock wealth than on access to political or commercial advantages. This has proved quite critical at the historical moment when pastoralist resources are being allocated, such as when influential Maasai have gained enormous land allocations during privatisation, influential Tswana, including the Botswana's president, have acquired title to enormous private ranches or cattle posts in drier regions where ownership of crucial water points entails effective control of surrounding tracts of land, or Nuer have benefited from occupying political positions in the new Government of Southern Sudan. At the same time, as rural class formation has proceeded during the post-independence period, herding communities have come to represent a subordinate, labouring class within a larger regional system; especially where they have lost secure land rights, they resemble a rural proletariat, as has been pointed out for the Basarwa of Botswana (Wilmsen 1989).

Focusing on the spread of modern weaponry and the increase in rangeland violence, we must distinguish between widespread insurgencies that pit pastoralist communities against the state, and endemic conflicts that focus on resource scarcity, livestock raiding and banditry, which are endemic in many pastoralist areas. The Somali toppled one government but in seventeen years have failed to install another, while the Sudan People's Liberation Army – in which the Nuer have played a major role – fought the Sudan government to a standstill, and gained local autonomy and the right to determine the political future of southern Sudan through a referendum. The Tuareg rebellion in Mali and Niger, which was contained by the Sahara region, initially resulted in a negotiated settlement, but re-erupted in 2012 after the fall of the Gadaffi regime in Libya, leading to the take-over of northern Mali by Tuareg nationalists and Islamists. But less dramatic is the inexorable violence that unsettles many pastoral regions in a broad arc from Somalia through southern Ethiopia, northern Kenya from east to west, and northern Uganda, in which automatic weapons empower pastoralist communities clashing with one another over water, grazing and livestock, leading groups to curtail their cycles of pastoralist mobility and withdraw from customary territories or watering sources (Galaty 2005b; Eaton 2008). Less pervasive but nonetheless traumatic is conflict engendered by land loss, such as when Maasai or Fulbe pastoralists resist displacement from grazing territories or watering sources by cultivators, or by outsiders claiming title (Galaty 2008).

Most surprising, the pursuit of rangeland development projects has in most cases failed to achieve economic growth or enhanced livestock production and marketing in the pastoralist zone (Homewood, Kristjanson and Trench 2009). But through emphasising creation of ranching associations or groups (in Niger, Tanzania and Kenya), borehole syndicates (in Botswana), socialist state ranches (in Tanzania and Ethiopia), or individualised ranches (in Uganda and Kenya), development initiatives have often undermined customary systems of land use and tenure in the attempt to establish semi-sedentary ranching enterprises. While the developmental aim of establishing a 'modern' livestock industry on the basis of transformed pastoralism has almost nowhere succeeded, most importantly due to intrinsic problems of establishing static forms of husbandry in drylands where mobility is an ecological necessity, these initiatives have destabilised systems of pastoral territoriality, displacing herding communities and engendering resource conflicts (Galaty 1994a, 1999; Galvin et al. 2008; Homewood 2008). This has been the experience among Fulbe of Niger, Maasai in Kenya and Tanzania, and Tswana in Botswana. It remains to be seen whether schemes of pastoralist land reform can be formulated which would derive some of the benefits that tenure security in theory promises (De Soto 2000; North 2005) while retaining the scale of grazing lands and flexibility of movement that pastoral land use in the African rangelands seems to demand (Scoones 1995).

The cases reviewed above demonstrate how ubiquitous globalising influences are, such that we can no longer speak of pastoralist livelihoods solely as adaptations to arid land constraints or simply framed by customary norms and institutions. Rather, we should think about herder decisions and the pastoral systems they practise as guided by two simultaneous lines of sight, towards the local distribution of grazing resources that they 'track' through herd movements and towards the upheavals emerging from shifts in power, property and population dynamics that have created a new political terrain for rangeland pastoralism. But the imposition of structures of state and market and the increasingly pervasive presence of global culture can best be viewed as providing tools with which pastoralists may seek to build modern lives rather than as rigid moulds into which they are pressed. As Sahlins states, 'global homogeneity and local differentiation have developed together'; world culture is not just stamped with uniformity but by, in Hannerz's words (cited by Sahlins), 'an organization of diversity' (Sahlins 1999: ix). The pastoralist indigenisation of modernity includes the refashioning of tenure systems to fit with arid land constraints, the novel use of state institutions to sustain local livelihoods, and the use of new technologies in transport and telecommunications to make livestock production more efficient. Burned when planned development has sought to displace local strategies and techniques in favour of the latest donor fashion or a solution in theory to perceived inefficiency, thus often exacerbating rather than relieving poverty, pastoralists adopt innovations with caution.

But they often seek to modernise the indigenous, using local knowledge of dryland husbandry, networks of local collaboration, forms of local authority, systems of rights in resources, and seasonal patterns of territorial perception and use, to pursue modern aims, of educating youth, competing politically, building up capital through the routine marketing of livestock, building permanent homes, and accessing national information. New types of territoriality often represent a compromise between national land-use programs and local recognition of resource rights and patterns of use. Maasai settlements are increasingly dispersed and nucleated, as households exploit sub-division to claim lands through building permanent structures and reducing the size of homesteads. But despite the greater autonomy of households, they still seek help among neighbours, exchange information, gather for collective rituals, celebrations and fund raising, mobilise for community defence when applicable, and avoid isolation by using markets and small towns, where social services and commodities can be found, as sites of important political discussion and sociability.

With the expansion of states and the pervasiveness of political influence, the growing role of government in the formal allocation of land for commercial purposes, the advancement of sedentarisation through small-town urbanisation and settlement, and the spread of agriculture, forms of pastoralist mobility have evolved in several ways to meet modern challenges. First, there has been an increase in the sort of bilateral negotiations that have long characterised Fulbe relations with their neighbours, confirming seasonal access to grazing or post-harvest stubble, sale of mown hay, or accommodation of entire herds on leased land. Second, we find much more spontaneous and dynamic movement of livestock into diverse areas where grass can be found under the customary principle that unused pasture is a collective resource and that pre-emption is a herder's right. So with increasing constraints due to the 'modernisation' of land rights, herders are responding creatively, through innovatory negotiations and enterprise.

Douglas North's insistence that institutional innovations may hold the key to increased economic productivity (North 1990, 2005) is an important complement to exclusive emphasis on new, productivity-enhancing technologies as the key to modernisation. Certainly there do not seem to be any system-transforming technological innovations for animal husbandry under dryland conditions, though with sufficient capital modestly enhanced livestock breeding, use of veterinary medicine, labour-saving fencing and modified feed systems hold promise. The record of major institutional innovations through privatised rangeland and the provision of credit, thought by some to hold the promise of significant economic growth (De Soto 2000), has not been positive where national institutions charged with adjudicating land and providing agricultural credit have been inefficient, politically motivated and corrupt, as in Kenya and elsewhere. But we must ask whether, even with efficient and incorruptible national institutions, the creation of ecologically non-viable units of rangeland production in arid lands through subdivision and privatisation, or encouraging herders to seek credit with land titles laid down as collateral where returns on livestock marketing fluctuate radically under conditions of disease and drought, make sense. New forms of mobility and territoriality represent practical innovations both of technology and institutions that achieve modest economic gains under contemporary constraints. The creative enterprise shown by pastoralists cum ranchers in accessing new resources while their own land bases are being eroded, devising novel institutional strategies that link local, national and global structures, should be seen as an embrace of the modern, even as 'modernisation' via planned development so often fails to secure or enhance arid-land livelihoods.

Notes

1. Research in Kenya was supported by the Social Sciences and Humanities Research Council of Canada (SSHRC) and McGill University, and carried out through affiliation with the Ethnography Unit of the National Museums of Kenya and the Mainyoito Pastoralist Integrated Development Organisation (MPIDO). I acknowledge the assistance of Joseph Ole Simel, Amos Ole Kaitei, and Justus Lesiamon in Kenya, and Julia Bailey, Stephen Moiko and Anne-Elise Keen at McGill, and the provision of institutional support by the Max Planck Institute for Social Anthropology in Halle and the Centre for Society, Technology and Development at McGill during the writing of this chapter.
2. This split in southern forces re-emerged in 2009 with the eruption of civil strife in the south.

References

Archambault, C. 2011. 'Ethnographic Empathy and the Social Context of Rights: "Rescuing" Maasai Girls from Early Marriage', *American Anthropologist* 113(4): 632–43.

Bernus, E. 1981. *Touregs Nigeriens: Unité, Culturelle et Diversité Regionale d'un Peuple Pasteur.* Paris: ORSTOM.

Berry, S. 1993. *No Condition is Permanent: The Social Dynamics of Agrarian Change in Sub-Saharan Africa.* Madison: University of Wisconsin Press.

Bollig, M. 2006. *Risk Management in a Hazardous Environment: A Comparative Study of Two Pastoral Societies.* New York: Springer.

Bonte, P. 1991. '"To Increase Bows, God Created the King": The Function of Cattle in Intralacustrine Societies', in J. Galaty and P. Bonte (eds), *Herders, Warriors, and Traders: Pastoralism in Africa.* Boulder, CO: Westview Press, pp.62–88.

Bubenzer, O., A. Bolten and F. Darius (eds). 2007. *Atlas of Cultural and Environmental Change in Arid Africa.* Köln: Heinrich-Barth-Institut.

Campbell, D.J., et al. 2000. 'Land Use Conflict in Kajiado District, Kenya', *Land Use Policy* 17: 337–48.

Casimir, M. 1992. 'The Dimensions of Territoriality: An Introduction', in M. Casimir and A. Rao (eds), *Mobility and Territoriality: Social and Spatial Boundaries among Foragers, Fishers, Pastoralists, and Peripatetics.* Oxford: Berg, pp.1–26.

Collier, P. 2007. *The Bottom Billion: Why the Poorest Countries are Failing and What Can Be Done about It.* Oxford: Oxford University Press.

Deng, F. 1995. *War of Visions: Conflict of Identities in the Sudan.* Washington, DC: Brookings Institution.

De Soto, H. 2000. *The Mystery of Capital: Why Capitalism Triumphs in the West and Fails Everywhere Else.* New York: Basic Books.

Eaton, D. 2008. 'The Business of Peace: Raiding and Peace Work along the Kenya–Uganda Border (Part I)', *African Affairs* 107: 89–110.

Evans-Pritchard, E.E. 1940. *The Nuer: A Description of the Modes of Livelihood and Political Institutions of a Nilotic People.* Oxford: Oxford University Press.

Fratkin, E., and E. Roth (eds). 2005. *As Pastoralists Settle: Social, Health, and Economic Consequences of Pastoral Sedentarization in Marsabit District, Kenya.* New York: Kluwer Academic.

Galaty, J. 1993. 'Maasai Expansion and the New East African Pastoralism', in T. Spear and R. Waller (eds), *Being Maasai: Ethnicity and Identity in East Africa*. London: James Curry, pp.61–86.

——— 1994a. 'Rangeland Tenure and African Pastoralism', in E. Fratkin, K. Galvin and E. Roth (eds), *African Pastoralist Systems: An Integrated Approach*. Boulder, CO: Lynne Rienner, pp.185–204.

——— 1994b. 'Ha(l)ving Land in Common: The Subdivision of Maasai Group Ranches in Kenya', *Nomadic People* 34/35: 109–22.

——— 1999.'Grounding Pastoralists: Law, Politics, and Dispossession in East Africa', in E. Fratkin and T. McCabe (eds), *Social Change among East African Pastoralists. Nomadic Peoples* (special issue) 3(2): 56–73.

——— 2005a. 'Double-voiced Violence in Kenya', in V. Broch-Due (ed.), *Violence and Belonging: The Quest for Identity in Post-colonial Africa*. London: Routledge, pp.173–94.

——— 2005b. 'States of Violence: Ethnicity, Politics and Pastoral Conflict in East Africa', *Geography Research Forum* 25: 105–27.

——— 2007. 'Mobiles for the Mobile: Global Technologies and New Political Strategies in Pastoral Africa', paper presented at the session '"Wired Nomads": Globalizing Links and Focal Problems', AAA Annual Meeting, 28 November to 2 December 2007, Washington, DC.

——— 2008. 'Violence and its Mediations: Local Organizations, Community Conflict and the State in East Africa', in A. Bellagamba and G. Klute (eds), *Beside the State: Emerging Power in Contemporary Africa*. Munich: Lit Verlag, pp.1–11.

Galvin, K., et al. (eds). 2008. *Fragmentation in Semi-arid and Arid Landscapes: Consequences for Human and Natural Systems*. New York: Springer.

Grayzel, J. 1990. 'Markets and Migration: A Fulbe Pastoral System in Mali', in J. Galaty and D. Johnson (eds), *The World of Pastoralism: Herding Systems in Comparative Perspective*. London: Guilford Press, pp.35–67.

Hitchcock, R. 1990. 'Water, Land and Livestock: The Evolution of Tenure and Administrative Patterns in the Grazing Areas of Botswana', in J. Galaty and D. Johnson (eds), *The World of Pastoralism: Herding Systems in Comparative Perspective*. London: Guilford Press, pp.216–98.

Homewood, K. 2008. *Ecology of African Pastoralist Societies*. Oxford: James Currey.

Homewood, K., P. Kristjanson and P.C. Trench (eds). 2009. *Staying Maasai: Livelihoods, Conservation and Development in East African Rangelands*. New York: Springer.

Hughes, L. 2006. *Moving the Maasai: A Colonial Misadventure*. Houndsmills: Palgrave McMillan.

Hutchinson, S. 1996. *Nuer Dilemmas: Coping with Money, War, and the State*. Berkeley: University of California Press.

Lesorogol, C. 2008. *Contesting the Commons: Privatizing Pastoral Lands in Kenya*. Ann Arbor: University of Michigan Press.

McCabe, T. 1994. 'Mobility and Land Use Among African Pastoralists: Old Conceptual Problems and New Interpretations', in E. Fratkin, K. Galvin and E. Roth (eds), *African Pastoralist Systems*. Boulder, CO: Lynne Rienner, pp.69–90.

McPeak, J.G., and P.D. Little. 2006. *Pastoral Livestock Marketing in Eastern Africa: Research and Policy Challenges*. Bourton-on-Dunsmore: Intermediate Technology Publications.

Mwangi, E. 2007. *Socioeconomic Change and Land Use in Africa: The Transformation of Property Rights in Maasailand*. New York: Palgrave MacMillan.

North, D. 1990. *Institutions, Institutional Change, and Economic Performance*. Cambridge: Cambridge University Press.

——— 2005. *Understanding the Process of Economic Change*. Princeton, NJ: Princeton University Press.

Pelican, M. 2009. 'Complexities of Indigeneity and Autochthony: An African Example', *American Ethnologist*, 36(1): 52–65.

Peters, P. 1994. *Dividing the Commons: Politics, Policy and Culture in Botswana*. Charlottesville: University of Virginia Press.

Sahlins, M. 1999. 'What is Anthropological Enlightenment? Some Lessons of the Twentieth Century', *Annual Review of Anthropology* 28: i–xxiii.

Salzman, P.C. 2004. *Pastoralism: Equality, Hierarchy, and the State*. Boulder, CO: Westview Press.

Scoones, I. 1995. 'New Directions in Pastoral Development in Africa', in I. Scoones (ed.), *Living with Uncertainty: New Directions in Pastoral Development in Africa*. London: Intermediate Technology Publications, pp.1–36.

Sen, A. 1993. 'Capability and Well-Being', in M. Nussbaum and A. Sen (eds), *The Quality of Life*. Oxford: Clarendon Press, pp.30–53.

Schneider, H. 1979. *Livestock and Equality in East Africa: The Economic Basis for Social Structure*. Bloomington: Indiana University Press.

Spear, T., and R. Waller. 1993. *Being Maasai: Ethnicity and Identity in East Africa*. London: James Currey.

Sperling, L., and J. Galaty. 1990. 'Cattle, Culture and Economy: The Dynamics of East African Pastoralism', in J. Galaty and D. Johnson (eds), *The World of Pastoralism: Herding Systems in Comparative Perspective*. London: Guilford Press, pp.69–98.

Thompson, M., and K. Homewood. 2002. 'Entrepreneurs, Elites, and Exclusion in Maasailand: Trends in Wildlife Conservation and Pastoralist Development', *Human Ecology* 10(1): 107–37.

Wilmsen, E. 1989. *Land Filled with Flies: A Political Economy of the Kalahari*. Chicago: University of Chicago Press.

Worby, E. 1988. 'Livestock Policy and Development Ideology in Botswana', in D. Attwood, T. Bruneau and J. Galaty (eds), *Power and Poverty: Development and Development Projects in the Third World*. Boulder, CO: Westview Press, pp.155–80.

Notes on Contributors

Susanne Berzborn is currently an independent consultant. Her PhD, received from Cologne University, deals with rural livelihoods in South Africa. She specialises in Southern Africa and her research focuses on land rights, natural resource management, social networks and identity. Recently, she worked on volunteer services in South Africa. Her publications include *Ziegen und Diamanten: Sicherung des Lebensunterhalts im ländlichen Südafrika* (2006).

Michael Bollig is Professor of Social and Cultural Anthropology at the University of Cologne. He received his PhD from the University of Tübingen. He is the author of *Risk Management in a Hazardous Environment: A Comparative Study of Two Pastoral Societies* (2006) and has co-edited *The Practice of War: Production, Reproduction and Communication of Armed Violence* (2007) and *African Landscapes: Interdisciplinary Approaches* (2009). He is currently finishing a book on the history of human–environment relations in north-western Namibia from a political ecology perspective.

Christo Botha is Associate Professor of History in the Department of Geography, History and Environmental Studies at the University of Namibia. He obtained his PhD from the University of Pretoria, South Africa. He has conducted research on environmental issues and land settlement in colonial Namibia and investigated colonial development policies in an African homeland (Damaraland). His research interests also extend to the field of German–Afrikaner relations in Namibia and the expanding field of studies of South Africa's wars in Southern Africa. He is currently investigating the role of individuals who contributed towards the establishment of a culture of tolerance and liberal values in Namibia.

Ute Dieckmann is currently working as a research coordinator for the Land, Environment and Development Project at the Legal Assistance Centre in Windhoek, Namibia, mainly working on marginalised communities, such as the San, and the current land reform process in Namibia. She received her PhD from the University of Cologne, Germany, and is the author of *The Hai||om from the Etosha Region: A History of Colonial Settlement, Ethnicity and Nature Conservation* (2007).

John G. Galaty studied anthropology in Chicago and Paris, and now teaches anthropology and African Studies at McGill University, Canada. He is co-editor of *The Future of Pastoral Peoples* (1981), *A World of Pastoralism: Herding Systems in Comparative Perspective* (1990), *Herders, Warriors and Traders: Pastoralists in Africa* (1991), and *Uprooted: Dispossession in Africa* (1999). He is currently writing a book on demarcations of Maasai land, livelihoods and identities.

Clemens Greiner is a post-doctoral researcher at the Institute of Social and Cultural Anthropology, University of Cologne. He received his PhD from Hamburg University for a multi-sited ethnography of rural–urban migration and livelihoods in Namibia. His current research focuses on changing land use, boundary conflicts and the transformation of socio-ecological systems in East Pokot, Kenya.

Dag Henrichsen is a Namibian historian and archivist at the Basler Afrika Bibliographien (Namibia Resource Centre and Southern Africa Library) in Switzerland. He received his PhD from the University of Hamburg and has published widely on Namibia's (visual) history in the nineteenth and twentieth century. He ist the author of *Herrschaft und Alltag im vorkolonialen Zentralnamibia: Das Herero- und Damaraland im 19. Jahrhundert* (2011) and editor of *Israel Goldblatt, Building Bridges: Namibian Nationalists Clemens Kapuuo, Hosea Kutako, Samuel Witbooi, Brendan Simbwaye* (2010).

Babett Jánszky is a social anthropologist and former member of the ACACIA research project at the University of Cologne. She studied Social Anthropology, Psychology and Communication- and Media sciences at Leipzig and Aix-Marseille. Ms. Jánszky works on her doctoral thesis on household economy and risk management among Tama peasants in eastern Chad. Her research interests focus on households in violent conflict and adaptation strategies to ecological and political crises.

Friederike Jesse obtained her PhD from the Institute of Prehistoric Archaeology at the University of Cologne with a thesis on early ceramics in the Wadi Howar region. From 2000 to 2007 she was responsible for the archaeological work of the Wadi Howar Project at the University of Cologne's collaborative research centre, ACACIA. From 2008 to 2012 she was then directing a research project at the Kushite fortress of Gala Abu Ahmed in lower Wadi Howar (northern Sudan).

Grit Jungstand is a social anthropologist and former member of the ACACIA research project at the University of Cologne. She studied social anthropology, political science and economics at Göttingen, Oklahoma City and Berlin. Her research concerns Africa, mainly the Sahelian zone, with a focus on economic and social change as well as the anthropology of conflict. She is currently working as a consultant in the field of development cooperation.

Birgit Keding gained her PhD in prehistory from the University of Cologne. Within the framework of ACACIA project, she conducted extensive fieldwork in the Eastern Sahara, in particular in the Wadi Howar, Sudan, and the Ennedi Mountains, Chad. Her major research interests include the archaeology of pastoralism and the spread of food production, especially in Africa, as well as in the relationship between cultural and environmental change at a worldwide level. Presently she is working in Kenya on cultural contacts between northern and eastern Africa during the early to late Holocene.

Rudolph Kuper is former head of the African Archaeology Unit at the University of Cologne and founder and current chairman of the Heinrich Barth Institute for African Archaeology and Environmental History e.V.. He received his PhD from Cologne University for his work on Neolithic landscape archaeology in western Germany. Since 1980 he directed extended field research into the inter-relation between climatic change and cultural development in the Eastern Sahara (projects B.O.S. and ACACIA). Presently he is leading an interdisciplinary project into rock art and landscape archaeology at Wadi Sura, Gilf Kebir, south-west Egypt.

Paul Lane is Senior Lecturer in the Department of Archaeology, University of York, where he currently directs the 'Historical Ecologies of East African Landscapes' project. He is also an Honorary Research Fellow of the School of Geography, Archaeology and Environmental Studies at the University of the Witwatersrand. His research interests include the role of memory in the past and its material expressions, social constructions of space and time, the archaeology of pastoralism and mobility, and the transition to food production in eastern and southern Africa. He is a former Director of the British Institute in Eastern Africa, Nairobi and has also taught archaeology at the University of Botswana and the University of Dar es Salaam, Tanzania.

Tilman Lenssen-Erz is head of rock-art research at the Institute of Prehistoric Archaeology, University of Cologne. His research concerns a contextual approach to rock art in Namibia and the Eastern Sahara, where he has guided many expeditions. He has compiled the six volumes on '*The Rock Paintings of the Upper Brandberg* and has published widely on spatial and ecological issues in rock art.

Veerle Linseele is postdoctoral research fellow of the Research Foundation (FWO), Flanders, and is affiliated with the Centre for Archaeological Sciences at the Katholieke Universiteit Leuven and the Royal Belgian Institute of Natural Sciences. She is also part-time lecturer in Environmental Archaeology at Ghent University. She is the author of *Archaeofaunal Remains from the Past 4,000 years in Sahelian West Africa: Domestic Livestock, Subsistence Strategies and Environmental Changes* (2007). She is an expert in the study of archaeological animal remains, including bones from all groups of vertebrates and shells. For her current postdoctoral project on divergent pathways in early food production and early complex societies, she is analysing archaeozoological remains from Egypt and Sudan.

Peter D. Little is Chair and Professor of Anthropology and Director of the Development Studies Program, Emory University, USA. He received his PhD from Indiana University and is the author or editor/co-editor of nine books, including *Elusive Granary: Herder, Farmer and State in Northern Kenya* (1992), *Somalia: Economy Without State* (2003) and *Risk and Social Change in an African Rural Economy: Livelihoods in Pastoralist Communities* (2011). Currently he is completing a book on neoliberalism and the anthropology of reform in sub-Saharan Africa.

Meike Meerpohl studied social anthropology and Islamic science at the University of Muenster, Germany, and in Syria and Jordan. As a researcher at the University of Cologne, she conducted ethnographic research about trade and trade routes in the arid zones of Sudan, Chad and Libya. She is currently finishing a book on the Tima of the Nuba Mountains of Sudan, and works on intercultural communication in tourism at the Centre of Mediterranean Studies at the University of Bochum.

Matthias Österle holds a PhD in cultural and social anthropology from the University of Cologne. He did extensive fieldwork with Pokot pastoralists in northern Kenya. His postdoctoral research focused on the consequences of armed conflict among Turkana, Samburu and Pokot communities in Kenya. Since 2009 he has been working in Karamoja, Uganda, in the field of conflict transformation and peace building.

Julia Pauli is Professor in the Institute of Social and Cultural Anthropology, Hamburg University, Germany. Her main research interests are gender and kinship studies, demographic anthropology, transnational migration and class formation processes. She has done extensive fieldwork in Mexico (since 1995) and Namibia (since 2003).

Nadja Pöllath is an archaeozoologist, and currently employed as scientific assistant at the Institute of Palaeoanatomy and the History of Veterinary Medicine, Munich University. Until 2007 she was in charge of the archaeozoological remains excavated in Egypt and Sudan by the ACACIA research project. Her focus is on archaeofaunas from north-eastern Africa and the Near East.

Heiko Riemer is research assistant at the University of Cologne, from which he holds a PhD in prehistoric archaeology. He has been working on landscape and environmental archaeology in the deserts of Egypt, Libya and Sudan for more than fifteen years. He is the author of *El Kharafish: The Archaeology of Sheikh Muftah Pastoral Nomads in the Desert around Dakhla Oasis* (2011), and co-editor of *Desert Animals in the Eastern Sahara* (2009). Currently he is co-director of the Wadi Sura project, conducting field research on rock art and landscape archaeology in south-west Egypt.

Karim Sadr is Head of the School of Geography, Archaeology and Environmental Studies at the University of the Witwatersrand, Johannesburg, South Africa. He received his PhD from Southern Methodist University in Dallas, Texas. He is the author of *The Development of Nomadism in Ancient North-east Africa* (1991) and has published several papers on the archaeology of hunter-gatherers and herders in southern Africa. He is currently finishing a monograph on his archaeological surveys at Kasteelberg, on the west coast of South Africa.

Michael Schnegg is Professor of Social and Cultural Anthropology at the University of Hamburg, Germany. He holds an MA from the University of Cologne and a PhD from the University of Hamburg. His research focuses on understanding how people in sub-Saharan Africa value, use and govern ecosystem services in an increasingly deterritorialised world.

Martin Solich is a research associate and doctoral candidate at the Institute of Social and Cultural Anthropology at the University of Cologne. Currently, he is part of a collaborative research project on 'Our Way to Europe' and is modelling mobility patterns of nomadic foragers.

Hans-Peter Wotzka is Professor of Prehistoric Archaeology at the University of Cologne. He holds an M.A. degree from Southampton University and received his PhD from the University of Hamburg. He specialises in African archaeology, and his books include: *Studien zur Archäologie des zentralafrikanischen Regenwaldes* (1995); *Aridity, Change and Conflict in Africa* (2007, with Michael Bollig, Olaf Bubenzer and Ralf Vogelsang); *Proceedings of the Third International Conference on the Archaeology of the Fourth Nile Cataract* (2012). His current research brings together archaeology, archaeobotany and historical linguistics to generate new field data and theoretical perspectives on environmental and cultural change in the African Inner Congo Basin during the Iron Age.

Index

www.ingramcontent.com/pod-product-compliance
Ingram Content Group UK Ltd.
Pitfield, Milton Keynes, MK11 3LW, UK
UKHW021835100926
8556IPUK00004B/357

9 780857 4590